RENEWABLE ENERGY: RESEARCH, DEVELOPMENT AND POLICIES

BIOETHANOL AND BEYOND

ADVANCES IN PRODUCTION PROCESS AND FUTURE DIRECTIONS

RENEWABLE ENERGY: RESEARCH, DEVELOPMENT AND POLICIES

Additional books in this series can be found on Nova's website under the Series tab.

Additional e-books in this series can be found on Nova's website under the eBooks tab.

RENEWABLE ENERGY: RESEARCH, DEVELOPMENT AND POLICIES

BIOETHANOL AND BEYOND

ADVANCES IN PRODUCTION PROCESS AND FUTURE DIRECTIONS

MICHEL BRIENZO
EDITOR

nova
science publishers
New York

Library of Congress Cataloging-in-Publication Data

ISBN: 978-1-53613-478-0

Published by Nova Science Publishers, Inc. † New York

Contents

PREFACE

This book approaches the advances in the process of producing bioethanol, and future directions regarding its uses as fuel (or fuel additive) and the process integration with co-generation and biorefinery. Bioethanol, in this book, is reported as being produced from first to fourth generation, including macroalgal derivatives. A part of the book is dedicated to possibilities in bioethanol production (chapters 1 and 2), fermentation process contamination (chapter 3), analytical standards applied to quality control (chapter 4), and analysis of the fermentation process tanks through mathematical models (chapter 5). When it comes to substrate and material, ozonolysis is presented as a potential pretreatment process (chapter 6), and microalgae (chapter 7) and macroalgae (chapter 8), which are noteworthy as substrates. Covering the enzymatic hydrolysis advance, there is an overview on enzymes properties and technologies (chapter 9), cellulases production from invertebrate animals (chapter 10), and production by solid-state fermentation (chapter 11). The uses of ethanol were approached in chemical looping (chapter 12), use in diesel engines (chapter 13), and for hydrogen production (chapter 14). The bioethanol production chain is evaluated as being integrated with the energy co-generation using densification process (chapter 15).

Bioethanol is produced worldwide in an average of 25 billion gallons, according to the Renewable Biofuels Association. From this amount, the USA produce 14, and Brazil, 7 billion gallons. Bioethanol is produced in other countries in lower amounts: the European countries produce 1.3 billion gallons, and China, 0.8 billion gallons. The bioethanol production has been a concern for several countries, due to economic and environmental aspects. Among many positive points, ethanol is characterized as renewable for being produced from crops such as sugarcane (sucrose) and corn (starch) (1G). Currently, aiming at a more sustainable production, researches have been dedicated to deconstructing the plant biomass residues (sugarcane bagasse and straw, corn stover, wood, etc) into fermentable sugars, which could be fermented into ethanol (2G). As another category, bioethanol can be produced from microalgae (3G) and from macroalgae residues (4G).

The book begins by presenting an overview of the possibilities of producing bioethanol, authored by Felipe A. F. Antunes, Anuj K. Chandel, Júlio C. Santos, Talita M. Lacerda, Kelly J. M. Dussán, Débora D. V. Silva, Gerônimo V. Tagliaferro, Thaís S. S. Milessi, Paulo R. F. Marcelino, Larissa P. Brumano, Ruly Terán-Hilares, and Silvio S. da Silva. The authors detailed the current technologies that are used worldwide for bioethanol production. A description of the consolidate technologies for ethanol 1G and innovative approaches for the development of the 2G ethanol have been performed. For the 2G ethanol production, residual materials from agroindustry activities, such as sugarcane bagasse and leaves, corn cobs, woods, and others are commonly used. However, there are still challenges for the conversion of residual biomass into bioethanol. Technology processes steps have been described at length: the availability of biomass and the bioprocess including pretreatments, detoxification, enzymatic hydrolysis, fermentation of pentose and hexose sugars, as well as different process configuration, such as Separate (or Sequential) Hydrolysis and Fermentation (SHF), Simultaneous Saccharification and Fermentation – (SSF) and Simultaneous saccharification and co-fermentation (SSCF). Moreover, the process feasibility has been discussed for the technologies.

In chapter 2, the production of bioethanol from fermentable sugars from the plant cell walls is discussed. The authors, Vimal K. Balasubramanian, Ning Yuan, and Venugopal Mendu, focused on economically feasible points for the production of 2G ethanol. The importance of the biomass production with minimal inputs has been discussed, as well as the reduction of the pretreatment costs required to break down the plant cell walls in order to access the sugars. In fact, it has been demonstrated that an efficient breakdown is a key in the biomass conversion feasibility. Moreover, the use of all the polysaccharides from the plant biomass can contribute, e.g. cellulose, hemicellulose, and pectin. A solution to contribute for overcoming the recalcitrance is suggested, based on developing biomass varieties, resulting in higher biomass yields and also facilitating the enzymatic saccharification.

The third chapter focused on a problem related to the fermentation process and its contaminations. The chapter is authored by Mariane S. Raposo; Renata M. Christofoleti-Furlan; Camila de S. Varize; Cauré B. Portuga, and Luiz C. Basso. The bioethanol fermentation process in Brazil is conducted with yeast cell recycling, and the process is not in full aseptic condition. As a consequence, microbial contamination is frequent with yeast and bacteria. The contamination results in a direct competition, with industrial yeast, for carbohydrates and nutrients, with negative impact in the fermentation process yield (efficiency). Toxic problems can be caused due to the production of organic acids by the contaminating microorganisms. The authors detailed techniques for controlling the contamination by applying antibiotics, acid treatments, and chemical agents, as well as new approaches on natural compounds.

In chapter four, the kinetic models in bioreactor for the production of bioethanol are discussed. The authors Hugo I. Valázques-Sanchez, and Rafael Maya-Yescas focused on analyzing equivalence trajectories. The control or the knowledge of the reaction rates makes it possible to design, optimize, and control the biochemical reactor. As a challenging issue, the authors discussed the cells growth modeling complexity, based on biochemical reactions/pathways, substrate assimilation, and product generation. Theoretical models, based on empiric equations, considered the cell growth rates as uniform, ignoring the internal dynamic process of the cells. Chemostat model prediction varies according to the unstructured growth model and biomass yield. However, by using several kinetic models it is possible to analyze the dynamic behavior of the process in the bioreactor. The authors suggested a prediction of the maximum dilution rate from the Han-Levenspiel growth model compared to other unstructured models.

Chapter five is dedicated to showing the Brazilian technical analytical standards applied to the quality control of bioethanol, authored by Ana C.G. Panegossi; Rodrigo Sequinel; Sandra I. Maintinguer, and Danilo L. Flumignan. The authors described the fundamental aspects of bioethanol quality assurance through standardized specifications, the official technical standards' analysis, the analytical methods, and the certified reference materials from regulatory authorities. Quality control is important to gain commercialization for the bioethnaol in the external market. It is fundamental to advertise the bioethanol as a reliable product, and overcome technical barriers for its commercialization. The bioethanol quality in Brazil is supported by three federal agencies: the Brazilian Technical Standards Association (ABNT), responsible for technical standardization; the National Agency of Petroleum, Natural Gas, and Biofuels (ANP), responsible for the definition of standardized specifications and quality control; and the National Institute of Metrology, Quality, and Technology (Inmetro), responsible for metrology standards.

Chapter six is dedicated to another important step in bioethanol production: the pretreatment, necessary because of the residues from agroindustry. The authors, Olavo M. Perrone, Sidnei E. Bordignon, Mauricio Boscolo, Eleni Gomes, and Roberto da Silva, presented ozonolysis applied to lignocellulosic biomass. The lignocellulosic biomass is recognized as an important source of green carbon for bioethanol production. However, the lignocellulosic material is naturally recalcitrant and needs pretreatments to overcome this resistance and allow enzymes to act on the cellulose in order to release fermentable sugars. In this scenery, ozonolysis outstands as an advanced pretreatment with high specificity in the delignification under mild operation conditions. Moreover, ozonolysis can be combined with other pretreatment techniques to improve the process feasibility and increase the yields.

Chapter seven introduces the production of third generation ethanol (3G) from microalgae. The chapter is authored by Eduardo B. Sydney, Ana M. V. Escallon, Alessandra C. N. Sydney, Juliana V. M. Bittencourt, Júlio C. de Carvalho, and Carlos R.

Soccol. Microalgae can accumulate lipids, proteins, and carbohydrates, which are valuable feedstock for the production of many value-added products. In a biorefinery context, the microalgae components can be extracted and converted into different products. Moreover, the carbohydrate fraction of the microalgae biomass can be used to produce bioethanol, which defines the 3G ethanol. Focused on bioethanol, microalgae species with the ability to accumulate high amounts of carbohydrtes are under research. An advantage of this carbohydrate source is that the fermentable sugars are directly released by enzymatic hydrolysis. According to the authors, 1,7 ton of microalgae (dray base) is necessary to produce 1 m^3 of ethanol.

The next chapter introduces the fourth generation ethanol (4G) concept with the use of carragena and glucan from macroalgae. This chapter is authored by Fernando Masarin, Fernando R. P. Cedeno, Eddyn G. S. Chavez, Levi E. O, Valéria C. Gelli, and Rubens Monti. According to the authors, 3G bioethanol is produced by using macroalgae biomass as feedstock; and 4G bioethanol is produced by using by-products or residues generated in the processing of macroalgae biomass. Marine resources can gain importance with the need of increasing the bioethanol production, when the conditions for 1G and 2G ethanol are limited. Moreover, a diverse energy matrix is desired. Macroalgae have the advantage of being rich in carbohydrates and devoid of lignin. The microalgae cell wall is formed by an amorphous mucilage matrix, containing polysacharides such as carratenan, agar, and starch. As strucutual carbohydrate, there are the cellulose fibres. The authors emphasize the lower recalcitrance of the macroalgae residues compared to lignocellulosic biomass and also its higher productivity.

Chapter nine, authored by Marcos P. Vicentim, is an overview of the cellulose enzymatic hydrolysis and its technological applications. An enzymatic hydrolysis after the pretreatment step is recognized as a good strategy for releasing fermentable sugars from lignocellulosic materials. However, cellulose has structural and physicochemical properties that impact the process slowdown and rate. Factors such as lignin content, cellulose crystallinity, and particle size limit the cellulase access to cellulose. Enzymes with action of hydrolases and oxidases are described as tools to break down the cellulose fibres. The author reviewed the starting producing cellulases purpose in 1980s, following several industrial applications that gained over the time. Moreover, the bioethanol production comes out as the most potential industrial application of cellulases.

In chapter ten the researches for discovering new enzymes from different model invertebrate animals, endogenous, and exogenous enzymes in the microflora are shown. This chapter is authored by Youssef B. Sade, Daniela T. Moraes, Paulo J. Beltrão, and Marcos P. Vicentim. As part of the efforts to develop the enzymatic hydrolysis process, an efficient enzymatic cocktail should contain enzymes/proteins with specials characteristics related to catalysis, synergism, and stability. Structurally different enzymes have been found in animals, compared to enzymes from fungi and bacteria. The microflora of invertebrate animals is rich in microrganisms that produce enzymes to degrade lignocellulosic materials. Besides endogenous enzymes being found, the importance of microrganisms lignocellulosic-degrading enzymes is demonstrated.

The next chapter, authored by Fernanda P. Casciatori, and Priscila A. Casciatori-Frassatto, approaches the production of enzymes by solid-state cultivation. The enzymatic hydrolyis process had its cost reduced in the last decade. However, it still restricts the process feasibility. The enzymatic hydrolysis of the biomass can require cellulases, hemicellulase, ligninases, and auxiliary enzymes, with the advantage of operating under mild conditions, since that, with specific action, there are no degradation products. The authors defend the idea of using agro-industrial residues as substrates for enzyme production. Authors also suggest that solid state cultivation (SSC), using filamentous fungi, outcome as a feasible and sustainable process for the 2G ethnaol production.

Chapter twelve is dedicated to reviewing Bioethanol Chemical Looping (CL) as a form of CO_2 capture technology, showing economic and energetic costs. The chapter is authored by F. García-Labiano, E. García-Díez, L.F. de Diego, A. Abad, P. Gayán, and J. Adánez. The authors describe the current status of the use of bioethanol in two processes: Chemical Looping Combustion (CLC) for energy production (heat and electricity), and Chemical Looping Reforming (CLR) for syngas/H2 generation. The chapter includes the basic concepts of the technology with a focus on the use of bioethanol and potential oxygen carriers and operational experience in continuous units. According to the authors, CL is a promising form of CO_2 capture, based on the use of oxygen as carrier in order to transfer the oxygen from the air to the fuel.

Chapter thirteen is focused on bioethanol usage in diesel engines, deeding direct use of ethanol. The chapter is authored by Breda Kegl and Marko Kegl. The authors describe blends of bioethanol and mineral diesel, as well as bioethanol and biodiesel. Among several physicochemical properties, the authors compared stability, density, viscosity, cetane index, cetane number, cloud point, pour point, cold filter plugging point, flash point, corrosiveness, lubricity, FT–IR spectrum, carbon–hydrogen–nitrogen composition, water content, and sound of velocity. As a function of the bioethanol effects on fuel injection, the analysis has been extended towards injection pressure, duration, and timing, and on fuel injection system components, like surface roughness of injection elements and nozzle hole discharge coefficient. Authors concluded that 15% bioethanol in mineral and biodiesel is appropriate.

Chapter fourteen reports the use of bioethanol on mixed oxide catalysts to produce hydrogen. The chapter is authored by Lidiya Y. Dolgikh, Yuri I. Pyatnitsky, and Peter E. Strizhak. Hydrogen is a promising biofuel due to its lack of emission of air pollutants and greenhouse gases. Among the resources for hydrogen production, sustainable ways for producing it are currently in focus by several researches. Ethanol and water stand out as resources for a friendly way to produce hydrogen. The authors present a thermodynamic analysis of the steam reforming of the bioethanol process. Results of studies with several catalysis such as Cu and Cu-M (M = Ni, Co, Fe) on several types of zirconia supports and MFe_2O_4 (M = Ni, Co, Fe, Mn, Cu, Mg, Zn) ferrites have been presented. Authors concluded that the $FeFe_2O_4$, $MnFe_2O_4$, and $MgFe_2O_4$ ferrites could be considered as durable catalysts for hydrogen production via the SRE reaction. The $FeFe_2O_4$ ferrite has also been pointed out as efficient for the ethanol conversion into acetone, suggesting a two-step process involving an intermediate formation of acetone.

Chapter fifteen, the last chapter, proposes an integration between ethanol (1G and 2G) production and energy co-generation. The chapter is authored by Andrea C. De Conti, Cláudio De Conti, and Nestor P. Perez. Gasification process efficiently work with densificated (energetical) lignocellulosic residues. The densified residue can be used in gasifiers for energy co-generation. The authors focused on describing the energetic densification, biomass characterization, kinds of energetic densification (briquettes and pellets), and machinery. Moreover, this chapter presented the concepts of gasification and the kinds of gasifiers used for energy co-generation. The energetic densification is a result of compacting the material, which increases the energetic density. According to the authors, the densification makes it viable to transport the biomass to long distances, as well as storing it. From the bioethnaol production plant, solid fuels can be generated and used outside the company, collaborating for a more diverse energetic matrix.

In: Bioethanol and Beyond
Editor: Michel Brienzo
ISBN: 978-1-53613-478-0

Chapter 1

BIOETHANOL: AN OVERVIEW OF PRODUCTION POSSIBILITIES

Felipe A. F. Antunes[1,**], Anuj K. Chandel[1], Júlio C. Santos[1], Talita M. Lacerda[1], Kelly J. M. Dussán [2], Débora D. V. Silva[2], Gerônimo V. Tagliaferro[1], Thaís S. S. Milessi[1], Paulo R. F. Marcelino[1], Larissa P. Brumano[1], Ruly Terán-Hilares[1] and Silvio S. da Silva[1]

[1]Department of Biotechnology, School of Engineering,
São Paulo-University, São Paulo, Lorena, SP, Brazil
[2]Department of Biochemistry and Chemical Technology,
Institute of Chemistry, Universidade Estadual Paulista, Araraquara, SP, Brazil

ABSTRACT

Bioethanol production mainly from sugarcane juice and corn, called first generation ethanol (1G), has already presented consolidated worldwide technology in the current scenario. Within this context, novel approaches are pivotal for development of technologies regarding bioethanol production from different renewable biomass. In this chapter, besides a global review of the 1G ethanol production from different carbon sources, the all steps and challenges for the ethanol 2G production, based on use of lignocellulosic materials (such as sugarcane bagasse and leaves, corn cobs, woods, agricultural wastes and others) are presented in detail. Among these steps, the biomass availability and bioprocess including pretreatments, detoxification, enzymatic hydrolysis, fermentation of pentose and hexose sugars, as well as different process configuration such as Separate (or Sequential) Hydrolysis and Fermentation (SHF), Simultaneous Saccharification and Fermentation – (SSF) and Simultaneous saccharification and co-fermentation (SSCF) are presented. Moreover, the third generation ethanol production

* Corresponding Author Email: felipeantunes22@gmail.com

(3G), from microalgae are also described. Additionally, new advances and emerging technologies such as strain improvement, biomass improvement (genetically modified) are elucidated, highlighting their real potential. Future prospects and economical scenario are also presented. In all sections of this chapter, content are discussed according to their real feasibility for Bioethanol production.

Keywords: bioethanol, biomass, ethanol 2G, ethanol 3G, sacharification, fermentation

INTRODUCTION

In the last years, many policies have been globally established aiming to reduce environmental impacts and greenhouse gas emission due to the use of fossil fuels (Posada et al., 2013). The development of sustainable and clean energy sources has become a worldwide issue, especially after the Conference of the Parties (COP 21) realized in Paris in 2015 where a global agreement was stablished by 195 countries to avoid the increase of world temperature in 2°C until 2020. For this, a significantly reduction on greenhouse gas emission must be achieved.

The transport sector contributes with 19% of world's CO_2 emissions and more than 70% of emitted carbon monoxide (Aditiya et al., 2016). Within this context, biofuels represent a clean and sustainable alternative of energy to substitute fossil fuels, which offers environmental and socioeconomic advantages, such as the increase of farm income of developing countries and the reduction of greenhouse gas emissions (Hamelinck; Hooijdonk; Faaij, 2005).

The bioethanol is a compound with a high octane number and high oxygen content, being an excellent additive for gasoline, allowing cleaner combustion as increasing the oxygen content of the fuel (Goldemberg; Coelho; Guardabassi, 2008). This biofuel can be obtained from different raw material sources of sugary, starches and lignocellulosic origin (Gírio et al., 2010). The main difference between ethanol production from these raw materials is in fermentable sugars release step. While sugary sources only need one stage for sugar extraction, ethanol production from starches sources, such as corn, requires a step for hydrolysis of the carbohydrate chains to obtain a sugar syrup (Mussatto et al., 2010). On the other hand, lignocellulosic biomass sources require one more step called pretreatment, aimed to release fermentable carbohydrates from recalcitrant biomass matrix (Nigam; Singh, 2011).

The bioethanol is classified according to the raw material in first or second generation. The first-generation ethanol (1G) is produced on large scale from food and oil crops such as sugarbeet, sugarcane and corn (Brennan; Owende, 2010), and has a well-established and known production process with viable and consolidated economic levels. However, there are some concerns about environmental impacts that limit first generation ethanol production increase. The food-versus-fuel debate is the main issue of this biofuel, which attributes the food prices rising to ethanol 1G production.

The second generation ethanol (2G), on the other hand, is produced from lignocellulosic biomass, such as agricultural and forest residues, e.g., sugarcane bagasse (Aditiya et al., 2016). Its large-scale application has already been in development, with many bottlenecks for its economic viability. However, despite the remaining challenges, this technology is a promising sustainable alternative, since lignocellulosic biomass has large availability, low cost and does not compete for land with food production (Nigam, Singh, 2011).

In addition, 2G ethanol production can increase productivity per hectare of planted biomass, contributing to the reduction of cultivated area for production (Pereira et al. 2015).

The lignocellulosic biomass potential is linked to its composition once it is mainly composed by cellulose, hemicellulose and lignin (Fengel; Wegener, 1989). The cellulose and hemicellulose are rich in fermentable sugars that can be released from biomass structure through a pretreatment step and used as carbon source for ethanol production (Canilha et al., 2012; Gírio, 2010).

Nowadays, the 2G ethanol production technologies are targets of several public and private companies, which have been investing in researches in this field. It is a process with high complexity since it involves steps that do not subsist in conventional sugar and starch fermentation. The biomass pretreatment, hydrolysate detoxification and enzymatic hydrolysis steps have already needed to be optimized in order to achieve efficient and competitive process level (Naik et al., 2010; Nigam; Singh, 2011). Additionally, the breakdown of hemicellulose fraction releases mainly xylose, which makes this pentose sugar conversion into ethanol, another challenge for this process due to low availability of efficient microorganisms that assimilate C5 sugars (Canilha et al., 2012; Carvalho et al., 2013).

Currently, the new technologies have gained attention of scientific community in the search of an economically competitive biofuel production process. Thus, the use of microalgae in biodiesel and bioethanol production as well as genetic manipulation of microorganisms and plants have been studied by many research groups. Actually, the microalgae have a great carbohydrate level in its composition and some advantages, such as fast growth rate and high biomass productivity. The bioethanol from microalgae is called "third generation (3G) ethanol" and its production requires a pretreatment step before fermentation, once most of carbohydrates are inside microalgae cell and cell wall (Harun; Danquah, 2011).

The genetic manipulation of microorganisms and plants is an approach used by several research groups around the world to improve ethanol 1G production and to develop the ethanol 2G as a viable process. The genetic modification of microorganisms has been extensively studied and *Saccharomyces cerevisae* yeasts is the most modified microorganism in this field, aimed to improve its tolerance to stress conditions of 1G ethanol process as well as to express genes of xylose isomerase enzyme in order to assimilate xylose in 2G ethanol production (Swinnen et al., 2012; Demeke et al., 2013). The genetic modification of plants, on the other hand, mainly aims to increase biomass sugar content to enhance pretreatment step (Furtado et al., 2014).

Taking this into account, the bioethanol is a biofuel worthy of attention in the search for clean and sustainable energies, presenting many advantages compared to fossil fuels once it can be obtained from several raw materials. However, there are bottlenecks that still have to be overcoming, highlighting several studies and new technologies that have been carried out in worldwide scale that will be discussed throughout this chapter.

FIRST GENERATION ETHANOL

The biofuels produced from renewable resources are alternatives to minimize the impacts of world dependence on fossil fuels and CO_2 emissions. Thus, the bioethanol produced by renewable sources has received investment in many countries. Currently, the bioethanol is the main biofuel used in the world and plays a significant role in the global energy matrix. (Lin; Tanaka, 2006; Manochio et al., 2017). The ethanol, also called ethylic alcohol, has been historically produced by the fermentation of carbohydrates derived from vegetables. Presently, majority of world ethanol manufacturing is obtained by this biotechnological process (called first generation – 1G), in a most feasible and sustainable method than the same product obtained by chemical route, from ethylene derived from petroleum (Bastos, 2007).

For the ethanol 1G production, three main steps are required: release and availability of fermentable sugars from plants; sugars fermentation by microorganisms and separation and purification of ethanol, generally carried out by distillation, rectification and dehydration (Dias et al., 2011).

In industrial scale, sugars or starch are commonly used as substrates in the process, obtained from sources such as sugarcane and sugar beet (sucrose sources), or corn, wheat and cassava (starch sources). The choice of each source takes into account the land availability for the crop, country climate, soil quality and other local geographic concerns. In addition, sugars obtained from sugarcane, sugar beets, molasses, and fruits can be converted in ethanol driven by microrganisms, once they are in monomeric form, profitable for microorganism assimilation. On the other hand, carbohydrates from starches, e.g., corn, cassava, potatoes, and root crops need to be hydrolyzed in monomeric sugars, usually by action of enzymes, e.g., amylases that depolymerise polysaccharide in glucose monomers, requiring an additional step in the process. (Lin and Tanaka, 2006).

The sugar-based ethanol plants are predominantly produced in Brazil, European Union and India from sugarcanes and beets while the US European Union Countries,

China and Canada dominate the starch-based ethanol, generally produced from corn and grains (Lennartsson et al., 2014). Although corn, sugar beet and sugarcane have been described as the most productive crops worldwide (either in terms of productivity per area as well as in ethanol yields), other biomasses rich in fermentable sugars can also be used such as sweet sorghum, cassava, wheat and rye (Manochio et al., 2017). The characteristics of different crops for ethanol production is presented in Table 1.

Table 1. Characteristics of sugarcane, corn and sugarbeet for ethanol production

	Sugarcane (in Brazil)	**Corn (In USA)**	**Sugarbeet (in Europe)**	**Reference**
Productivity per area	60 – 120 ton/hectare	15 – 20 ton/hectare*	13 25– ton/hectare for non-irrigated farmland 50-100 ton/hectare for irrigated farmland	BNDES, 2008
Energy balance*	8.1-10	1.4	2.0	Goldemberg and Guardabassi, 2009
Production cost (US$ L^{-1} ethanol)	0.24–0.42	0.50–0.87	0.46–0.77	Manochio et al., 2017
Greenhouse gas (GHG) emissions (kg CO2eq L^{-1} ethanol)	0.25	1.70	0.41	Manochio et al., 2017
Avoided emissions of GHG (%)	69–89	30–38	35–56	Manochio et al., 2017
Total production (billion liters) (D)	22.5	34	2.7	Goldemberg and Guardabassi, 2009
Area cultivated (million hec.) (E)	3.4	8.13	0.49	Goldemberg and Guardabassi, 2009
Yield (liter/hectare) (D/E)	6.471	4.182	5.500	Goldemberg and Guardabassi, 2009

*which 50% consists of dry matter, thus, only between 7.5 and 10 t per hectare of corn are really used for producing ethanol.

Regarding the fermentation step, there are several microorganisms that can convert glucose and sucrose to ethanol, such as the yeast *Saccharomyces cerevisae* (Naik et al., 2010), *Kluyveromyces marxianus* (Castro; Roberto, 2014), *Candida utilis* (Vallet et al. 1996), *Kluyveromyces fragilis* and *Pachysolen tannophilus* (Sanchez et al. 1999); the bacteria *Clostridium sporogenes*, *Clostridium indole* (pathogenic), *Clostridium sordelli* (pathogenic), *Zymomonas mobilis*, *Spirochaeta aurantia*, *Spirochaeta stenostrepta*, *Spirochaeta litoralis*, *Erwinia amylovora* (Miyamoto 1997) and mold *Mycelium* (Naik et al., 2010). Since decades, the most used microorganism is the yeast *Saccharomyces cerevisiae*, due to high ethanol conversion as well as generally recognized as safe (GRAS) as a food additive for human consumption, being secure for industrial use (Lin; Tanaka, 2006), followed by the bacteria *Zymomona mobilis*. *Sacharomyces cerevisiae*, similarly several other microrganisms, metabolizes glucose by the Embden–Meyerhoff (EM), while a different metabolism is used for ethanol production by *Zymomona mobilis*, drived by the Entner–Doudoroff pathway (ED). While in the EM pathway one glucose molecule lead to two Adenosine Tri-Phosphate (ATP), in ED pathway only one ATP molecule is generated by metabolized glucose. Thus, higher specific rate of sugar uptake and ethanol yields and productivity, lower biomass production are verified in ED pathway. On the other hand, the use of *Zymomona mobilis* present disadvantages such as the sucrose inability conversion by some industrial substrates. The main inhibitory factors include thermal, osmotic and ethanol stress (product inhibition) along with limited substrate range (Panesar et al., 2006). Thus, *S. cerevisiae* is still preferred by the industry due to features as stability and hardiness in bioprocess (Lin; Tanaka, 2006).

Theoretically, the maximum conversion efficiency of glucose to ethanol is 51% on the weight basis, which comes from a stoichiometric calculation:

$$C_{16}H_{12}O_2 \rightarrow 2\ C_6H_5OH + 2\ CO_2$$
(mol. wt. 180) (mol. wt. 2x46) (mol. wt. 2x44)

2 molecular wt of ethanol (2x46)/ molecular wt of glucose (180)x100 = 0.51.

However, the yeast commonly use some part of available glucose for cell mass production and other metabolic products, and hence, reducing the conversion efficiency (Lee et al., 2006; Naik et al., 2010).

After the fermentation step, the fermented liquid is distilled to separate and purify the ethanol, which is then dehydrated to concentrations above 99.7% for fuel applications. In the bottom of the distillation column accumulates a stillage that consist of about 10% of total solids, including residual substrate, yeast, and by-products. The removal of these solid particles is carried out by centrifugation and decanter, and the remaining thin stillage is sent to evaporation step. The cake resulted from centrifugation and the remained syrup from the evaporation are normally mixed to produce protein source, normally used as animal feed, playing a crucial role in the overall process economy (Lennartsson et al., 2014).

Since last decades, many policies to support the production and use of ethanol, have been globally established aimed to promote energy independence and decrease environmental impacts caused by the used of fossil fuels, such as gasoline and diesel (Posada et al., 2013). For example, Brazil established in 1975 the Brazilian Alcohol Program (PROÁLCOOL), aimed to reduce fossil fuels imports, due to the oil crises carried in 70's decade. In this policy, the sugarcane was choosing as feedstock for raw material, with intensified studies in agricultural crop as well as in process development, leading Brazil to a stand out position in ethanol 1G production technology (Manochio et al., 2017; Renó et al., 2014). Also, in the Organisation for economic co-operation and development (OECD), countries such as USA and EU, subsidies and policies support lead to an important growth in first generation ethanol production in the last years (FAO, 2008).

According to the Renewable Fuels Association (2016), the United States and Brazil, the two largest ethanol producers, are responsible for 85% of total world production, achieving more than 100,631 million liters in 2016. Figure 1 presents the world fuel bioethanol production profile according to main countries producing, and used raw material.

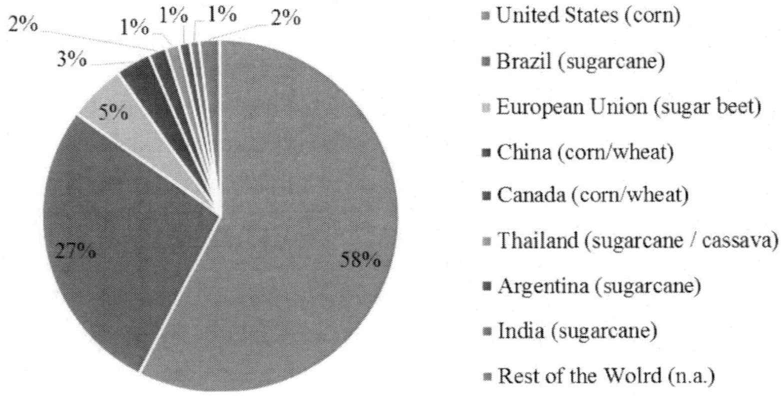

Figure 1. World fuel bioethanol production (millions of liters) and the respective main carbon source in 2016 (Renewable Fuels Association).

Actually, the first generation ethanol production is currently a well-known technology, however, the main disadvantage of first generation biofuels is the food-versus-fuel debate, which discusses the use of cultivated area and the use of crops that compete with the food production. Thus, new process and technologies, as second generation, by using lignocellulosic materials, and third generation, by using algae, have been explored aiming other sustainable alternatives.

Second Generation Ethanol: Main Steps for Production

The ethanol production from lignocellulosic biomass offers environmental benefits and in long-term will be economically viable compared to fossil fuels. The lignocellulosic biomass consists of cellulose, hemicellulose and lignin, as well as other components in smaller amounts. The fractions of cellulose and hemicellulose are carbohydrate polymers and a potential source of fermentable sugars.

The cellulose is one of the major structural components of the cell wall, and provides mechanical resistance and chemical stability to plants. It is a homopolymer formed by β-1,4 glycosidic bonds of repeated D-glucose units, found in both crystalline and non-crystalline structure (Fengel; Wegener, 1989). It is the most abundant organic compound in the biosphere and about 75 billion tons of pulp are processed worldwide (NRC, 2000). Hemicellulose is a branched heteropolysaccharide (polysaccharide comprising more than one type of monomer) of different sugars C5 and C6, which is also part of the structure of the plant cell wall. It is mainly composed by arabino-xylans, gluco-mannans and galactans (Raven et al., 1992). Among the most relevant aspects of hemicellulose structure and composition are the lack of crystalline structure, mainly due to the highly branched structure, and the presence of acetyl groups bound to the polymer chain (Scheller; Ulvskov, 2010). Additionally, lignin is an aromatic compounds-rich structure that forms a protective layer in plant cell wall. Lignin consists of three- dimensional polymeric bonds between phenylpropane units with various types of bonds between the monomers, resulting in a complex polyphenolic structure, which cannot be converted into fermentable sugar (Calvo-Flores; Dobado, 2010).

The fuel ethanol production from lignocellulosic biomass includes steps such as pretreatment (cellulose/hemicellulose hydrolysis), hexose/pentose fermentation, separation and distillation (Ojeda; Kafarov, 2009). Intense efforts have been applied in recent years aimed to develop efficient technologies for the pretreatment of biomass, enzymes for saccharification of cellulose/hemicellulose and suitable technologies for the fermentation of both C6 and C5 sugars.

Biomass Availability, Composition, Pretreatment, Detoxification, Enzymatic Hydrolysis and Processes Configuration

Biomass Availability and Its Composition

The lignocellulosic materials represent a viable, inexpensive, and renewable raw material for the production of second-generation bioethanol. Besides, these materials have no competition in food production. Different kind of wood and residues or co-products from agroindustry have been studied as raw material for bioethanol production in different countries, according to their availability.

In recent years, estimates of worldwide availability of agricultural wastes have been published. According to Saini et al. (2015), around 354, 731, 128 and 180 million tons of wheat straw, rice straw, corn straw and sugarcane bagasse respectively are produced per year, which have an estimated bioethanol production potential of 104, 205, 58 and 51 gigaliters, respectively. Additionally, Sarkar et al. (2012) analyzed the availability of these biomasses in the different regions of the planet, and concluded while in Asia there is a greater amount of rice straw, in America the highlights are corn straw and sugarcane bagasse, and wheat straw in Europe. The authors also presented that the contribution of Africa and Oceania to production of these materials are very small. Specifically, it is possible to verify that in studies for bioethanol production, corn residues and switchgrass are more evaluated in USA (Bansal et al., 2016; Kim; Dale, 2016), sorghum and miscanthus in Asian countries (Zhang et al., 2010; Shi-wei et al., 2017) and sugarcane co-products in Indian and Latin America countries (Sukumaran et al., 2010; Bezerra; Ragauskas, 2016; Marrugo et al., 2016).

Due to the typical cell wall constituents of these materials, mainly cellulose, hemicellulose, high concentrations of glucose and xylose can be obtained from cellulosic (C6) and hemicellulosic (C5) fractions. Table 2 shows the content of cellulose, hemicellulose and lignin in different lignocellulosic materials.

Table 2. Cellulose, hemicellulose and lignin content in different plant biomass

Lignocellulosic Biomass	Cellulose (%)	Hemicellulose (%)	Lignin (%)	Reference
Beech wood residual	43.10	20.20	24.2	Katsimpouras et al. (2017)
Cotton stalk	40.10	13.60	29.40	Keshav et al. (2017)
Corncob	45.60	36,90	15,10	Brar et al. (2016)
Corn stover	36.50	22.10	18.80	Yang et al. (2016)
Miscanthus	45.30	25.30	25.5	Auxenfans et al. (2017)
Palm press fiber	25.45	17.35	36.00	Brito et al. (2017)
Rice straw	35.10	25.94	19.86	Castro and Roberto (2015)
Sugarcane bagasse	34.60	18.90	27.60	Dussan et al. (2016)
Sugarcane straw	31.70	27.00	31.10	Hernández-Pérez et al. (2016)
Switchgrass	31.90	20.20	20.70	Parthasarathi et al. (2016)
Wheat straw	39.10	29.40	24.60	Auxenfans et al. (2017)

Other lignocellulosic materials have also been evaluated as raw material for the production of bioethanol, such as agave leaves (Avila-Gaxiola et al., 2017), mango stem barks residues (Carrillo-Nieves et al., 2017), Poaceae biomass (Djajadi et al., 2017), chili post-harvest residue (Sindhu et al., 2017), poplar wood chips (Yu and Christopher, 2017), Juncus maritimus (Smichi et al., 2016), prickly pear cladodes (Santos et al., 2016), fiber sorghum (Forte et al., 2016), switchgrass (Luque et al., 2016) and others.

Pretreatment

The pretreatment is a prerequisite step to solubilize fermentable sugars present in lignocellulosic materials structure, which can be converted, by microorganisms, in bioproducts such as ethanol.

Table 3. Different pretreatments methods: Conditions, advantages and disadvantages

Pretreatments	Conditions, Advantages and Disadvantages	References
	Physical and Physical-Chemical	
Mechanical	Promotes the reduction of polymerization degree	Sun and Cheng (2002); Chang and Holtzapple (2000)
	The total hydrolysis yield increases from 5-25%.	
	Reduces digestion time by 23-59%.	
	No produce inhibitors	
	Requires high power consumption	
Thermal	Temperatures above 180°C.	Avila-Gaxiola et al., 2016; Djajadi et al., 2017; Smichi et al., 2016.
	Extothermic reaction	
	Solubilization of hemicellulose and lignin	
	Produces phenolic compounds	
	Temperatures above 250°C should be avoided to prevent pyrolysis reaction.	
Pyrolysis	Requires high temperatures (300-600°C)	Luque et al., 2016; Khiyami et al., 2005
	Cellulose decomposes rapidly into gases	
	Formation of volatile products	
	Produces bio-oil and char	
Extrusion	Expansion of the material resulting in an increase of the specific area	Zheng and Rehmann, 2014; Lamsal et al., 2010
	No degrade products	
	Increased accessibility of cellulose to enzymatic action	
	Combined with other pretreatments to improve sugar recovery	
Steam Explosion	Requires high temperature (160-290ºC) and pressure (0.5-7 MPa).	Auxenfans et al., 2017; Pielhop et al., 2016; Pielhop et al., 2017; Yang et al., 2016; Yuan et al., 2017; Zhang et al., 2016
	Partial degradation of hemicellulose and forms acids	
	Produces inhibitors such as furfural, 5-HMF and soluble phenolic compounds	
	No corrosion problems	
	High energy demand	
	Produces inhibitors such as furfural, 5-HMF and soluble phenolic compounds	
	No corrosion problems	
	High energy demand	
Liquid hot water	pH should be maintained between 4 and 7 to prevent formation of inhibitors	Bhagia et al., 2016; Shibuya et al., 2017; Rogalinski et al., 2008; Perez et al., 2008
	Requires temperatures between 170-230ºC	
	Catalytic degradation of sugars occurs in a series of reaction that are difficult to control and results in undesirable side products.	
	High amount of solubilized products. Partial solubilization of lignin.	
Ammonia fiber explosion	Temperatures between 70-170ºC with pressure of 1.0-2.0 MPa	Balan et al., 2009; Bradshaw et al., 2007; Murnen et al., 2007; Li et al., 2010; Chandel et al., 2013
	Requires ammonia recovery	
	Solubilization of hemicellulose and lignin	
	No produce inhibitors	
	Environmental problems due to ammonia	

Ozonolysis	Use of ozone at room temperature and normal pressure No produce inhibitors Degrades hemicellulose and lignin Large amount of ozone are required	Barrera-Martínez et al., 2016; Garcia-Cubero et al., 2009; Travaini et al., 2013
CO_2 explosion	Pressures up to 7 MPa and low temperatures (30-50°C) No produce inhibitors Uses supercritical CO_2 Low environmental impact No degrade sugars Increases surface area of material	Zheng et al., 1995; Zhang and Wu 2014; Daza Serna et al., 2016
	Chemicals	
Acid	Uses acids such as H_2SO_4, HCl and HNO_3 Hydrolysis temperatures of diluted acids of 120- 160°C and for concentrated acids of 170-190°C The solubilization of hemicellulose and the precipitation of lignin are more pronounced with strong acid compared to dilute acid Produces degradation products The use of strong acid is not attractive for the ethanol production because there is a risk in the formation of inhibiting compounds and equipment wear. The use of diluted acid is considered to be one of the most promising methods because the side reactions that occur during the pretreatment can be avoided	Brar et al., 2016; Brito et al., 2017; Castro et al., 2015; Costa et al., 2017; Keshav et al., 2016; Lu et al., 2017; Mesa et al., 2017; Santos et al., 2016; Sun et al., 2016
	Corrosion problems	
Alkaline	The first reactions that occur are solvation and saponification. Uses alkalis as NaOH, KOH, $Ca(OH)_2$ The temperature is maintained between 60-120°C. Lower molecular weight compounds are formed and there is the risk of degradation and loss carbon. The crystalline state of the cellulose can be modified and these effects may decrease the positive effects of the removal of lignin and cellulose. Solubilization of hemicellulose and lignin	Carrillo-Nieves et al., 2017; Kamoldeen et al., 2017; Lu et al., 2017; Ranganathan et al., 2017; Sindhu et al., 2017; Sun et al., 2016
Oxidative	Uses oxidizing compounds like hydrogen peroxide and peracetic acid. Low temperatures (30-60°C) In many cases, it is not selective oxidant and, therefore, there are losses of hemicellulose and cellulose. Peracetic acid is very selective with lignin and not present significant carbohydrate losses.	Katsimpouras et al., 2017; Lu et al., 2017
Ionic Liquids	Temperatures up to 110°C Complete solubilization of the lignocellulosic biomass by the breakdown in the polysaccharides of the H bonds by the anions of the ionic liquids. The process cannot be applied at industrial level because of the high cost of ionic liquids. Requires recovery of ionic liquids	Kim et al., 2017; Li et al., 2016; Parthasarathi et al., 2016; Yamada et al., 2017

Table 3. (Continued)

OrganoSolv	Organic solvents such as methanol, ethanol, acetone, ethylene glycol	Zhao et al., 2017; Mesa et al., 2011; Zhao et al., 2009
	Temperatures between 150-200ºC and low pH (2.0-3.4)	
	High cost of solvents and requires solvent recovery	
	Solubilization of hemicellulose and lignin	
Biological		
Fungal	White-rot, Brown-rot and Soft-rot fungi	Sindhu et al., 2016; Wan and Li 2012; Sánchez 2009; Singh et al., 2008
	Pleurotus ostreatus, Phanerochaete sordida, Pycnoporus cinnabarinus, Sporotrichum pulverulentum	
	Solid State Processes	
	Fungi produce enzymes that degrade hemicellulose and lignin	
	Very slow processes	
Bacteria	*Sphingomonas paucimobilis, Bacillus circulans, Anaerocellum thermophilum, Clostridium thermocellum*	Sindhu et al., 2016; Kurakake et al., 2007; Singh et al., 2008
	Very slow processes	
	The products are lactate, ethanol, acetate, CO2, and hydrogen	
Combinations		
Thermal-acid	The external acid catalyzes and solubilizes hemicellulose.	Ra et al., 2017; Sharma et al., 2016; Zaafouri et al., 2017
	The external acid reduces the optimum temperature and gives a better enzymatic substrate.	
	Extraction of hemicellulose and lignin can reorient cellulose to a more crystalline form.	
	Considerable production of furfural	
Thermal-alkaline	Temperatures of 100-150ºC with addition of Ca(OH)2.	Sharma et al., 2016; Gandi et al., (1997), Chang et al., (1998); Chang et al., (2001)
	Increases digestibility of biomass containing low lignin	
Thermal-oxidative	Uses peracetic acid and treated steam at 231°C.	Ando et al., (1988); Klinke et al., (2002)
	The soluble sugars produced are mainly polymers.	
	Phenolic monomers are not final products during oxidation, but are further degraded to carboxylic acids.	
	The production of furfural and HMF is low during oxidation.	
Thermal, alkaline and oxidative	Low sugar degradation as a result of using the relatively low temperature (150°C).	Brienzo et al., 2016; Chang et al., (2001)
	The digestibility of the treated biomass is 13 times greater than the untreated biomass.	
Microwave-assisted ionic liquids	The lignocellulosic material is directly solubilized in an ionic liquid in the presence of microwaves.	Kostas et al., 2017; Wang et al., (2011); Wu et al., (2012)
	The cellulose precipitates in the presence of water.	
	The lignin remains in the solution.	
	Similar results to those obtained in steam explosion and chemical pretreatments.	
Ammonia and CO2 Explosion	Ammonium loadings around 1: 1 (kg of ammonium / kg dry biomass) at room temperature for 10-60 days, or at temperatures up to 120°C for a few minutes.	Kim e Hong (2001); Alizadeh et al., (2005)
	Carbon dioxide can be applied as supercritical carbon dioxide (35°C, 73 bar), increasing yield.	
Bio-OrganoSolv	*Ceriporiopsis subvermispora* grown for up to 8 weeks	Itoh et al., 2003; Muñoz et al., 2007; Monrroy et al., 2010

From an economic view, the pretreatment is a pivotal step in the bioconversion process, once it can improve the separation between the components of cell wall, avoiding formation of compounds that inhibit hydrolysis and subsequent fermentation processes. The pre-treatments can be physical, chemical, biological or a combination of these different processes (Table 3). These pretreatment have been applied in different biomass, for example, alfalfa fiber, alfalfa newsprint, aspen wood chips bagasse, barley straw, Bermuda grass, big bluestem, brassica carinata residue, cane leaves, coastal Bermuda grass, corn stalk, corn stover, cotton straw, eucalyptus softwood bagasse, green hay, hardwood, loblolly pine, miscanthus, olive pulp, olive stones, peanut, pine, poplar wood bagasse, prairie cord grass, reed canary grass, rice hulls, rice straw, rye straw, sawdust, soybean hull, straw, sugarcane bagasse, sweet sorghum bagasse, switchgrass, timothy grass, waste cotton, wheat bran, wood (Kumar et al., 2009; Sarkar et al., 2012; Saini et al., 2015).

Usualy, there are two main aims for pretreatments. The first one is when pretreatment is used in order to obtain hemicellulosic hydrolysate, rich in xylose, pentose that can be converted into bioethanol by some microorganisms, mainly yeasts. For example, Brito et al. (2017) reported that the diluted sulfuric acid-catalyzed hydrolysis (30% of dry biomass in 5% H_2SO_4 at 121°C for 60 min) was effective to removal of 88.4% of the hemicellulose of palm press fiber (PPF), a co-product generated during the palm oil milling process, obtaining hemicellulosic hydrolysate. This hydrolysate was fermented by *Scheffersomyces stipitis* NRRLY 7124, producing 12.1 L of ethanol per ton of palm press fiber. Also, Keshav et al. (2017) evaluated bioethanol production by coculture of *Saccharomyces cerevisiae* VS3 and *Pichia stipitis* NCIM3498 in cotton stalk hydrolysate. Cotton stalk was pre-treated with NaOH (3% w/v; room temperature for 24 h) with lignin removal of 52.48%, followed by two-stage dilute acid hydrolysis (1% H_2SO_4; 110°C for 30 min in stage-I and 3% H_2SO_4, 130°C for 60 min in stage-II), releasing 29.40 g/L of sugars and holocellulose hydrolysis of 63.50%. Batch fermentation of detoxified hydrolysate resulted in 11.64 g/L of bioethanol (yield, 0.47 g/g). The second aim is when pretreatment is used only for fractionation of lignocellulosic material with removal of the C5 fraction and lignin, exposing the cellulose for glucose solubilization. For example, Barrera-Martínez et al. (2016) pretreated the sugarcane using ozone as oxidant agent to remove lignin, with subsequent acid or enzymatic hydrolysis. The lignin oxidation with increase in total sugar (30.83% and 18.41%, respectively) was observed with reducing sugars (27.17% and 13.43%, respectively) produced after acid or enzymatic hydrolysis. Moreover, Auxenfans et al. (2017) evaluated combined steam explosion pretreatment with dilute sulfuric acid impregnation in three different biomasses: *Miscanthus* × *giganteus*, poplar and wheat straw, to promote glucose release by enzymatic hydrolysis. The authors observed that pretreatment resulted in removal of the hemicellulosic fraction, in significant modifications in the lignin structure and its redistribution, favoring the accessibility of cellulose. Despite behaviour, the differences observed between woody and non-woody biomasses, according to the authors, the changes in syringyl/guaiacyl ratio and cellulose crystallinity were not a relevant factor to enzymatic digestibility. In addition, pretreatment of beech wood residual biomass by using acetone/water oxidation (hot acetone, water, and oxygen) was reported by Katsimpouras et al. (2017). In that work, the optimized conditions that resulted in separation of lignin were: acetone/water ratio 1:1, 40 atm initial pressure of 40 vol% O_2 gas (20°C) and 64 atm at reaction temperature of 175°C for 2 h incubation. The obtained hydrolysate (55.1% cellulose conversion) was fermented by *Saccharomyces cerevisiae* strain Ethanol Red®, resulting in 64.6% ethanol yield (72.2 g/L ethanol).

Presently, new developments of biomass pretreatment have been investigated in order to enhance the efficiency of sugars releasing. For example, hydrodynamic cavitation was simultaneously employed with alkaline solution, in order to improve the efficiency of sugarcane bagasse pretreatment (Téran-Hilares et al., 2016; 2017). In optimum conditions, up to 94% yield of enzymatic hydrolysis of cellulose and hemicellulose were obtained within 30 min of pretreatment, corresponding to significantly better results when compared to alkaline and combined ultrasound assisted alkaline processes. The use of ionic liquids also corresponds to a potential alternative and efficient pretreatment method for the removal of lignin and hemicelluloses, and for the reduction of cellulose crystallinity prior to its enzymatic hydrolysis. Recently, Elgharbawy and co-workers (2016) published a very detailed review on this subject, considering also ionic liquids cellulase-compatible systems for simultaneous pretreatment and saccharification of biomass. Moreover, Xu et al. (2016) reported a compatible cellulase-ionic liquid system derived from *Penicillium oxalicum*, improving the saccharification rate of rice straw from 47% to 83%. Ionic liquids, mainly 1-ethyl-3-methylimidazolium acetate (Bian et al. 2014), are known to be non-volatility, thermal-stable and very effective in biomass delignication and cellulose solubilization (Bezerra; Ragauskas, 2016). In addition, Hashmi and co-workers (2017) recently compared the properties of sugarcane bagasse pre-treated via autohydrolysis, with the one from 1-butyl-3-methylimidazolium acetate pre-treatment, verifying more amorphous cellulosic structure, and hence, enhanced digestibility during enzymatic hydrolysis. The wide variety of strategies for improving the accessibility of enzymes to polysaccharides is a clear indicative of the challenges that still need to be overcome for the conversion of biomass into ethanol. However, the unquestionable importance of the full exploitation of renewable resources for the production of chemicals and fuels still motivates researchers in this field.

Detoxification

During some pretreatment types, some by-products that negatively interfere in the fermentation process can be released and/or generated. For example: acetic acid can be formed by hydrolysis of acetyl group of hemicellulose; formic and levulinic acids are sugar degradation products; phenolic compounds are formed mainly by the partial degradation of lignin; and formaldehydes or aldehydes furans (furfural and 5-hydroxymethylfurfural) are formed by degradation of pentoses and hexoses, respectively.

All these compounds as well as metal ions released from the hydrolysis equipment are potent inhibitors of microbial metabolism (Kelly et al., 2008; Silva et al., 2004; Palmqvist; Hahn-Hängerdal, 2000b; Felipe et al., 1995).

The presence of inhibitory compounds in the hydrolysates often imposes the need for a detoxification step before using as fermentation medium aiming to reduce/remove the concentration of these inhibitors (Silva et al., 2014) that are partially removed/reduced or converted in inactive compounds (Chandel et al., 2012). The choice of the detoxification method depends on the type of hydrolysate and the microorganism tolerance to inhibitors.

Different physical-chemical treatments or biological treatments are reported to detoxify hydrolysates. Neutralization methods, adsorption by ion exchange resins, adsorption by activated carbon and overliming with calcium hydroxide, and microbiological have been proposed as potential detoxification strategies, individually applied or in combination (Table 4).

Among different methodologies, the overliming is considered the most widely used technique due to the high effiency in removal of toxic inhibitors, such as furfural and hydroxymethylfurfural (Cardona et al., 2010). During overliming, the acids are removed from the hydrolysate by addition of calcium hydroxide (or other hydroxide), resulting in a pH adjustment, condition under which a precipitate is formed (Alves et al., 1998; Marton et al., 2006). This precipitate consists mainly of calcium salts of low solubility, such as calcium sulfate, which is capable to precipitate acidic compounds. Moreover, the overliming technique causes other beneficial effects, including removal of phenolic compounds, precipitation of metal ions, and conversion of furfural to furfurylic acid (Marton et al., 2006).

Enzymatic Hydrolysis of Lignocellulosic Biomass

The carbohydrate fraction of pretreated biomass can be enzymatically hydrolyzed in order to release monomeric sugars as glucose, xylose, arabinose and cellobiose, which are used in subsequent fermentation steps for different biomolecules production (Cardona et al., 2010; Terán Hilares et al., 2017). The enzymatic hydrolysis requires three steps: adsorption of enzyme into carbohydrate fraction, degradation of carbohydrate fraction into monomeric sugars, and desorption of enzyme from substrate (Sant'Ana da Silva et al., 2016).

Table 4. Different detoxification methods: Conditions, advantages and disadvantages

Detoxification Methods	Conditions, Advantages and Disadvantages	References
	Physical	
Evaporation	Separations of volatiles and non-volatiles compounds	Palmqvist & Hahn-Hägerdal (2000a)
Extraction	Organic solvents, 3:1 organic phase: aqueous phase (v/v).	Palmqvist e Hahn-Hägerdal (2000a)
Adsorption	Supercritic solvents in co-current with hydrolysate, 20 MPa, 40 °C; followed by depressurization	Alves et al. (1998), Lee et al. (1999), Marton et al. (2006)
	Activated charcoal. Addition 2.5 - 10% (m/v).	
	Removal of furans, phenols and acetic acid.	
	Chemical	
Alkaline (Overliming and liming)	Addition of Ca(OH)2 ou CaO until pH 7 - 10.5 followed by pH adjustement with acid (pH 5.5 - 6.5).	Alves et al. (1998), Martinez et al. (2000), Chandel et al. (2007); Larsson et al., 1999
	Removal of furfural, HMF and phenolic compounds.	
Combined alkaline	Addition of KOH until pH 10, followed by pH adjustment until 6.5 with HCl and addition of sodium sulfite 1% at 90°C.	Palmqvist e Hahn-Hägerdal (2000a)
	Removal of ketones, aldehydes and volatile compounds.	
Ions exchange resins	Ratio resin/hydrolysate 1:10. Agitation for 1 h at room temperature.	Larsson et al., 1999; Chandel et al., (2007); anilha et al., (2010)
	Removal of furans, phenols and acetic acid.	
	Biological	
Microbiological	Trichoderma reesei, Pseudomonas putida ou Streptomyces setonii.	Palmqvist e Hahn-Hägerdal (2000a), Cheng et al. (2008) & Fonseca et al. (2011)
	Issatchenkia occidentalis	
	Removal of phenols, furfural and acetic acid. 5% loss of sugars.	
Enzymatic	Lacase from Cyathus stercoreus.	Larsson et al., 1999; Moreno et al., 2012
	Lacase and lignin peroxidase from Trametes versicolor	
	Rotary agitation at 100 rpm for 4 - 12 h at 30°C.	
	Removal of phenols.	
	Without removal of furans and acetic acid.	
	Minimal loss of sugars.	

For complete hydrolysis of cellulose in glucose monomers, some synergetic action enzymes are required. Among them, the hydrolases are the main biocatalysts, including exo-1,4-β- glucanases (EC 3.2.1.91 and EC 3.2.1.176), endo-1,4-β-glucanases (EC 3.2.1.4) and β-glucosidases (EC 3.2.1.21) (Maeda et al., 2011; Van Dyk and Pletschke, 2012).

Each enzyme presents specific activity: exo-glucanases act in the reducing and non-reducing ends of cellulose chains; endo-glucanases cleave internal β-glycosidic bonds in the cellulose chains, reducing the degree of polymerization; and the β-glucosidases release glucose units from cellobiose (Van Dyk; Pletschke, 2012). Also, other enzymes can increase the efficiency of classical hydrolytic cellulases acting on the surfaces of the insoluble substrate, where they promote chain breaks in the polysaccharide molecules in the crystalline matrix. These enzymes correspond to previous known Carbohydrate Binding Modules (CBM33) and Glycoside Hydrolases (GH61), reclassified in the CAZy database to AA9 and AA10 respectively. They are the copper-dependent oxidative enzymes called Lytic polysaccharide monooxygenases (LPMOs), with specific characteristics and action mechanism (Walton; Davies, 2016; Vaaje-Kolstad et al., 2017).

Moreover, for the other carbohydrate macromolecular fraction, the hemicellulose, it is required a more complex of enzymes capable to cleave the internal glycoside linkages (xylanases) and branch point-degrading enzymes (α-L-arabinofuranosidase, acetylxylan esterase, and α-D-glucuronidase) (Meng; Ragauskas, 2014; Moreira; Filho, 2016).

Actually, the hydrolysis yield of carbohydrate fraction is influenced by different factors, such as solid composition, enzyme loading, composition of blend enzyme, temperature of process, crystallinity, degree of polymerization, composition and structure of biomass, efficiency of pretreatment and presence of surfactants (Eriksson et al., 2002; Yang et al., 2011; Moreira; Filho, 2016). A key factor for the process is the enzyme accessibility to substrate; therefore, one of the objectives of pretreatment is to increase the enzyme accessibility, which can be obtained by lignin removal, reduction in crystallinity and polymerization degree, and hence, increasing in the porosity of the material (Zhao et al., 2012; Sun et al., 2016; Terán Hilares et al., 2016). An increase in porosity of the pretreated material has been reported for different types of pretreatment. For example, in the technical reports of company E-PIC srl. (2016), an increase in the porosity, surface area and volume in rice straw after pretreatment with hydrodynamic cavitation was reported, which were attributed to mechanical (microjets and shockwaves) and chemical (lignin removal) effects. By using a similar approach, higher enzymatic digestibility yield values (more than 90%) for sugarcane bagasse pretreated was reported by Terán Hilares et al. (2017). In other pretreatment as steam explosion, hemicellulose is partial solubilized and lignin is reorganized, exposing the cellulose surface for enzymes attack (Zhang et al., 2017). Commonly, enzymatic hydrolysis in order to evaluate the efficiency of pretreatment are usually carried out in Erlenmeyer flasks loaded with 2-5% of solids (Ahmed et al., 2016; Zhang et al., 2017).

However, to make biomass enzymatic conversion economically feasible by producing high sugars concentration after hydrolysis, the use of high solid loading in the reactor is required (10-15% or above); however, the increase in hydrolysis yield is not linear with increasing in solids content (Wang et al., 2011). When complex substrates as lignocelluloses are used, the adsorption of enzyme is not correlated with efficiency of conversion because some fractions of enzymes are adsorbed into lignin, reducing the efficiency of hydrolysis of carbohydrate fractions, thus requiring additional enzyme loading (Huang; Penner, 1991; Eriksson et al., 2002). Also, when high solids loading is used, long process time is usually required, turning the process more expensive (Modenbach; Nokes, 2013).

The enzyme activity can also be reduced or inactivated by irreversible adsorption of cellulases on cellulose (Converse, 1998; Ma et al., 2008). In order to avoid this phenomenon, the use of surfactants is an alternative, favoring the desorption of enzyme from biomass, avoiding the unproductive adsorption of enzyme onto lignin (Eriksson et al., 2002; Lin et al., 2017). Additionally, surfactants were reported as stabilizing compounds for enzymes, preventing their denaturation during the hydrolysis (Yoon; Robyt, 2005). Besides, they can change the nature of the substrate, increasing the cellulose surface by residual lignin removal (Helle et al., 1993; Kristensen et al., 2007). As recent report by Agrawal et al. (2017), an increase in 43.4% of enzymatic hydrolysis yield of dilute acid pretreated wheat straw was observed by the addition of surfactant tween 80 in 5% v/v. In other study reported by Zhang et al. (2017), the addition of surfactant (bovine serum albumin – BSA and Tween 80) during enzymatic hydrolysis of sugarcane bagasse pretreated with ferric chloride (160°C, 10 min, and 0.1 M $FeCl_3$) was evaluated. In that study, commercial enzyme cellulose (Cellic CTec2, Novozymes) was used and different enzymes loading (5, 10 and 20 FPU/g substrate) were added to biomass. In the same study, a control assay using 5 FPU/g substrate (without surfactant addition) resulted in 22.7% of hydrolysis yield in 24 h, value increased to 61.6% and 62.2% by addition of BSA and Tween 80, respectively. Additionally, using 10 and 20 FPU/g substrate, no difference was observed in hydrolysis yield.

The use of adequate reactors is other pivotal factor to be taking into account to favor enzymatic hydrolysis of biomass (Humbird et al., 2010; Fockink et al., 2017). During enzymatic hydrolysis with high solid loading, the heat and mass transfer limitations can be observed, including boundary layer resistance and resistance to internal enzyme diffusion (Fockink et al., 2016). Additionally, the concentration of end products of hydrolysis as glucose and cellobiose is higher, resulting in inhibition of enzyme activity (Sun; Cheng, 2002). Thus, the saccharification and fermentation process (SSF) has been developed. Specifically, released sugars are simultaneously converted into ethanol or other biomolecules, reducing the concentration of end products of hydrolysis in the medium (Rodrigues et al., 2016; Terán Hilares et al., 2017).

Fermentation of C5 and C6 and Different Process Configurations: SSF, SHF, CBP

The released sugars after acid or enzymatic hydrolysis can be converted in several products by using yeast, filamentous fungi or bacteria. Ethanol is the most known fermented product obtained from C6 by using the yeast *Saccharomyces cerevisiae* (that drives glucose the glycolytic pathway). Different strategies to improve the ethanol productivity have been reported, e.g, batch, feed batch and continuous process

(Laopaiboon et al., 2007). However, this yeast specie is not capable of fermentation of C5 sugars in its wild form.

C5 sugars (e.g., xylose) can be fermented in ethanol by other yeasts such as *Scheffersomyces stipitis*, *Scheffersomyces shehatae* (that present specific enzymes that drive pentose compound to specific pathways of conversion); however, usually low conversion efficiency is observed compared to *S. cerevisiae* C6 fermentation. Currently, different new-engineered strains of *S. cerevisiae* have been reported, which convert efficiently C5 and C6 into ethanol. For example, in the work of Ko et al. (2016), glucose and xylose were efficiently co-fermented into ethanol by engineered *S. cerevisiae*, achieving 50 g/L of ethanol (yield of 0.43 g of ethanol/ g of sugars) from 70 g/L of glucose and 40 g/L of xylose present in enzymatic hydrolysate. Additionally, around 18-21 g/L of ethanol (0.43-0.46 g of ethanol/g of sugar) was produced from acid hydrolysate of rice straw, using the same engineered strain.

Different alternatives for fermentation process have been developed, considering the possibility of integration with hydrolysis step and co-fermentation of C5 and C6 sugars. These alternatives include Separate Hydrolysis and Fermentation process (SHF), Simultaneous Saccharification and Fermentation (SSF) or Co-fermentation (SSCF) and Consolidated bioprocess (CBP), as schematically show in Figure 2 for ethanol production from pretreated biomass.

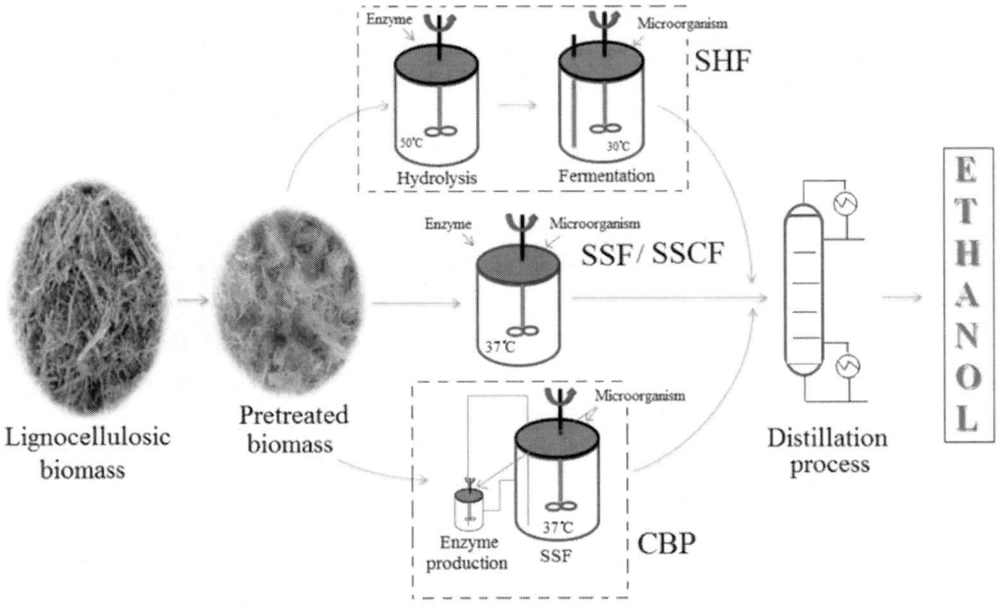

Figure 2. Schematic representation of different configurations for ethanol production from lignocellulosic biomass. SHF - Separate Hydrolysis and Fermentation process, SSF - Simultaneous Saccharification and Fermentation or SSCF - Co-fermentation and CBP - Consolidated bioprocess. SSF/SSCF and CBP alternatives can be performed in single or interconnected reactors.

Fermentation is usually carried out after hydrolysis steps, which is known as Separate Hydrolysis and Fermentation process (SHF). This configuration presents an advantage to the possibility to carry out the hydrolysis and fermentation steps in optimized temperatures and pH conditions (Singh et al., 2013). However, the main disadvantage of SHF corresponds to inhibition of enzyme activity by accumulation of product (glucose and cellobiose) in hydrolysis process and the need of different vessels, increasing the number of required equipment (Dahnum et al., 2015). As an alternative, Simultaneous Saccharification and Fermentation (SSF) process is proposed. In this process, the released sugars are simultaneously converted into ethanol as they are hydrolyzed from the material. When C5 is co-fermented with C6, this alternative is called Simultaneous Saccharification and Co-Fermentation (SSCF). This process present advantages compared to SHF, allowing reducing the number of vessel in the process, the fermentation process time and fermentation cost, avoiding the accumulation of inhibitory product of hydrolysis (Wingren et al., 2003). For example, in the work of Alfani et al. (2000), the fermentation time required in a SSF process was shorter (≈30 h) compared to the SHF process (96 h). Additionally, in that work the SSF alternative resulted in large increase in ethanol productivity (0.837 g/L.h for SSF compared to 0.313 g/L.h for SHF). The main disadvantages of SSF/SSCF process correspond to development of the process in non-optimized temperature and pH condition (Zhu et al., 2012; Dahnum et al., 2015) once is carried out using a same vessel. Alternative studies were proposed in order to improve the efficiency of process kepting the optimal temperatures for each step. For example, in the study of Viola et al. (2013), a two-chamber reactor separated by a membrane (acrylic tissue with 10 μm pores) was used to improve enzymatic hydrolysis and fermentation of lignocellulosic materials. In that work, by using the proposed system, an increase in 20% in ethanol yield was reached compared to the value obtained using only one vessel. A different alternative using interconnected column reactors process was recently reported by Terán Hilares et al. (2017). In that study, hydrodynamic cavitation pretreated SCB (pretreatment conditions: 3 bar of inlet pressure, 0.3 M of NaOH, 20 min at 60°C) was enzymatically hydrolyzed (hydrolysis conditions: 11% solid loading, 20 FPU/g of biomas) and the sugars released was simultaneously fermented into ethanol by immobilized *Sheffersomyces stipitis*. Two different columns were used: the hydrolysis was carried out in a reactor at 50°C loaded with pretreated bagasse and the fermentation was carried out at 30°C using cells immobilized in calcium alginate. In that process, 17.26 g/L of ethanol (0.48 g of ethanol/g of sugars consumed) was produced and 62.33% of carbohydrate fraction was hydrolyzed. Other recent work was also reported by Akhtar et al. (2017). In that study, microwave-alkali-acid pretreated rice straw was used. Initially, biomass was pretreated with 1% of NaOH during 3 min at 850 W and 150°C, followed by acid pretreatment using 1% (v/v) H_2SO_4 to achieve a liquid-to-solid ratio of 10:1 (v/w) for 1 h at 121°C. Under optimized SSF conditions (substrate loading, 11% (w/v); enzyme concentration, 0.5% (v/v); inoculum size, 8% (v/v); pH, 4.5 at 30°C after

24 h), 25.2 g/L of ethanol was produced using a lab-scale bioreactor, corresponding to ethanol yield of 0.38 g/g.

Consolidated bioprocess (CBP) is another interesting option, which integrates biological events required for biomass conversion (enzyme production, hydrolysis of carbohydrate fraction and fermentation of C5 and C6) in a single reactor. For example, a CBP process for ethanol production was reported by Liu et al. (2015). In that work, the alkaline-pretreated sugarcane bagasse was supplemented with wheat bran, and used as substrate to produce enzymes by *Trichoderma* and *Penicillium*, releasing fermentable sugars, achieving 20 g/L in 72 h. The obtained sugars were fermented by *Saccharomyces cerevi*siae in ethanol, verifying 5.8 g/L (40.84% of theoretical yield) after 24h of fermentation.

Biomass Genetic Manipulation and Alternative Pretreatment Methods for Improved Ethanol Production

Currently, the technology for the production of first generation ethanol is well established and is widely exploited in several countries around the world. However, even considering the high yields obtained by the current methods, any improvement that optimizes the carbon utilization of raw materials is desirable from the economic point of view, since this step corresponds to 60% of the total costs of manufacturing (Wisselink, 2007). On the other hand, the production of second-generation ethanol is still an emerging technology whose success depends on several factors, e.g., the biomass source and its further pretreatment method, the fermentation microorganisms, the conditions of the fermentative process, among others. Several drawbacks still need to be overcome in order to convert it into an efficient and economic viable route, which is the main objective of several research groups mainly focused on genetic manipulation of microorganisms and plants, and on the search of novel, non-conventional, biomass pretreatment methods. This section aims to show some of the tools currently used to improve bio-based ethanol production.

Genetic Manipulation of Microorganisms

Several research groups around the world have been relying on molecular biology, genetics and metabolic engineering to improve the production of first-generation ethanol and to convert second-generation ethanol into a technically and economically viable procedure. Researches focusing on the genetic modification of microorganisms have been intensively carried out in the last 20 years, specially in the case of *Saccharomyces cerevisae* yeasts, widely used in the production of first-generation ethanol, aiming to improve their tolerance in the adverse conditions normally considered in general bioprocesses, such as the high temperature that the fermentation tanks can reach during the bioprocesses, the alcohol content during fermentation, the substrate concentrations and possible bacterial contaminations. Within this context, authors have been studying improved production of ethanol with the aid of microbial genetic tools, by different tools. For example, Kim and co-workers (2011) identified several genes (CMP2, IMD4, SSK2, PPG1, DLD3, PAM1, and MSN2) in *S. cerevisiae* strains that make them alcohol- and thermo-tolerant yeasts, resisting to concentrations of ethanol up to 15% and temperatures around 42°C. Also, Yang and co-workers (2011) used "global transcriptional machinery engineering" (gTME), compared the transcriptional profiles of two strains of *S. cerevisiae*, and found eighteen genes associated with ethanol tolerance, from which they created an alcohol-tolerant mutant. Further studies, as the one reported by Vanegas et al. (2012), elucidated the role of ergosterol and unsaturated lipids present in the yeasts cellular membrane, which allowed building alcohol- and thermo-tolerant mutants. Through metabolic engineering, it was also possible to produce osmotolerant *S. cerevisae* strains, by mitigating or replacing the glycerol production pathway by reducing acetate (Papapetridis et al., 2017).

In addition to the production of first-generation ethanol, several *S. cerevisiae* strains were genetically modified, making it able to convert xylose to ethanol, improving their metabolic flux. Moreover, Traff et al. (2002), Toivari et al. (2004) and Chang et al. (2007), used modified *S. cerevisiae* strain for metabolizing pentoses, mainly xylose, through overexpression of aldose reductase and xylitol dehydrogenase native genes. Additionally, in order to overcome the limitations related to the fact that *S. cerevisiae* does not have a specific xylose transport system, Hamacher et al. (2002) and Sedlak et al. (2004) showed that Hxt7, Hxt5, Gal2, Hxt1, and Hxt4 systems are permissive for xylose, which allows, by expressing these genes, to improve the absorption and metabolization of C5 sugars by the yeasts and, consequently, the ethanol productivity through the concepts of evolutionary engineering (Hector et al., 2008; Runquist et al., 2010; Diao et al., 2013; Gonçalves et al., 2014).

For the production of second-generation ethanol, some interesting studies reported on the development of mutant strains of *S. cerevisiae* tolerant to the toxic compounds commonly present in the hydrolysates, such as furans and their derivatives. Within this context, Almeida et al. (2009) identified genes encoding reductases that act on the oxidation of furans, and the overexpression of these genes enhanced *S. cerevisiae* growth and fermentation rate in culture medium containing 5-hydroxymethylfurfural. More recently, some wild strain species of the genus *Saccharomyces* yeasts, used in industrial fermentations, were tested in culture media with different concentrations of furfural, in order to select them for the production of second-generation ethanol (Field et al., 2015; Kim et al., 2017). Once engineered *S. cerevisiae* strains were used, the intracellular spermidine (polyamine) content increased cell tolerance against various fermentation inhibitors, including furans and organic acids.

Despite the technological success of using mutant strains of *S. cerevisiae* for better performances in fermentation processes for the production of first and second-generation ethanol, this strategy is still limited by the negative public opinion (Lennartsson et al., 2014), as the use of GMOs in industrial bioprocesses may cause environmental and public health risks. The utilization of antibiotics to control microbial contaminants in industrials fermenters or as transformation markers can generate and release into the environment antibiotic-resistant organisms, corresponding to relevant examples of this scenario (Kang et al., 2014).

Therefore, bioprospecting studies on microorganisms with potential characteristics to be used in the production of ethanol have been highlighting in recent years. From bacteria with versatile metabolisms such as *Escherichia coli*, *Klebsiella oxytoca* and *Zymomonas mobilis*, which can ferment pentoses and hexoses to ethanol, to yeasts that ferment only xylose, such as *Pichia*, *Candida*, *Schizosaccharomyces* and *Pachysolen genus*, many studies demonstrated their capacity of considerably improve ethanol production (Dien et al., 2003; Hahn-Hägerdal et al., 2007; Mussato et al., 2012). Thermo-tolerant species such as some *Kluyveromyces marxianus* strains have also been evaluated for second-generation ethanol production (Castro; Roberto, 2014). Additionally, secretomic analysis of filamentous fungi that produce hydrolases (such as cellulases and xylanases, important in the production of second-generation ethanol), have recently been carried out. These studies were mainly conducted with the *Penicillium echinulatum*, *Trichoderma reesei* and *Aspergillus niger* (Ribeiro et al., 2012; Borin et al., 2015; Schneider et al., 2016).

Genetic Manipulation of Biomass

Many plants, such as corn, rice, sugarcane and beet, can also be genetically modified for the efficient production of ethanol. This strategy is less common, however also relevant, the modification of some perennial and woody plants with rapid growth were reported (Knauf; Moniruzzaman, 2004; Sticklen, 2004), highlighting the broadness and potential of this strategy.

The ethanol obtained from genetically modified plants, either to increase the sugar content in their composition or to decrease the corresponding lignin content, are considered major advances in bioenergy. In addition, some genetically modified plants are engineered to resist to agricultural pests. The main example to be mentioned is *Saccharum officinarum* species produced by the Sugarcane Technology Center (CTC, Brazil), proved to be resistant to *Diatraea saccharalis* (known as "broca da cana-de-açúcar"), responsible for considerable damages to sugarcane fields in Brazil (ÚNICA, 2017).

Despite having a complex genome and a large phenotypic variation of germplasm, the variety of current genomic technologies will enable better understanding of the critical differences in the composition of biomass to identify targets for the improvement of sugarcane and to increase the productivity of biofuels (Hoang et al., 2015).

The main genetic modifications of sugarcane species are carried out in their cell walls, by increasing the cellulose content and reducing its crystallinity, and/or altering the composition of non-cellulosic polysaccharides or lignin, making possible to modify the chemical bonds present in the biomass components in order to facilitate its deconstruction (Furtado et al., 2014).

Many studies are driven to reduce the biomass recalcitrance, by reducing the lignin content and changing its structure and composition, improving the ethanol production yield. Some studies in this regard were carried out by manipulating hemicellulose biosynthesis through the expression of microbial hydrolases at high levels, promoting the autohydrolysis of biomass. These studies indicated that if genetic manipulation could be successful, ethanol yield at low costs and high yield could be achivied, without affecting the integrity of the cell wall or the plant development (Li et al. 2014).

Although the genetic modification of plants in order to achieve better yields of ethanol production is already a reality, much is still to be developed. The lignin biosynthesis and its role in plant biology are not completely elucidated (Poovaiah et al., 2014), and are in the target of several researchers around the world that constantly contribute to the implementation of ethanol as a viable and accessible fuel.

Third Generation Bioethanol from Microalgae

In the last years, microalgae have received great attention as a renewable source for biofuels, mainly for biodiesel and bioethanol production, due to composition rich in lipids and carbohydrates (Odjadjare; Mutanda; Olaniran 2015; Singh et al. 2016; Shokrkar; Ebrahimi; Zamani 2017). Some species from genus *Chlorella*, *Dunaliella*, *Chlamydomonas*, *Scenedesmus*, and *Spirulina* are known by their particularly high carbohydrate content that can reach over 50% of the dry cell weight under specific culture conditions (John et al. 2011). Thus, microalgae are a versatile raw material for biofuels. Actually, their use has some advantages, as fast growth rate and high biomass productivity, requiring small physical spaces for cultivation when compared to other crop. Besides, the microalgae can use wastewater as culture medium for biomass production, removing chemical and organic contaminants, heavy metals and pathogens and thus, acting concomitantly as effluent treatment method (Ansari et al. 2017; Brennan; Owende 2010).

The ethanol from microalgae is called " third generation (3G)" and its production can be performed through the alcoholic fermentation of the carbohydrates present in the biomass, requiring a physical, chemical or biological pretreatment before fermentation (Harun; Danquah 2011). Disruption of microalgae cells for fermentable sugars release is required to produce bioethanol, considering the most of carbohydrates are present inside the cell wall, or intracellularly as energy storage in the form of starch. Moreover, hydrolysis of polysaccharides is required to use the monosaccharides for the fermentation step. Different methods have been studied for cell rupture and hydrolysis of carbohydrates of the cell wall, such as: acid and alkaline hydrolysis, autoclaving hydrolysis, microwaves and enzymatic hydrolysis (Hernández et al. 2015). The carbohydrate production and hydrolysate composition is influenced by the type of microalgae or hydrolysis method (Table 5).

Table 5. Sugars profile of hydrolysates obtained from different microalgae

Microalga	Pretreatment	Sugars profile (%)*	Total carbohydrate content (%)	References
C. sorokiniana	Acid hydrolysis (H_2SO_4, 30 min, 121°C)	Glu (70.8), Xyl (13.8), Mal (7.7), Rha (6.5), Fuc (1.0)	18.2	Hernández et al. (2015)
S. almeriensis		Glu (52.2), Xyl (33.4), Mal (n.d.), Rha (6.2), Fuc (7.5)	14.5	
N. gaditana	Enzymatic hydrolysis	Glu (59), Xyl (28.8), Mal (n.d.), Rha (6.5), Fuc (4.6)	11.2	Kim et al. (2014)
C. vulgaris	Enzymatic hydrolysis	Glu (48.2), Xyl (4.0), Mal (6.3), Rha (13.8), Man (6.3), Gal (23.7)	22.4	
P. tricornutum	Acid hydrolysis (TFA, 90 min, 121°C)	Glu (5.9), Gal (24.4), Rha (3.6), Man (37.8), Xyl (5.8), Fuc (4.9)	8.9	Ortiz-Tena et al. (2016)
N. salina		Glu (65.4), Gal (13.5), Rha (5.6), Man (1.9), Xyl (4.7), Fuc (5.0)	9.5	
P. Purpureum		Glu (71.0), Gal (14.3), Rha (n.d.), Man (n.d.), Xyl (12.9), Fuc (n.d.)	49.5	
C. vulgaris		Glu (76.5), Gal (11.4), Rha (3.0), Man (1.7), Xyl (2.1), Fuc (n.d.)	29.5	

* Glu – glucose, Xyl – xylose, Gal – galactose, Mal – maltose, Man – Mannose, Rha – rhamnose, Fuc – Fucose.

According to Ortiz-Tena et al. (2016), starch constitutes the majority of total biomass carbohydrates fraction with 46.5%, followed by mono- and oligosaccharides with 26.4%, and structural polysaccharides (cell wall) with 6.0%. A precise determination of the monosaccharide composition of crude microalgae is essential to use biomass for bioethanol production, especially to choose pretreatment and fermentation strategies. Carbohydrates from microalgae can be found in different forms such as storage (starch), structural (cell wall, for example, cellulose) or signaling (glycolipids or glycoproteins). Identification of sugars origin provides valuable information for the development of a strategy, for example, using hydrolytic enzymes. Besides, the microorganism selection for ethanol production in the fermentation step of the process is dependent of sugar profile of hydrolysate, considering e.g., characteristics as possibility of pentose fermentation.

The production of third generation bioethanol is influenced by several factors intrinsic in the microalgal biomass culture process (Figure 3), such as the culture medium, the bioreactor and process conditions, method of carbohydrates extraction and hydrolysis, among other factors. Microalgae can be stimulated to grow under three different metabolism conditions based on the carbon source: by using CO_2 as the sole carbon source (autotrophic); By using organic nutrients as carbon source, such as glucose, wastewater, leached and others (heterotrophic); and with both inorganic and organic carbon sources (mixotrophic).

For heterotrophic/mixotrophic cultivation, different mediums have been used, including carbon sources as glucose, glycerin, glycine, mannitol, sodium acetate, sodium bicarbonate (Cao et al. 2014; Bumbak et al. 2011). For example, Xu, Miao, and Wu (2006) studied the heterotrophic cultivation of the microalga *Chlorella protothecoides* addition 10 g/L of glucose in the culture medium. The results showed an increase of 45% of carbohydrate content in the cell. In other work, Guldhe et al. (2017) has shown the use of agricultural wastewater with carbon organic (COD - 96 mg/L), as a heterotrophic cultivation medium, has increased the productivity of carbohydrates in the *C. sorokiniana* biomass, allowing the production of biofuels. In mixotrophic culture, Ramsundar et al. (2017) studied the urea supplemented mixotrophic cultivation. The medium was prepared of urea (1500 mg/L) and domestic wastewater (185.5 mg/L) for *C. sorokiniana* cultivation. The results showed that urea supplemented mixotrophic cultivation further enhanced biomass and carbohydrate in 162.5 mg/L.d and 20.1 mg/L.d, respectively.

Autotrophic cultivation medium is usually composed mainly by nitrogen source and minerals. For marine microalgae, e.g., Guillard medium f/2 is usually applied (33.3 g/L NaCl, 75.0 mg/L $NaNO_3$, 5.0 mg/L $NaH_2PO_4 \cdot H2O$, 3.15 mg/L $FeCl_3 \cdot 6H_2O$, 22.2 g/L $ZnSO_4 \cdot 7H_2O$, 180 g/L $MnCl_2$, 6.3 g/L $Na_2MoO_4 \cdot 2H_2O$, 10 g/L $CoCl_2 \cdot 6H_2O$, 9.8 g/L $CuSO_2 \cdot 5H_2O$, 100 g/L Thiamine (B1), 0.5 g/L Cyanocobalamin (B12) and 0.5 g/L Biotin (B7)) (Guillard, 1975). The variety and concentration, depletion or repletion of nitrogen source influence the amount of carbohydrates in the microalgal cell, affecting the bioethanol productivity (Kim et al. 2014; Kim; Bae; Lee, 2016). According to Ho et al. (2013b), nitrogen depletion in the culture medium of the microalga *Chlorella vulgaris* FSP-E promoted the accumulation of carbohydrate in the microalgal cell, reaching values of 51% of biomass containing 93% of glucose in carbohydrate fraction.

The commonly systems used for microalgae biomass production in phototrophic cultivation are classified in two types: photosynthetic ponds (raceway pond) formed by large shallow tanks constructed of concrete, glass fiber or polycarbonate and mechanically agitated; or advanced photobioreactors which correspond to closed equipment constructed of flat or serpentine panels, spirals or cylinders made of glass, plastic or polycarbonate (transparent for light input), with mechanical stirring, bubble and airlift systems or static. Closed photobioreactors present advantages compared to ponds due to low contamination, better control of gas-liquid transfer rate, higher biomass density, greater exposure to light intensity, high productivity per area and low harvest cost (Olivieri; Salatino; Marzocchella 2014).

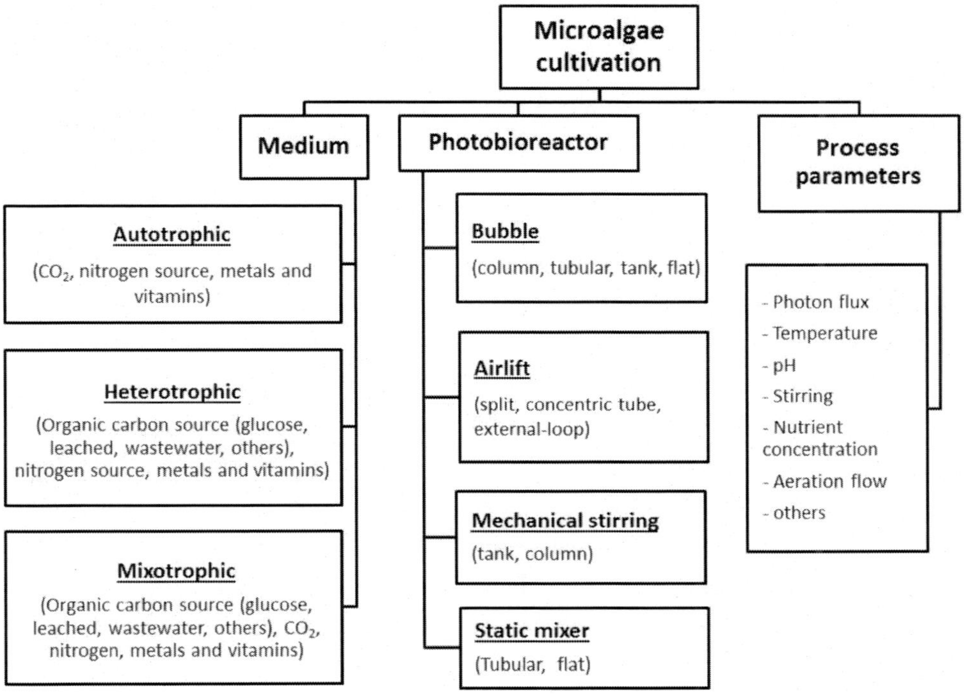

Figure 3. Factors influencing in cultivation and composition microalgal.

The production of bioethanol from microalgae biomass is certainly a sustainable and environmentally friendly approach (John et al., 2011). However, the study of new strains (Bahadar; Bilal Khan 2013), bioreactors and raceway (Olivieri; Salatino; Marzocchella 2014; Mendoza et al. 2013), culture medium (Kim et al. 2014), harvest (Gerardo et al., 2015), carbohydrate extraction methods (Lee; Lewis; Ashman 2012) and genetic engineering (Ghosh et al., 2016; Singh et al. 2016) have been carried out and new ones are required to increase productivity of the biomass with high carbohydrate content in the cell resulting in high productivity in the bioethanol production.

Conclusion

Bioethanol as a transportation fuel has a central role in transitioning of petroleum based products to biobased products under the umbrella of sustainable biorefinery. Geopolitical concerns worldwide have thrusted to switch over to sustainable green biomass based products from conventional petroleum refinery based fuels or chemicals.

Bioethanol has set an example in USA and Brazil as it has been successfully employed as renewable liquid transportation fuel. Other countries in Europe, Asia (China and India) are also making policies to blend ethanol in petrol to some extent which will surely impact on environment and economics as well. Ethanol currently is being blended to petrol in first generation ethanol, which is being produced either from corn (in USA) and sugarcane juice (in Brazil). However, in future lignocellulosic ethanol or so called second generation ethanol will take place central role in order to make biorefinery more viable and sustainable for energy and biochemical production on biorefinery concept. While first generation ethanol production is a well-established process, second generation ethanol process is still not mature. Biomass pretreatment and hydrolysis with desired yields under industrial parameters are the major bottlenecks. Enzyme cost and availability of a robust fermenting microorganisms which can utilize all forms of sugars from biomass in presence of cell wall derived inhibitors and giving similar ethanol yield and productivity as 1G ethanol are also major concerns. Process consolidation (hydrolysis and fermentation simultaneously) is an important necessity. In this line, simultaneous saccharification and co-fermentation or consolidated bioprocessing could play a revolutionary role using higher total biomass solids or sugars derived from biomass. Cost of commercially available cellulase is a major impeding factor for the successful commercial operations for 2G ethanol production. In-house and cost competitive cellulase production and subsequently its use to biomass saccharification with desired sugars yield can bring a breakthrough in 2G ethanol production. Simultaneous sugars concentration and detoxification by using specific membranes is possible. However, membrane cost is an important factor for their deployment in biorefineries. Integration of 2G ethanol production with 1G ethanol production or sugarcane processing mills is necessary to bring down the net ethanol production cost as both the process can use similar machineries and equipment saving operational and capital expenses. The lignin recovered after cellulosic sugars extraction should be sent to boilers for energy generation. However, regarding current studies, in a near future, lignin will be able to be converted in high value based products.

Third generation (3G) biofuels production is still in an early stage and needs more realistic approaches at commercial level. Micro and macroalga have potential to serve as an excellent feedstock not only for biofuels production but also high commercial value based bioproducts of health and nutritional value. Judicious technological exploration of algal biomass could bring paradigm shift in biobased economy in near future. However, for the successful realization of biofuels at commercial scale still needs more scientific and technological improvements. All the industries who have started 2G ethanol program are still struggling because of process maturity and higher production cost. Concerted efforts are required at every front in engineering (robust process engineering, process consolidation) and science (in-house cellulase production, strain improvement for cellulase and ethanol production or even energy crops development) for making bioethanol a reality in future.

ACKNOWLEDGMENTS

We thank the Research Council for the State of São Paulo (FAPESP) (Award Number 2014/27055-2 and 16/10636-8), Brazilian National Council for Scientific and Technological Development (CNPq), and Brazilian Federal Agency for the Support and Evaluation of Graduate Education (CAPES).

REFERENCES

Aditiya, H. B., Mahlia, T. M. I., Chong, W. T., Nur, H., Sebayang, A. H. 2016. "Second generation bioethanol production: A critical review." *Renewable and Sustainable Energy Reviews*, 66:631-653.

Agrawal, R., Satlewal, A., Kapoor, M., Mondal, S., Basu, B. 2017. "Investigating the enzyme-lignin binding with surfactants for improved saccharification of pilot scale pretreated wheat straw." *Bioresource Technology*, 224:411–418.

Ahmed, M. A., Seo, Y. H., Terán-Hilares, R., Rehman, M. S. U. R., Jong-In Han, J-I. 2016. "Persulfate based pretreatment to enhance the enzymatic digestibility of rice straw." *Bioresource Technology*, 222:523–526.

Akhtar, N., Goyal, D., Goyal, A. 2017. "Characterization of microwave-alkali-acid pre-treated rice straw for optimization of ethanol production via simultaneous saccharification and fermentation (SSF)." *Energy Conversion and Management*, 141:133–144.

Alfani, F., Gallifuoco, A., Saporosi, A., Spera, A., Cantarella, M. 2000. "Comparison of SHF and SSF processes for the bioconversion of steam-exploded wheat straw." *Journal of Industrial Microbiology and Biotechnology*, 25: 184–192.

Alizadeh, H.; Teymouri, F.; Gilbert, T. I.; Dale, B. E. 2005. "Pretreatment of Switchgrass by Ammonia Fiber Explosion (AFEX)." In *Twenty-Sixth Symposium on Biotechnology for Fuels And Chemicals*, 1133-114. New York: Humana Press.

Almeida, J. R. M., Bertilsson, M., Hahn-Hägerdal, B., Lidén, G., Gorwa-Grauslund, M.-F. 2009. "Carbon fluxes of xylose-consuming Saccharomyces cerevisiae strains are affected differently by NADH and NADPH usage in HMF reduction." *Applied Microbiology and Biotechnology* 84:751-61.

Alves, L., Felipe, M. G. A., Silva, J. B. A. E., Prata, A. M. R. 1998. "Pretreatment of sugarcane bagasse hemicellulose hydrolysate for xylitol production by C*andida guilliermondii*." *Applied Biochemistry and Biotechnology* 70:89-98.

Ando, S., Kakimoto, T., Itoh, K., Arai, I., Kiyoto, K., Hanai, S. 1998. "Increased digestibility of cedar by pretreatment with peracetic acid and steam explosion." *Biotechnology and Bioengineering* 31:802-804.

Auxenfans, T., Crônier, D., Chabbert, B., Paës, G. 2017. "Understanding the structural and chemical changes of plant biomass following steam explosion pretreatment." *Biotechnology for Biofuels* 10:1-36.

Avila-Gaxiola, E., Avila-Gaxiola J., Velarde-Escobar O., Ramos-Brito F., Atondo-Rubio G, Yee-Rendon C. 2017. "Effect of drying temperature on agave tequilana leaves: a pretreatment for releasing reducing sugars for biofuel production." *Journal of Food Process Engineering* 40:1-8.

Balan, V., Bals, B., Chundawat, S. P., Marshall, D., Dale, B. E. 2009. "Lignocellulosic biomass pretreatment using AFEX." *Biofuels: methods and protocols* 58:61–77.

Bansal, A., Illukpitiya, P., Tegegne, F., Singh, S. 2016. "Energy efficiency of ethanol production from cellulosic feedstock." *Renewable and Sustainable Energy Reviews* 58:141-146.

Barrera-Martínez, I., Guzman, N., Pena, E., Vazquez, T., Ceron-Camacho, R., Folch, J., Salazar, J. O. H., Aburto, J. 2016. "Ozonolysis of alkaline lignin and sugarcane bagasse: Structural changes and their effect on saccharification." *Biomass and Bioenergy* 94:167-172.

Bastos, V. D. 2007. *Ethanol, Alcohol Chemistry and Biorefinery*. Rio de Janeiro: Editora Setorial B.

Bezerra, T. L., and Ragauskas, A. J. 2016. "A review of sugarcane bagasse for second-generation bioethanol and biopower production." *Biofuels, Bioproducts and Biorefining* 10:634-647.

Bhagia, S., Li, H., Gao, X., Kumar, R., Wyman C. E. 2016. "Flowthrough pretreatment with very dilute acid provides insights into high lignin contribution to biomass recalcitrance." *Biotechnology for Biofuels* 9:1-15.

Bian, J., Peng, F., Peng, X. P., Xiao, X., Peng, P., Xu, F., Sun, R. C. 2014. "Effect of [Emim] Ac pretreatment on the structure and enzymatic hydrolysis of sugarcane bagasse cellulose." *Carbohydrate Polymers* 100:211-17.

BNDES, GCEE. 2008. *Bioethanol from sugarcane: energy for sustainable development*. Rio de Janeiro: Editora Senac Rio.

Borin, G. P., Sanchez, C. C., de Souza, A. P., de Santana, E. S., de Souza, A. T., Paes Leme, A. F., Squina, F. M., Buckeridge, M., Goldman, G. H., and Oliveira, J. V. 2015. "Comparative secretome analysis of *Trichoderma reesei* and *Aspergillus niger* during growth on sugarcane biomass." *PloS One* 1-20.

Bradshaw, T. C., Alizadeh, H., Teymouri, F., Balan, V., Dale, B. E. 2007. "Ammonia fiber expansion pretreatment and enzymatic hydrolysis on two different growth stages of reed canary grass." *Applied Biochemistry and Biotechnology* 140:395–405.

Brar, K. K., Kaur, S., Chadha, B. S. 2016. “A novel staggered hybrid SSF approach for efficient conversion of cellulose/hemicellulosic fractions of corncob into etanol.” *Renewable Energy* 98:16-22.

Brennan, L., Owende, P. 2010. “Biofuels from microalgae – A review of technologies for production, processing, and extractions of biofuels and co-products.” *Renewable and Sustainable Energy Reviews* 14:557-577.

Brienzo, M., Abud, Y., Ferreira, S., Corrales, C. N. R., Ferreira-Leitão, V. S., Souza, W., Sant’Anna, C. “Characterization of anatomy, lignin distribution, and response to pretreatments of sugarcane culm node and internode.” *Industrial Crops and Products* 84:305–313.

Brito, P. L., Ferreira, C. M. A., Silva, A. F. F., Pantoja, L. A., Nelson, D. L., Santos, A. S. 2017. “Hydrolysis, Detoxification and Alcoholic Fermentation of Hemicellulose Fraction from Palm Press Fiber.” *Waste and Biomass Valorization.* doi: 10.1007/ s12649-017-9882-4.

Calvo-Flores, F. G. and Dobado, J. A. 2010. “Lignin as Renewable Raw Material.” *Chem Sus Chem*, 3:1227-1235.

Canilha, L., Carvalho, W., Felipe, M. G. A., Silva, J. A., Giulietti, M. 2010. “Ethanol production from sugarcane bagasse hydrolysate using *Pichia stipitis*.” *Applied Biochemistry and Biotechnology* 161:84-92.

Canilha, L., Chandel, A. K., Milessi, T. S. S., Antunes, F. A. F., Freitas, W. L. C., Felipe, M. G. A., Silca, S. S. 2012. “Bioconversion of sugarcane biomass into ethanol: an overview about composition, pretreatment methods, detoxification of hydrolysates, enzymatic saccharification and ethanol fermentation.” *Journal of Biomedicine and Biotechnology*. doi: 10.1155/2012/989572.

Cardona, C. A., Quintero, J. A., Paz, I. C. 2010. “Production of bioethanol from sugarcane bagasse: status and perspectives.” *Bioresource Technology* 101: 4754 - 4766.

Carrillo-Nieves D., Ruiz H. A., Aguilar C. N., Ilyina A., Parra-Saldivar R., Torres J. A., Hernández J. L. M. 2017. “Process alternatives for bioethanol production from mango stem bark Residues.” *Bioresource Technology* 239:430–436.

Carvalho, A. F. A., Neto, P. O., Silva, D. F., Pastore, G. M. 2013. “Xylo-oligosaccharides from lignocellulosic materials: Chemical structure, health benefits and production by chemical and enzymatic hydrolysis.” *Food Research International* 51:75-85.

Castro R. C. A, Roberto I. C. 2015. “Effect of nutrient supplementation on ethanol production in different strategies of saccharification and fermentation from acid pretreated rice straw.” *Biomass and Bioenergy* 78:156-163.

Castro, R. C. A., Roberto, I. C. 2014. “Selection of a thermotolerant *Kluyveromyces marxianus* strain with potential application for cellulosic ethanol production by simultaneous saccharification and fermentation.” *Applied Biochemistry and Biotechnology* 172:1553-1564.

Chandel, A. K., Antunes, F. A. F., Silva, M. B., Silva, S. S. 2013. "Unraveling the structure of sugarcane bagasse after soaking in concentrated aqueous ammonia (SCAA) and ethanol production by *Scheffersomyces (Pichia) stipitis*." *Biotechnology for Biofuels* 6:1-11.

Chandel, A., Mpabanga, T. P., Silva, S. S., Singh, O. V. 2012. "Detoxification strategies applied to lignocellulosic hydrolysates for improved xylitol production." In *D-Xylitol fermentative production, application and commercialization*, edites by Silva, S. S., Chandel, A. K., 63-82. Heidelberg: Springer.

Chandel, A. K., Kapoor, R. K., Singh, A., Kuhad, R. C. 2007. "Detoxification of sugarcane bagasse hydrolysate improves ethanol production by *Candida shehatae* NCIM 3501." *Bioresource Technology* 98:1947-1950.

Chang, Q., Griest, T., Harter, T., and Petrash, J. 2007. "Functional studies of aldo-keto reductases in *Saccharomyces cerevisiae*." *Biochimica et Biophysica Acta* 1773:321-29.

Chang, V., Holtzapple, M. 2000. "Fundamental factors affecting biomass enzymatic reactivity." *Applied Biochemistry and Biotechnology* 84:5-37.

Chang, V., Nagwani, M., Holtzapple, M. 1998. "Lime pretreatment of crop residues bagasse and wheat straw." *Applied Biochemistry and Biotechnology* 74:135-159.

Chang, V., Nagwani, M., Kim, C. H., Holtzapple, M. 2001. "Oxidative lime pretreatment of high-lignin biomass." *Applied Biochemistry and Biotechnology* 94:1-28.

Chang, V. S., Kaar, W. E., Burr, B., Holtzapple, M. T. 2001. "Simultaneous saccharification and fermentation of lime-treated biomass." *Biotechnology Letters* 23:1327-1333.

Cheng, K., Cai, B., Zhang, J., Ling, H., Zhou, Y., Ge, J., Xu, J. 2008. "Sugarcane bagasse hemicellulose hydrolysate for ethanol production by acid recovery process." *Biochemical Engineering Journal* 38:105-109.

Converse, A. O., Matsuno, R., Tanaka, M., Taniguchi, M., 1988. "A model for enzyme adsorption and hydrolysis of microcrystalline cellulose with slow deactivation of the adsorbed enzyme." *Biotechnology and Bioengineering* 32:38-45.

Costa R. M. F., Pattathil S., Avci U., Lee S. J., Hazen S. P., Winters A., Hahn M. G., Bosch M. 2017. "A cell wall reference profile for Miscanthus bioenergy crops highlights compositional and structural variations associated with development and organ origin." *New Phytologist* 213:1710–1725.

Dahnum, D., Tasum, S. O., Triwahyni, E., Nurdin, M., Abimanyu, H. 2015. "Comparison of SHF and SSF Processes Using Enzyme and Dry Yeast for Optimization of Bioethanol Production from Empty Fruit Bunch." *Energy Procedia* 68:107-116.

Daza Serna, L. V., Orrego Alzate, C. E., Cardona Alzate, C. A. 2016. "Supercritical fluids as a green technology for the pretreatment of lignocellulosic biomass." *Bioresource Technology* 199:113-120.

Demeke, M. M., Dietz, H., LI, Y., Foulquié-Moreno, M. R., Mutturi, S., Deprez, S., Abt, T. D., Bonini, B. M., Liden, G., Dumortier, F., Verplartse, A., Boles, E., Thevelein, J. M. 2013b. "Development of a D-xylose fermenting and inhibitor tolerant industrial Saccharomyces cerevisiae strain with high performance in lignocellulosic hydrolysates using metabolic and evolutionary engineering" *Biotechnology for Biofuels*, 6:1-24.

Diao, L., Liu, Y., Qian, F., Yang, J., Jiang, Y., Yang, S. 2013. "Construction of fast xylose-fermenting yeast based on industrial ethanol-producing diploid Saccharomyces cerevisiae by rational design and adaptive evolution." *BMC Biotechnology* 13:110-18.

Dias, M. O. S., Cunha, M. P., Jesus, C. D. F., Rocha, G. J. M., Pradella, J. C. C., Rossell, C. E. V., Filho, R. M.; Bonomi, A. 2011. "Second generation ethanol in Brazil: Can it compete with electricity production?" *Bioresource Technology* 102:8964–8971.

Dien, B. S., Cotta, M. A., and Jeffries, T. W. 2003. "Bacteria engineered for fuel ethanol production: current status." *Applied Microbiology and Biotechnology* 63:258-66.

Dimarogona, M., Topakas, E., Christakopoulos, P. 2012. "Cellulose degradation by oxidative enzymes." *Computational and Structural Biotechnology Journal*. 3:1-8. doi: 10.5936/csbj.201209015.

Djajadi D. T., Hansen A. R., Jensen A., Thygesen L. G., Pinelo M., Meyer A. S., Jørgensen H. 2017. "Surface prope rties correlate to the digestibility of hydrothermally pretreated lignocellulosic Poaceae biomass feedstocks." *Biotechnology for Biofuels* 10:1-15. doi: 10.1186/s13068-017-0730-3.

Dussán, K. J., Silva, D. D. V., Perez, V. H., da Silva, S. S., 2016. "Evaluation of oxygen availability on ethanol production from sugarcane bagasse hydrolysate in a batch bioreactor using two strains of xylose-fermenting yeast." *Renewable Energy* 87:703-710.

Elgharbawy, A. A., Alam, M. Z., Moniruzzaman, M., and Goto, M. 2016. "Ionic liquid pretreatment as emerging approaches for enhanced enzymatic hydrolysis of lignocellulosic biomass." *Biochemical Engineering Journal* 109:252-67.

E-PIC S. R. L., 2016. *Biomass pretreatment (BIOGAS)* [Internet]. Mongrando (BI). Year [cited 2017, Jun 12]. Available from: https://www.epic-srl.com/en/cavitation-technologies/biomass-pretreatment-biogas.

Eriksson, T., Borjesson, J., Tjemeld, F. 2002. "Mechanism of surfactant effect in enzymatic hydrolysis of lignocellulose." *Enzyme and Microbial Technology* 31:353-364.

FAO. Food and Agriculture Organization of the United Nations. 2008. *The state of food and agriculture*. Rome:FAO.

Felipe, M. G. A., Vieira, D. C., Vitolo, M., Silva, S. S., Roberto, I. C., Mancilha, I. M. 1995. "Effect of acetic acid on xylose fermentation to xylitol by *Candida guilliermondii*." *Journal of Basic Microbiology* 35:171-177.

Fengel, D., Wegener, G. 1989. *Wood chemistry, ultraestruture, reactions*. Bellin:Walter de Gruyter.

Field, S. J., Ryden, P., Wilson, D., James, S. A., Roberts, I. N., Richardson D. J., Waldron, K. W. 2015. "Identification of furfural resistant strains of *Saccharomyces cerevisiae* and *Saccharomyces paradoxus* from a collection of environmental and industrial isolates." *Biotechnology for Biofuels* 8:33. doi: 10.1186/s13068-015-0217-z.

Fockink, D. H., Urio, M. B., Chiarello, L. M., Sánchez, J. H., Ramos, L. P. 2016. "Principles and Challenges Involved in the Enzymatic Hydrolysis of Cellulosic Materials at High Total Solids." In Soccol, C. R., Brar, S. K., Faulds, C., Ramos, L. P. *Green Fuels Technology*, New York: Springer International Publishing.

Fockink, D. H., Urio, M. B., Sánchez, J. H., Ramos, L. P. 2017. "Enzymatic Hydrolysis of Steam-Treated Sugarcane Bagasse: Effect of Enzyme Loading and Substrate Total Solids on Its Fractal Kinetic Modeling and Rheological Properties." *Energy Fuels* 31:6211–6220.

Fonseca, B. G., Moutta, R. D. O., Ferraz, F. D. O., Vieira, E. R., Nogueira, A. S., Baratella, B. F., Rodrigues, L. C., Hou-Rui, Z., Silva, S. S. 2011. "Biological detoxification of different hemicellulosic hydrolysates using *Issatchenkia occidentalis* CCTCC M 206097 yeast." *Journal of Industrial Microbiology & Biotechnology* 38:199-207.

Forte, A., Zucaro, A., Fagnano, M., Fierro, A. 2017. "Potential environmental impact of bioethanol production chain from fiber sorghum to be used in passenger cars." *Science of the Total Environment* 598:365–376.

Furtado, A., Lupoi, J. S., Hoang, N. V., Healey, A., Singh, S., Simmons, B. A., Henry, R. J. 2014. "Modifying plants for biofuel and biomaterial production." *Plant Biotechnology Journal* 12:1246-58.

Gandi, J., Holtzapple, M. T., Ferrer, A., Byers, F. M., Turner, N. D., Nagwani, M., Chang, S. 1997. "Lime treatment of agricultural residues to improve rumen digestibility." *Animal Feed Science and Technology* 68:195-211.

Garcia-Cubero, M. A., Gonzalez-Benito, G., Indacoechea. I., Coca, M., Bolado, S. 2009. "Effect of ozonolysis pretreatment on enzymatic digestibility of wheat and rye straw." *Bioresourse Technology* 100:1608–1613.

Goldemberg, J., Guardabassi, P. 2010. "Perspective: the potential for first generation ethanol production from sugarcane." *Biofuels, Bioproducts and Biorefining* 4:17–24.

Gonçalves, D. L., Matsushika, A., Belisa, B., Goshima, T., Bon, E. P., Stambuk, B. U. 2014. "Xylose and xylose/glucose co-fermentation by recombinant Saccharomyces cerevisiae strains expressing individual hexose transporters." *Enzyme and Microbial Technology* 63:13-20.

Hahn-Hägerdal, B., Karhumaa, H. B. K., Fonseca, C., Spencer-Martins, I., and M. F. Gorwa-Grauslund. 2007. "Towards industrial pentose-fermenting yeast strains." *Applied Microbiology and Biotechnology* 74:937-53.

Hamacher, T., Becker, J., Gárdonyi, M., Hahn-Hägerdal, B., and Boles, E. 2002. "Characterization of the xylose-transporting properties of yeast hexose transporters and their influence on xylose utilization." *Microbiology* 148:2783-2788.

Hashmi, M., Sun, Q., Tao, J., Wells Jr., T., Shah, A. A., Labbé, N., and Ragauskas, A. J. 2017. "Comparison of autohydrolysis and ionic liquid 1-butyl-3-methylimidazolium acetate pretreatment to enhance enzymatic hydrolysis of sugarcane bagasse." *Bioresource Technology* 224:714-720.

Hector, R. E., Qureshi, N., Hughes, S. R., and Cotta, M. A. 2008. "Expression of a heterologous xylose transporter in a Saccharomyces cerevisiae strain engineered to utilize xylose improves aerobic xylose consumption." *Applied Microbiology and Biotechnology* 80:675-84.

Helle, S. S., Duff, S. J. B., Cooper, D. G. 1993. "Effect of surfactants on cellulose hydrolysis." *Biotechnology and Bioengineering*, 42, 611–617.

Hernández-Pérez, A. F., Costa, I. A. L., Silva, D. D. V., Dussán, K. J., Villela, T. R., Canettieri, E. V., Carvalho Jr, J. A., Soares Neto, T. G., Felipe, M. G. A. 2016. "Biochemical conversion of sugarcane straw hemicellulosic hydrolyzate supplemented with co-substrates for xylitol production." *Bioresource Technology* 200:1085-1088.

Hoang, N. V., Furtado, A., Botha, F. C., Simmons, B. A., and Henry, R. J. 2015. "Potential for genetic improvement of sugarcane as a source of biomass for biofuels." *Frontiers in Bioengineering and Biotechnology* 3:1-15. doi: 10.3389/fbioe.2015. 00182.

Horn, S. J., Vaaje-Kolstad, G., Westereng, B., Eijsink, V. 2012. "Novel enzymes for the degradation of cellulose." *Biotechnology for Biofuels*, 45:1-12. doi: 10.1186/1754-6834-5-45.

Huang, X. L., Penner, M. H., 1991. "Apparent substrate inhibition of the Trichoderma reesei cellulase system." *Journal of Agricultural and Food Chemistry* 39: 2096–2100.

Humbird, D., Mohagheghi, A., Dowe, N., Schell, D. J. 2010. "Economic impact of total solids loading on enzymatic hydrolysis of dilute acid pretreated corn stover." *Biotechnology Progress* 26:1245-1251.

Itoh, H., Wada, M., Honda, Y., Kuwahara, M., Watanabe, T. 2003. "Bioorganosolve pretreatments for simultaneous saccharification and fermentation of beech wood by ethanolysis and white-rot fungi." *Journal of Biotechnology* 103:273–280.

Kamoldeen, A. A., Lee, C. K., Abdullah, W. N. W., Leh, C. P. 2017. "Enhanced ethanol production from mild alkali-treated oil-palm empty fruit bunches via co-fermentation of glucose and xylose." *Renewable Energy* 107:113-123.

Katsimpouras, C., Kalogiannis, K. G., Kalogianni A., Lappas A. A., Topakas E. 2017. "Production of high concentrated cellulosic ethanol by acetone/water oxidized pretreated beech wood." *Biotechnology for Biofuels* 10:1-16. doi: 10.1186/s13068-017-0737-9.

Kelly, C., Jones, O., Barnhart, C., Koti, S., Linga, V. R. 2008. "Effect of furfural, vanillin and syringaldehyde on Candida guilliermondii growth and xylitol biosynthesis." *Applied Biochemistry and Biotechnology* 148:97–108.

Keshav, P. K., Shaik, N., Koti, S., Linga, V. R. 2016. "Bioconversion of alkali delignified cotton stalk using two-stage dilute acid hydrolysis and fermentation of detoxified hydrolysate into ethanol." *Industrial Crops and Products* 91:323–331.

Khiyami M. A., Pometto, A. L., Brown, R. C. 2005. "Detoxification of corn stover and corn starch pyrolysis liquors by *Pseudomonas putida* and *Streptomyces setonii* suspended cells and plastic compost support biofilms." *Journal of Agricultural and Food Chemistry* 53:2978–2987.

Kim, K. H., Dutta, T., Ralph, J., Mansfield, S. D., Simmons. B. A., Singh, S. 2017. "Impact of lignin polymer backbone esters on ionic liquid pretreatment of poplar" *Biotechnology for Biofuels*10:1-10. doi: 10.1186/s13068-017-0784-2.

Kim, H. S., Kim, N. R., Yang, J., Choi, W. 2011. "Identification of novel genes responsible for ethanol and/or thermotolerance by transposon mutagenesis in *Saccharomyces cerevisiae.*" *Applied Microbiology and Biotechnology* 62:1569-72.

Kim, K. H.; Hong, J. 2001. "Supercritical CO_2 pretreatment of lignocellulose enhances enzymatic cellulose hydrolysis." *Bioresource Technology* 77:139-144.

Kim, S., Dale, B. E. 2016. "A distributed cellulosic biorefinery system in the US Midwest based on corn stover." *Biofuels, Bioproducts and Biorefining* 10:819–832.

Kim, S. K., Jo, J. H., Jin, Y. S., Seo, J. H. 2017. "Enhanced ethanol fermentation by engineered *Saccharomyces cerevisiae* strains with high spermidine contents." *Bioprocess and Biosystems Engineering* 40:683-91.

Klinke, H. B., Ahring, B. K., Schmidt, A. S., Thomsen, A. B. 2002. "Characterization of degradation products from alkaline wet oxidation of wheat straw." *Bioresource Technology* 82:15-26.

Knauf, M., and Moniruzzaman, M. 2004. "Lignocellulosic biomass processing: A perspective." *International Sugar Journal* 106:147-50.

Ko, J. K., Um, Y., Woo, H. M., Kim, K. H., Lee, S. M. 2016. "Ethanol production from lignocellulosic hydrolysates using engineered *Saccharomyces cerevisiae* harboring xylose isomerase-based pathway." *Bioresource Technology* 209: 290-296.

Kostas E. T., Beneroso D., Robinson J. P. 2017. "The application of microwave heating in bioenergy: A review on the microwave pretreatment and upgrading technologies for biomass." *Renewable and Sustainable Energy Reviews* 77: 12–27.

Kristensen, J. B., Borjesson, J., Bruun, M. H., Tjermeld, F., Jorgensen, H. 2007. "Use of surface active additives in enzymatic hydrolysis of wheat straw lignocellulose." *Enzyme and Microbial Technology*, 40:888-895.

Kumar, P., Barrett, D. M., Delwiche, M. J., Stroeve, P. 2009. "Methods for Pretreatment of Lignocellulosic Biomass for Efficient Hydrolysis and Biofuel Production." *Industrial & Engineering Chemistry Research* 48:3713– 3729.

Kurakake, M., Ide, N., Komaki, T. 2007. "Biological pretreatment with two bacterial strains for enzymatic hydrolysis of office paper." *Current Microbiology* 54:424–428.

Lamsal, B., Yoo, J., Brijwani, K., Alavi, S. 2010. "Extrusion as a thermomechanical pretreatment for lignocellulosic ethanol." *Biomass & Bioenergy*. 34:1703-1710.

Laopaiboon, L., Thanonkeo, P., Jaisil, P., Laopaiboon, P. 2007. "Ethanol production from sweet sorghum juice in batch and fed-batch fermentations by *Saccharomyces cerevisiae*." *World Journal of Microbiology and Biotechnology*, 23:1497–1501.

Larsson, S., Reimann, A., Nilvebrant, N. O., Jönsson L. J. 1999. "Comparison of different methods for the detoxification of lignocellulose hydrolyzates of spruce." *Applied Biochemistry and Biotechnology*. 77:91–103.

Lee, S., Speight, J. G., Loyalka, S. K. 2007. *Handbook of alternative fuel technologies*. USA: CRC Taylor and Francis Group.

Lee, W., Lee, J., Shin, C., Park, S., Chang, H., Chang, Y. 1999. "Ethanol production using concentrated oak wood hydrolysates and methods to detoxify." *Applied Biochemistry and Biotechnology* 78:547-559.

Lennartsson, P. R., Erlandsson, P., Taherzadeh, M. J. 2014. "Integration of the first and second generation bioethanol processes and the importance of by-products." *Bioresource Technology* 165:3–8.

Li, B. Z., Balan, V., Yuan, Y. J., Dale, B. E. 2010. "Process optimization to convert forage and sweet sorghum bagasse to ethanol based on ammonia fiber expansion (AFEX) pretreatment." *Bioresource Technology*, 101:1285-1292.

Li, H. Y., Chen, X., Wang. C. Z., Sun, S. N., Sun, R. C. 2016. "Evaluation of the two-step treatment with ionic liquids and alkali for enhancing enzymatic hydrolysis of Eucalyptus: chemical and anatomical changes." *Biotechnology for Biofuels* 9:166. doi: 10.1186/s13068-016-0578-y.

Li, Q., Song, J., Peng, S., Wang, J. P., Qu, G. Z., Sederoff, R. R. and Chiang, V. L. 2014. "Plant biotechnology for lignocellulosic biofuel production." *Plant Biotechnology Journal* 12:1174-1192.

Lin, X., Lou, H., Qiu, X., Pang, Y., Yang, D. 2017. "Effect of sodium dodecyl sulfate and cetyltrimethylammonium bromide catanionic surfactant on the enzymatic hydrolysis of Avicel and corn stover." *Cellulose* 24:669-676.

Lin, Y.; Tanaka, S. 2006. "Ethanol fermentation from biomass resources: current state and prospects." *Applied Microbiology and Biotechnology* 69:627-642.

Liu, Y., Zhang, Y., Xu, J., Sun, Y., Yuan, Z., Xie, J. 2015. "Consolidated bioprocess for bioethanol production with alkali-pretreated sugarcane bagasse." *Applied Energy* 157:517-522.

Lü, F., Chai, L., Shao, L., He, P. 2017. "Precise pretreatment of lignocellulose: relating substrate modification with subsequent hydrolysis and fermentation to products and by‑products." *Biotechnology for Biofuels* 10:1-15. doi: 10.1186/s13068-017-0775-3.

Luque, L., Oudenhoven, S., Westerhof, R., Rossum, G., Berruti, F., Kersten, S., Rehmann L. 2016. "Comparison of ethanol production from corn cobs and switchgrass following a pyrolysis‑based biorefinery approach." *Biotechnol Biofuels* 9:1-14. doi: 10.1186/s13068-016-0661-4.

Ma, A., Hu, Q., Qu, Y., Bai, Z., Liu, W., Zhuang, G. 2008. "The enzymatic hydrolysis rate of cellulose decreases with irreversible adsorption of cellobiohydrolase I." *Enzyme and Microbial Technology* 42:543-547.

Maeda, R. N., Serpa, V. S., Rocha, V. A. L., Mesquita, R. A. A., Santa Anna, L. M., Castro, A. M., Driemeier, C. E., Pereira Jr. N., Polikarpov, I. 2011. "Enzymatic hydrolysis of pretreated sugar cane bagasse using Penicillium funiculosum and Trichoderma harzianum cellulases." *Process Biochemistry* 46:1196-1201.

Manochio, C., Andrade, B. R., Rodrigues, R. P., Moraes, B. S. 2017. "Ethanol from biomass: A comparative overview." *Renewable and Sustainable Energy Reviews* 80:743–755.

Marrugo, G.; Valdés, C. F., Chejne, F. 2016. "Characterization of Colombian Agroindustrial Biomass Residues as Energy Resources." *Energy Fuels* 30:8386–8398.

Martinez, A., Rodriguez, M. E., York, S. W., Preston, J. F., Ingram, L. O. 2000. "Effects of $Ca(OH)_2$ treatments ("overliming") on the composition and toxicity of bagasse hemicellulose hydrolysates." *Biotechnology and Bioengineering* 69:526-536.

Marton, J. M., Felipe, M. G. A., Silva, J. B. A., Pessoa Jr., A. "Evaluation of the activated charcoals and adsorption conditions used in the treatment of sugarcane bagasse hydrolysate for xylitol production." *Brazilian Journal of Chemical Engineering*, 23:9-21.

Meng, X., Ragauskas, A. J. 2014. "Recent advances in understanding the role of cellulose accessibility in enzymatic hydrolysis of lignocellulosic substrates." *Current Opinion in Biotechnology* 27:150–158.

Mesa, L., González, E., Cara, C., González, M., Castro, E., Mussatto, S. I. 2011. "The effect of organosolv pretreatment variables on enzymatic hydrolysis of sugarcane bagasse." *Chemical Engineering Journal* 168:1157–1162.

Mesa, L., Martínez, Y., Barrio, E., González, E. 2017. "Desirability function for optimization of Dilute Acid pretreatment of sugarcane straw for ethanol production and preliminary economic analysis based in three fermentation configurations." *Applied Energy* 198:299–311.

Miyamoto, K. 1997. *Renewable biological systems for alternative sustainable energy production*. Osaka: FAO.

Modenbach, A. A., Nokes, S. E. 2013. "Enzymatic hydrolysis of biomass at high-solids loadings – A review." *Biomass and Bioenergy* 56:526-544.

Monrroy M., Ibañez J., Melin V., Baeza J., Mendonça R. T., Contreras D., Freer J. 2010. "Bioorganosolv pretreatments of by a brown rot fungus (*Gloephyllum trabeum*) and ethanolysis." *Enzyme and Microbial Technology* 47:11-16.

Moreira, L. R., Filho, E. X. 2016. "Insights into the mechanism of enzymatic hydrolysis of xylan." *Applied Microbiology and Biotechnology* 100:5205-1524.

Moreno, A. D., Ibarra, D., Fernandez, J., Ballesteros, M. 2012. "Different laccase detoxification strategies for ethanol production from lignocellulosic biomass by the thermotolerant yeast *Kluyveromyces marxianus* CECT 10875." *Bioresource Technology* 106:101-109.

Muñoz, C., Mendonça, R., Baeza, J., Berlin, A., Saddler, J. and Freer, J. 2007. "Bioethanol production from bio-organosolv pulps of *Pinus radiata* and *Acacia dealbata*." *Journal of Chemical Technology and Biotechnology* 82: 767–774.

Murnen, H. K., Balan, V., Chundawat, S. P. S., Bals, B., Sousa, L. d. C. and Dale, B. E. 2007. "Optimization of Ammonia Fiber Expansion (AFEX) Pretreatment and Enzymatic Hydrolysis of *Miscanthus* x *giganteus* to Fermentable Sugars." *Biotechnol Progress* 23: 846–850.

Mussato, S. I., Machado, E. M. S., Carneiro, L. M., and Teixeira, J. A. 2012. "Sugar metabolism and ethanol production by different yeast strains from coffee industry wastes hydrolysates." *Applied Energy* 92:763-8.

Naik, S. N.; Goud, V. V.; Rout, P. K.; Dalai, A. K. 2010. "Production of first and second generation biofuels: a comprehensive review." *Renewable and Sustainable Energy Reviews* 14:578-597.

NRC - National Research Council. 2000. *Biobased Industrial Products: Research and Commercialization Priorities*. Washington: The National Academies Press. doi: 10.17226/5295.

Ojeda, K.; Kafarov, V. 2009. "Energy analysis of enzymatic hydrolysis reactors for transformation of lignocellulosic biomass to bioethanol." *Chemical Engineering Journal* 154:390-395.

Palmqvist, E., Hahn-Hägerdal, B. 2000a. "Fermentation of lignocellulosic hydrolysates. I: inhibition and detoxification." *Bioresource Technology* 74:17-24.

Palmqvist, E., Hahn-Hägerdal, B. 2000b. "Fermentation of lignocellulosic hydrolysates. II: inhibitors and mechanisms of inhibition." *Bioresource Technology* 74:25-33.

Panesar, P. S., Marwaha, S. S., Kennedy, J. F. 2006. "*Zymomonas mobilis*: an alternative ethanol producer." *Journal of Chemical Technology and Biotechnology* 81:623–635.

Papapetridis, I., Van Dijk, M., Van Maris, A. J. A., Pronl, J. T. 2017. "Metabolic engineering strategies for optimizing acetate reduction, ethanol yield and osmotolerance in *Saccharomyces cerevisiae*." *Biotechnology for Biofuels* 10:107. doi: 10.1186/s13068-017-0791-3.

Parthasarathi, R., Sun, J., Dutta, T., Sun, N., Pattathil, S., Konda, N. V. S. N. M., Peralta, A. G., Simmons, B. A., Singh, S. 2016. "Activation of lignocellulosic biomass for higher sugar yields using aqueous ionic liquid at low severity process conditions." *Biotechnology for Biofuels* 9:160. doi: 10.1186/s13068-016-0561-7.

Perez, J. A., Ballesteros, I., Ballesteros, M., Saez, F., Negro, M. J., Manzanares, P. 2008. "Optimizing liquid hot water pretreatment conditions to enhance sugar recovery from wheat straw for fuel-ethanol production." *Fuel* 87:3640–3647.

Pielhop, T., Amgarten, J., Rohr, P. R., Studer, M. H. 2016. "Steam explosion pretreatment of softwood: the effect of the explosive decompression on enzymatic digestibility." *Biotechnology for Biofuels* 9:152. doi: 10.1186/s13068-016-0567-1.

Pielhop, T., Amgarten, J., Studer, M. H., Rohr, P. R. 2017. "Pilot‑scale steam explosion pretreatment with 2‑naphthol to overcome high softwood recalcitrance." *Biotechnology for Biofuels* 10:130. doi: 10.1186/s13068-017-0816-y.

Poovaiah, C. R., Nageswara-Rao, M., Soneji, J. R., Baxter, H. L., and Stewart, C. N. 2014. "Altered lignin biosynthesis using biotechnology to improve lignocellulosic biofuel feedstocks." *Plant Biotechnology Journal* 12:1163-73.

Posada, J. A., Patel, A. D., Roes, A., Blok, K., Faaij, A. P. C.; Patel, M. K. 2013. "Potential of bioethanol as a chemical building block for biorefineries: Preliminary sustainability assessment of 12 bioethanol-based products." *Bioresource Technology* 135:490-499.

Ra, C. H., Kim, M. J., Jeong, G. T., Kim, S. K. 2017. "Efficient utilization of *Eucheuma denticulatum* hydrolysates using an activated carbon adsorption process for ethanol production in a 5-L fermentor." *Bioprocess and Biosystems Engineering* 40:373–381.

Ranganathan, A., Smith O. P., Youssef, N. H., Struchtemeyer, C. G., Atiyeh, H. A., Elshahed, M. S. 2017. "Utilizing Anaerobic Fungi for Two-stage Sugar Extraction and Biofuel Production from Lignocellulosic Biomass." *Frontiers in Microbiology* 8:635. doi: 10.3389/fmicb.2017.00635.

Raven, P. H., Evert, R. F., Eichhorn, S. E. 1992. *Biology of Plants*. New York: W. H. Freeman & Company.

Renó, M. L. G., Olmo, O. A. D., Palacio, J. C. E., Lora, E. E. S., Venturini, O. J. 2014. "Sugarcane biorefineries: Case studies applied to the Brazilian sugar–alcohol industry." *Energy Conversion and Management* 86:981-991.

Ribeiro, D. A., Cota, J., Alvarez, T. M., Brüchli, F., Bragato, J., Pereira, B. M. P., Pauletti, B. A., Jackson, G., Pimenta, M. T. B., Murakami, M. T., Camassola, M., Ruller, R., Dillon, A. J. P., Pradella, J. G. C., Leme, A. F. P., Squina, F. M. 2012. "The Penicillium echinulatum secretome on sugarcane bagasse." *PLoS One* 7:e50571. doi: 10.1371/journal.pone.0050571.

Rodrigues, T. H., de Barros, E. M., da Sá Brigido, J., da Silva, W. M., Rocha, M. V., Gonçalves, L. R. 2016. "The Bioconversion of Pretreated Cashew Apple Bagasse into Ethanol by SHF and SSF Processes." *Applied Biochemistry and Biotechnology* 178:1167-1183.

Rogalinski, T., Ingram, T., Brunner, G. 2008. "Hydrolysis of lignocellulosic biomass in water underelevated temperatures and pressures." *The Journal of Supercritical Fluids* 47:54–63.

Runquist, D., Hahn-Hagerdal, B., and Radstrom, P. 2010. "Comparison of heterologous xylose transporters in recombinant *Saccharomyces cerevisiae*." *Biotechnology for Biofuels* 17:3-5. doi: 10.1186/1754-6834-3-5.

Saini, J. K.; Saini, R.; Tewari, L. 2015. "Lignocellulosic agriculture wastes as biomass feedstocks for second-generation bioethanol production: Concepts and recent developments." *3 Biotech* 5:337-353.

Sánchez, C. 2009. "Lignocellulosic residues: biodegradation and bioconversion by fungi." *Biotechnology Advances* 27:185–194.

Sanchez, S., Bravo, V., Castro, E., Moya, A. J., Camacho. F. 1999. "Comparative study of the fermentation of D-glucose/D-xylose mixtures with *Pachysolen tannophilus* and *Candida shehatae*." *Bioprocess Engineering* 21:525–532.

Sant'Ana da Silva, A., Souza, M. F., Ballesteros, I., Manzanares, P., Ballesteros, M., Bon, E. P. S. 2016. "High-solids content enzymatic hydrolysis ofhydrothermally pretreated sugarcane bagasse using a laboratory-made enzyme blend and commercial preparations." *Process Biochemistry* 51:1561-1567.

Santos T. N., Dutra E. D., Prado A. G., Leite F. C. B., Souza R. F. R., Santos D. C., Abreu C. A. M., Simões D. A., Morais M. A., Menezes R. S. C. 2016. "Potential for biofuels from the biomass of prickly pear cladodes: Challenges for bioethanol and biogas production in dry areas." *Biomass and Bioenergy* 85: 215-222.

Sarkar, N., Ghosh, S. K., Baneerjee, S., Aikat, K. 2012. "Bioethanol production from agricultural wastes: An overview." *Renewable Energy*, 37:19-27.

Scheller, H. V., Ulvskov, P. 2010. "Hemicelluloses." *Annual Review of Plant Biology* 61:263-289.

Schneider, W. D. H., Gonçalves, T. A., Uchima, C. A., Couger, M. B., Prade, R., Squina, F. M., Dillon, A. J. P., Camassola, M. 2016. "*Penicillium echinulatum* secretome analysis reveals the fungi potential for degradation of lignocellulosic biomass." *Biotechnology for Biofuels* 9:66. doi: 10.1186/s13068-016-0476-3.

Sedlak, M., Ho, N. W. 2004. "Characterization of the effectiveness of hexose transporters for transporting xylose during glucose and xylose co-fermentation by a recombinant *Saccharomyces yeast*." *Yeast* 21:671-84.

Sharma S., Sharma V., Kuila A. 2016. "Cellulase production using natural medium and its application on enzymatic hydrolysis of thermo chemically pretreated biomass." *3 Biotech* 6:139. doi: 10.1007/s13205-016-0465-z.

Shibuya M., Yasukawa M., Sasaki K., Tanaka Y., Takahashi T., Kondo A., Matsuyama H. 2017. "Up-concentration of sugars in pretreated-rice straw by an osmotic pressure-driven method." *Biochemical Engineering Journal* 121:13–16.

Shi-wei, H., Lei-ming, W., Persson, S., Liang-cail, P., Sheng-qiul, F. 2017. "Sweet sorghum and Miscanthus: Two potential dedicated bioenergy crops in China." *Journal of Integrative Agriculture*, 16:1236–1243.

Silva, D. D. V., Arruda, P. V., Dussán, K. J., Felipe, M. G. A. 2014. "Adaptation of *Scheffersomyces stipitis* Cells as a strategy to the improvement of ethanol production from sugarcane bagasse hemicellulosic hydrolysate." *Chemical Engineering Transactions* 38: 427-432.

Silva, D. D. V., Felipe, M. G. A., Mancilha, I. M., Luchese, R. H., Silva, S. S. 2004. "Inhibitory effect of acetic acid on bioconversion of xylose in xylitol by *Candida guilliermondii* in sugarcane bagasse hydrolysate." *Brazilian Journal of Microbiology* 35:248-254.

Sindhu, R., Binod, P., Mathew, A. K., Abraham, A., Gnansounou E., Ummalyma, S. B., Thomas L., Pandey A. 2017. "Development of a novel ultrasound-assisted alkali pretreatment strategy for the production of bioethanol and xylanases from chili post harvest residue." *Bioresource Technology* 242:146-151.

Sindhu, R., Binod, P., Pandey, A. 2016. "Biological pretreatment of lignocellulosic biomass – An overview." *Bioresource Technology* 199:76–82.

Singh, P., Suman, A., Tiwari, P., Arya, N., Gaur, A., Shrivastava, A. K. 2008. "Biological pretreatment of sugarcane trash for its conversion to fermentable sugars." *World Journal of Microbiology and Biotechnology* 24:667–673.

Singh, A., Sharma, P., Saran, A. K., Singh, N., Bishnoi, N. 2013. "Comparative study on ethanol production from pretreated sugarcane bagasse using immobilized *Saccharomyces cerevisiae* on various matrices." *Renewable Energy* 50:488-493.

Smichi, N., Messaoudi, Y., Moujahed, N., Gargouri, M. 2016. "Ethanol production from halophyte *Juncus maritimus* using freezing and thawing biomass pretreatment." *Renewable Energy* 85:1357-136.

Sticklen, M. B. 2004. *Proceedings of the 2nd International Ukrainian Conference on Biomass for Energy* 133-136 (Ukraine National Academy of Sciences, Kiev).

Sukumaran, R. K., Surender, V. J., Sindhu, R., Binod, P., Janu, K. U., Sajna, K. V., Rajasree, K. P., Pandey, A. "Lignocellulosic ethanol in India: Prospects, challenges and feedstock availability." *Bioresource Technology* 101:4826–4833.

Sun S., Chen W., Tang J., Wang B., Cao X., Sun S., Sun R. C. 2016. "Synergetic effect of dilute acid and alkali treatments on fractional application of rice straw." *Biotechnol Biofuels* 9:217. doi: 10.1186/s13068-016-0632-9.

Sun, Y., Cheng, J. 2002. "Hydrolysis of lignocellulosic materials for ethanol production: A review." *Bioresource Technology* 83:1–11.

Sun, S., Sun, S., Cao, X., Sun, R. 2016. "The role of pretreatment in improving the enzymatic hydrolysis of lignocellulosic materials." *Bioresource Technology* 199: 49-58.

Sun, Y., Cheng, J. 2002. "Hydrolysis of lignocellulosic materials for ethanol production: A review." *Bioresource Technology*, 83: 1-11.

Terán Hilares, R., dos Santos, J. C., Ajaz Ahmed, M., Hwan Jeon, S., Silva, S. S, Han, J. 2016. "Hydrodynamic cavitation-assisted alkaline pretreatment as a new approach for sugarcane bagasse biorefineries." *Bioresource Technology* 214: 609– 614.

Terán Hilares, R., Almeida, G. F, Ahmed, M. A., Antunes, F. A. F., Silva, S. S., Han, J., dos Santos, J. C. 2017. "Hydrodynamic cavitation as an efficient pretreatment method for lignocellulosic biomass : A parametric study." *Bioresource Technology* 235:301–308.

Terán Hilares, R., Ienny, J. V., Marcelino, P. F., Ahmed, M. A., Antunes, F. A. F., Silva, S. S., Santos, J. C. 2017. "Ethanol production in a simultaneous saccharification and fermentation process with interconnected reactors employing hydrodynamic cavitation-pretreated sugarcane bagasse as raw material." *Bioresource Technology* 243:652-659.

Terán-Hilares, R., Almeida, G. F., Ahmed, M. A., Antunes, F. A. F., Silva, S. S., Han, J. - I., Santos, J. C. 2017. "Hydrodynamic cavitation as an efficient pretreatment method for lignocellulosic biomass: A parametric study." *Bioresource Technology* 235:301-3088.

Terán-Hilares, R., Santos, J. C., Ahmed, M. A., Jeon, S. H., Silva, S. S., Han, J. -I. 2016. "Hydrodynamic cavitation-assisted alkaline pretreatment as a new approach for sugarcane bagasse biorefineries." *Bioresource Technology* 214:609-14.

Toivari, M. H., Salusjarvi, L., Ruohonen, L., Penttilä, M. 2004. "Endogenous xylose pathway in *Saccharomyces cerevisiae*." *Applied and Environmental Microbiology* 70:3681-3686.

Traff, K. L., Jonsson, L. J., Hahn-Hägerdal, B. 2002. "Putative xylose and arabinose reductases in *Saccharomyces cerevisiae*." *Yeast* 19:1233-1241.

Travaini, R., Otero, M. D. M., Coca, M., Da-Silva, R., Bolado, S. 2013. "Sugarcane bagasse ozonolysis pretreatment: effect on enzymatic digestibility and inhibitory compound formation." *Bioresource Technology* 133:332–339.

UNICA, *Sugarcane Industry Union*. 2017. Accessed July 11. http://www.unica.com.br/noticia/16900437920320868796/cana-geneticamente-modificada-desenvolvida-pelo-ctc-e-aprovada-na-ctnbio/.

Vaaje-Kolstad, G., Forsberg, Z., Loose, J. S. M., Bissaro, B., Vincent G. H., Eijsink, V. G. H. 2017. "Structural diversity of lytic polysaccharide monooxygenases." *Current Opinion in Structural Biology*, 44, 67–76.

Vallet, C. Said, R.; Rabiller, C.; Martin, M. L. 1996. "Natural abundance isotopic fractionation in the fermentation reaction: influence of the nature of the yeast." *Bioorganic Chemistry* 24:319–330.

Van Dyk, J. S., Pletschke, B. I. 2012. "A review of lignocellulose bioconversion using enzymatic hydrolysis and synergistic cooperation between enzymes—Factors affecting enzymes, conversion and synergy." *Biotechnology Advances* 30:1458-1480.

Vanegas, J. M., Contreras, M. F., Faller, R., and Longo, M. L. 2012. "Role of unsaturated lipid and ergosterol in ethanol tolerance of model yeast biomembranes." *Biophysical Journal* 1023:507-16.

Viola, E., Zimbardi, F., Valerio, V., Nanna, F., Battafarano, A. 2013. "Use of a two-chamber reactor to improve enzymatic hydrolysis and fermentation of lignocellulosic materials." *Applied Energy* 102:198-203.

Walton, P. H., Davies, G. J. 2016. "On the catalytic mechanisms of lytic polysaccharide monooxygenases." *Current Opinion in Chemical Biology* 31: 195– 207.

Wan, C., Li, Y. 2012. "Fungal pretreatment of lignocellulosic biomass." *Biotechnology Advances* 30:1447–1457.

Wang, P., Yu, H., Zhan, S., Wang, S. 2011. "Catalytic hydrolysis of lignocellulosic biomass into 5-hydroxymethylfurfural in ionic liquid." *Bioresource Technology* 102:4179-4183.

Wang, W., Kang, L., Wei, H., Arora, R., Lee, Y. Y. 2011. "Study on the Decreased Sugar Yield in Enzymatic Hydrolysis of Cellulosic Substrate at High Solid Loading." *Applied Biochemistry and Biotechnology*, 164:1139–1149.

Wingren, A., Galbe, M., Zacchi, G. 2003. "Techno-Economic Evaluation of Producing Ethanol from Softwood: Comparison of SSF and SHF and Identification of Bottlenecks." *Biotechnology Progress* 19:1109–1117.

Wisselink, H. W., Toirkens, M. J., Berriel, M. D. F., Winkler, A. A., Van Dijken, J. P., Pronk, J. T., Van Maris, A. J. A. 2007. "Engineering of Saccharomyces cerevisiae for efficient anaerobic alcoholic fermentation of L-arabinose." *Applied and Environmental Microbiology* 73:4881-91.

Wu, Y., Zhang, C., Liu, Y., Fu, Z., Dai, B., Yin, D. 2012. "Biomass char sulfonic acids (BC-SO3H)-catalyzed hydrolysis of bamboo under microwave irradiation." *Bio Resources* 7:5950-5959.

Xu, J., Wang, X., Liu, X., Xia, J., Zhang, Tong, and Xiong, P. 2016. "Enzymatic in situ saccharification of lignocellulosic biomass in ionic liquids using an ionic liquid-tolerant cellulases." *Biomass and Bioenergy* 93:180-6.

Yamada, R., Nakashima, K., Asai-Nakashima, N., Tokuhara, W., Ishida, N., Katahira, S., Kamiya, N., Ogino, C., Kondo, A. 2017. "Direct Ethanol Production from Ionic Liquid-Pretreated Lignocellulosic Biomass by Cellulase-Displaying Yeasts." *Applied Biochemistry and Biotechnology* 182:229–237.

Yang, S., Zhang, Y., Yue, W., Wang, W., Wang, Y. Y., Yuan, T. Q., Sun R. C. 2016. "Valorization of lignin and cellulose in acid-steam-exploded corn stover by a moderate alkaline ethanol post-treatment based on an integrated biorefinery concept." *Biotechnology for Biofuels* 9:238. doi:10.1186/s13068-016-0656-1.

Yang, B., Dai, Z., Ding, S-Y., Wyman, C. E. 2011. "Enzymatic hydrolysis of cellulosic biomass." *Biofuels* 2:421–450.

Yang, J., Bae, J. Y., Lee, Y. M., Kwon, H., Moon, H. Y., Kang, H. A., Yee, S. B., Kim, W., Choi, W. 2011. "Construction of Saccharomyces cerevisiae strains with enhanced ethanol tolerance by mutagenesis of the TATA binding protein gene and identification of novel genes associated with ethanol tolerance." *Biotechnology and Bioengineering* 108:1776-87.

Yoon, S-H., Robyt, J. F. 2005. "Activation and stabilization of 10 starch-degrading enzymes by Triton X-100, polyethylene glycols, and polyvinyl alcohols." *Enzyme and Microbial Technology* 37:556-562.

Yu, Y and Christopher, L. P. 2017. "Detoxification of hemicellulose-rich poplar hydrolysate by polymeric resins for improved ethanol fermentability." *Fuel*, 203:187–196.

Yuan S. F., Guo G. L., Hwang W. S. 2017. "Ethanol production from dilute-acid steam exploded lignocellulosic feedstocks using an isolated multistress-tolerant *Pichia kudriavzevii* strain." *Microbial Biotechnology*. doi: 10.1111/1751-7915.12712.

Zaafouri K., Ziadi M., Hassen-Trabelsi A. B., Mekni S., Assi B., Alaya M., Bergaoui L., Hamdi M. 2017. "Optimization of Hydrothermal and Diluted Acid Pretreatments of *Tunisian Luffa cylindrica* (L.) Fibers for 2G Bioethanol Production through the Cubic Central Composite Experimental Design CCD: Response Surface Methodology." *Bio Med Research International*. doi: 10.1155/2017/9524521.

Zhang, H. D., Wu, S. B. 2014. "Enhanced enzymatic cellulose hydrolysis by subcritical carbon dioxide pretreatment of sugarcane bagasse." *Bioresource Technology* 158:161– 165.

Zhang, J., Song, Y., Wang, B., Zhang, X., Tan, T. 2016. "Biomass to bio-ethanol: The evaluation of hybrid *Pennisetum* used as raw material for bio-ethanol production compared with corn stalk by steam explosion joint use of mild chemicals." *Renewable Energy* 88:164-170.

Zhang, C., Xie, G., Li, S., Ge, L., He, T. 2010. "The productive potentials of sweet sorghum ethanol in China." *Applied Energy* 87:2360-2368.

Zhang, H., Ye, G., Wei, Y., Li, X., Zhang, A., Xie, J. 2017. "Enhanced enzymatic hydrolysis of sugarcane bagasse with ferric chloride pretreatment and surfactant." *Bioresource Technology* 229:96–103.

Zhang, X., Yuang, Q., Cheng, G. 2017. "Deconstruction of corncob by steam explosion pretreatment: Correlations between sugar conversion and recalcitrant structures." *Carbohydrate Polymers* 156:351–356.

Zhao, X., Cheng, K., Liu, D. 2009. "Organosolv pretreatment of lignocellulosic biomass for enzymatic hydrolysis." *Applied Microbiology and Biotechnology* 82:815–827.

Zhao, X., Li, S., Wu, R., Liu, D. 2017. "Organosolv fractionating pretreatment of lignocellulosic biomass for efficient enzymatic saccharification: chemistry, kinetics, and substrate structures." *Biofuels, Bioproducts and Biorefining*. 11:567–590.

Zhao, X., Zhang, L., Liu, D. 2012. "Biomass recalcitrance. Part I: the chemical compositions and physical structures affecting the enzymatic hydrolysis of lignocellulose." *Biofuels Bioproducts and Biorefinering* 6:465–482.

Zheng, J., Rehmann, L. 2014. "Extrusion Pretreatment of Lignocellulosic Biomass: A Review." *International Journal of Molecular Science*. 15:18967-18984.

Zheng, Y., Lin, H. M., Wen, J. 1995. "Supercritical carbon dioxide explosion as a pretreatment for cellulose hydrolysis." *Biotechnology Letters* 14:845–850.

Zhu, M., Li, P., Gong, X., Wang, J. 2012. "A comparison of the production of ethanol between simultaneous saccharification and fermentation and separate hydrolysis and fermentation using unpretreated cassava pulp and enzyme cocktail." *Bioscience, Biotechnology, and Biochemistry* 76:671–678.

In: Bioethanol and Beyond
Editor: Michel Brienzo
ISBN: 978-1-53613-478-0

Chapter 2

Bioethanol Production from Fermentable Sugars of Plant Cell Walls

Vimal Kumar Balasubramanian, Ning Yuan and Venugopal Mendu*

Fiber and Biopolymer Research Institute, Department of Plant & Soil Science, Texas Tech University, Lubbock, TX, US

Abstract

Plant cell walls encompass reservoir of complex polymeric sugars such as cellulose, hemicelluloses and pectins, which are arranged in a highly sophisticated and orderly manner that account for more than 60% of the cell wall by weight. Plant cell wall sugars are the focus for second generation bioethanol production via biochemical/enzymatic deconstruction followed by microbial fermentation process which outcompetes the first-generation biofuels made from starch/simple sugars from food crops. In comparison to fossil fuels, plant biomass based bioethanol is still at primitive stage for industrial scale production due to economic infeasibility. The cost associated with plant biomass production, collection, pretreatment and fermentation are prohibitive with the current technologies. To make bioethanol production economically feasible, it is important to produce biomass in marginal lands with minimal inputs, reduce pretreatment costs involved in breaking down of plant cell walls to access the sugars and fermentation. In addition, efficient breakdown and utilization of total fermentable sugars from cellulose, hemicellulose and pectin will result in economically feasible and environmental friendly plant biomass based bioethanol. Here, we focus on the cell wall based fermentable sugars, conversion methodologies and various challenges associated with the conversion process and recent scientific advances in this field.

* Corresponding Author Phone: 806-834-6327. Email: venugopal.mendu@ttu.edu.

INTRODUCTION

Utilization of natural carbon sources as biofuels is ancient as the human civilization. Use of these energy resources helped the mankind for their survival, establishment and evolution on this planet. The forms and kinds in which the natural carbon resources are used by the human beings changed with novel innovations in energy extraction methods; began with crude burning of the wood material to the highly refined fossil fuels obtained through industrial processing of crude oil derived from dead and decomposed organic matter [1]. Further, the non-renewable and environmental hazardous nature of fossil fuels led to the exploration of carbon based alternative fuels that are renewable and environmental friendly. Photosynthetic green plants fix the atmospheric carbon using sunlight and synthesize the most abundant and renewable Sugar Related Biopolymers (SRPs) on earth such as cellulose, hemicellulose and pectin. Conversion of these SRPs into biofuels is eco-friendlier than combustion of fossil fuel as the SRPs are fixed from the atmosphere and are put back into the atmosphere.

Plant cell walls are composed of SRPs that play a major role in plant growth, development and defense. The SRPs, together with lignin form a highly fortified secondary cell wall which stands as a magnificent scripture resulted from 700 million years of terrestrial plant evolution. The focus of second generation biofuels is to produce bioethanol from the glucose molecules derived from cellulose. Though, the commercialization of cellulosic ethanol has been accomplished to date [2], it is still not cost-effective in comparison to fossil fuels. This can be attributed to the logistics involved in the production of bioethanol from plant biomass such as cost associated with the biomass production, collection, processing, separation of cellulose from other polymers via pretreatment and fermentation of the derived sugars to produce bioethanol. Recent advances in plant cell wall research with the aim of obtaining liquid transportation fuel from plant cell walls made significant progress on reducing the biomass recalcitrance. However, there are several challenges in making the technology cost-effective and economically feasible for commercial production. Here in this chapter, the latest developments in the field of plant cell wall research along with the challenges and future directions towards an affordable cellulosic ethanol production from plant cell walls are discussed.

THE EMERGENCE OF CELLULOSIC ETHANOL

Majority of global energy requirements were met by coal in nineteenth century while crude oil in the twentieth century. Currently, following Saudi Arabia and Russia, USA is the third largest oil producing country in the world. There is a rapid depletion in the number of high quality oil reserves and overall production levels, posing a risk for more than 180 countries which import oil for its energy needs [3]. Moreover, use of coal and oil for electricity generation and transportation uses contributed 78% of the total global greenhouse gas emissions increase from 1970-2011 (https://www.epa.gov/ghgemissions). The concern on the adverse effects of coal and fossil fuel usage on environment led to the use of bioethanol as an alternative environmental friendly transportation fuel. The first-generation bioethanol was produced from the simple sugars derived from the starch rich plant based feedstocks such as corn, potato, sugarcane, etc. which was used in static and automobile engines. Meanwhile, the fall in global oil price due to enormous supply from the oil producing countries nearly bankrupted the ethanol industries. Later in 2005, to reduce petroleum dependence, the US government created a policy to blend ethanol in gasoline for automobiles. This resulted in an increase in price of corn due to the diversion of corn for ethanol production leading to food vs fuel debate.

The Food Conservation and Energy Act of 2008 provided incentives to companies that uses non-food based resources for ethanol production also known as second generation biofuels. This led to an increased investment in technologies that derive ethanol from biomass based feedstocks and the first commercial cellulosic ethanol plant (POET-DSM, Iowa) was built in 2013. Currently, approximately 78% of bio-ethanol is made from first generation feedstock such as corn and sugarcane; whereas biomass based second generation fuel contributes only 4% of the world's total bioethanol production [4]. To make the biomass based bioethanol economically feasible, among several factors, the feedstock bioprocessing and effective utilization of biomass are two very important factors to be addressed. Significant research has been conducted on the development of feedstocks with reduced recalcitrance and efficient downstream bioprocessing technologies with improved pretreatment, saccharification and fermentation. In addition, the other fermentable SRPs from the plant cell wall biomass presents an excellent opportunity to improve the economic feasibility of bioethanol production. Currently, the technologies that can utilize various fermentable sugars are limited and it is essential to improve the technologies for efficient use of all the available sugars. This book chapter focused on the availability and nature of fermentable cell wall sugars and the processing technologies to produce bioethanol. The topic is covered under two sub-headings; (1) Fermentable sugars of plant cell wall (2) Bioprocessing of cell walls into ethanol.

Fermentable Sugars of the Plant Cell Wall

Plant cell walls are important structural components that help the plants to with stand the mechanical pressures, protect the plants from biotic and abiotic stress factors. Plant cell wall is dynamic in nature allowing the active cell elongation while maintaining the rigidity to withstand the internal turgor pressure of the cells. The type of cell walls and composition varies among the various plant species, among plant organs and tissues within the organs. For example, parenchymal and collenchymal cell walls in leaves develop only primary cell walls, whereas the sclerenchymal cell present in stem fibers and endodermal cells in roots develop both primary and secondary cell walls and an additional tertiary cell wall is present in tension wood of trees [5]. The plant cell walls are mainly composed of SRPs such as cellulose, hemicellulose and pectin in addition to the phenolic lignin polymer. The composition of SRPs varies in the cell walls of land plants based on the species as well as tissue types and the average values range from 15-50% cellulose, 5-50% hemicellulose, 7-20% lignin and 30-50% pectin of the total cell wall weight [6, 7]. These SRPs are derived from primary metabolites such as hexoses (glucose, fructose, mannose, galactose) and pentoses (arabinose, xylose) that are derived from Calvin cycle and pentose phosphate pathway respectively [8, 9]. Cellulose is an unbranched homopolymer of glucose while hemicelluloses and pectins are mixed and branched polymers composed of hexoses, pentoses, deoxy-hexose (Fucose) and Uronic acids (galacturonic acid and glucuronic acid) (Figure 1) [10]. The biosynthesis of hexoses and pentoses are followed by activation into their respective nucleotide sugars such as UDP-glucose, UDP-xylose, etc. [9]. The activated nucleotide sugars are then transferred to the growing polymer chain by specific glycosyl transferases (GTs) and other additional enzymes as listed in CAZy database [11]. Apart from cell wall polymer synthesis, these activated monosaccharides are also used in nucleotide, protein and secondary metabolite biosynthetic process [12]. In the following sections, different cell wall polymers that can serve as a source for bioethanol production are discussed.

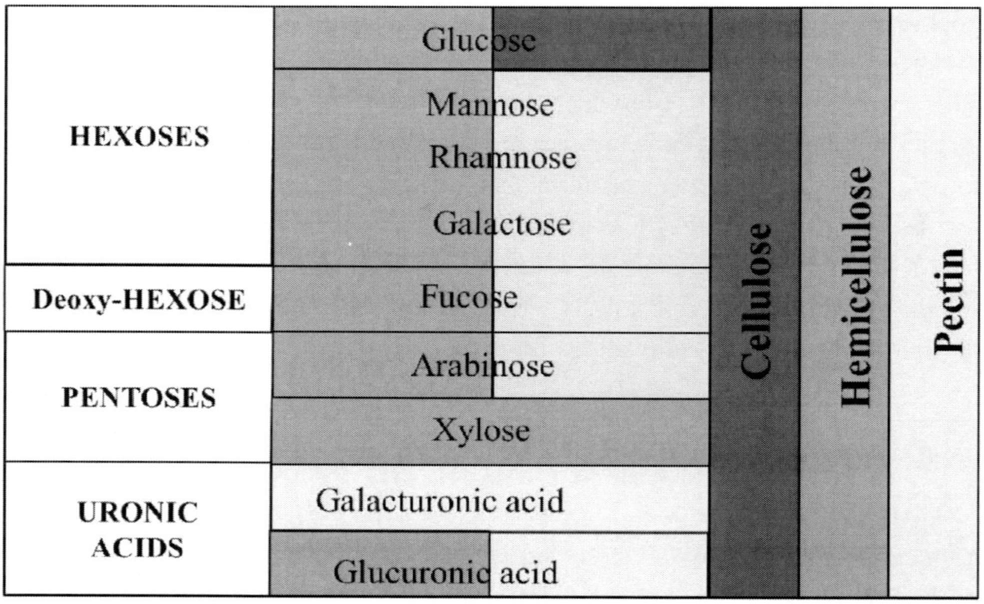

Figure 1. Composition of hexoses and pentoses in major cell wall polysaccharides (Black-cellulose, Gray-Hemicellulose and Light gray-Pectin).

Cellulose

Cellulose is the most abundant biopolymer on earth and is composed of beta 1-4 linked glucose units (Figure 1). The repeating unit in cellulose is cellobiose, not glucose as the adjacent glucose units are arranged at an angle of 180 degree to each other. The homogenous nature of cellulose composition i.e., glucose, which is a direct substrate for ethanol production, makes it an attractive polymer for bioethanol production. The cellulose content of individual cells varies and can range from 45% in wood tissue to 96% in cotton fibers [13]. The cellulose is synthesized by plasma membrane localized Cellulose Synthase A (CESA) protein complexes [14]. During the biosynthetic process, the individual microfibrils engage in inter and intra molecular hydrogen bonding which results in the production of highly crystalline cellulose fibers. The crystalline nature of cellulose makes it recalcitrant against enzymatic/mild acid hydrolysis to break down to glucose monomers for ethanol production by fermentation process [15]. Therefore, the genes that control the crystallinity of cellulose microfibrils need to be discovered and characterized as they determine the efficiency of cellulose hydrolysis process. Apart from *CESA* genes involved in the biosynthesis of crystalline cellulose, few other genes such as *COBRA* [16] and *CHL/CHI* [17] are found to regulate the cellulose crystallinity. Mutation in *CESA/COBRA/CHL/CHI* genes reduce the crystallinity of cellulose as seen with *cesa1cesa3* Arabidopsis mutant [18]. The reduction in crystallinity results in increased saccharification [19], resulting in increased bioethanol yield from the biomass and reduced pretreatment costs. In addition to reducing crystallinity, increasing the amount of cellulose deposited in the cell walls also can increase the amount of bioethanol production from the same amount of plant biomass. Increase in the total cellulose content was achieved by genetic engineering of both monocot and dicot plants. Overexpression of Sucrose Synthase (SuSy) genes resulted in 2-6% increase in cellulose content in poplar [20]. Also, total cellulose content was increased by overexpressing native UDP-glucose pyrophosphorylase in jute [21] and by overexpressing an Arabidopsis transcription factor AtSHINE in rice [22]. Discovery of the genes involved and mechanism of cellulose biosynthesis in dedicated bioenergy crops is needed to tailor the biomass with higher cellulose content with less crystallinity to improve saccharification efficiency for sustainable and profitable bio-ethanol production.

Hemicellulose

Next to cellulose, hemicellulose ranks as the second most abundant carbohydrate based polymer in the plant cell wall. Amount of hemicellulose ranges from 28-37% in the cell walls of different crop plants [10]. The hemicellulose is a complex heteropolymer synthesized by cellulose synthase like A (CSLA) proteins along with specific GTs associated with each polymer biosynthesis [23]. While the cellulose is synthesized by the plasma membrane localized cellulose synthase complexes [24], the hemicellulose biosynthesis occurs in Golgi vesicles, transported across the plasma membrane and deposited in the apoplast [23]. Unlike cellulose, hemicellulose is a complex heteropolymer comprised of hexoses, pentoses, deoxy-hexose and uronic acids (Figure 1). For instance, xyloglucan is the most predominant polymer in hemicellulose, which is made up of glucose backbone with xylose, fucose and galactose as the sidechains. Apart from xyloglucan, hemicelluloses are also comprised of glucuronoarabinoxylan, galactomannan, arabinoxylan and glucomannan. The structural complexity and composition of hemicellulose creates recalcitrance in several ways in utilizing the sugars for bioethanol production. Hemicelluloses require different set of enzymes for their hydrolysis into monosaccharides which ultimately results in higher bioethanol production costs [25]. Further, the major sugar present in hemicellulose i.e., xylose (Figure 2) acts as inhibitor for most of the fermenting microorganisms by forming inhibitory compound xylitol and as a consequence the utilization of xylose is reduced when glucose is present as main substrate [26]. This problem can be overcome by increasing the hexose content of hemicellulose (e.g., glucose, mannose, galactose) by tissue targeted higher expression of hexose biosynthetic genes [10]. It is possible to manipulate the composition of hemicellulose as some tree species such as pine and spruce which contain higher hexose content than other plants (Figure 2).

In addition, hemicelluloses (e.g., glucuronoxylan) contain acetyl side groups that interact with lignin in cell wall thereby creating a recalcitrant matrix posing additional problems in polymer separation for sugar extraction. This recalcitrance was reduced by mutating Cinnamyl-alcohol dehydrogenase gene involved in lignin biosynthesis which resulted in reduced interaction with hemicelluloses, resulting in reduced biomass recalcitrance [27]. Moreover, the acetyl groups are released as acetic acid during fermentation process that reduces the efficiency of fermentation [28]. Transgenic plants with reduced acetyl transferase activity resulted in reduced release of acetic acid during fermentation process leading to higher ethanol yield [28]. In addition to modifying the composition, increasing the total hemicellulose content also creates a positive impact on biomass saccharification process as they contain more SRPs and release more fermentable sugars [29]. Until now, the hemicellulose biosynthesis is one of the poorly understood fields of cell wall research due to its complexity and warrants further research to engineer the composition of hemicellulose for efficient bioethanol production.

Pectin

Pectin is the third SRP that is present mainly in the primary cell walls and middle lamellae regions of the cells and constitutes up to 35% of primary cell wall that varies by species [30]. As a component of cell wall, pectin provides mechanical strength and involves in cell wall integrity signaling during pathogen attack, cell wall extension and organogenesis [31]. Pectin is primarily made up of galacturonic acid residues (Figure 1) and 60% of the pectin present in cell wall is composed of the polymer homogalacturonan (HG) [32]. Apart from HG, other polymers such as rhamnogalacturonan (RG 1 and RG2), apigalacturonan and xylogalacturonan are also part of pectin polysaccharides [30]. Similar to hemicellulose, pectin is another complex heteropolysaccharide, made up of many different monosaccharides (Figure 1). Similar to hemicellulose, the biosynthesis of pectin takes place in Golgi vesicles which is exported and deposited in the growing primary cell walls. It is estimated that, there are about 67 glycosyl transferases involved in biosynthesis of pectin polysaccharide [30]. Similar to other cell wall polymers, pectin forms a complex matrix in plant cell walls and acts as a recalcitrant factor in bioethanol production. The recalcitrance of pectin comes from methylesterification of pectin, which is a process of removing the methyl groups present in pectin side chain. The methyl esterification is catalyzed by the enzyme pectin methyl esterase (PME), which facilitates higher interpolymer linkage ability of the pectin molecules. Inhibition of PME in Arabidopsis plant resulted in enhanced saccharification of the cell wall sugars [33]. In general, pectin is water soluble in nature and pectin rich sources such as fruit and vegetable waste, potato peel and sugar beet pulp can be used as biofuel feedstocks which require minimal pretreatment and lesser enzyme loading for bioethanol production [34]. There is a tremendous interest in bioethanol industry to increase the pectin content due to its lack of tight association with lignin in the cell wall which results in reduced pretreatment costs. Similar to hemicellulose, the research on pectin is still an emerging field and crucial genes involved in biosynthesis and modification needs to be identified and functionally characterized. Also, the pectin composition needs to be carefully altered as it directly influences the cell wall integrity and defense mechanism which ultimately governs plant growth, survival and yield [35].

ENGINEERING NEXT-GENERATION BIOETHANOL FEEDSTOCKS

Due to high pretreatment costs, creating less recalcitrant plant cell wall is one of the prime factors involved in developing next generation feedstock for affordable biomass based ethanol production. In case of first-generation biofuel feedstock such as starch (corn) and sucrose (sugarcane), two main factors are of major focus; overall crop yield and sugar/starch content. However, the second-generation biofuels are mainly plant biomass based hence, the total biomass, composition of different cell wall polymers (lignocellulosic material) and recalcitrance associated with their extraction are of a major focus. Since the cell walls are important for the plant growth, development and survival under different environmental conditions, any alteration in the composition may have deleterious effect on the overall plant development, survival and biomass yield. A careful balance must be maintained to produce tailored biomass without compromising the biomass yield and performance. This can be achieved by a complete understanding of the molecular and biochemical mechanism involved in the cell wall polymer biosynthesis and composition.

Of all the cell wall polymers, cellulose biosynthesis is the most investigated process, probably due to its less complicated composition compared to other cell wall polymers. Several breakthroughs have been made in cellulose biosynthesis which made it possible to achieve an increased cellulose content [36], biomass content [37], improved carbon allocation to wood [38] etc. In case of hemicelluloses, there are still many unanswered questions due to its complex composition, branching and modifications. Despite lack of complete understanding, several advancements have been made to tailor the biomass to improve the bioethanol production. For example, the acetyl content of hemicellulose has been reduced in transgenic poplar through RNA interference technology by targeting reduced wall acetylation (RWA) genes which resulted in higher sugar release [28]. In case of pectin, another complex cell wall polymer, progress has been made in improving pectin content of biomass. Reduction in the PME gene expression improved the disease resistance [39] while enhanced pectin content improved phosphorous remobilization [40], which are very important traits in generating biomass crops with improved disease resistance and higher nutrient use efficiency. Another major focus of the second-generation biomass engineering research is to reduce the lignin content which was considered as a major hindrance for the extraction of cell wall sugars for bioethanol production. Successful discoveries such as brown mid rib (*bmr*) mutant lines in corn and sorghum showed a reduced lignin content due to mutation in lignin biosynthetic genes that ultimately result in higher saccharification efficiency and improved digestibility of the cattle fodder [41].

Recent developments in lignin research in model and crop species led to creation of transgenic poplar trees with higher sugar release [42] and Arabidopsis with higher saccharification of biomass [43] with no compromise in plant growth and development. Despite benefits of reduced lignin content for bioethanol production, the plants with reduced lignin content result in compromised plant defense of the engineered plants [44]. Instead of reduced lignin content, altered lignin composition to reduce the biomass recalcitrance will offer a better solution for sustainable production of biofuel crops with increased saccharification. Recent breakthroughs in basic cell wall polymer research are advancing the knowledge about cell wall biosynthetic process and offers an excellent opportunity for producing tailor made feedstocks for sustainable bioethanol production in near future [45-49].

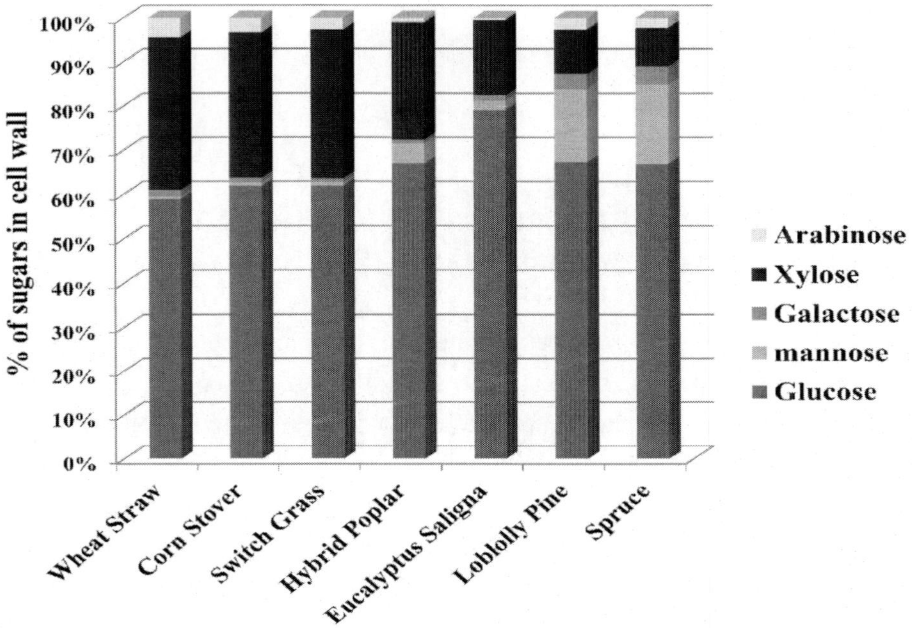

Figure 2. Percentage of major hexose (glucose, mannose & galactose) and pentose sugars (Xylose & Arabinose) in cell wall of different feedstocks. Data derived from [25].

BIOPROCESSING OF CELL WALLS INTO ETHANOL

Biomass represents a renewable and humongous biochemical battery that is made up of energy rich cell wall polymers. Currently, biomass derived fuels are the promising alternative source for producing environmental friendly liquid transportation fuels. Biomass can be converted into fuel through thermochemical and biochemical methods. Thermochemical process involves thermal gasification of the biomass to produce syngas that is primarily composed of carbon monoxide and hydrogen that are used to produce energy or pyrolyzing the biomass which involves anaerobic combustion to produce bio oil [50]. While, the biochemical process involves careful extraction of fermentable sugars from the plant biomass and conversion to bioethanol through fermentation technology. Unlike thermochemical based energy production, when it comes to fermentation based bioethanol production, multiple stages are involved in the production process. These stages include biomass pretreatment to reduce the recalcitrance associated with the cell wall, extraction of the useful cell wall polymers, and hydrolysis of cell wall polymers into simple sugars and fermentation of sugars into ethanol. Among these stages, the biomass type, the choice of pretreatment method, hydrolysis and fermentation techniques determine the ultimate cost of bio-ethanol production. Hence, it is important not only to understand the biomass composition to reduce the processing costs, it is also important to determine the biomass processing technologies for the production of cost effective bioethanol. Significant progress has been made in bioprocessing technologies for efficient polymer extraction and bioethanol production which are discussed in the following section in detail.

Biomass Pretreatment: The First Step in Fermentable Ethanol Production

Pretreatment of biomass is a very important step in the bioethanol production pipeline as the efficiency of pretreatment has a direct effect on the downstream processing. Broadly, pretreatment is defined as the process of reducing the association between different cell wall polymers thereby reducing the crystallinity of the cell wall. The ultimate aim of pretreatment process is the reduction of biomass particle size and increase in the porosity of cell wall for better enzymatic access. The cost of pretreatment accounts up to 19% of the total production costs in bioethanol generation process [51]. The biomass pretreatment methods can be broadly classified into four types; physical, chemical, biological and solvent fractionation [52]. The physical pretreatment process involves use of mechanical force to reduce the particle size and crystallinity of the biomass [53]. In chemical pretreatment method, acids (dilute sulfuric acid), steam, alkali (ammonia based) and oxidation based techniques are used. In this method, chemical hydrolysis of hemicellulose and lignin occurs thereby releasing the cellulose for enzymatic saccharification. The chemical method operates under high temperature and pressure and often produces downstream inhibitory compounds that require extensive washing of the residual cellulose slurry which is neither cost effective nor ecofriendly [52]. Various downstream inhibitory compounds formed during the pretreatment process, the problems associated and the strategies to overcome the issues are reviewed in detail elsewhere [54]. The biological pretreatment method employs biological organisms which produce cell wall modifying enzymes for the biomass deconstruction. In biological pretreatment method, fungal species such as basidiomycetes and actinomycetes are used which produces cell wall degrading enzymes such as cellulases and laccases to break down the cell wall. But this pretreatment method is highly time consuming (up to several days) and the fungal cultures feed on the released cell wall sugars. In solvent fractionation process, organic solvents such as alcohols and phosphoric acids are used under high temperature and pressure to dissolve non-cellulosic polymers thereby leaving cellulose for enzymatic saccharification [52]. The major problem associated with organic solvent fractionation method is the economical way of recovering the used organic solvents after the pretreatment process. The latest and currently most used organic solvent process is the ionic liquid based pretreatment and it is showing promising results with the biomass pretreatment process.

Ionic Liquid Pretreatment: Highly Promising Next Gen Technology in Biomass Pretreatment

An efficient pretreatment method which requires less energy input, operates at room temperature and recyclable is required for sustainable and economical production of bioethanol. Ionic liquids appear to be a promising pretreatment option for sustainable bioethanol production. Ionic liquids (IL) are solvents composed of salts (anions, cations) that are liquid at room temperature and are used nowadays in the pretreatment process for dissolving different cell wall polymers such as cellulose, hemicellulose and lignin [55]. Several ILs that are functional at room temperature have been discovered which avoid the pretreatment at higher temperatures and there by saves energy input [56]. When compared to organic acids and alcohols, ILs show low vapor pressure and high temperature stability. ILs have several advantages over inorganic acids in pretreatment process. During the pretreatment, the inorganics causes decomposition of polysaccharides resulting in the formation of microbial inhibitors such as furfural, hydroxymethyl furfural while the ILs do not produce these inhibitors [54]. The research on solubility of carbohydrates by ILs dates back several decades and currently there are many ILs discovered to be a good solvent for all kinds of carbohydrates [55, 57]. Various ILs are good solvents for lignin extraction [56], which is a phenolic polymer of the cell wall and especially, 1-Ethyl -3- Methylimidazolium acetate acts as a good solvent, which showed up to 63% lignin extraction from maple wood flour and reduced the cellulose crystallinity and boosted saccharification of cellulose up to 93% [58]. Further, combination of 1-ethyl-3-methylimidazolium cation along with a combination of two anions such as alkylbenzenesulfonates and xylenesulfonates were used to pre-treat sugarcane bagasse that recovered 93% of the total lignin [59]. The residual cellulose obtained after the pretreatment can be fermented for ethanol production while, the dissolved lignin can be recovered using an anti-solvent and used for other industrial applications. Also, many renewable chemicals can be produced from ionic liquid pre-treated lignocellulosic biomass [60]. By maintaining a higher IL to biomass ratio, it is possible to dissolve the biomass completely and the dissolved polymers can be regenerated using antisolvent for simultaneous saccharification and fermentation process [61]. Overall, the ILs offers an excellent platform for biomass pretreatment and product recovery for efficient bioethanol production.

Cell Wall Hydrolysis of Pretreated Biomass

The biomass pretreatment reduces the crystallinity and interpolymer associations however, it does not depolymerize the cell wall polymers into its monomeric units. In order to ferment the cell wall sugars in to ethanol, these polymers need to be converted into their simplest monomeric units through a process called saccharification as the fermenting microorganisms utilize only the monosaccharide (glucose/xylose) form of sugars. Saccharification is a hydrolytic process in which chemical bonds between the sugar units are broken down with the help of hydrolytic reaction with water as the donor of hydrogen and hydroxyl groups for the resulting monosaccharides. Cell wall saccharification can be carried out by chemicals or by hydrolytic enzymes. In chemical method, dilute and concentrated acids are used to depolymerize the cell wall polymers, whereas in enzymatic method, cell wall hydrolyzing enzymes such as cellulase, hemicellulase, pectinases, etc. are used. The ultimate aim of the two methods is to produce simpler monomeric sugars for the downstream fermentation process.

Chemical Hydrolysis

In chemical hydrolysis, the mild acid pretreated biomass (the residual fibers or crystalline polysaccharides) is treated with concentrated acids such as sulfuric acid and hydrochloric acids [62]. These strong acids break the interpolymer and intrapolymeric bonds resulting in the breakdown of polymers into monomeric sugar units. This method of acid hydrolysis has been used to quantify the monomeric sugar content of plant cell wall polysaccharides [63, 64] and in commercial scale ethanol production [62]. Apart from plant biomass, the mild acid hydrolysis is commonly used for treating agricultural and municipal wastes that are rich in polymeric sugars [65]. Strong acid hydrolysis has several advantages such as compatibility with diverse feedstocks, higher yield of monomeric sugars and requirement of mild reaction conditions. But, corrosion of hydrolytic reaction chambers and recycling of acids were among the technical difficulties associated with this method. To overcome these issues, a combination of ionic liquid and mild acid based hydrolysis was used to depolymerize cellulose and convert them into monomeric sugars, which resulted in poor yields compared to strong acid hydrolysis methods [66, 67]. By adjusting the composition of ionic liquid and the amount of water used in the process, up to 70% of glucose and 79% of xylose was solubilized into the ionic liquid and out of which 94% glucose and 88% xylose were recovered for the fermentation process [68]. By this way of using a combination of ionic liquid and dilute acid hydrolysis, the use of enzyme based method for hydrolysis of cell wall polymers can be avoided which is very expensive stage in bioethanol production [69, 70].

Enzymatic Hydrolysis

The alternative method to chemical hydrolysis is by using cell wall degrading enzymes (CWDE) produced primarily from microbes such as bacteria and fungi. In nature, these heterotrophic microorganisms feed upon the plant material for the metabolic needs. Since plants contain recalcitrant cell walls, these microbes evolved to produce plant CWDEs to degrade the cell walls. These CWDEs represent a cocktail of different enzymes such as exo and endo-cellulases, hemicellulase, pectinase and laccases that degrade different components of the plant cell walls. In commercial bioethanol production, the enzymatic hydrolysis step involves use of CWDEs from specific bacteria (e.g., *E. Coli*) and fungi (e.g., *S. Cerevisiae*) for cell wall hydrolysis process. Among the cell wall polymers, hemicelluloses and pectin require different class of enzymes to convert them into monomeric sugars (Figure 1), whereas cellulose and lignin require less number of enzymes [25, 71]. The CWDEs cleave the cellulose either in the middle (e.g., endoglucanase) and/or from the ends (e.g., exoglucanase). In case of hemicellulose, endo-1,4-beta xylanase that belong to GH10 family is abundantly present in fungi that is involved in breaking the xylan backbone of xyloglucan polymer. In case of pectin, polygalacturonases and pectin lyases are involved in cleaving the glycosidic bond present in the pectic sugars. The list of CWDEs and their mode of action has been reviewed elsewhere [71, 72].

After the hydrolysis by CWDEs, the resulting hexose end product is channeled into glycolysis pathway of yeast to form glyceraldehyde 3-phosphate which first is reduced into acetaldehyde and then to ethanol. Whereas in case of pentoses, the pentose phosphate pathway intermediate D-xylulose-5-phophate gets converted into glyceraldehyde 3-phosphate which is then reduced to ethanol [73]. Nowadays, the CWDEs are immobilized on silica nanoparticles that boosts their stability, recovery and catalytic activity thereby increasing bioethanol production [74]. The rate of saccharification of cell wall polymers by these CWDEs are determined by several factors such as substrate particle size, composition, pretreatment process and hydrolytic conditions such as temperature and batch time [75, 76]. Currently, biotechnological approaches were used to engineer these CWDEs to be produced within plants endogenously thereby using plants as bioreactors for enzyme production [77, 78]. This led to cut down the cost associated with the enzyme production and purification and resulted in a much easier processing of plant biomass in the saccharification process. In addition, the advancements in the downstream processing technologies such as consolidated bioprocessing (CBP) involves the use of recombinant fungi and bacteria as a one-step hydrolysis and fermentation step for bioethanol production [79] which results in the reduced cost of bioethanol production.

Fermentation of Hydrolyzed Sugars into Ethanol

Fermentation is the final step in the bioethanol production pipeline, which involves the use of microorganisms such as yeast that utilize glucose to synthesize Adenosine

Triphosphate (ATP) and ethanol is produced as a byproduct. This process can take place under both aerobic and anaerobic conditions. In first generation feedstock based ethanol production (e.g., corn and sugarcane), the starch/sugars obtained from the plant is converted into glucose and fermented directly by bacteria (*Zymomonas mobilis*) or fungi (*Saccharomyces cerevisiae*) in a batch or continuous fermenter [80]. In case of biomass based second generation biofuels, different fermentation technologies are used such as (1) separate hydrolysis and fermentation (SHF) (2) simultaneous saccharification and co-fermentation (SSF) (3) consolidate bioprocessing (CBP). The aim of these different types of fermentation methodologies is reducing the number of stages in the ethanol production process thereby reducing the associated costs. In SHF method, as the name indicates, the saccharification and fermentation process are performed separately in two different chambers. This is advantageous in removing the inhibitory compounds formed during saccharification before the start of fermentation process and flexibility to use different temperature required for hydrolytic enzymes (~50°C) and yeast (~32°C). Despite the advantages, this method has several disadvantages such as the end product inhibition of saccharification, high capital cost and chances of substrate/product cross contamination [81]. In case of SSF, both saccharification and fermentation are carried out in the same chamber, which means both hydrolyzing enzymes and fermenting yeast are used simultaneously. The SSF method have several advantages such as low substrate inhibition, low capital cost, higher ethanol yield [82]. This SSF method is also considered to be the first-generation consolidated bioprocessing (CBP) technique, where the hydrolysis and fermentation are conducted in a single step process. In second generation CBP, the pretreatment of biomass is avoided, thereby directly using the biomass in a single step process to generate bioethanol [83]. This is achieved by considerable research in the development of competitive strains that can take-on the biomass recalcitrance without any additional pretreatments. Using this approach, the cost associated with pretreatment is avoided, thereby considerable reduction in the price of ethanol produced [84]. The developments in CBP technique will be further described in following section as it is the latest bioprocessing technology.

Consolidated Bioprocessing of Biomass: The Next-Generation of Bioprocessing and Fermentation Technology

The improvements in biological processing techniques such as CBP are providing a unique opportunity to overcome the recalcitrance associated with lignocellulosic biomass to produce affordable bioethanol in near future. The CBP technology includes novel advancements in research in both process engineering and biotechnological sectors. In the latter sector, innovations have been made in bacterial and fungal strains, to utilize the biomass with/without pretreatment to convert them directly into bioethanol. In general, fungal strains (e.g., *S. cerevisiae* or baker's yeast) are used to ferment glucose into ethanol, but the fungal species lacks the ability to utilize pentose sugars due to missing pentose transporters in their system [85]. Recent advances in biotechnological research in fungi led to the utilization of xylose [86, 87], which is the second most abundant sugar in most of plant biomass. Due to the ability to use pentose sugars and withstand higher temperatures and inhibitory compounds, bacterial strains were used for fermentation of lignocellulosic biomass [88]. In case of second generation CBP, fungal strains that are cellulolytic (*C. thermocellum*) in nature are engineered to withstand higher ethanol levels produced in the fermentation process, whereas, the strains that can withstand the ethanol (*S. cerevisiae*) are engineered to improve cellulolytic properties [89]. In case of bacterial strains, recent research in *C. thermocellum* and *C. bescii* is providing promising strains that can withstand high temperatures and directly utilize the crystalline cellulose rich biomass for conversion into ethanol [83, 90]. More recombinant approaches that are involved in developing CBP suitable strains and their efficiency in ethanol fermentation are reviewed in detailed elsewhere [79]. In the future, metabolic engineering is required to identify/develop/improve novel strains that can maximize ethanol yield by CBP technology which will result in economical bioethanol production.

The Current Status and Future Perspectives of the Bio-Ethanol Industry

Despite the promising future of the bioethanol as an environmental friendly transportation fuel, it is economically not sustainable with the current technologies. There is an immediate need to make rapid advancement in biomass manipulation through biotechnology and improve the efficiency of downstream processing technologies thereby making bioethanol production more economical. From the research and development perspective, a significant progress has been made in the field of biomass manipulation and processing technologies. Understanding the plant cell wall biosynthesis, development and composition led to the generation of plants with less recalcitrant biomass while and breakthrough discoveries in downstream process development significantly improved the efficiently of polymer separation and bioethanol production. Field trials with low lignin biomass crops such as poplar, switchgrass and maize shows promising results for enhanced saccharification thereby boosting ethanol yield [91-93]. In case of process developments, several cutting-edge research discoveries were made by exclusive organizations such as Joint Bioenergy Institute (https://www.jbei.org/research/ publications/), Great lake bioenergy research center (https://www.glbrc.org/ research/technologies), and Bioenergy science center (http://bioenergycenter.org/besc/ industry/intellectual-property.cfm). There is an intense research on developing better varieties with higher biomass yield, developing better strains of bacteria and fungi for biomass saccharification and developing better bioprocessing techniques to efficiently depolymerize the biomass. Recently, the US government sanctioned $40 million for as a part of 5-year fiscal period beginning 2018 towards development of advanced biomass based energy and products which boosts the research and development in the sector (https://energy.gov/articles/department-energy-provides-40-million-4-doe-bioenergy-research-centers). In addition, partnerships between energy related companies and universities are in place to advance research and development to create next generation biofuel from the plant biomass. With this progress in place, the future of cellulosic bioethanol could benefit from (1) creating novel strategies to address biomass recalcitrance without compromising the plant growth and survival (2) developing efficient consolidated approach for biomass processing (3) increasing the public/producers' awareness on next generation feedstock cultivation (4) government-private ventures that can establish more bioethanol plants using newer technologies and (5) government policies on promoting/utilizing the second generation bioethanol in the automobile sector. With this level of biomass based bioethanol research, the stage is set for key inventions that can transform the cellulosic ethanol production more economical in the near future.

REFERENCES

[1] Council, NR. *Oil in the Sea: Inputs, Fates, and Effects*. 1985, Washington, DC: The National Academies Press. 601.

[2] Reuters. *TABLE-Open and planned US cellulosic ethanol plants*. 2009, [cited 2017 August 9], Available from: http://libanswers.liverpool.ac.uk/faq/49511.

[3] Hall, C; et al., Hydrocarbons and the evolution of human culture. *Nature*, 2003, 426(6964), p. 318-322.

[4] Balan, V. Current Challenges in Commercially Producing Biofuels from Lignocellulosic Biomass. *ISRN Biotechnology*, 2014, p. 31 pages.

[5] Mellerowicz, EJ; Gorshkova, TA. Tensional stress generation in gelatinous fibres: A review and possible mechanism based on cell-wall structure and composition. *Journal of Experimental Botany*, 2012, 63(2), p. 551-565.

[6] Welker, C; et al., Engineering Plant Biomass Lignin Content and Composition for Biofuels and Bioproducts. *Energies*, 2015, 8(8), p. 7654.

[7] Harholt, J; Suttangkakul, A; Vibe Scheller, H. Biosynthesis of Pectin. *Plant Physiology*, 2010, 153(2), p. 384-395.

[8] Sillero, A; Selivanov, VA; Cascante, M. Pentose phosphate and calvin cycles: Similarities and three-dimensional views*. *Biochemistry and Molecular Biology Education*, 2006, 34(4), p. 275-277.

[9] Yin, Y; et al., Evolution of Plant Nucleotide-Sugar Interconversion Enzymes. *PLOS ONE*, 2011, 6(11), p. e27995.

[10] Pauly, M; Keegstra, K. Cell-wall carbohydrates and their modification as a resource for biofuels. *The Plant Journal*, 2008, 54(4), p. 559-568.

[11] Cantarel, BL; et al., The Carbohydrate-Active EnZymes database (CAZy): An expert resource for Glycogenomics. *Nucleic Acids Research*, 2009, 37 (Database issue), p. D233-D238.

[12] Bar-Peled, M; O'Neill, MA. Plant Nucleotide Sugar Formation, Interconversion, and Salvage by Sugar Recycling*. *Annual Review of Plant Biology*, 2011, 62(1), p. 127-155.

[13] Abidi, N; Cabrales, L; Haigler, CH. Changes in the cell wall and cellulose content of developing cotton fibers investigated by FTIR spectroscopy. *Carbohydrate Polymers*, 2014, 100, p. 9-16.

[14] Kumar, M; Turner, S. Plant cellulose synthesis: CESA proteins crossing kingdoms. *Phytochemistry*, 2015, 112, p. 91-99.

[15] Hall, M; et al., Cellulose crystallinity-a key predictor of the enzymatic hydrolysis rate. *FEBS J*, 2010, 277.

[16] Ben-Tov, D; et al., COBRA-LIKE2, a Member of the Glycosylphosphatidylinositol-Anchored COBRA-LIKE Family, Plays a Role in Cellulose Deposition in Arabidopsis Seed Coat Mucilage Secretory Cells. *Plant Physiology*, 2015, 167(3), p. 711-724.

[17] Sánchez-Rodríguez, C; et al., CHITINASE-LIKE1/POM-POM1 and Its Homolog CTL2 Are Glucan-Interacting Proteins Important for Cellulose Biosynthesis in Arabidopsis. *The Plant Cell*, 2012, 24(2), p. 589.

[18] Harris, DM; et al., Cellulose microfibril crystallinity is reduced by mutating C-terminal transmembrane region residues CESA1(A903V) and CESA3(T942I) of cellulose synthase. *Proceedings of the National Academy of Sciences of the United States of America*, 2012, 109(11), p. 4098-4103.

[19] Harris, D; DeBolt, S. Synthesis, regulation and utilization of lignocellulosic biomass. *Plant Biotechnology Journal*, 2010, 8(3), p. 244-262.

[20] Coleman, HD; Yan, J; Mansfield, SD. Sucrose synthase affects carbon partitioning to increase cellulose production and altered cell wall ultrastructure. *Proceedings of the National Academy of Sciences of the United States of America*, 2009, 106(31), p. 13118-13123.

[21] Zhang, G; et al., Overexpression of UDP-glucose pyrophosphorylase gene could increase cellulose content in Jute (*Corchorus capsularis* L.). *Biochemical and Biophysical Research Communications*, 2013, 442(3), p. 153-158.

[22] Ambavaram, MMR; et al., Coordinated Activation of Cellulose and Repression of Lignin Biosynthesis Pathways in Rice. *Plant Physiology*, 2011, 155(2), p. 916-931.

[23] Pauly, M; et al., Hemicellulose biosynthesis. *Planta*, 2013, 238(4), p. 627-642.

[24] Bashline, L; Li, S; Gu, Y. The trafficking of the cellulose synthase complex in higher plants. *Annals of Botany*, 2014, 114(6), p. 1059-1067.

[25] Álvarez, C; Reyes-Sosa, FM; Díez, B. Enzymatic hydrolysis of biomass from wood. *Microbial Biotechnology*, 2016, 9(2), p. 149-156.

[26] Sharmin, F; Huygens, F; Hargreaves, M. A review of pentose utilizing bacteria from bagasse hemicellulose. *African Journal of Microbiology Research*, 2013, 7(22), p. 2694-2700.

[27] Carmona, C; et al., Why genetic modification of lignin leads to low-recalcitrance biomass. *Physical Chemistry Chemical Physics*, 2015, 17(1), p. 358-364.

[28] Pawar, PMA; et al., Downregulation of RWA genes in hybrid aspen affects xylan acetylation and wood saccharification. *New Phytologist*, 2017, 214(4), p. 1491-1505.

[29] Xu, N; et al., Hemicelluloses negatively affect lignocellulose crystallinity for high biomass digestibility under NaOH and H2SO4 pretreatments in Miscanthus. *Biotechnology for Biofuels*, 2012, 5(1), p. 58.

[30] Caffall, KH; Mohnen, D. The structure, function, and biosynthesis of plant cell wall pectic polysaccharides. *Carbohydrate Research*, 2009, 344(14), p. 1879-1900.

[31] Kohorn, BD; et al., Pectin activation of MAP kinase and gene expression is WAK2 dependent. *The Plant Journal*, 2009, 60(6), p. 974-982.

[32] Ridley, BL; O'Neill, MA; Mohnen, D. Pectins: structure, biosynthesis, and oligogalacturonide-related signaling. *Phytochemistry*, 2001, 57(6), p. 929-967.

[33] Lionetti, V; et al., Engineering the cell wall by reducing de-methyl-esterified homogalacturonan improves saccharification of plant tissues for bioconversion. *Proceedings of the National Academy of Sciences*, 2010, 107(2), p. 616-621.

[34] Lesiecki, M; Białas, W; Lewandowicz, G. Enzymatic hydrolysis of potato pulp. *Acta Sci Pol Technol Aliment*, 2012, 11(1), p. 53-9.

[35] Xiao, C; Anderson, CT. Roles of pectin in biomass yield and processing for biofuels. *Frontiers in Plant Science*, 2013, 4, p. 67.

[36] Yang, Y; et al., Overexpression of a Domain of Unknown Function 266-containing protein results in high cellulose content, reduced recalcitrance, and enhanced plant growth in the bioenergy crop Populus. *Biotechnology for Biofuels*, 2017, 10(1), p. 74.

[37] Jin, YL; et al., Overexpression of Populus trichocarpa CYP85A3 promotes growth and biomass production in transgenic trees. *Plant Biotechnology Journal*, 2017, 15(10), p.1309-1321.

[38] Unda, F; et al., Altering carbon allocation in hybrid poplar (*Populus alba* × *grandidentata*) impacts cell wall growth and development. *Plant Biotechnology Journal*, 2017, 15(7), p. 865-878.

[39] Lionetti, V; et al., Overexpression of Pectin Methylesterase Inhibitors in Arabidopsis Restricts Fungal Infection by *Botrytis cinerea*. *Plant Physiology*, 2007, 143(4), p. 1871-1880.

[40] Zhu, XF; et al., Pectin enhances rice (*Oryza sativa*) root phosphorus remobilization. *Journal of Experimental Botany*, 2015, 66(3), p. 1017-1024.
[41] Godin, B; et al., Improved sugar yields from biomass sorghum feedstocks: comparing low-lignin mutants and pretreatment chemistries. *Biotechnology for Biofuels*, 2016, 9(1), p. 251.
[42] Cai, Y; et al., Enhancing digestibility and ethanol yield of Populus wood via expression of an engineered monolignol 4-O-methyltransferase. *Nature Communications*, 2016, 7, p. 11989.
[43] Yang, F; et al., Engineering secondary cell wall deposition in plants. *Plant Biotechnology Journal*, 2013, 11(3), p. 325-335.
[44] Sattler, S; Funnell-Harris, D. Modifying lignin to improve bioenergy feedstocks: strengthening the barrier against pathogens? *Frontiers in Plant Science*, 2013, 4(70).
[45] Ivakov, A; et al., Cellulose Synthesis and Cell Expansion Are Regulated by Different Mechanisms in Growing Arabidopsis Hypocotyls. *The Plant Cell*, 2017, 29(6), p. 1305-1315.
[46] Shinohara, N; et al., The plant cell-wall enzyme AtXTH3 catalyses covalent cross-linking between cellulose and cello-oligosaccharide. *Scientific Reports*, 2017, 7, p. 46099.
[47] Simmons, TJ; et al., Folding of xylan onto cellulose fibrils in plant cell walls revealed by solid-state NMR. *Nature Communications*, 2016, 7, p. 13902.
[48] Sechet, J; et al., Xyloglucan Metabolism Differentially Impacts the Cell Wall Characteristics of the Endosperm and Embryo during Arabidopsis Seed Germination. *Plant Physiology*, 2016, 170(3), p. 1367-1380.
[49] Xiao, C; et al., Xyloglucan Deficiency Disrupts Microtubule Stability and Cellulose Biosynthesis in Arabidopsis, Altering Cell Growth and Morphogenesis. *Plant Physiology*, 2016, 170(1), p. 234-249.
[50] De Filippis, P; et al., Biomass Gasification Plant and Syngas Clean-up System. *Energy Procedia*, 2015, 75, p. 240-245.
[51] Wooley, R; et al., Process Design and Costing of Bioethanol Technology: A Tool for Determining the Status and Direction of Research and Development. *Biotechnology Progress*, 1999, 15(5), p. 794-803.
[52] da Costa Sousa, L; et al., 'Cradle-to-grave' assessment of existing lignocellulose pretreatment technologies. *Current Opinion in Biotechnology*, 2009, 20(3), p. 339-347.
[53] Silveira, MHL; et al., Current Pretreatment Technologies for the Development of Cellulosic Ethanol and Biorefineries. *Chem Sus Chem*, 2015, 8(20), p. 3366-3390.
[54] Jönsson, LJ; Martín, C. Pretreatment of lignocellulose: Formation of inhibitory by-products and strategies for minimizing their effects. *Bioresource Technology*, 2016, 199, p. 103-112.

[55] Hou, Q; et al., Pretreatment of Lignocellulosic Biomass with Ionic Liquids and Ionic Liquid-Based Solvent Systems. *Molecules*, 2017, 22(3), p. 490.

[56] Mora-Pale, M; et al., Room temperature ionic liquids as emerging solvents for the pretreatment of lignocellulosic biomass. *Biotechnology and Bioengineering*, 2011, 108(6), p. 1229-1245.

[57] Zakrzewska, ME; Bogel-Łukasik, E; Bogel-Łukasik, R. Solubility of Carbohydrates in Ionic Liquids. *Energy & Fuels*, 2010, 24(2), p. 737-745.

[58] Lee, SH; et al., Ionic liquid-mediated selective extraction of lignin from wood leading to enhanced enzymatic cellulose hydrolysis. *Biotechnology and Bioengineering*, 2009, 102(5), p. 1368-1376.

[59] Tan, SSY; et al., Extraction of lignin from lignocellulose at atmospheric pressure using alkylbenzenesulfonate ionic liquid. *Green Chemistry*, 2009, 11(3), p. 339-345.

[60] Varanasi, P; et al., Survey of renewable chemicals produced from lignocellulosic biomass during ionic liquid pretreatment. *Biotechnology for Biofuels*, 2013, 6(1), p. 14.

[61] Zavrel, M; et al., High-throughput screening for ionic liquids dissolving (ligno) cellulose. *Bioresource Technology*, 2009, 100(9), p. 2580-2587.

[62] Groenestijn, JWV; Hazewinkel, O; Bakker, RR. Pretreatment of lignocellulose with biological acid recycling (Biosulfurol process). *Sugar industry*, 2006, 131(9), p. 639-641.

[63] Hoebler, C; et al., Rapid acid hydrolysis of plant cell wall polysaccharides and simplified quantitative determination of their neutral monosaccharides by gas-liquid chromatography. *Journal of Agricultural and Food Chemistry*, 1989, 37(2), p. 360-367.

[64] Pettolino, FA; et al., Determining the polysaccharide composition of plant cell walls. *Nat. Protocols*, 2012, 7(9), p. 1590-1607.

[65] Li, C; et al., Scale-up and process integration of sugar production by acidolysis of municipal solid waste/corn stover blends in ionic liquids. *Biotechnology for Biofuels*, 2017, 10, p. 13.

[66] Li, C; Wang, Q; Zhao, ZK. Acid in ionic liquid: An efficient system for hydrolysis of lignocellulose. *Green Chemistry*, 2008, 10(2), p. 177-182.

[67] Sievers, C; et al., Ionic-Liquid-Phase Hydrolysis of Pine Wood. *Industrial & Engineering Chemistry Research*, 2009, 48(3), p. 1277-1286.

[68] Binder, JB; Raines, RT. Fermentable sugars by chemical hydrolysis of biomass. *Proc Natl Acad Sci USA*, 2010, 107.

[69] Sun, N; et al., Production and extraction of sugars from switchgrass hydrolyzed in ionic liquids. *Biotechnology for Biofuels*, 2013, 6(1), p. 39.

[70] Jiang, Lq; et al., Production of 2,3-butanediol from cellulose and Jatropha hulls after ionic liquid pretreatment and dilute-acid hydrolysis. *AMB Express*, 2013, 3, p. 48-48.

[71] Van Dyk, JS; Pletschke, BI. A review of lignocellulose bioconversion using enzymatic hydrolysis and synergistic cooperation between enzymes-Factors affecting enzymes, conversion and synergy. *Biotechnology Advances*, 2012, 30(6), p. 1458-1480.

[72] Kubicek, CP; Starr, TL; Glass, NL. Plant Cell Wall–Degrading Enzymes and Their Secretion in Plant-Pathogenic Fungi. *Annual Review of Phytopathology*, 2014, 52(1), p. 427-451.

[73] Dellomonaco, C; Fava, F; Gonzalez, R. The path to next generation biofuels: successes and challenges in the era of synthetic biology. *Microbial Cell Factories*, 2010, 9(1), p. 3.

[74] Min, K; Yoo, YJ. Recent progress in nanobiocatalysis for enzyme immobilization and its application. *Biotechnology and Bioprocess Engineering*, 2014, 19(4), p. 553-567.

[75] Leu, SY; Zhu, JY. Substrate-Related Factors Affecting Enzymatic Saccharification of Lignocelluloses: Our Recent Understanding. *Bio Energy Research*, 2013, 6(2), p. 405-415.

[76] FF; Shastri, Y. Optimal control of enzymatic hydrolysis of lignocellulosic biomass. *Resource-Efficient Technologies*, 2016, 2, p. S96-S104.

[77] Liu, W; Yuan, JS; Stewart, Jr. CN. Advanced genetic tools for plant biotechnology. *Nat Rev Genet*, 2013, 14(11), p. 781-793.

[78] Willis, JD; Mazarei, M; Stewart, CN. Transgenic Plant-Produced Hydrolytic Enzymes and the Potential of Insect Gut-Derived Hydrolases for Biofuels. *Frontiers in Plant Science*, 2016, 7, p. 675.

[79] Olson, DG; et al., Recent progress in consolidated bioprocessing. *Current Opinion in Biotechnology*, 2012, 23(3), p. 396-405.

[80] Zabed, H; et al., Bioethanol Production from Fermentable Sugar Juice. *The Scientific World Journal*, 2014, 2014, p. 11 pages.

[81] Dahnum, D; et al., Comparison of SHF and SSF Processes Using Enzyme and Dry Yeast for Optimization of Bioethanol Production from Empty Fruit Bunch. *Energy Procedia*, 2015, 68, p. 107-116.

[82] Olofsson, K; Bertilsson, M; Lidén, G. A short review on SSF-an interesting process option for ethanol production from lignocellulosic feedstocks. *Biotechnology for Biofuels*, 2008, 1(1), p. 7.

[83] Chung, D; et al., Direct conversion of plant biomass to ethanol by engineered *Caldicellulosiruptor bescii*. *Proceedings of the National Academy of Sciences*, 2014, 111(24), p. 8931-8936.

[84] Lynd, LR; et al., How biotech can transform biofuels. *Nat Biotech*, 2008, 26(2), p. 169-172.

[85] Gírio, FM; et al., Hemicelluloses for fuel ethanol: A review. *Bioresource Technology*, 2010, 101(13), p. 4775-4800.

[86] Zha, J; et al., Enhanced expression of genes involved in initial xylose metabolism and the oxidative pentose phosphate pathway in the improved xylose-utilizing *Saccharomyces cerevisiae* through evolutionary engineering. *Journal of Industrial Microbiology & Biotechnology*, 2014, 41(1), p. 27-39.

[87] Mohd Azhar, SH; et al., Yeasts in sustainable bioethanol production: A review. *Biochemistry and Biophysics Reports*, 2017, 10, p. 52-61.

[88] Dien, BS; Cotta, MA; Jeffries, TW. Bacteria engineered for fuel ethanol production: current status. *Applied Microbiology and Biotechnology*, 2003, 63(3), p. 258-266.

[89] Ali, SS; et al., Fungal-mediated consolidated bioprocessing: the potential of *Fusarium oxysporum* for the lignocellulosic ethanol industry. *AMB Express*, 2016, 6, p. 13.

[90] Singh, N; et al., Cellulosic ethanol production via consolidated bioprocessing by a novel thermophilic anaerobic bacterium isolated from a Himalayan hot spring. *Biotechnology for Biofuels*, 2017, 10(1), p. 73.

[91] Van Acker, R; et al., Improved saccharification and ethanol yield from field-grown transgenic poplar deficient in cinnamoyl-CoA reductase. *Proceedings of the National Academy of Sciences of the United States of America*, 2014, 111(2), p. 845-850.

[92] Li, M; et al., Study of traits and recalcitrance reduction of field-grown COMT down-regulated switchgrass. *Biotechnology for Biofuels*, 2017, 10(1), p. 12.

[93] Voorend, W; et al., Overexpression of GA20-OXIDASE1 impacts plant height, biomass allocation and saccharification efficiency in maize. *Plant Biotechnology Journal*, 2016, 14(3), p. 997-1007.

In: Bioethanol and Beyond
Editor: Michel Brienzo

ISBN: 978-1-53613-478-0

Chapter 3

CONTAMINATION IN BIOETHANOL FERMENTATION PROCESSES

Mariane Soares Raposo**, Renata Maria Christofoleti-Furlan, Camila de Souza Varize, Cauré Barbosa Portugal and Luiz Carlos Basso

Departament of Biological Science, Escola Superior de Agricultura Luiz de Queiroz, University of São Paulo, Piracicaba, São Paulo, Brazil

ABSTRACT

The industrial bioethanol fermentation process is usually not conducted in fully aseptic conditions and, moreover, successive yeast cell recycling is a common practice in the Brazilian industry. In this scenario, microbial contamination is frequent and expected, and may involve both bacteria and yeasts. The homo- and heterofermentative lactic acid bacteria (LAB) strains are the major contaminants in bioethanol distilleries, affecting fermentation in different ways. Additionally, wild yeasts, including *Saccharomyces* or non-*Saccharomyces*, may also emerge during the process. These microorganisms can compete with selected industrial yeasts for sugars and nutrients, endangering the efficiency of ethanol fermentations. In addition, bacteria divert sugars to organic acids production, which in combination to ethanol make the environment toxic to the fermenting yeasts. Therefore, yeast flocculation, excessive foam formation and operational difficulties are common handicaps in the process. The control of contamination is critical to minimize the growth and proliferation of undesirable microorganisms, and the main strategies currently employed include acid treatment, and addition of antibiotics and other chemical agents. The use of antibiotics has become an object of concern due to residue in yeast biomass destined for animal feed, and emergence of resistant bacteria. Therefore, some industrial unities are opting for the use of alternative antimicrobials. Other approaches regard the search for natural compounds to control contamination in bioethanol fermentations.

* Corresponding Author Email: mraposo@usp.br.

INTRODUCTION

All over the world, microbial contamination is one of the most challenging problems faced by commercial bioethanol production processes, independently on the feedstock employed (Bayrock and Ingledew 2004; Broda and Leja 2010; Beckner et al. 2011) and contamination by exogenous microorganisms is usually unavoidable (Brexó and Sant'Ana 2017). The main factors favoring undesirable microorganisms proliferation in the ethanol industry comprise the low quality of feedstock, difficulties on adopting sterilization practices for large volumes of substrate and water, and the extensive practice of recycling (Basso et al. 2011; Lisboa et al. 2015).

Bacteria and wild yeasts are the main contaminants impairing bioethanol fermentation processes (Beckner et al. 2011; Lopes et al. 2016; Brexó and Sant'Ana 2017). The infections by those microorganisms can lead to reduced ethanol yield, as well as decrease in the efficiency and productivity of industrial fermentations. Furthermore, other associate problems may appear, including accumulation of polysaccharides and organic acids, gummy biofilm formation, yeast flocculation, decreased yeast cell viability, excessive foam production, longer fermentation time and other operational complications (Skinner-Nemec et al. 2007; Beckner et al. 2011; Amorim et al. 2011; Brexó and Sant'Ana 2017).

Lactic acid bacteria (LAB) commonly emerge as the major contaminants affecting the ethanol fermentation processes (Beckner et al. 2011; Lucena et al. 2010; Lopes et al. 2016). *Lactobacillus* has been found as the most abundant genus endangering the process (Gallo 1989; Lucena et al. 2010; Costa et al. 2015; Brexó and Sant'Ana 2017). Lactobacilli are well adapted to the harsh conditions of the fermentative environment as high ethanol concentration, low pH and low oxygenation, which may probably be related to their adaptive success (Beckner et al. 2011). These microorganisms reduce fermentation yield by directly competing for sugars and nutrients and by producing inhibitory compounds such as lactate and/or acetate, which can be harmful to yeast metabolism (Bayrock and Ingledew 2004; Beckner et al. 2011).

The LAB strains are traditionally classified in three distinct groups, obligate homofermentative, obligate heterofermentative and facultative heterofermentative, based on the pathway of hexose sugars fermentation and the respective by-products (Kandler 1983; Kandler and Weiss 1986; Fugelsang and Edwards 2007). These groups seem to differently affect yeast and ethanol yield in conditions of fuel ethanol production using sugarcane-based substrates (Vital 2013; Basso et al. 2014).

Wild yeasts may also easily contaminate fermentations and compete with the selected starter yeast (Amorim et al. 2011), and both *Saccharomyces* and non-*Saccharomyces* species may emerge as potential contaminants of the process (Basso et al. 2008; Basílio et al. 2008; Amorim et al. 2011; Lopes et al. 2016).

Wild *S. cerevisiae* strains have been reported as the main yeasts infecting fermentations (Basso et al. 2008), although *Dekkera bruxellensis* has been frequently reported as a common contaminant, and related to acute contamination episodes (Arunachalam and Ricke 2010). Other frequent yeast genera are *Candida* and *Pichia* (Basílio et al. 2008).

The control of microbial contaminants in the industrial bioethanol fermentation is mandatory, most likely because of the productivity and economical losses they drive (Muthaiyan et al. 2011). Different approaches as acid treatment or the use of chemical agents and antibiotics have been applied especially for bacterial control (Beckner et al. 2011). The use of antibiotics may be critical because they are commonly cost expensive, induce the selection of resistant strains, and can generate residues in yeast biomass and other by-products (Eguchi 2007). In this sense, biofuel production unities worldwide are adopting measures to reduce their usage in the process (Lopes et al. 2016) and searching for other natural, safer and cost- effective strategies.

Environmental variables and microbial interactions in the ethanol fermentation processes are complex, but it is mandatory to properly identify the factors that influence the selection and dissemination of contaminating microorganisms (Beckner et al. 2011).

This chapter addresses the key scientific findings within the scope of microbial contamination in the ethanol production industry. It comprises the understanding of the fermentation environment, the contaminating microbes in the system, the possible methods of control, as well as recent research and advances in this field.

MICROBIAL CONTAMINATION

Lactic Bacteria

The ethanol production is a highly sensitive process considering microbial spoilage. Environmental factors, feedstock quality and management, large volumes of processed substrates and continuous recycling of yeast cells, make it impossible to work under aseptic conditions. The process is normally carried out under certain levels of contamination (Skinner and Leathers 2004; Basso et al. 2011; Lopes et al. 2016), and the continuous risk of bacterial proliferation is the main cause of yield drop in distilleries.

Bacteria directly affect the fermentation yield because they preferably divert fermentable carbohydrates into different organic acids, which may be toxic to yeasts and lead to decrease in cell viability. Another factor that contributes to the reduction of fermentation efficiency is the competition for nutrients between the bacteria and the industrial yeast strains (Bayrock and Ingledew 2004; Skinner and Leathers 2004; Eggleston et al. 2006a).

These contaminants enter the process mainly by feedstock (Solomon 2009; De Souza et al. 2016) and can be found in pipes, heat exchangers, tanks, and other equipments (Reed and Nagodawithana 1991; Sharmin et al. 2013, Costa et al. 2015). Soil and sugarcane associated microbiota have also been recently reported as a source of contamination (Costa et al. 2015; De Souza et al. 2016).

Many studies have identified the bacteria that contaminate the process (Table 1), contributing to our recent knowledge on the diversity of such microorganisms commonly installed in the process. Lactic acid bacteria (LAB) are the main contaminant agents in fermentation (Gallo 1989; Narendranath et al. 1997; Lucena et al. 2010; Dellias 2014; Lopes et al. 2016), and the most recurrent genera are *Lactobacillus*, *Pediococcus*, *Leuconostoc* and *Weisella* (Lushia 2005; Heist 2009).

Table 1. Recurrent contaminant lactic acid bacteria in alcoholic fermentation (Beckner et al. 2011)

Organism	Reference
Bacteroides forsythus	Skinner and Leathers (2004)
Bifidobacterium sp., *B. adolecentis*, *B. angulatum*, unidentified *Bifidobacterium* sp.	
Clostridium sp., *C. aerotolerans*, *C. clostridiiforme*	
Eubacterium biforme	
Fusobacterium nucleatum	
Lactobacillus sp., *L. acidophilus*, *L. amylovorus*, *L. brevis*, *L. buchneri*, *L. casei*, *L. crispatus*, *L. delbrueckii* subsp. *delbrueckii*, *L. delbrueckii* subsp. *lactis*, *L. diolivorans-like*, *L. ferintoshensis*, *L. fermentum*, *L. gasseri*, *L. helveticus*, *L. hilgardii*, *l. lindneri*, *L. manihotivorans*, *L. mucosae*, *L. nagelii*, *L. paracasei* subsp. *paracasei*, *L. pentosus*, *L. plantarum*, *L. reuteri*, *L. rhamnosus*, *L. salivarius*, *L. vini*, unidentified *Lactobacillus* sp.	Skinner and Leathers (2004) Lucena et al. (2010) Chang et al. (1995)
Lactococcus sp., *L. lactis* subsp. *lactis*, *L. raffinolactis*	Skinner and Leathers (2004)
Leuconostoc sp., *L. carnosum*, *L. citreum*, *L. lactis* subsp. *lactis*, *L. mesenteroides* subsp. *cremoris*	
Oenoccocuss kitaharae-like	Lucena et al. (2010)
Pediococcus sp., *P. acidilactici*, *P. damnosus*, *P. parvulus*, *P. pentosaceus*, unidentified *Pediococcus* sp.	Skinner and Leathers (2004)
Propionibacterium granulosum	
Weisella sp., *W. confusa*, *W. paramesenteroides*, *W. viridescens*	Skinner and Leathers (2004) Lucena et al. (2010)

These microorganisms show rapid growth, can survive and overcome those fermentation environments characterized by low pH, and high temperature and ethanol concentrations (Kandler 1983; Kandler and Weiss 1986; Skinner and Leathers 2004). The LAB are not able to perform phosphorylation in association with electron transfer but, alternatively, they perform phosphorylation at the substrate level. They are tolerant anaerobes, most of which obtain energy from sugar metabolism, and they are nutrient demanding microorganisms (Madigan et al. 2004).

Lactobacillus has been reported as the most common genus contaminating the fermentation processes (Gallo 1989; Skinner and Leathers 2004; Lucena et al. 2010; Beckner et al. 2011; Costa et al. 2015), wherein *L. fermentum*, *L. vini*, *L. plantarum and L. delbrueckii* appear as the most prominent species (Gallo 1989; Skinner and Leathers 2004; Lucena et al. 2010; Vital 2013; Dellias 2014).

Gallo (1989) investigated bacterial contamination at the fermentation stage in an industrial unity in Brazil. Three hundred and thirty-four (334) strains were isolated and classified using probability matrices, 98.52% were Gram-positive, 87.76% had rod morphology, and 73.95% was non-spore forming. The most common genera identified in such cases were *Lactobacillus* (59.75%) and *Bacillus* (26.58%). Costa et al. (2015) also reported the presence of *Lactobacillus* as the predominant genus (62.6%) in fermented sugarcane must in a Brazilian distillery using culture independent methods. Species of this genus were likewise identified as the main contaminants of fermentations of other substrates other than sugarcane such as starch-based feedstocks as corn and tapioca (Chang et al. 1995; Skinner and Leathers 2004).

The consumption of hexoses and the production of associated by-products by LAB allow classifying them into three distinct metabolic groups (Kandler 1983; Kandler and Weiss 1986). The first one congregates the obligate homofermentative, producing only lactate during fermentation (Figure 1). These microorganisms convert glucose in pyruvate via the Embden-Meyerhof-Parnas pathway (EMP or Glycolysis). NADH (Nicotinamide-Adenine-Dinucleotide), produced by the oxidation of glyceraldehyde-3-phosphate to 1,3-bisphosphoglycerate, is reoxidized to NAD^+ in lactate formation from pyruvate. This reaction occurs through the action of the enzyme lactate dehydrogenase (LDH), which is involved in the production of D-lactic acid, L-lactic acid, or both (Fugelsang and Edwards 2007).

These bacteria produce 2 moles of lactic acid and 2 moles of ATP from 1 mole of hexose (Kandler and Weiss 1986), which is metabolized by the action of the enzyme aldolase. This enzyme catalyzes the conversion of 1 mole of fructose-1,6-bisphosphate into 1 mole of glyceraldehyde-3-phosphate and 1 mole of dihydroxyacetone-phosphate (Fugelsang and Edwards 2007).

When lactate production pathway is hampered, the homofermentatives can produce mannitol as an end-product of glucose catabolism. This is not common, but when it occurs, small amounts of mannitol can be produced (Chalfan et al. 1975; Loesche and

Kornman 1976; Ferain et al. 1996, Wisselink 2002). The reaction starts with the glycolysis intermediate fructose 6-phosphate reduction catalyzed by mannitol 1-phosphate dehydrogenase into mannitol 1-phosphate, which in turn is dephosphorylated to mannitol (Chakravorty 1964; Ferain et al. 1996; Wisselink et al. 2002). Moreover, the homofermentatives can also consume the mannitol present in the medium. This metabolite can be transported into the cell by a specific mannitol transporter, phosphorylated and oxidized to fructose 6-phosphate (Kandler and Weiss, 1986; Neves et al. 2000; Wisselink et al. 2002).

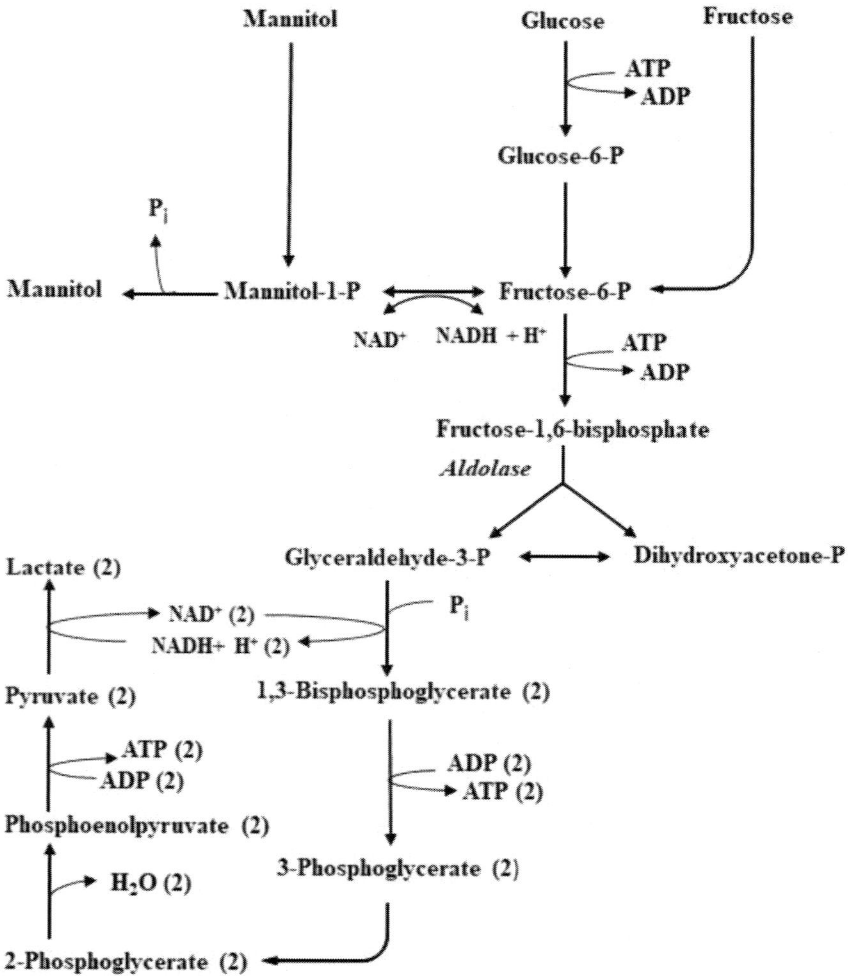

Figure 1. Metabolism of obligate homofermentative LAB with production of lactate (Adapted from Wisselink et al. 2002; Fugelsang and Edwards 2007).

The second class assembles the obligate heterofermentative lactic bacteria (Figure 2), mainly characterized by the absence of the enzyme aldolase. Such condition establishes the inability to cleave fructose biphosphate into triose phosphate. Alternatively, these organisms oxidize glucose-6-phosphate into 6-phosphoglycolate, which is then decarboxylated to generate the pentose phosphate that will be cleaved by the enzyme phosphoketolase to form acetyl-phosphate and triose phosphate (Madigan et al. 2004; Fugelsang and Edwards 2007). Under anaerobic conditions, equimolar amounts of lactate, ethanol and carbon dioxide are produced with only 1 mole of ATP yield from 1 mole of hexose (Wisselink et al. 2002). An additional ATP can be produced by converting acetyl-phosphate to acetate, rather than ethanol (Fugelsang and Edwards 2007).

When the obligate heterofermentative bacteria grow in aerobiosis the oxygen serves as final acceptor of electrons, and under anaerobic or oxygen-limited conditions fructose may be reduced to mannitol (von Weymarn et al. 2002). This reaction is catalyzed by the enzyme NAD-linked mannitol dehydrogenase (Sakai and Yamanaka 1968a, b; Wisselink et al. 2002). Strains from the genera *Lactobacillus*, *Leuconostoc* and *Oenococcus* can produce mannitol directly from fructose (Saha 2003).

In *Lactobacillus brevis*, the production of mannitol is induced by fructose. By growth on this sugar, rather than glucose, it was observed high activity of the enzyme mannitol dehydrogenase (Martinez et al. 1963). In the case of *Leuconostoc pseudomesenteroides*, the activity of this enzyme is not dependent on the carbohydrate source (Grobben et al. 2001).

Only heterofermentative bacteria can produce higher amounts of mannitol and, therefore, this metabolite is indicative of heterofermentative's contamination (Alves 1994; Narendranath et al. 1997; Eggleston et al. 2004; Eggleston et al. 2006b; Borges et al. 2015).

In addition, the heterofermentative bacteria are able to metabolize the glycerol produced by yeasts. This compound can be oxidized or reduced. When the metabolite is oxidized by glycerol dehydrogenase and dihydroxy-acetone kinase enzymes, it is converted into dihydroxy-acetone phosphate (DHAP), which finally reaches the glycolytic pathway (Lonvaud-Funel 2002; Garai-Ibabe et al. 2008)

When the glycerol is reduced it is converted into 3-hydroxypropionaldehyde (3-HPA) by coenzyme B_{12}-dependent glycerol and diol dehydratases (Talarico and Dobrogosz 1990; Sauvageot et al. 2002; Garai-Ibabe et al. 2008). The 3-HPA can be subsequently reduced to 1,3-propanediol (1,3-PDL) by a NADH-linked 1,3-propanediol dehydrogenase (Talarico et al. 1990; Veiga da Cunha and Foster 1992; Garai-Ibabe et al. 2008), or can be oxidized to 3-hydroxypropionic acid (3-HP) (Lonvaud-Funel 2002; Garai-Ibabe et al. 2008).

Finally, the third group of lactic bacteria gathers the facultative heterofermentative organisms, which show similar metabolism traits in comparison to the homofermentative ones. However, under glucose limiting conditions these microorganisms generate acetic acid or ethanol as by-products (Kandler 1983; Kandler and Weiss 1986; Calasso and Gobbetti 2011).

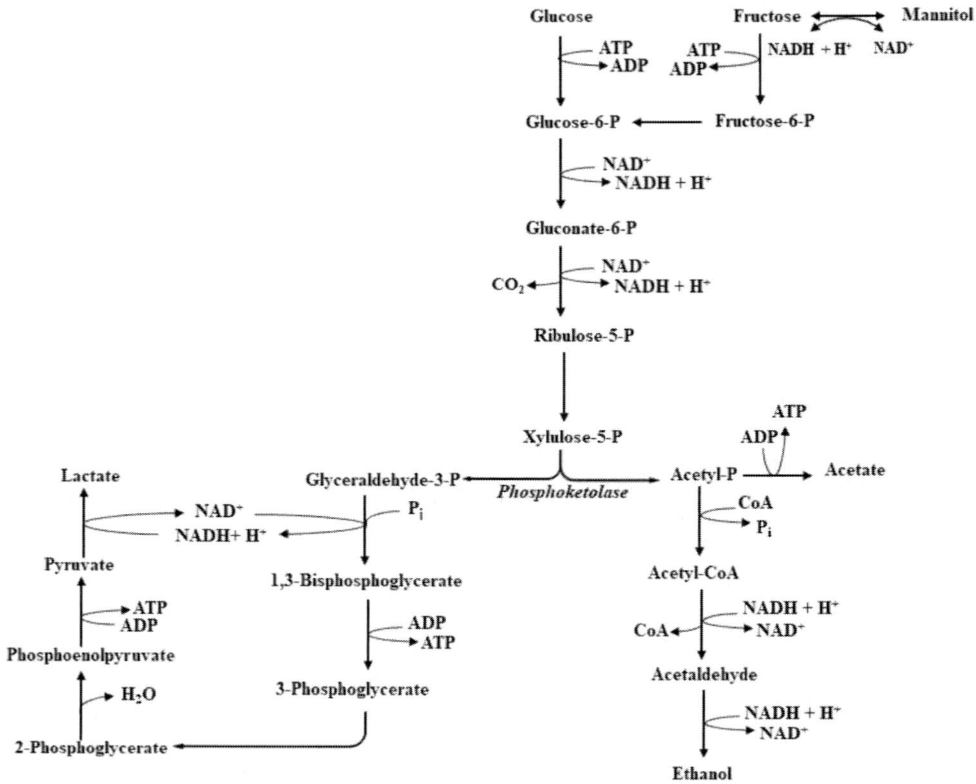

Figure 2. Metabolism of obligate heterofermentative LAB with production of lactate, ethanol, acetic acid and CO_2 (Adapted from Wisselink et al. 2002; Fugelsang and Edwards 2007).

Owing to all the microorganisms and metabolites present in fermentation, the ethanol industry environment becomes very complex. This circumstance, together with the stressing conditions imposed by the fermentative process, may establish challenging situations to industrial yeasts. In this scenario, it is imperative to better understand the microbial interactions between yeast and bacteria.

The interactions between industrial yeast and LAB directly interfere on fermentation. One such example is the inhibition of fermenting yeasts by some species of the genus *Lactobacillus*. Competition for nutrients and the production of organic acids by these bacteria are the main factors affecting ethanol production by yeasts (Narendranath et al. 1997; Narendranath 2003; Gomes 2009; Vital 2013; Basso et al. 2014).).

Basso et al. (2014) evaluated the interaction between homo and heterofermentative lactic bacteria in mixed cultures with yeasts during the fermentation process. The authors considered *Lactobacillus fermentum* (heterofermentative), *L. plantarum* (homofermentative), and the yeast strain CAT-1 (Basso et al. 2008) that is largely used in Brazilian distilleries.

Such condition simulates the microbial interactions occurring in industries, and it was found that the homofermentative strain preferably metabolizes glucose, while the heterofermentative strain displays a fructophilic behavior. Considering this condition, it is expected that heterofermentative bacteria would be ecologically favored in the process, since yeasts generally prioritize to metabolize glucose (Vital 2013; Basso et al. 2014).

When simulating the industrial process with cell recycling, Basso et al. (2014) also observed that heterofermentative population increased with the recycles, whereas homofermentative one remained at a lower concentration. Besides that, heterofermentatives seem to be more competitive during fermentation, leading to decreased yeast cell viability along the cell recycle. Such competitiveness may be related to the presence of lactic and acetic acid and ethanol in the medium, which can make the fermentative environment more toxic to yeast cells.

However, the prevalence of hetero- rather than homofermentative bacteria in industrial fermentation processes is not observed. The ratio between these two metabolic profiles is quite more equitable (Costa et al. 2008; Vital 2013). It is possible that in such complex environments, a synergistic interaction may be established between these groups of microorganisms, and homofermentative bacteria population remains active in the process (Vital 2013).

Bacterial lactate and acetate lead to a pronounced growth decrease of the yeast strain employed in the process, besides influencing the efficiency of carbohydrate conversion and increasing the medium acidification (Maiorella et al. 1983).

Bayrock and Ingledew (2004) reported that the presence of lactate in the medium may lead to yeast's glucose consumption. In such conditions, yeasts may boost sugar metabolism due to the need for more energy (ATP), obtained from the catabolism of glucose, to expel H^+ ions resulting from the acid flow of non-dissociated acid lactic that changes inner environment pH (Bayrock and Ingledew 2004).

Although lactate and acetate seem to be the main inhibitors influencing yeast ethanol production, Nobre (2005) reported that the isolated presence of bacterial cell metabolites was not able to directly affect yeast cell viability. Currently, it is known that many other by-products may be involved on inhibition of yeast growth and fermentation yield loss. Other compounds are actually known to be implicated, including diacetyl, reuterin, hydrogen peroxide, phenylacetic acid, protein compounds, 3-hydroxy fatty acids, cyclic dipeptides, caproic acid and carbon dioxide (Beckner et al. 2011).

When yeast and *Lactobacillus* interact during ethanol fermentations, the divertion of sugars to glycerol formation seems to be lower (Alves 1994; Thomas et al. 2001). The produced glycerol may occur as a response to the osmotic stress experienced by the cell in the presence of lactic and acetic acid (Meikle et al. 1988).

Vital (2013) and Basso et al. (2014) observed that the production of glycerol is higher when yeasts are in contact with heterofermentative bacteria. On the contrary, glycerol production seems to be lower under homofermentative bacteria interactions. Studies are still needed to understand which factors really drive lower glycerol production in such cases.

On the other hand, synergism is also reported between contaminating bacteria and industrial yeasts. The LAB have specific nutritional demands, requiring amino acids and vitamins that are not commonly available in such environments, so when yeast cells undergo autolysis, the cell content is released to the extracellular medium and such compounds are then used by bacteria. Furthermore, some fermentation products may also contribute to the growth of bacteria, as is the case of sucrose breakdown by yeasts, releasing monosaccharides that are metabolized by bacteria (Selby et al. 1975; Essia-ngand et al. 1992; King and Beelman 1986).

Wild Yeasts

Although lactic acid bacteria commonly emerge as the major contaminants in the fermentative environments, different wild yeasts may usually affect the efficiency of ethanol fermentations (Beckner et al. 2011). Virtually, any different species or strain other than the selected starter can be considered as a possible contaminant. Industrial ethanol fermentation process is usually carried out under non-aseptic conditions due to the difficulty on adopting sterilization practices to large volumes of sugarcane juice and water. In addition, due to the extensive cell recycling during the whole crop season, wild yeasts may easily contaminate the medium and compete with selected starter strains (Amorim et al. 2011). Both *Saccharomyces* and non-*Saccharomyces* species may emerge as hidden contaminants of the ethanol production process (Basso et al. 2008; Basílio et al. 2008; Amorim et al. 2011; Lopes et al. 2016).

The wild *S. cerevisiae* strains seem to be the most frequent contaminants found in Brazilian ethanol production process (Cabrini e Gallo 1999; Silva-Filho et al. 2005; Basso et al. 2008; Basílio et al. 2008). Basso et al. (1993) showed that, during two consecutive crop seasons in bioethanol distilleries in the Southeast region of Brazil, the wild *S. cerevisiae* strains may overcome the starter yeasts in up to 40 days along cell recycling procedures. Cabrini and Gallo (1999) also reported the presence of contaminant yeasts, predominantly belonging to the genus *Saccharomyces*, in an ethanol plant in São Paulo State. Da Silva-Filho et al. (2005a) and Basso et al. (2008) confirmed the rapid emergence of wild yeasts, with some of them eventually replacing the starter strain and persisting in the process along yeast recycling.

Up to 80% of contaminant *S. cerevisiae* can display undesirable traits for efficient ethanol fermentations, such as cells flocculation or flotation, excessive foam formation, lengthy fermentation time and higher levels of residual sugars (Basso et al. 2008). The rough colonies of wild *S. cerevisiae* detected in synthetic medium are usually related to the formation of cellular chains (pseudohyphal growth) and high sedimentation rate, some traits that have been observed in a large number of isolates (Basso et al. 2008). The yeast strains displaying such characteristics seem to present lower fermentation performance in comparison to other typical morphotypes (smooth, bright colonies with dispersed cells), showing longer fermentation time, high levels of residual sugars and lower ethanol yield (Reis et al. 2013). Such strains seem to be more tolerant to osmotic and ethanol stress (Reis et al. 2017), traits that can be related to the robustness and dominance of some wild strains in the fermentation process, thus enabling their eventual success in displacing the starter yeast used in the process. The dominance of strains with tendency to sedimentation or flocculation may directly impair the centrifugation steps in industrial processes, resulting in a less-concentrated yeast cream. This condition is associated to larger wine volumes, with increased buffer capacity, that will need higher sulfuric acid consumption to attain the desirable pH during the acid treatment. In such situation, yeast cells face a more stressful condition that will impact on their physiological activity (Amorim et al. 2011).

Despite the high frequency of yeasts formerly regarded as contaminants, it is worth mentioning that some of them may eventually overcome and persist in the process. Some of these strains may display some desirable attributes (stress tolerance, high fermentative capacity, and high ethanol yields), and can be selected for later application in ethanol production unities (Cabrini e Gallo 1999; Da Silva-Filho et al. 2005b; Basso et al. 2008; Della-Bianca et al. 2013).

In addition, it is not uncommon to detect non-*Saccharomyces* species as contaminants in fermentation processes. Some studies have reported the incidence of 12% to 30% of different species in distilleries of northeastern Brazil, with higher rates in those units that use raw sugarcane juice for fermentation (Da Silva-Filho et al. 2005a; Basílio et al. 2008). On the other hand, another study comprising distilleries of Southern-central Brazil identified less than 5% of the contaminants detected as non-*Saccharomyces* strains (Basso et al. 2008). The differences in the incidence of these populations are especially attributed to the employment of raw sugarcane juice and to the continuous fermentation system, most widespread in the northeastern distilleries. Among the non-*Saccharomyces* strains found in bioethanol process, *Dekkera bruxellensis* has currently been reported as one of the most frequent and critical contaminant (Beckner et al. 2011).

The prevalence and persistence of *D. bruxellensis* have been reported mainly in continuous system fermentation processes that recycle the yeast biomass (De Souza Liberal et al. 2007; Passoth et al. 2007; Pereira et al. 2012). High prevalence of *D. bruxellensis* in association to decreased ethanol yields have been also observed in continuous fermentation systems that employ crude sugarcane juice in Brazilian distilleries in Northeast region (De Souza Liberal et al. 2007; Basílio et al. 2008). In fact, populational fluctuations between *S. cerevisiae* and *D. bruxellensis* can occur in industrial fermentations during a single crop season (De Souza Liberal et al., 2007; Pereira et al., 2012).

In such cases, *D. bruxellensis* was detected in greater proportion in some periods, coming to overlap *S. cerevisiae* population, even with the introduction of new periodical batches of pure starter culture. In Sweden, Passoth et al. (2007) also detected *D. bruxellensis* dominating the fermentation process of wheat-based mashes in stable ethanol-producing consortium with *Lactobacillus vini*. Similarly, different strains of the genus *Dekkera* were also identified in spoilage corn mash fermentations of fuel-ethanol process (Abbott et al. 2005).

The procedures of cell recycling seem to play a major role on the proliferation of *Dekkera* spp., since these microorganisms are continuously reintroduced and recycled in those systems. Pereira et al. (2012) and Meneghin et al. (2013) observed expressive growth of *D. bruxellensis* during cell recycling processes of sugarcane wort, demonstrating that the proliferation of this microorganism is commonly associated to lower ethanol production. It is worth noting that even under low cell density (10 cells ml^{-1}) and in co-culture with *S. cerevisiae*, the population of *D. bruxellensis* can grow significantly during successive cycles.

Initial insights postulated that the competitiveness of *Dekkera* spp. would be correlated with its acidogenic character (De Miniac 1989). However, the production of acetic acid by this yeast is strongly influenced by aerobic conditions (Ciani and Ferraro 1997; Aguilar Uscanga et al. 2003; Abbot et al. 2005), and this trait does not seem the main reason for *Dekkera* success in industrial processes (Abbot et al. 2005; De Souza Liberal et al. 2007; Pereira et al. 2012).

Slow growth rate is also a common characteristic of *Dekkera* species in the fermentative environment (Abbot et al. 2005; Blomqvist et al. 2010; Galafassi et al. 2011), which could mean a competitive disadvantage. However, some studies have noticed that *D. bruxellensis* may display higher biomass yields and lower glycerol production (Blomqvist et al. 2010; Pereira et al. 2012), as well as ethanol yields equivalent to or higher than *S. cerevisiae* (Blomqvist et al. 2010; Galafassi et al. 2011; Meneghin et al. 2013), which suggest that *D. bruxellensis* is more energy efficient under anaerobic conditions (Blomqvist et al. 2010). *Dekkera* is considered a low-rate ethanol producer under oxygen-limited conditions, i.e., it lasts a long time to completely consume and ferment the available sugars in the medium (Blomqvist et al. 2010; Galafassi et al. 2011; Meneghin et al. 2013). This kind of behavior is strictly related to the Custer's effect, which weakens sugar assimilation ability in anaerobic environments, consequently inhibiting the fermentative metabolism (Wijsman et al. 1984).

Another important factor regarding the competitive edges of *D. bruxellensis* may be related to nutritional aspects and stress tolerance. Some *D. bruxellensis* strains show similar or even higher resistance to ethanol (Renouf et al. 2006; Pereira et al. 2012) and to organic acids (Abbot et al. 2005) than *S. cerevisiae*. One such possibility linked to these traits is that *D. bruxellensis* may alternatively use other nutrient sources found in the medium. De Barros Pita et al. (2011) showed that the adaptive success of *D. bruxellensis* over *S. cerevisiae* in sugarcane fermentations is mostly related to its ability to assimilate nitrate, favoring its installation and proliferation. The assimilation of nitrate in these cases seems to inhibit the Custer's (negative Pasteur) effect, improving its growth and the fermentative metabolism under anaerobic conditions (Galafassi et al. 2013). Such a physiological condition could explain why occasional contamination episodes concerning this species are not correlated to ethanol yield drops in industrial processes (Meneghin et al. 2013).

Basílio et al. (2008) also detected the presence of *Pichia galeiformis* and *Candida tropicalis* associated to acute contamination events. Other non-*Saccharomyces* yeasts have already been identified as contaminants of biofuel fermentations, including *Schizosaccharomyces pombe*, *Candida krusei* (Basso et al. 2008), *C. entomophila*, *Torulopsis* sp., *Pichia* sp. (Cabrini e Gallo 1999), *Hansenula*, *Kloeckera*, *Kluyveromyces*, *Oidium*, *Rhodotorula*, *Torula*, *Trichosporon*, and *Cryptococcus* (Oliva-Neto et al. 2013). The genera *Candida* and *Meyerozyma* appear as the prevalent taxa in sugarcane fermentations in Brazilian distilleries. But a high percentage of non-classified fungi are frequently observed, suggesting the existence of a fungal community still unknown in association to sugarcane ethanol processes (Costa et al. 2015).

Contaminant yeasts can be introduced in the process by different ways. Feedstock, soil impurities, as well as cutting and grinding equipments are intrinsically associated to contamination (Costa et al. 2015). The water can also be considered another important source, and evidences of *D. bruxellensis* spoilage have already been related in alcohol-producing units (Meneghin et al. 2013). Acute contaminations in such cases can lead to the need of a complete biomass removal and replacement within the fermentation tanks (De Souza Liberal et al. 2007; Beckner et al. 2011), meaning a costly, laborious practice in industrial processes. The early detection of microbial contamination is imperative to control these handicaps, allowing distilleries to prevent and avoid acute contamination episodes and higher economical losses (De Souza Liberal et al. 2007; Basílio et al. 2008). The use of safe, viable, cost-effective fungicides and antimicrobial agents that do not damage the selected starter yeast is even more difficult, and to date, it is still an object of research (Beckner et al. 2011).

Effects of Lactic Bacteria in Bioethanol Fermentation

Contamination by lactic acid bacteria (LAB) is a frequent problem in the ethanol industry, affecting production yield in distilleries and causing constant inefficiency in the process. The continual influx of bacteria into the fermentation systems is a critical factor that may limit or even completely inhibit ethanol production. Competition for nutrients between LAB and yeasts, as well as organic acid release into the medium by these bacteria, constitute the main biological conditions modulating such relationships (Beckner et al. 2011).

The ethanol yield decrease depends on populational density of contaminants, the species/strain of dominant bacteria, and the selected yeast used in the fermentation (Narendranath et al. 1997; Gomes 2009). Some estimates indicate that when *Lactobacillus* population reaches 10^8-10^9 cells ml^{-1}, the yield losses may reach 27% (Narendranath et al. 1997; Bischoff et al. 2009.). An example of that was demonstrated by Gomes (2009), who assessed four yeast strains frequently used in industrial fermentations, and concluded that only *S. cerevisiae* CAT-1 exerts a larger antibacterial action, being more tolerant to the bacterial presence.

The influence of organic acids produced by LAB in ethanol production is another critical point in the process control. The undissociated forms of lactate and acetate can diffuse through the cell membrane, and acidify the cytoplasm (Narendranath et al. 2001), impacting on the cell growth, and leading to sluggish fermentation (Rasmussen et al. 1995; Huang et al. 1996; Edwards et al. 1999).

However, the inhibitory effect of these compounds observed in laboratory conditions seems not to display the same effects in the industrial environment. Such differences mainly rely on the higher complexity and the variable conditions observed, with intricate nutritional and non-nutritional factors influencing yeast growth and sensitiveness (Graves et al. 2007).

The inhibitory action of these organic acids is inversely correlated to the pH, so that the higher the pH value, the lower the undissociated acid concentration in the medium (Narendranath and Power 2005). In steady-state fermentation, raising the pH of the medium to 6.0, the authors observed a 4.4-log increase in *Lactobacillus* population after a 3-day period, followed by a decrease of 83% in the yeast population. Therefore, in this case, the production of ethanol reduced by 44%, and lactate concentration increased by 2% (w/v).

Some works show that higher concentrations of the undissociated form of lactic acid lead to an increase in glucose consumption, which may be explained by the necessity of yeast cell to produce ATP to drive out H^+ ions from the cell. The presence of lactic and acetic acid acting synergistically can result in increased glycerol production, and reduced yeast growth and cell viability (Narendranath et al. 2001; Bayrock and Ingledew 2004; Beckner et al. 2011; Basso et al. 2014).

Glycerol is the metabolite released by yeasts in greater quantity in anaerobiosis, and its production can be influenced by LAB by the release of organic acids in the medium. In those cases, the yeast cell diverts carbohydrates to produce glycerol, a compound that is correlated to the intracellular osmotic and redox balance (Scanes et al. 1998). The yeast cell changes the carbon flux to stay alive, increasing the synthesis of glycerol, and consequently decreasing the production of ethanol (Dong et al. 2015).

The proliferation of LAB in the fermentation environment has direct consequences on yeast viability. *Lactobacillus fermentum*, for example, may lead to critical viability levels of yeast cells, slowing down the fermentation kinetics and leading to much residual sugar in the medium (Thomas et al. 2001). Competition for trace nutrients (vitamins and amino acids such as thiamine and glutamic acid, respectively) is another obstacle causing yeast inhibition (Chang 1995; Bayrock and Ingledew 2004). The lack of available nutrients for the yeasts, in combination with the organic acids released by those bacterial populations, lead to the fall in viability and cell density (Nobre 2005).

The flocculation of yeast cells is another fact that represents a handicap with bacterial contamination. This kind of situation occurs because the bacteria have a coat of protein that allows them to adhere to yeast cell wall, thus forming bigger agglomerates (Gallo and Canhos 199a1). These cell clusters reduce the contact surface between the available substrates and the cell, directly impacting on the fermentation kinetics and leading to higher amount of residual sugars at the end of fermentation (Amorim and Oliveira 1982; Oliva-Neto and Yokoya 1997). *L. fermentun*, *L. fructosus*, *L. buchneru* and *L. plantarum* are some species that induce yeast flocculation (Alcarde 2001).

THE CONTROL OF MICROBIAL CONTAMINANTS

In the ethanol fermentation process, both agricultural and industrial practices should be monitored to control microbial contaminants. Infestation of sugarcane by borer, for example, leads to fungal infections and directly increases microbial contamination of the juice (Amorim et al. 2000; Cheavegatti-Gianotto et al. 2011). The post-harvesting variables related to cane deterioration such as ambient temperature, cane variety, period of storage, maturity status, methods of harvesting, etc. also influence microbial contamination (Solomon 2009).

Delayed times between unit operations as harvesting and milling also contribute to microbial proliferation. Furthermore, other uncontrolled factors, like air humidity, raise the amount of soil particles adhered to the vegetative partitions of the plant and bring typical bacterial populations of these ecosystems into the process (Amorim et al. 2000).

Some practices that may contribute to minimize microbial growth and proliferation include: planting varieties of sugarcane resistant to borer (Dinardo-Miranda et al. 2012); improving plant quality at harvesting; quickly processing of the harvested plants, avoiding long-term storage; carrying out adequate plant cleaning to remove soil dirtiness, and regular sanitation of equipments (Solomon 2009; Misra et al. 2017).

Acceptable levels of bacterial contamination must not exceed 10^5 cells mL^{-1} (Alcarde et al. 2003). Anyway, several antimicrobial agents, such as chemicals and antibiotics, have been studied and are currently used in the fuel ethanol industry. The antibiotics generally display a selective action, and compounds like monensin, virginiamycin, penicillin, streptomycin and tetracycline have been the most commonly used antibiotics for bacterial control in fuel ethanol fermentations (Muthaiyan et al. 2011). In addition to penicillin, virginiamycin is now the main antibiotic employed, since it is still active at lower pH values and is considered one of the better choices by the industry (Hynes et al. 1997). The antibiotic monensin (commercially named Kamoran) exerts both bactericidal and bacteriostatic activities. It is obtained from fermentation processes employing *Streptomyces cinnamonensis*, which produces the crystalline sodium monensin, and acts preventing the transport of cations in the membrane of Gram-positive bacteria (Alcarde et al. 2001).

Once yeast cell recycle is a common practice in Brazilian distilleries, the most frequent control strategy for bacterial contamination is the acid treatment. In this process yeast cells are collected from the fermentation tank by centrifugation as a yeast suspension containing ca. 70% cell (wet basis), diluted with equal volume of water and treated with sulfuric acid (pH 2.0-2.5). After one or two hours the cells return to the tank for the next fermentation (Gallo and Canhos 1991b; Basso et al. 2008; Lopes et al. 2016). This treatment can reduce the bacterial population by up to 44.5% (Gallo 1989). Nevertheless sulfuric acid treatment is not efficient enough, and antibiotics are used to achieve a better control.

Some studies have pointed out that the growth of lactic acid bacteria can be expressively inhibited in the presence of sulfur dioxide (Reguant et al. 2005). The bactericidal effect of this compound is based on the free fraction that permeates the cell by diffusion, and is converted in bisulfite ion (HSO^{3-}) in the cytosol. The HSO^{3-} reacts with cellular proteins, nucleic acids and some cofactors, inhibiting enzymes, and consequently inducing the cell death (Romano and Suzzi 1993). Although sulfur dioxide is intentionally used in wine fermentation this is not practiced in fuel ethanol production. Sulfur dioxide is found in molasses in variable titers, since this compound is used in the clarification step of sugarcane juice prior sucrose crystallization. While sulfur dioxide exerts yeast inhibitory effect at 200 mg/L, at a lower level in the substrate (100 mg/L) it displays a desirable antibacterial action without affecting yeast physiology (Alves, 1994).

Some compounds like urea hydrogen peroxide and ammonia used as nitrogen source in industrial fermentations also exert bactericidal activity against lactic acid bacteria and other microbial contaminants (Narendranath et al. 2000; Tajkarimi et al. 2008; Broda and Grajek, 2009).

The main controlling contamination strategies employed in biofuel production unities worldwide seek to reduce the use of antibiotics in the process (Lopes et al. 2016). Reasons for that include antibiotic residues in yeast biomass and other co-products (mainly distillers dried grains with soluble - DDGS) from ethanol production processes, which preclude their usage for animal feed, development and selection of antibiotic-resistant bacteria, and concerns about safety of animal, human and environmental health

(Muthaiyan et al. 2011; FAO 2012). This way, the search for alternative antimicrobials to control contamination in bioethanol fermentations is necessary.

In this field, ozone was recently evaluated for treatment of uncooked corn mash to reduce bacterial contamination prior to ethanol fermentation with positive results (Rasmussen et al. 2015). Some ethanol plants have been using chlorine dioxide in the gaseous form (*in situ* generated). This compound has action against bacteria inhibiting the growth of *Lactobacillus*, *Leuconostoc* and *Bacillus*. However, it has action against fungi, but in concentration until 50 ppm it does not inhibit the growth of industrial yeasts (Meneghin et al. 2008; Lopes et al. 2016).

Regarding the widespread use of antibiotics and the emergence of resistant bacteria, a promising approach relies on search for natural compounds prospected from natural sources and biologically active with antimicrobial properties, including plant compounds and extracts, chitosan, and bacteriocins (Muthaiyan et al. 2011). Biocides of natural origin do not cause bacterial resistance and can replace antibiotics and hazardous chemicals. These compounds enable antibiotic-free products, comply with regulations for additives in animal feed, and can increase yeast performance improving fermentation (Muthaiyan et al. 2011).

Some ethanol distilleries have used plant extracts, such as those obtained from *Humulus lupulus* (hops). Hop extract contains humulones (α-acids) and lupulones (β-acids) that show antimicrobial activity on Gram-positive bacteria (Langezaal et al., 1992; Rückle and Senn 2006; Muthaiyan et al. 2011; Leite et al. 2013; Lopes et al. 2016).

The usage of bacteriocins has also brought out positive results for the control of bacterial contamination during ethanol fermentations. Nisin was able to reduce the lactobacilli metabolism and the lactic acid production (Peng et al. 2012). Bacteriophage lytic enzymes (Roach et al. 2013) and secondary metabolites produced by *Pseudomonas* (Matsuoca Góis et al. 2013) likewise showed potential to be applied in controlling such handicaps (Roach et al. 2013). Furthermore, the use of bacteriophages as antimicrobial agents has also been evaluated in view of controlling bacterial contamination in biofuel fermentation processes (Silva and Sauvageau 2014; Wroley-Morse et al. 2015; Liu et al. 2015).

Indeed, yet in an approach of antibiotic-free strategies, a genetically-engineered *S. cerevisiae* strain expressing endolysins (bacteriophage lytic enzymes) was constructed and assessed for reducing *Lactobacillus* contamination during ethanol fermentations. The modified yeast strain producing the lytic enzymes was able to reduce *L. fermentum* growth, and likewise lactate and acetate levels (Khatibi et al. 2014).

Natural anti-microbial agents seem to be less harmful to the environment and cost-effective in comparison to traditional antibiotics currently employed in the ethanol industry, and therefore should be better explored (Langezaal et al., 1992; Rückle and Senn 2006; Muthaiyan et al. 2011; Leite et al. 2013; Lopes et al. 2016).

CONCLUSION

Bioethanol production process entails a complex environment, in which a great diversity of microorganisms may be found (Gallo 1989; Cabrini and Gallo 1999; Basso et al. 2008; Basílio et al. 2008; Lucena et al. 2010; Costa et al. 2015).

Studies concerning the microbial communities of sugarcane and ethanol production process through culture-independent methods have revealed an unexplored biodiversity associated to such environments. Costa et al. (2015) assessed the microbial diversity of contaminants of a Brazilian ethanol distillery, and showed for the first time, the presence of Archaea in sugarcane juice and mixed juice used for fermentation. They confirmed the presence of different Fungi, Bacteria, and Archaea, demonstrating the existence of a great microbial diversity in these environments. It was revealed changes in the microbial population as the stages of ethanol production advance. Such alterations have already been observed earlier by traditional microbiological techniques (Gallo 1989; Cabrini and Gallo 1999; Lucena et al. 2010), however the use of cultivation independent methods have made it possible to assess unknown microorganisms until then and to expand our knowledge of different microorganisms associated with this process.

The genus *Lactobacillus* although present throughout the process steps only dominates in the wine during fermentation (Gallo 1989; Costa et al. 2015), meaning that process outputs exert a positive selective pressure on these microorganisms. Not only the fermentation itself favors *Lactobacillus*, but specifically the sulfuric acid treatment contributes to increase the population of some species belonging to that genus (Gallo 1989).

Investigations about the microbiome of sugarcane have revealed a highly unknown fungal and bacterial diversity associated to the plant different organs (roots, stalks and leaves) (De Souza et al. 2016). The native soil that shelters the plants influences the formation of such microbiota. The diversity of microorganisms associated to the stalks, which are the main part of the plant by which sugarcane juice is obtained, may influence the microbial communities of biofuel productions. Regarding yeast diversity, it is likely that yeast population succession observed in the fermentative process along cell recycling (Da Silva-Filho et al. 2005a; Basso et al. 2008) is related to intrinsic factors of each producing unit and to the employed feedstock.

A relevant aspect to be considered nowadays that is probably influencing specially the yeast population in the fermentation vat is the recently introduced mechanical harvesting in sugarcane plantation. Not only more soil impurities, but more tips and leaves from the cane are collected, contributing to an increase in starch, phenols and other compounds in cane juice, which requires a more severe juice clarification treatment, prior to sucrose chrystallization. Meanwhile, due to the higher price of sugar in relation to ethanol, Brazilian distilleries have diverted more sugarcane for sucrose production, and today only molasses (a by-product of sugar industry) is used to compose the ethanol substrate. This higher proportion of molasses imposes a more stressful condition to the fermenting yeast, and it has been observed that wild *S. cerevisiae* strains presenting rough colony morphology has been dominating and persisting the whole crop season, overcoming the *S. cerevisiae* starter yeast (presenting smooth colony morphology) yet in the initial period of the season.

Other recent contribution concerning contaminants assessed by culture-independent techniques are about bacteria associated with biofilm formation in the Brazilian bioethanol fermentative process. Dellias (2014) observed that the main inhabitants of the bacterial community found in fermentation vats are lactic acid bacteria, predominantly from the genus *Lactobacillus*, and that the biofilm structure is composed by a mixed culture of bacteria and yeasts. Costa et al. (2015) identified in the ethanol process, after the stages of juice clarification and evaporation, bacteria of the genera *Tatumella* and *Paenibaccilus*, which are also able to form biofilms. The relevance of biofilm formation in the fermentative process is mainly related to the resistance of such biological structure to cleaning and disinfection procedures, including the use of antibiotics and other hostile environmental conditions like ethanol, temperature, acids, etc. (Kubota et al. 2008; Yang et al. 2012). Such resistance contributes to contamination reoccurrence and persistence, and the inherent problems for its control (Skinner-Nemec et al. 2007). Studies concerning microbial populations inhabiting biofilms and interactions among them could contribute to better understanding the resistance mechanisms to hostile environment conditions. New insights in this way probably would collaborate with novel approaches to control contamination in industrial process.

Although several research works have been carried out in the last years, the microbiological universe of alcoholic fermentation, even though kept inside a fermenter, is quite complex. Variations may occur in response to external and internal interferences, sometimes very far from the fermentation vat, as well as in the feedstock employed in the process. It is evidenced that both the feedstock and the soil may influence microbial communities permeating the ethanol producing process. This is an indicative that the environment where the distillery is located, its production practices, and the feedstock geographical region, as well as the biotic and abiotic conditions, can influence the microbial communities of each ethanol producing unit, likewise the interactions between bacteria and fungi in these environments. This could partially explain the events of microbial population succession observed in the process along the crop season, but also in the way a specific population settles in each producing unit.

Interactions between microorganisms, likewise with the environment around them, remain unknown to some extent. Further research works should be carried out to understand the cellular mechanisms underlying the physiological responses and competitiveness related to such interactions. In the same way, the diversity of microorganisms in bioethanol processes also should be considered. New contributions in this sense assertively will bring important information to the development of more sustainable strategies - which are cost effective and equally safe for environment, humans and animals - of contaminants control, and will help the industrial sector to better cope with such constraint.

References

Abbott, D. A. & Ingledew, W. M. (2004). Buffering capacity of whole corn mash alters concentrations of organic acids required to inhibit growth of Saccharomyces cerevisiae and ethanol production. *Biotechnology Letters*, *26*, 1313–1316.

Abbott, D. A., Hynes, S. H. & Ingledew, W. M. (2005). Growth rates of *Dekkera/ Brettanomyces* yeasts hinder their ability to compete with *Saccharomyces cerevisiae* in batch corn mash fermentation. *Applied Microbiology and Biotechnology*, *66*(6), 641-647.

Aguilar Uscanga, M. G., Délia, M. L. & Strehaiano, P. (2003). *Brettanomyces bruxellensis*: effect of oxygen on growth and acetic acid production. *Applied Microbiology and Biotechnology*, *61*(2), 157-162. doi:10.1007/s00253-002-1197-z.

Alcarde, A. R., Walder, J. M. M. & Horii, J. (2001). Comparison between gamma radiation and kamoran hj in the decontamination of sugarcane must. *Journal of food processing and preservation*, *25*(2), 137-147. doi: 10.1111/j.1745-4549.2001. tb00449.x.

Alcarde, V. E. (2001). *Avaliação de parâmetros que afetam a floculação de leveduras e bactérias isoladas de processos industriais de fermentação alcoólica*. Universidade de São Paulo. Tese de Doutorado. [*Evaluation of parameters affecting the flocculation of yeasts and bacteria isolated from industrial processes of alcoholic fermentation*. University of Campinas. PhD thesis].

Alcarde, A. R., Walder, J. M. M. & Horii, J. (2003). Fermentation of irradiated sugarcane must. *Scientia Agricola*, *60*(4), 677-681. doi: 10.1590/S0103-90162003000400011.

Alves, D. M. G. (1994). *Fatores que afetam a formação de ácidos orgânicos, bem como outros parâmetros da fermentação alcoólica*. MSc Diss., University of São Paulo, Piracicaba. [*Factors affecting the formation of organic acids, as well as other parameters of alcoholic fermentation*].

Amorim, H. B. & Oliveira, A. J. (1982). Infecção na fermentação: como evitá-la. *Álcool e Açúcar*, São Paulo., *2*, 12-18. [Infection in fermentation: how to avoid it. *Alcohol and Sugar*].

Amorim, H. V., Oliveira, A. J., Silva, L. F. L. F., Bernardino, C. D., Godoy, A. & Alves, D. M. G. (2000). Impact of sugar cane quality on sugar and alcohol yields. *International Sugar Journal*, *102*(1214), 86-88.

Amorim, H. V., Lopes, M. L., Oliveira, J. V. C., Buckeridge, M. S. & Goldman, G. H. (2011). Scientific challenges of bioethanol production in Brazil. *Applied Microbiology and Biotechnology*, *91*(5), 1267-1275. doi:10.1007/s00253-011-3437-6.

Arunachalam, M. & Ricke, S. (2010). Current perspectives on detection of microbial contamination in bioethanol fermentors. *Bioresource Technology*, *101*(13), 5033-5042. doi: 10.1016/j.biortech.2009.11.005.

Basílio, A. C. M., de Araújo, P. R. L., de Morais, J. O. F., da Silva Filho, E. A., de Morais, Jr. M. A. & Simões, D. A. (2008). Detection and identification of wild yeast contaminants of the industrial fuel ethanol fermentation process. *Current Microbiology*, *56*(4), 322-326. doi:10.1007/s00284-007-9085-5.

Basso, L. C., Oliveira, A. J., Orelli, V. F. D. M., Campos, A. A., Gallo, C. R. & Amorim, H. V. (1993). Dominância das leveduras contaminantes sobre as linhagens industriais avaliada pela técnica de cariotipagem. Trabalho apresentado no *V Congresso Nacional da STAB*, Águas de São Pedro, São Paulo, *1*, 246-250. [Dominance of contaminant yeasts over industrial strains evaluated by karyotyping technique. Paper presented at *V STAB National Congress*, Águas de São Pedro, São Paulo, *1*, 246-250].

Basso, L. C., Amorim, H. V., Oliveira, A. J. & Lopes, M. L. (2008). Yeast selection for fuel ethanol in Brazil. *FEMS Yeast Research*, *8*(7), 1155-1163. doi: 10.1111/j.1567-1364.2008.00428.xç.

Basso, L. C., Basso, T. O. & Rocha, S. N. (2011). Ethanol production in Brazil: the industrial process and its impact on yeast fermentation. In: SANTOS BERNARDES, M.A. dos (Ed.). *Biofuel Production - Recent Developments and Prospects*, *5*, 85-100. doi: 10.5772/17047.

Basso, T. O., Gomes, F. S., Lopes, M. L., Amorim, H. V., Eggleston, G. & Basso, L. C. (2014). Homo- and heterofermentative lactobacilli differently affect sugarcane-based fuel ethanol fermentation. *Antonie Van Leeuwenhoek*, *105*(1), 169-177. doi: 10.1007/s10482-013-0063-6.

Bayrock, D. P. & Ingledew, W. M. (2004). Inhibition of yeast by lactic acid bacteria in continuous culture: nutrient depletion and/or acid toxicity? *Journal of Industrial Microbiology and Biotechnology*, *31*(8), 362-368. doi: 10.1007/s10295-004-0156-3.

Beckner, M., Ivey, M. L. & Phister, T. G. (2011). Microbial contamination of fuel ethanol fermentations. *Letters in Applied Microbiology*, *53*, 387-394. doi:10.1111/j. 1472-765X.2011.03124.x.

Bischoff, K. M., Liu, S., Leathers, T. D., Worthington, R. E. & Rich, J. O. (2009). Modeling bacterial contamination of fuel ethanol fermentation. *Biotechnology and Bioengineering*, *103*, 117–122. doi: 10.1002/bit.22244.

Blomqvist, J., Eberhard, T., Schnürer, J. & Passoth, V. (2010). *Fermentation characteristics of Dekkera brucellensis strains. Applied Microbiology and Biotechnology*, *87*, 1487-1497.

Borges, E. P., Lopes, M. L., Bernardino, C., Godoy, A., Ré, F. E., Cherubin, R. A., Paulillo, S. C. L. & Amorim, H. V. (2015). *The benefits of applied research: 37 years of discoveries, adaptations and solutions*. Lecture presented by Prof. Dr. Henrique V. Amorim, winner of SPRI Science and Technology Award at Sugar Processing Research Institute (SPRI), 2014. Ribeirão Preto, São Paulo, August 31 to September 3.

Brexó, R. P. & Sant`Ana, A. S. (2017). Impact and significance of microbial contamination during fermentation for bioethanol production. *Renewable and Sustainable Energy Reviews*, *73*, 423-434. doi: 10.1016/j.rser.2017.01.151.

Broda, M. & Grajek, W. (2009). Ammonia disinfection of corn grains intended for ethanol fermentation. *Acta Scientiarum Polonorum Technologia Alimentaria*, *8*(4), 33-38.

Broda, M. & Leja, K. (2010). The microbiological situation of distilleries in the Wielkopolska region in Poland. *Polish Journal of Environmental Studies*, *19*(5), 901-906.

Cabrini, K. T. & Gallo, C. R. (1999). Identificação de leveduras no processo de fermentação alcoólica em usina do Estado de São Paulo, Brasil. [Identification of yeasts in the process of alcoholic fermentation in a plant in the State of São Paulo, Brazil.] *Scientia Agricola*, *56*(1), 1-9. doi:10.1590/S0103-90161999000100028.

Calasso, M. & Gobbetti, M. (2011). *Lactobacillus spp.: Other Species*. Elsevier Ltd., 125-131.

Chakravorty, M. (1964). Metabolism of mannitol and introduction of mannitol 1-phosphate dehydrogenase in Lactobacillus plantarum. *Journal of Bacteriology, 87,* 1246–1248.

Chalfan, Y., Levy, R. & Mateles, R. I. (1975). Detection of mannitol formation by bacteria. *Applied Microbiology*, *30*(3), 476.

Chang, I. S., Kim, B. H., Shin, P. K. & Lee, W. K. (1995). Bacterial contamination and its effects on ethanol fermentation. *Journal of Microbiology and Biotechnology, 5,* 309–314.

Cheavegatti-Gianotto, A., de Abreu, H. M. C., Arruda, P., Bespalhok Filho, J. C., Burnquist, W. L., Creste, S. & César Ulian, E. (2011). Sugarcane (Saccharum X officinarum): a reference study for the regulation of genetically modified cultivars in Brazil. *Tropical Plant Biology*, *4*(1), 62–89. doi: 10.1007/s12042-011-9068-3.

Ciani, M. & Ferraro, L. (1997). Role of oxygen on acetic acid production by *Brettanomyces/Dekkera* in winemaking. *Journal of the Science of Food and Agriculture*, *75*(4), 489-495. doi:10.1002/(SICI)1097-0010(199712)75:4<489::AID-JSFA902>3.0.CO;2-9.

Costa, V. M., Basso, T. O., Angeloni, L. H. P., Oetterer, M. & Basso, L. C. (2008). Production of acetic acid, ethanol and optical isomers of lactic acid by Lactobacillus strains isolated from industrial ethanol fermentations. *Ciência e Agrotecnologia*, *32*, 503–509.

Costa, O. Y. A., Suto, B. M., Tupinambá, D. D., Bergmann, J. C., Kyaw, C. M., Kruger, R. H., Barreto, C. C. & Quirino, B. F. (2015). Microbial diversity in sugarcane ethanol producion in a Brazilian distillery using a culture-independent method. *Journal of Industrial Microbiology and Biotechnology*, *42*(1), 73-84. doi:10.1007/s10295-014-1533-1.

Da Silva-Filho, E. A., dos Santos, S. K. B., Resende, A. M., de Morais, J. O. F., de Morais, Jr. M. A. & Simões, D. A. (2005a). Yeast population dynamics of industrial fuel-ethanol fermentation process assessed by PCR-fingerprinting. *Antonie van Leeuwenhoek*, *88* (1), 13-23. doi:10.1007/s10482-004-7283-8.

Da Silva-Filho, E. A., de Melo, W. F., Antunes, D. F., dos Santos, S. K. B., Resende, A. D., Simões, D. A. & de Morais, Jr. M. A. (2005b). Isolation by genetic and physiological characteristics of a fuel-ethanol fermentative *Saccharomyces cerevisiae* strain with potential for genetic manipulation. *Journal of Industrial Microbiology and Biotechnology*, *32*(10), 481-486. doi:10.1007/s10295-005-0027-6.

De Barros Pita, W., Leite, F. C., de Souza Liberal, A. T., Simões, D. A. & de Morais, J. R. M. A. (2011). The ability to use nitrate confers advantage to *Dekkera bruxellensis* over *S. cerevisiae* and can explain its adaptation to industrial fermentation processes. *Antonie van Leeuwenhoek*, *100*(1), 99-107. doi:10.1007/s10482-011-9568-z.

Dellias, M. T. F. (2014). *Comunidade bacteriana dos biofilmes da fermentação alcoólica: estrutura, composição, suscetibilidade aos antimicrobianos e formação de biofilme em culturas pura*, Tese Doutorado, Universidade de São Paulo. [*Bacterial community of biofilms from alcoholic fermentation: structure, composition, susceptibility to antimicrobials and biofilm formation in pure cultures*. PhD thesis, University of São Paulo].

De Miniac, M. (1989). Contamination des fermentations alcooliques industrialles par les levures du genre: *Brettanomyces*. *Industries Alimentaires et Agricoles*, *106*(7-8), 559-563. [The contamination of industrial alcoholic fermentations by the yeasts of genus: *Brettanomyces*. *Food and Agricultural Industries*, *106*(7-8), 559-563].

De Souza, R. S. C., Okura, V. K., Armanhi, J. S. L., Jorrín, B., Lozano, N., da Silva, M. J., González-Guerrero, M., de Araújo, L. M., Verza, N. C., Bagheri, H. C., Imperial, J. & Arruda, P. (2016). Unlocking the bacterial and fungal communities assemblages of sugarcane microbiome. *Scientific Reports*, *6*, 28774. doi:10.1038/srep28774.

De Souza Liberal, A. T., Basílio, A. C. M., do Monte Resende, A., Brasileiro, B. T. V., da Silva-Filho, E. A., de Morais, J. O. F., Simões, D. A. & de Morais, Jr. M. A. (2007). Identification of *Dekkera bruxellensis* as a major contaminant yeast in continuous fuel ethanol fermentation. *Journal of Applied Microbiology*, *102*(2), 538-547. doi:10.1111/j.1365-2672.2006.03082.x.

Della-Bianca, B. E., Basso, T. O., Stambuk, B. U., Basso, L. C. & Gombert, A. K. (2013). What do we know about the yeast strains from the Brazilian fuel ethanol industry? *Applied Microbiology and Biotechnology*, *97*, 979-991. doi:10.1007/s00253-012-4631-x.

Dinardo-Miranda, L. L., Anjos, I. A., Costa, V. P. & Fracasso, J. V. (2012). Resistance of sugarcane cultivars to Diatraea saccharalis. *Pesquisa Agropecuária Brasileira*, *47*(1). doi: 10.1590/S0100-204X2012000100001.

Dong, S., Lin, X. & Li, H. (2015). Regulation of Lactobacillus plantarum contamination on the carbohydrate and energy related metabolisms of Saccharomyces cerevisiae during bioethanol fermentation. *The International Journal of Biochemistry and Cell Biology, 68,* 33-41. doi: 10.1016/j.biocel.2015.08.010.

Edwards, C. G., Reynolds, A. G., Rodriguez, A. V., Semon, M. J. & Mills, J. M. (1999). Implications of acetic acid in the induction of slow/stuck grape juice fermentations and inhibition of yeast by *Lactobacillus* sp. *American Journal of Enology and Viticulture*, *50,* 204–210.

Eggleston, G., Legendre, B. L. & Tew, T. (2004). Indicators of freeze damaged sugarcane varieties which can predict processing problems. *Food Chemistry*, *87*(1), 119-133.

Eggleston, G., Basso, L. C., de Amorim, H. V., Paulillo, S. C. de, L. & Basso, T. O. (2006a). Mannitol as a sensitive indicator of sugarcane deterioration and bacterial contamination in fuel alcohol production. *Sugar Industry*, *132*, 33-39.

Eggleston, G. & Harper, W. (2006b). Determination of sugarcane deterioration at the factory: development of a rapid, easy and inexpensive enzymatic method to measure mannitol. *Food Chemistry*, *98*, 366-372. doi:10.1016/j.foodchem.2005.07.031.

Eguchi, J. Y. (2007). Antimicrobial assets used in industry. *Brazilian Society of Contamination Control Magazine*, *22*, 35-39. http://www.sbcc.com.br/sumario_22.htm.

Essia-ngand, T. J. E., Wolniewicz, E., Letourneau, F. & Villa, P. (1992). Stimulation of lactobacilli during alcoholic fermentation: action of sucrose hydrolysis by yeast. *Biotechnology Letters*, *14*, 741-746.

FAO [Food and Agriculture Organization]. (2012). *Biofuel co-products as livestock feed – Opportunities and challenges.* http://www.fao.org/docrep/016/i3009e/i3009e.pdf.

Ferain, T., Schanck, A. N. & Delcour, J. (1996). 13C nuclear magnetic resonance analysis of glucose and citrate end products in an ldhLldhD double-knockout strain of *Lactobacillus plantarum*. *Journal of Bacteriology*, *178*, 7311–7315.

Fugelsang, K. C. & Edwards C. G. (2007). Lactic acid bacteria. In *Wine Microbiology – Practical Applications and Procedures*. 29-44. Springer US.

Galafassi, S., Merico, A., Pizza, F., Hellborg, L., Molinari, F., PiSkur, J. & Compagno, C. (2011). *Dekkera/Brettanomyces* yeasts for ethanol production from renewable sources under oxygen-limited and low-pH conditions. *Journal of Industrial Microbiology and Biotechnology*, *38*(8), 1079-1088. doi:10.1007/s10295-010-0885-4.

Galafassi, S., Capusoni, C., Moktaduzzaman, M. & Compagno, C. (2013). Utilization of nitrate abolishes the "Custers effect" in *Dekkera bruxellensis* and determines a different pattern of fermentation products. *Journal of Industrial Microbiology and Biotechnology*, *40*(3-4), 297-303. doi:10.1007/s10295-012-1229-3.

Gallo, C. R. (1989). *Determinação da microbiota bacteriana de mosto e de dornas de fermentação alcoólica*. PhD Thesis, University of Campinas. [*Determination of bacterial contaminants from must and form alcoholic fermentation tanks*]. http://repositorio.unicamp.br/handle/REPOSIP/255723?mode=full.

Gallo, C. R. & Canhos, V. P. (1991a). Contaminantes bacterianos na fermentação alcoólica: revisão. [Bacterial contaminants in alcoholic fermentation: review.] *STAB: Açúcar, Álcool e Subprodutos, Piracicaba*, *9*, 15-40.

Gallo, C. R. & Canhos, V. P. (1991b). Effect of acid treatment on yeast on bacterial microbiota contaminating alcoholic fermentation. Stab: Açúcar, *Alcohol and Byproducts*, *9*(6), 35-37.

Garai-Ibabe, G., Ibarburu, I., Berregi, I., Claisse, O., Lonvaud-Funel, A., Irastorza, A. & Duenas, M. T. (2008). Glycerol metabolismo and bitterness producing lactic acid bacteria in cider making. *International Journal of Food Microbiology*, *121*, 253-261. doi: 10.1016/j.ijfoodmicro.2007.11.004.

Gomes, F. S. (2009). *Antagonismo entre leveduras e bactérias láticas na fermentação alcoólica*. MSc Diss., University of São Paulo, Piracicaba. [*Antagonism between yeasts and lactic bacteria in alcoholic fermentation*].

Graves, T., Narendranath, N. V., Dawson, K. & Power, R. (2007). Interaction effects of lactic acid and acetic acid at different temperatures on ethanol production by *Saccharomyces cerevisiae* in corn mash. *Applied Microbiology and Biotechnology*, *73*, 1190-1196. doi: 10.1007/s00253-006-0573-5.

Grobben, G. J., Peters, S. W. P. G., Wisselink, H. W., Weusthuis, R. A., Hoefnagel, M. H. N., Hugenholtz, J. & Eggink, G. (2001). Spontaneous formation of a mannitol-producing variant of *Leuconostoc pseudomesenteroides* grown in the presence of fructose. *Applied and Environmental Microbiology*, *67*, 2867–2870. doi: 10.1128/AEM.67.6.2867-2870.2001.

Heist, P. (2009). Identifying, controlling the most common microbial contaminants. *Ethanol Producer Magazine*. http://www.ethanolproducer.com/articles/5464/identifying-controlling-the-most-common-microbial-contaminants/.

Huang, Y. C., Edwards, C. G., Peterson, J. C. & Haag, K. M. (1996). Relationship between sluggish fermentations and the antagonism of yeast by lactic acid bacteria. *American Journal of Enology and Viticulture*, *47*, 1–10.

Hynes, S. H., Kjarsgaard, D. M., Thomas, K. C. & Ingledew, W. M. (1997). Use of virginiamycin to control the growth of lactic acid bacteria during alcohol fermentation. *Journal of Industrial Microbiology and Biotechnology*, *18*(4), 284-291.

Kandler, O. (1983). Carbohydrate metabolism in lactic-acid bacteria. *Antonie van Leeuwenhoek*, *49*, 209–224.

Kandler, O. & Weiss, N. (1986). "Genus *Lactobacillus"*. In *Bergey's manual of systematic bacteriology*, Vol. 2, edited by Williams, S.T., Sneath, P. H. A., Mair N. S., Sharpe, M. E., Holt. J. G., 1209-1234. Baltimore: Williams and Wilkins.

Khatibi, P. A., Roach, D. R., Donovan, D. M., Hughes, S. R. & Bischoff, K. M. (2014). *Saccharomyces cerevisiae* expressing bacteriophage endolysins reduce *Lactobacillus* contamination during fermentation. *Biotechnology for Biofuels*, *7*, 104. doi:10.1186/ 1754-6834-7-104.

King, S. W. & Beelman, R. B. Metabolic interaction between Saccharomyces cerevisiae and Leuconostoc oenos in a model grape juice/wine system. 1986. *American Journal os Enology and Viticulture*, *37*, 53-69.

Kubota, H., Senda, S., Nomura, N., Tokuda, H. & Uchiyama, H. (2008). Biofilm formation by lactic acid bacteria and resistance to environmental stress. *Journal of Bioscience and Bioengineering*, *106*(4), 381-386. doi:10.1263/jbb.106.381.

Langezaal, C. R., Chandra, A. & Scheffer, J. J. (1992). Antimicrobial screening of essential oils and extracts of some Humulus lupulus L. cultivars. *Pharmaceutisch Weekblad. Scientific Edition.*, *14*(6), 353-6.

Leite, I. R., Faria, J. R., Marquez, L. D. S., Reis, M. H. M., de Resende, M. M., Ribeiro, E. J. & Cardoso V. L. Evaluation of hop extract as a natural antibacterial agente in contaminated fuel ethanol fermentations. 2013. *Fuel Processing Technology*, 611-618.

Lisboa, L. A. M., Pascoaloto, I. M., Viana, R. S., Lopes, P. R. M. & Figueiredo, P. A. M. (2015). Contaminations in sugarcane juice: a time evaluation. *Brazilian Journal of Biosystems Engineering*, *9*(3), 279-288. doi: 10.18011/bioeng2015v9n3p279-288.

Liu, M., Bischoff, K. M., Gill, J. J., Mire-Criscione, M. D., Berry, J. D., Young, R. & Summer, E. J. (2015). Bacteriophage application restores ethanol fermentation characteristics disrupted by *Lactobacillus fermentum. Biotechnology for Biofuels*, *8*, 132. doi:10.1186/s13068-015-0325-9.

Loesche, W. J. & Kornman, K. S. (1976). Production of mannitol by *Streptococcus mutans*. *Archives of Oral Biology*, *21*, 551–553.

Lopes, M. L., Paulillo, S. C. L., Godoy, A., Cherubin, R. A., Lorenzi, M. S., Giometti, F. H. C., Bernardino, C. D., Amorim Neto, H. B. & Amorim, H. V. (2016). Ethanol production in Brazil: a bridge between science and industry. *Brazilian Journal of Microbiology*, *47S*, 64-76. doi:10.1016/j.bjm.2016.10.003.

Lonvaund-Funel, A. (2002). Lactic acid bacteria in winemaking: influence on sensorial and hygienic quality. In: Singh, V. P., Stapleton, R. D. (Rds.), Biotransformations:

bioremediation trchnology for health and environmental protection. *Progress in Industrial Microbiology*, vol. *36*. Elsevier Science, Amsterdam, pp. 231-262.

Lucena, B. T. L., dos Santos, B. M, Moreira, J. L. S., Moreira, A. P. B., Nunes, A. C., Azevedo, V., Miyoshi, A., Thompson, F. L. & Morais, Jr. M. A. (2010). Diversity of lactic acid bacteria of the bioethanol process. *Bio Med Central Microbiology*, *10*, 298–306. doi: 10.1186/1471-2180-10-298.

Lushia, W. & Heist, P. (2005). Antibiotic resistant bacteria in fuel ethanol fermentation. *Ethanol Producer Magazine*.

Madigan, M. T., Martinko, J. M. & Parker, J. (2004). Prokaryotic Diversity: Bacteria. In *Brock's Microbiology*., 333-420. São Paulo: Prentice Hall Brasil.

Maiorella, B., Blanch Harvey, W. & Wilke, C. R. (1983). By-product inhibition effects on ethanolic fermentation by *Saccharomyces cerevisiae*. *Biotechnology and Bioengineering*, *25*, 103–121. doi: 10.1002/bit.260250109.

Martinez, G., Barker, H. A. & Horecker, B. L. (1963). A specific mannitol dehydrogenase from *Lactobacillus brevis*. *Journal of Biological Chemistry*, *238*, 1598–1603.

Matsuoca Góis, C. G., Lopes-Santos, L., Oliveira Beranger, J. P., de Oliveira, A. G., Spago, F. R. & Andrade, G. (2013). The control of *Lactobacillus* sp. by extracellular compound produced by *Pseudomonas aeruginosa* in the fermentation process of fuel ethanol industry in Brazil. *Journal of Sustainable Bioenergy Systems*, *3*(3), 194-201. doi:10.4236/jsbs.2013.33027.

Meikle, A. J., Reed, R. H. & Gadd, G. M. (1988). Osmotic adjustments and accumulation of organic solutes in whole cells and protoplasts of *Saccharomyces cerevisiae*. *Journal of General Microbiology*, *134*, 3049-3060.

Meneghin, M. C., Bassi, A. P. G., Codato, C. B., Reis, V. R. & Ceccato-Antonini, S. R. (2013). Fermentative and growth performances of *Dekkera bruxellensis* in different batch systems and the effect of initial low cell counts in co-cultures with *Saccharomyces cerevisiae*. *Yeast*, *30*, 295-305. doi:10.1002/yea.2959.

Meneghin, S. P., Reis, F. C., Almeida, P. G. & Ceccato-Antonini. (2008). Chlorine dioxide against bacteria and yeasts from the alcoholic fermentation. *Brazilian Journal of Microbiology*, *39*, 337-343. doi:10.1590/S1517-83822008000200026.

Misra, V., Mall, A. K., Pathak, A. D., Solomon, S. & Kishor, R. (2017). Microorganisms affecting post-harvest sucrose losses in sugarcane. *International Journal of Current Microbiology and Applied Sciences*, *6*(7), 2554-2566. doi: 10.20546/ijcmas.2017.607.361.

Muthaiyan, A., Limayem, A. & Ricke, S. C. (2011). Antimicrobial strategies for limiting bacterial contaminants in fuel bioethanol fermentations. *Progress in Energy and Combustion Science*, *37*, 351-370. doi:10.1016/j.pecs.2010.06.005.

Narendranath, N. V., Hynes, S. H., Thomas, K. C. & Ingledew, W. M. (1997). Effects of lactobacilli on yeast-catalyzed ethanol fermentations. *Applied and Environmental Microbiology*, *63*(11), 4158-4163.

Narendranath, N. V., Thomas, K. C. & Ingledew, W. M. (2000). Urea hydrogen peroxide reduces the numbers of lactobacilli, nourishes yeast, and leaves no residues in the ethanol fermentation. *Applied and Environmental Microbiology*, *66*(10), 4187-4192.

Narendranath, N. V., Thomas, K. C. & Ingledew, W. M. (2001). Effects of acetic acid and lactic acid on the growth of Saccharomyces cerevisiae in a minimal medium. *Journal of Microbiology and Biotechnology*, *26*, 171–177.

Narendranath, N. V. (2003). Bacterial contamination and control in ethanol production. In: Jacques, K. A.; Lyons, T. P.; KelsalL, D. R. (Ed.). *The alcohol textbook*. 4th ed. Nottinghan: Nottinghan University Press. Chap., *20*, p. 287-298.

Narendranath, N. V. & Power, R. (2005). Relationship between pH and medium dissolved solids in terms of growth and metabolism of lactobacilli and Saccharomyces cerevisiae during ethanol production. *Applied Environmental Microbiology.*, *71*, 2239–2243.

Neves, A. R., Ramos, A., Shearman, C., Gasson, M. J., Almeida, J. S. & Santos, H. (2000). Metabolic characterization of Lactococcus lactis deficient in lactate dehydrogenase using *in vivo* 13C-NMR. *European Journal of Biochemistry*, *267*, 3859–3868.

Nobre, T. P. (2005). *Viabilidade celular de Saccharomyces cerevisiae cultivada em associação com bactérias contaminantes da fermentação alcoólica*. 90p. MSc Diss. Escola Superior de Agricultura "Luiz de Queiroz", Universidade de São Paulo, Piracicaba. [*Cell viability of Saccharomyces cerevisiae grown in association with contaminating bacteria from alcoholic fermentation*].

Oliva-Neto, P. & Yokoya, F. (1997). Effects of nutricional factors on growth of *Lactobacillus fermentum* mixed with *Saccharomyces cerevisiae* in alcoholic fermentation. *Revista de Microbiologia*, *28*, 25–31.

Oliva-Neto, P., Dorta, C., Carvalho, A. F. A., de Lima, V. M. G. & da Silva, D. F. (2013). The Brazilian technology of fuel ethanol fermentation – yeast inhibition factors and new perspectives to improve the technology. In *Materials and processes for energy: communicating current research and technological developments*, edited by Méndez-Vilas, A., 371-379. Badajoz: Formatex Research Center <http://www.formatex.info/energymaterialsbook/book/371-379.pdf>.

Passoth, V., Blomqvist, J. & Schnürer, J. (2007). *Dekkera bruxellensis* and *Lactobacillus vini* form a stable ethanol-producing consortium in a commercial alcohol production process. *Applied and Environmental Microbiology*, *73*(13), 4354-4356. doi:10.1128/AEM.00437-07.

Peng, J., Zhang, L., Gu, Z. H., Ding, Z. Y. & Shi, G. Y. (2012). The role of nisin in fuel ethanol production with *Saccharomyces cerevisiae*. *Letters in Applied Microbiology*, *55*(2), 128-134. doi:10.1111/j.1472-765X.2012.03275.x.

Pereira, L. F., Bassi, A. P. G., Avansini, S. H., Barbosa Neto, A. G., Brasileiro, B. T. R. V., Ceccato-Antonini, S. R. & de Morais, Jr. M. A. (2012). The physiological characteristics of the yeast *Dekkera bruxellensis* in fully fermentative conditions with cell recycling and in mixed cultures with *Saccharomyces cerevisiae*. *Antonie van Leeuwenhoek*, *101*(3), 529-539. doi:10.1007/s10482-011-9662-2.

Rasmussen, J. E., Schultz, E., Snyder, R. E., Jones, R. S. & Smith, C. R. (1995). Acetic acid as a causative agent in producing stuck fermentations. *American Journal of Enology and Viticulture*, *46,* 278–280.

Rasmusen, M. L., Koziel, J. A., Jane, J. & Pometto, A. L. (2015). Reducing bacterial contamination in fuel ethanol fermentations by ozone treatment of uncooked corn mash. *Journal of Agricultural and Food Chemistry*, *63*, 5239-5248. doi:10.1021/acs.jafc.5b00563.

Reed, G. & Nagodawithana, T. W. (1991). "Yeast derived products". New York: *Yeast Technology*, Van Nodtrand Reinhold, *2*, 225– 256.

Reis, V. R., Bassi, A. P. G., da Silva, J. C. G. & Ceccato-Antonini, S. R. (2013). Characteristics of *Saccharomyces cerevisiae* yeasts exhibiting rough colonies and pseudohyphal morphology with respect to alcoholic fermentation. *Brazilian Journal of Microbiology*, *44*(4), 1121-1131. doi: 10.1590/S1517-83822014005000020.

Reis, V. R., Antonangelo, A. T. B. F., Bassi, A. P. G., Colombi, D. & Ceccato-Antonini, S. R. (2017). Bioethanol strains of *Saccharomyces cerevisiae* characterized by microsatellite and stress resistance. *Brazilian Journal of Microbiology*, *48*(2), 268-274. doi:10.1016/j.bjm.2016.09.017.

Reguant, C., Carreté, R., Constantí, M. & Bordons, A. (2005). Population dynamics of *Oenococcus oeni* strains in a new winery and the effect of SO_2 and yeast strain. *FEMS Microbiology Letters*, *246*(1), 111-117. doi: 10.1016/j.femsle.2005.03.045.

Renouf, V., Falcou, M., Miot-Sertier, C., Perello, M. C., De Revel, G. & Lonvaud-Funel, A. (2006). Interaction between *Brettanomyces bruxellensis* and other yeast species during the initial stages of winemaking. *Journal of Applied Microbiology*, *100*, 1208-1219. doi:10.1111/j.1365-2672.2006.02959.x.

Roach, D. R., Khatibi, P. A., Bischoff, K. M., Hughes, S. R. & Donovan, D. M. (2013). Bacteriophage-encoded lytic enzymes control growth of contaminating *Lactobacillus* found in fuel ethanol fermentations. *Biotechnology for Biofuels*, *6*, 20. doi:10.1186/1754-6834-6-20.

Romano, P. & Suzzi, G. (1993). Sulphur dioxide and wine microorganisms. In Fleet G.H. (ed.). *Wine Microbiology and Biotechnology*, Switzerland. Harwood Academic Publishers, 373–393.

Rückle, L. & Senn, T. (2006). Hop acids can efficiently replace antibiotics in ethanol production. *International Sugar Journal*, *108*(1287), 139-147.

Saha, B. C. (2003). Production of mannitol by fermentation. In: Saha BC (ed) *Fermentation biotechnology*. American Chemical Society, Washington, DC, pp. 67–85.

Sakai, S. & Yamanaka, K. (1968a). Crystalline D-mannitol:NAD+ oxidoreductase from *Leuconostoc mesenteroides*. *Biochimica et Biophysica Acta*, *151*, 684–686.

Sakai, S. & Yamanaka, K. (1968b). Crystalline D-mannitol:NAD+ oxidoreductase from *Leuconostoc mesenteroides*: part II substrate and coenzyme specificity. *Agricultural and Biological Chemistry*, *32*, 894–899.

Sauvageot, N., Pichereaur, V., Louarme, L. C., Hartke, A., Auffray, Y., Laplace, J. M., 2002. Purification, characterization and subumits identification of diol dehydratase of *Lactobacillus collinoides*. *European Journal of Biochemistry* 269, 5731-5737.

Scanes, M. L. A., Hohmann, S. K. T. & Prior, B. A. (1998). Glycerol production by the yeast Saccharomyces cerevisiae and its relevance to wine: a review. *South African Journal of Enology and Viticulture*, *19*, 17–24.

Sharmin, F., Wakelin, S., Huygens, F. & Hargreaves, M. (2013). Firmicutes dominate the bacterial taxa within sugar-cane processing plants. *Scientific Reports 3*. Doi: 10.1038/srep03107.

Selby, S. J., Hillier, A. J. & Lee, G. J. (1975). The nature of the stimulation of the growth of Streptococcus lactis by yeast extract. *Journal of Dairy Reserarch*, *Cambridge*, *42*, 123-138.

Skinner, K. A. & Leathers, T. D. (2004). Bacterial contaminants of fuel ethanol production. *Journal of Industry Microbiology and Biotechnology, 31*, 401-408.

Skinner-Nemec, K. A., Nichols, N. N. & Leathers, T. D. (2007). Biofim formation by bacterial contaminants of fuel ethanol production. *Biotechnology Letters*, *29*(3), 379-383. doi:10.1007/s10529-0069250-0.

Silva, B. J. & Sauvageau, D. (2014). Bacteriophages as antimicrobial agents against bacterial contaminants in yeast fermentation processes. *Biotechnology for Biofuels*, *7*(1), 123. doi:10.1186/s13068-014-0123-9.

Silva-Filho EA, Santos SKB, Resende AM, Morais JOF, Morais MA Jr, Simões DA: Yeast population dynamics of industrial fuel-ethanol fermentation process assessed by PCR-fingerprinting. *Antonie Van Leeuwenhoek* 2005, 88:13-23.

Solomon, S. (2009). Post-harvest deterioration of sugarcane. *Sugar Tech*, *11*(2), 109-123. doi: 10.1007/s12355-009-0018-4.

Tajkarimi, M., Riemann, H. P., Hajmeer, M. N., Gomez, E. L., Razavilar, V. & Cliver, D. O. (2008). Ammonia disinfection of animal feeds - laboratory study. *International Journal of Food Microbiology*, *29*, 122(1-2), 23-28. doi: 10.1016/j.ijfoodmicro. 2007.11.040.

Talarico, T. L. & Dobrogosz, W. J. (1990). Purification and characterization of glycerol dehydrate from *Lactobacillus reuteri*. *Applied and Environmental Microbiology*, *56*, 1195-1197.

Thomas, K. C., Hynes, S. H. & Ingledew, W. M. (2001). Effect of lactobacilli on yeast growth, viability and batch and semi-continuous alcoholic fermentation of corn mash. *Journal of Applied Microbiology*, *90*, 819–828.

Veiga da Cunha, M. & Foster, M. A. (1992). Sugar-glycerol cofermentation in lactobacilli: the fate of lactate. *Journal of Bacteriology*, *174*, 1013-1019.

Vital, S. (2013). *Relações tróficas entre Saccharomyces cerevisiae e Lactobacillus sp. Nas condições da fermentação alcoólica industrial*. MSc Diss. University of São Paulo. [*Trophic relationships between Saccharomyces cerevisiae and Lactobacillus sp. under the conditions of industrial alcoholic fermentation*].

von Weymarn, N., Hujanen, M. & Leisola, M. (2002). Production of D-mannitol by heterofermentative lactic acid bacteria. *Process Biochemistry*, *37*, 1207-1213.

Wijsman, M. R., van Dijken, J. P., van Kleeff, B. H. & Scheffers, W. A. (1984). Inhibition of fermentation and growth in batch cultures of the yeast *Brettanomyces intermedius* upon a shift from aerobic to anaerobic conditions (Custer's effect). *Antonie Van Leeuwenhoek*, *50*(2), 183-192.

Wisselink, H. W., Weusthuis, R. A., Eggink, G., Hugenholtz, J. & Grobben, G. J. (2002). Mannitol production by lactic acid bacteria: a review. *International Dairy Journal*, *12*, 151-161.

Worley-Morse, T. O., Deshusses, M. A. & Gunsch, C. K. (2015). Reduction of invasive bacteria in ethanol fermentation using bacteriophages. *Biotechnology and Bioengineering*, *112*(8), 1544-1553. doi:10.1002/bit.25586.

Yang, L., Liu, Y., Wu, H., Song, Z., Hoiby, N., Molin, S. & Givskov, M. (2012). Combating biofilms. *FEMS Immunology and Medical Microbiology*, *65*(2), 146-157. doi:10.1111/j.1574-695X.2011.00858.

In: Bioethanol and Beyond
Editor: Michel Brienzo

ISBN: 978-1-53613-478-0

Chapter 4

AN ANALYSIS OF TRAJECTORY EQUIVALENCE IN STIRRED TANK BIOREACTORS USING UNSTRUCTURED KINETIC MODELS

Hugo Iván Velázquez-Sánchez[1], Rafael Maya-Yescas[2] and Ricardo Aguilar-López[3,*]

[1]Department of Biotechnology and Bioengineering, CINVESTAV-IPN, Mexico City, Mexico

[2]Chemical Engineering Department, Universidad Michoacana de San Nicolás de Hidalgo, Morelia, Michoacán, Mexico

[3]Department of Biotechnology and Bioengineering CINVESTAV-IPN, Mexico City, Mexico

ABSTRACT

In this chapter, the dynamic behavior of the most common state variables in stirred tank bioreactors is analyzed, considering different unstructured kinetic models under continuous operating mode. Analyses are performed based on the topological equivalence of trajectories framework applied to all the mathematical models selected for such systems. It is assumed that it is possible to find some diffeomorphism relating such models, which can determine under which conditions the selected mathematical models are (or not) dynamically equivalent. This analysis could help to explain the differences in performance between different kinetic models for the description of bioreactor dynamics occurring when shifting from batch to continuous operation, which is an important issue in biochemical reactor engineering. Numerical simulations are used to corroborate theoretical analysis and to set the necessary conditions to confirm that batch and continuous regime trajectories are equivalent.

* Corresponding Author Email: raguilar@cinvestav.mx.

Keywords: batch bioreactors, continuous bioreactors, topological equivalence, kinetic modelling, bioethanol

INTRODUCTION

It is well known that humanity is currently experiencing a major energetic crisis because the main source of it comes from a non-renewable resource in the form of fossil fuels. The reality of the oil sector indicates that reserves are at a critical level, therefore now research effort is focused into the study and development of processes aimed to consolidate alternative energy production and distribution technologies (Demirbas, 2009).

One of the key objectives set out for the use of alternative energy technologies is to reduce the environmental impact caused by both the exploitation and use of fossil fuel energy, especially considering the great environmental harm caused by the extraction, refining and use of oil derivatives, which are a threat to both terrestrial and marine ecosystems due to leakage of aromatic hydrocarbons or high weight molecules that due to their physicochemical properties are specially hard to remove from the physical site where they are discharged (Brown, 2003).

Therefore, when looking for strategies to address the need to exploit alternative energy sources it's not only needed to consider only technical or economic factors, but also take into account additional strategies to provide an added value to the process regarding either environmental benefits or side-product exploitation, which could help to increase the success rate of the process to be commercially competitive versus the traditional fossil fuel technologies (Dufey, 2006).

Biofuels, which can be defined as all compounds of organic nature derived from living beings and their metabolism that can potentially be utilized as fuels, lead a group of alternative energy sources aimed to provide solution to the issues raised above, as they seem more suitable to address specific issues that other technologies cannot overcome (Maddipati et al., 2011). Additionally, biofuels processes have extensive theoretical knowledge that support them, as fermentation technology is almost as ancient as humanity (Naik et al., 2010). However, current research and development of biofuel technologies cannot achieve reasonable production levels to make them attractive from the economic point of view due to both low conversion yields and final concentration after the biological fermentation stage of production. Another current issue with biofuel technology is feedstock availability and price, as traditional biotechnological processes require the use of monomeric or dimeric sugars as substrates for cell growth. Nonetheless, it is known that within the diversity of waste generated due to world trade and industrial activities there is a vast variety of hydrocarbonated components that can be utilized as raw materials for producing biofuels (Sharma et al., 2013).

Ethanol as Biofuel

One of the biofuel processes currently established at industrial-scale is the one for ethanol production, which can be used as an additive or substitute for gasoline in internal combustion engines. However, despite being one of the oldest fermentation processes, and counting with extensive knowledge derived from that vast experience, still requires additional efforts to become a solid long-term candidate to replace fossil fuels. Just to mention some of the current limitations of ethanol production it can be mentioned that the raw material used for its production in both Brazil and the United States, which are the largest producers of ethanol for its use as biofuel, comes from resources originally destined to both animal and human feedstock, such as sugar cane and corn respectively (Pimentel et al., 2007). Therefore, alternative substrates derived from crop residues such as rice straw, rice husk, corn stover, corn cobs or waste biomass like potato peel waste, orange peel waste, banana peel waste, food waste, livestock waste, paper waste and construction-derived wood residues are being proposed for its use as replacement substrates for ethanol production.

The use of agroindustrial waste has been the main objective of current trends for microbial bioethanol production. Nonetheless, further pretreatment processes must be applied to the feedstock to obtain fermentable sugars from such complex substrates, as these compounds are mainly constituted by sugars arranged into molecules of high molecular weight that exhibit a significant amount of radical branching, cyclization and other physicochemical properties that confer them a high biochemical stability and make them unsuitable to carry out a traditional fermentation process (Lo et al., 2008). Such pretreatment is known as saccharification and it may be based on either chemical or biological processes.

Saccharification Technologies Applied to Ethanol Production

Chemical based saccharification techniques are generally considered non environment-friendly, as they comprise the use of harsh acids and alkalis which leads to the generation of heat and unwanted chemical byproducts. The development of techniques for biological degradation of lignocellulosic material is a more recent research topic, which has emerged due to the large amount of knowledge generated in the areas of bioengineering and molecular biology that allows a more detailed study of the organisms that have the capacity to degrade the agroindustrial waste (Demirbas, 2009). These technologies promise higher conversion from polysaccharides to fermentable sugars due to the high specificity of the enzymes involved in the hydrolysis of said compounds, besides having the additional advantage of requiring less aggressive operating conditions such as ambient pressure, temperature and almost neutral pH compared to the chemical saccharification methods (Sandgren and Hiberg, 2005). Despite this, the cost of producing the enzymes required to perform said operation is relatively high and the organisms that natively produce proteins with such properties have complex nutritional requirements which hampers their viability (Pandey, 2003).

Separate hydrolysis and fermentation (SHF) processes had been successfully implemented for starch based ethanol production. In this process, starch is initially treated with amylolytic enzymes for liquefaction and then by glucoamylases for its saccharification.

The primary advantage of such configuration is that starch hydrolysis and sugar fermentation can be carried out separately which will minimize the interactions between these steps. However, α-amylase activities are normally inhibited by the accumulation of generated sugars during this process, which causes that SHF keeps being an expensive and time consuming process. Therefore, considerable efforts are being made to solve both the end-product inhibition and low conversion yield problems (Dey and Saggi, 2016).

Simultaneous Saccharification and Fermentation

As a way to solve the existing problems of bio-ethanol production by the SHF process the concept for simultaneous saccharification and fermentation (SSF) was developed. The sequence of the process for SSF is virtually the same as the SHF one, except that saccharification and fermentation steps are combined in a single stage into a single vessel, where the presence of ethanol in the mixture makes it less susceptible to contamination by unwanted microorganisms. Schematically in this process after liquefaction by α-amylase has been made both the glucoamylase and the fermentative microorganisms are added into the slurry (Bothast and Schlicher, 2005). Further enhancements have been proposed to make the process less time consuming by adding more than one organism species with synergistic relationships to be co-cultured together in the same vessel and achieve a much higher yield and a faster growth rate of the culture.

Simultaneous Saccharification and Co-Fermentation (SSCF) is another alternate process to SSF which facilitates pentose assimilation, which is a common issue in yeast based fermentations. Consolidated bioprocessing (CBP) is a similar approach which facilitates direct microbial conversion (DMC) and it integrates maximum biotransformation of biomass into ethanol in a single reactor by a single microorganism community (Carere et al., 2008).

MATHEMATICAL MODELLING AS A TOOL FOR THE DESIGN OF NOVEL ETHANOL PRODUCTION STRATEGIES

Mathematical modelling and simulation of biological systems can be used as a first step in novel process design and operation analysis, through the evaluation of the behavior of the studied system under different operating conditions and regimes. However, in reality, the construction of an adequate model that can reproduce the evolution of biological entities under study is not always possible, due to the high complexity of the metabolic pathways and regulation mechanisms exhibited by them. Nevertheless, the current state of society development and the arising environmental and energetic issues demand the implementation of novel biological-based production strategies that can satisfy the increasing demand of quality products while using the least quantity of the available resources.

The most important issue in biochemical reactor design and analysis is the adequate knowledge of the kinetics of the reaction rate (Froment et al., 2011); the satisfactory knowledge of the reaction rates allow designing, optimizing, controlling and operating satisfactorily these processes. Several methodologies have been considered to identify kinetic models, from classical differential and integral approaches (Levenspiel, 1969; Smith, 1970) to complex optimization techniques which allow to determinate a set of parameters to be fitting to experimental data sets (Froment et al., 2011).

In particular, a common practice followed in biochemical engineering to determine a kinetic model for a specific reacting system, is to obtain time series concentration data from batch experiments; these data are fitting using some class of unstructured kinetic models (Bailey and Ollis, 1986). Once the model parameters are tuned, the model with the best performance on the basis of experimental measures is selected. For the kinetic model discrimination, statistical and probabilistic criteria are generally considered to either accept or discard some models (Hosten and Froment, 1976).

Since cells growth involves various biochemical networks and multiple reactions, simultaneous modeling of biomass growth, substrate consumption and product formation by microorganisms is a challenging task. There are different approaches to model microbial kinetics, trying to describe the time evolution of these bioreactor states analytically, by developing unstructured kinetic models. These kind of kinetic models considers biomass as uniform (pseudo-homogeneous) quantity without internal dynamics (Zhang and Henson, 2001; Dong and Ma, 2013). Then, relationship between cell growth rates of microorganisms and substrate concentrations, which are valuable tools in bioengineering, is represented by a set of empirically derived equations referred as theoretical models (Bailey and Ollis, 1986), unfortunately.

In general, the procedure described above is useful to obtain adequate correlation between experimental and predicted data, generated by the selected kinetic model; however several kinetic model structures can produce similar results and it is difficult to discern which one is the most adequate to be scaled to continuous operating mode. Sometimes, a drastically different performance can be observed in continuous mode with respect to the batch one (Gómez-Acata et al., 2015).

As it can be noticed, existing bioreactor simulation models are strongly based on empiricism and restrictive assumptions. Reasons for these shortcomings include the lack of reliable and complete data sets as well as numerical solution constraints, since the governing equations are usually available. From the above, it is important to analyze the dynamic behavior of bioreactors operating under different scenarios, considering different available kinetic models, and determine the impact of the structure of such kinetic models on the analysis of process operation.

Analysis of dynamic trajectories of state variables could discern when some models are, or not, equivalent among them; therefore the expected behavior would be adequate. Several methodologies have been considered for the analysis of trajectories of nonlinear systems. Currently, sufficient and necessary conditions have been developed to demonstrate that different kinetic models are able of generate equivalent behavior, via a particular selection of the kinetic parameters in the corresponding mass balance equations (Craciun and Pantea, 2008). This property has been called confoundability, where the main issue lies in that the vector space spanned by the linear combination of the compounds corresponding to chemical reactants in batch or continuous operating modes is not empty. Additionally, two structurally different kinetic models are dynamical equivalent if they can generate equivalent equations for the dynamics of the mass balance (Szederkenyi et al., 2001). Another dynamic property has been called indistinguishability (Evans et al., 2004; Schnell et al., 2006), which establishes that two trajectories cannot be distinguished for distinct initial conditions lying at the domain of the vector state space generated by the differential equations describing mass balances.

As it is possible to notice, the above techniques to analyze dynamic equivalence are complex, mainly because of the employed advanced mathematical tools. Hence, in this work a SSF process by a culture of *Cellulomonas cellulans* under substrate inhibition conditions (Agarwal, Mahanty, & Dasu, 2009), is considered as application example. This process is taking place in a stirred tank bioreactor, and is modeled via classical mass balances for substrate and biomass concentrations employing unstructured kinetic models.

MATHEMATICAL BACKGROUND FOR THE ANALYSIS OF EQUIVALENCE OF TRAJECTORIES

First, consider the nonlinear system (1), which exhibits the general structure of mass balances in (bio) chemical reactors.

$$\dot{x} = f(x, u, \theta) \tag{1}$$

Here $x \in \mathcal{R}^n$ denotes the vector of the *n* dynamic states, which take values in $X \in \mathcal{R}^n$ as a connected manifold of dimension *n*. Function *f* will be, in general, assumed to be C^∞ on their arguments. $\theta \in \mathcal{R}^q$ denotes a set of vector parameters, taking values in some open subset Θ. *u* is the vector of control actions, mainly inlet flow variation in continuous systems.

An important topic to analyze in nonlinear systems is its structural stability, which means that different behaviors can occur if new periodic orbits and/or equilibrium points, caused by exogenous disturbances, appear; if this happens the system is considered as structurally unstable. In order to analyze structural stability with some detail, let us consider that there is a quoted function with quoted derivative (2) that allows representing the dynamic model of the system (1) as a control affine system (3). In model (1), control actions on flows are affecting the model in linear way, hence the control is affine (Maya-Yescas and Aguilar, 2003; Aguilar-López and Maya-Yescas, 2008). Therefore trajectories of both systems in (3) are topologically equivalents, also the system (1) is considered as the unperturbed system, and is said to be structurally stable.

$$g = \left\{ \frac{g(x)}{\|g(x)\|} < \infty, \ \left\| \frac{\partial g}{\partial x} \right\| < \infty \ \forall \ x \in \mathbb{R}^n \right\} \tag{2}$$

$$\dot{x} = f(x, u, \theta) \ and \ \dot{x} = f(x, \theta) + \ g(x)u \tag{3}$$

Now, in order to analyze the topologically equivalence relationships, consider the nonlinear system (1) near a hyperbolic critical point x_o, where $f(x_0, u, \theta) = 0$. If it is possible to linearize this system around this critical point using the Taylor linearization and via the Jacobian matrix of (1) evaluated in x_o, $J(x_o)$; then the linear tangent system (4) is obtained.

$$\dot{x} = J(x_o)x \tag{4}$$

System (4) is equivalent to system (3) in a neighborhood of x_o, therefore trajectories generated by (4) should be equivalent to trajectories generated by (3) in this neighborhood.

Let us consider two definitions and a theorem in order to formalize this relationship (Penko, 2001).

Definition 1

Two autonomous sets of differential equations modelling a dynamic system are topologically equivalent in the neighborhood of an equilibrium point, or exhibit the same qualitative structure near the equilibrium point, if there is a homomorphism mapping F from an open set M containing the origin onto an open set P containing the origin, which preserves their orientation by time in the sense that if a trajectory is directed from the state $x_1 \in M$ to the state $x_2 \in M$ of the system, then its image directs from $F(x_1) \in P$ to $F(x_2) \in P$. In this case, the two systems are said to be topologically conjugate in a neighborhood of the origin.

Hartman-Grobman Theorem

This theorem allows to represent the local phase portrait about certain types of equilibrium points in a nonlinear system by a similar, simpler linear system that is found by computing the system's Jacobian matrix at the equilibrium point. Formally, be G be an open subset of $\mathbb{R}^n$ containing the origin, let $f \in C^1(G)$ and let ϑ_t be the flow of the nonlinear system f(x). Assume that f(0) = 0 and the matrix J = Df(0) does not contain eigenvalues with zero real part. Then, there exist a homeomorphism transformation H of an open set containing the origin onto another open set V containing the origin. In fact, for each $x_0 \in U$, there is an open interval $I_0 \subset \mathbb{R}$ containing the origin in which for all $x_0 \subset U$ and $t \in I_0$, $H^\circ\vartheta_t(x_0) = e^{Jt}H(x_0)$. i.e., H maps trajectories in G near the origin onto trajectories in H near the origin, H is also 1-1, has continuous inverse and preserves the parameterization on time. Therefore, this homeomorphism may be thought of as continuously pushing the points around such that each point retains all of its original neighbors.

Now, let's consider two systems f_1, f_2, that are chosen to model the nonlinear dynamics of $\dot{x}_1 = f_1(x_1, u, \theta)$ and $\dot{x}_2 = f_2(x_2, u, \theta)$; $x_1, x_2 \in \mathbb{R}^n$.

Definition 2

(See Zhenting, et al., 2010). Assume that there exists a differential transformation $h \in \mathbb{R}^n$ that relates the two dynamic systems $x_1, x_2 \in \mathbb{R}^n$ by the Jacobian matrix J of the vector field f at the equilibrium point x_o (5). Then the systems given by x_1, x_2 are said to be *smoothly equivalent*.

$$f_1(x_o) = J^{-1}(x_o) f_2(h(x_o)) \tag{5}$$

Now, if equivalent dynamic behavior in these system exists, their evolution in time should be similar, $\dot{x}_1 \cong \dot{x}_2$ or $f_1(x_1) \cong f_2(x_2)$. If these systems obey the Hartman-Grobman theorem, then it is possible to find a derivative transformation (6) and it is possible to compare the state variables of both two systems (7). Moreover, from the linearized version (7) it is possible to find any point in the trajectory of system x_1 in terms of system x_2 (8).

$$f_1(x_1) \approx J_1 x_1, \; f_2(x_2) \approx J_2 x_2 \tag{6}$$

$$J_1 x_1 \cong J_2 x_2 \tag{7}$$

$$x_1 \cong J_1^{-1} J_2 x_2 \tag{8}$$

Now, if both systems are *fully equivalent*, it is easy to notice that their Jacobian matrices would be mutual inverses (9); in contrast if trajectory equivalence is not reached, a qualitative and/or quantitative index (□) of the *separation* between the systems' trajectories can be defined by comparing the difference between Jacobian matrices (10).

$$J_1^{-1} J_2 = I \tag{9}$$

$$J_1^{-1} J_2 = \Delta \neq I \tag{10}$$

In this work, the above concepts are considered and applied to analyze dynamics of mass balance equations in stirred tank bioreactors; growth rate of biomass use unstructured kinetic models developed by Aiba (1982), Andrew (Okpokwasili and Nweke, 2005), Haldane (Versyck, et al., 1997), Luong (1987), Han-Levenspiel (1988) and Moser (1981); all of them described in (Table A.1).

PROBLEM STATEMENT

The modelling of a continuous bioreactor considering that the coefficient related to yield of substrate ($Y_{X/S}$) could be either constant or a Gaussian type function. The basic model equations (11) are used as in previous cases; it is assumed that specific growth rate μ can be approached following different functions.

$$\Phi(\mathrm{x},\ \mathrm{S}) = \begin{cases} \dfrac{dx}{dt} = f_1 = -Dx + \mu x \\ \dfrac{dS}{dt} = f_2 = D(S_{in} - S) - \dfrac{\mu x}{Y_{X/S}} \end{cases} \tag{11}$$

Under classical mass balance modeling approach, the mathematical structures of the kinetic models are inherently structurally unstable; therefore continuous reactor models are able to predict state multiplicity and complex dynamic behavior under certain values of the convective input-output mass flows (Zhu and Huang, 2010).

Then, the Jacobian matrix (12) is calculated for any μ(X, S) system (11). Because these Jacobian matrix is not singular, its inverse matrix is computed in (23) for any μ_j(X, S) model, and the equivalence matrix is defined by (14).

Table 1. Unstructured Kinetic Models evaluated for the SSF process using *Cellulomonas cellulans*

Model	Equation	Parameters
Aiba	$\mu_{max}\frac{S}{(Ks+S)(1+\frac{S}{Ki})}$	$\mu_{max}=0.383$ $Ks=3.69$ $Ki=6.569$
Andrew	$\mu_{max}\frac{S}{(Ks+S)}\left(1-\frac{S}{S^*}\right)^n$	$\mu_{max}=0.412$ $Ks=2.40$ $S^*=2.5$
Haldane	$\mu_{max}\frac{S}{\left(Ks+S+{S^2}/{Ki}\right)}$	$\mu_{max}=0.2752$ $Ks=388.3$ $Ki=0.0256$
Han-Levenspiel	$\mu_{max}\frac{\left(1-\frac{S}{Cs}\right)^n S}{S+Kl\left(1-\frac{S}{Cs}\right)^m}$	$\mu_{max}=0.182$ $Cs=388.5$ $n=42.27$ $m=1202$
Luong	$\mu_{max}\frac{S^n}{Ks+S^n}$	$\mu_{max}=0.385$ $Ks=3.7$ $n=15718$
Moser	$\mu_{max}\frac{S}{Ks+S}e^{-S/Ki}$	$\mu_{max}=0.108$ $Ks=0.0015$ $Ki=2.66$

$$J(\Phi(x,S))=\frac{\partial(f_1,\ f_2)}{\partial(x,\ S)}=\begin{pmatrix}\frac{\partial f_1}{\partial x}=-D+\mu & \frac{\partial f_1}{\partial S}=x\frac{d\mu}{dS}\\ \frac{\partial f_2}{\partial x}=-\frac{\mu}{Y_{X/S}} & \frac{\partial f_2}{\partial S}=-D-x\frac{d({\mu}/{Y_{X/S}})}{dS}\end{pmatrix} \tag{12}$$

$$J(\Phi_j)^{-1}=\frac{1}{\left((-D+\mu_j)\left[-D-x_j\frac{d(\mu_j/Y_{X/S})}{dS_j}\right]+\frac{\mu_j x_j}{Y_{X/S}}\frac{d\mu_j}{dS_j}\right)}* \tag{13}$$

$$\begin{bmatrix} -D - x_j \dfrac{d\left(\mu_j/Y_{X/S}\right)}{dS_j} & -x_j \dfrac{d\mu_j}{dS_j} \\ \dfrac{\mu_j}{Y_{X/S}} & -D + \mu_j \end{bmatrix}$$

$$EM = J(\Sigma_1)^{-1} J(\Sigma_2) = \frac{1}{\left((-D+\mu_1)\left[-D - x_1 \dfrac{d(\mu_1/Y_{X/S})}{dS_1}\right] + \dfrac{\mu_1 x_1}{Y_{X/S}} \dfrac{d\mu_1}{dS_1}\right)} * \begin{bmatrix} (-D+\mu_2)\left(-D - x_1 \dfrac{d(\mu_1/Y_{X/S})}{dS_1} + \dfrac{\mu_1 x_2}{Y_{X/S}} \dfrac{d\mu_1}{dS_1}\right) & x_2 \dfrac{d\mu_2}{dS_2}\left(-D - x_1 \dfrac{d(\mu_1/Y_{X/S})}{dS_1}\right) - x_1 \dfrac{d\mu_1}{dS_1}\left(-D - x_2 \dfrac{d(\mu_2/Y_{X/S})}{dS_2}\right) \\ (-D+\mu_2)\dfrac{\mu_1}{Y_{X/S}} - (-D+\mu_1)\dfrac{\mu_2}{Y_{X/S}} & (-D+\mu_1)\left(-D - x_2 \dfrac{d(\mu_2/Y_{X/S})}{dS_2} + \dfrac{\mu_2 x_1}{Y_{X/S}} \dfrac{d\mu_1}{dS_1}\right) \end{bmatrix} \quad (14)$$

Again, if both systems are fully equivalent, then the Equivalence matrix $EM = I$. The conditions to reach such equivalent behavior between systems Φ_1 and Φ_2 are given by (15a-d).

$$\frac{1}{\left((-D+\mu_1)\left[-D - x_1 \dfrac{d(\mu_1/Y_{X/S})}{dS_1}\right] + \dfrac{\mu_1 x_1}{Y_{X/S}} \dfrac{d\mu_1}{dS_1}\right)} * (-D+\mu_2)\left(-D - x_1 \frac{d(\mu_1/Y_{X/S})}{dS_1} + \frac{\mu_1 x_2}{Y_{X/S}} \frac{d\mu_1}{dS_1}\right) = 1 \quad (15a)$$

$$x_2 \frac{d\mu_2}{dS_2}\left(-D - x_1 \frac{d\left(\frac{\mu_1}{Y_{X/S}}\right)}{dS_1}\right) - x_1 \frac{d\mu_1}{dS_1}\left(-D - x_2 \frac{d\left(\frac{\mu_2}{Y_{X/S}}\right)}{dS_2}\right) = 0 \quad (15b)$$

$$(-D+\mu_2)\frac{\mu_1}{Y_{X/S}} - (-D+\mu_1)\frac{\mu_2}{Y_{X/S}} = 0 \quad (15c)$$

$$\frac{1}{\left((-D+\mu_1)\left[-D - x_1 \dfrac{d(\mu_1/Y_{X/S})}{dS_1}\right] + \dfrac{\mu_1 x_1}{Y_{X/S}} \dfrac{d\mu_1}{dS_1}\right)} * (-D+\mu_1)\left(-D - x_2 \frac{d(\mu_2/Y_{X/S})}{dS_2} + \frac{\mu_2 x_1}{Y_{X/S}} \frac{d\mu_1}{dS_1}\right) = 1 \quad (15d)$$

The corresponding derivatives of the specific growth kinetic models are giving in Table 2.

Table 2. Derivatives of the Unstructured Kinetic Models evaluated for the SSF process using *Cellulomonas cellulans*

Kinetic Model	Derivative
Andrew	$\mu' = \mu_{max}\dfrac{1-\dfrac{S}{Ks+S}-\dfrac{S/_{Ki}}{1+S/_{Ki}}}{(Ks+S)(1+S/_{Ki})}$
Luong	$\mu' = \mu_{max}\dfrac{\left(1-\frac{S}{S^*}\right)^n\{(Ks+S)\left[1-\frac{nS}{S^*}\left(1-\frac{S}{S^*}\right)^{-1}\right]-S\}}{(Ks+S)^2}$
Han-Levenspiel	$\mu' = \mu_{max}\dfrac{-\left(\frac{-(S-Cs)}{Cs}\right)^n\left\{Kl[(m-n-1)S+Cs]\left[\frac{-(S-Cs)}{Cs}\right]^m-nS^2\right\}}{(S-Cs)\left[Kl\left(\frac{-(S-Cs)}{Cs}\right)^m+S\right]^2}$
Haldane	$\mu' = \mu_{max}\dfrac{Ks-S^2/_{Ki}}{\left(Ks+S+S^2/_{Ki}\right)^2}$
Moser	$\mu' = \mu_{max}\dfrac{nKsS^n}{S(Ks+S^n)^2}$

NUMERICAL RESULTS

Several numerical simulations where done in order to show the performance of the Despite any kinetic model of interest can be employed as comparison reference objective, Aiba's (1982) model is considered as reference in the case of constant yield coefficient; in contrast, Han-Levenspiel's (1988) model is considered as reference in the case of variable yield coefficient. Equivalence matrix (EM) was derived for each of the other models, after evaluation of the Jacobian matrices was performed. Some restrictions for the modelling are that it is bounded in the positive quadrant (mass concentrations cannot take negative values experimentally), it is governed by the principle of mass conservation, it is isothermal and homogeneous in the reactant concentrations in the entire vessel; no terms of death rate were considered (Sterner, 2012).

In the case of batch operation and constant yield coefficient, all the kinetic models exhibit different transient periods for substrate consumption kinetics (Figure 1). After 100 h of simulation, it can be noticed that Andrew's model reach the equilibrium point at 3 hours, Luong's model at 6 hours, Moser's model at 10 hours and Haldane's model at 16 hours; these four models reach equilibrium concentrations of *Sss* = 0 g/L and *Xss* = 4.35 g/L. In contrast, Han-Levenspiel's model reach equilibrium concentrations of *Sss* = 0.3 g/L and *Xss* = 4.25 g/L. Aiba's model shows an equivalent behavior in comparison with Luong's model, as predicted by the analysis above.

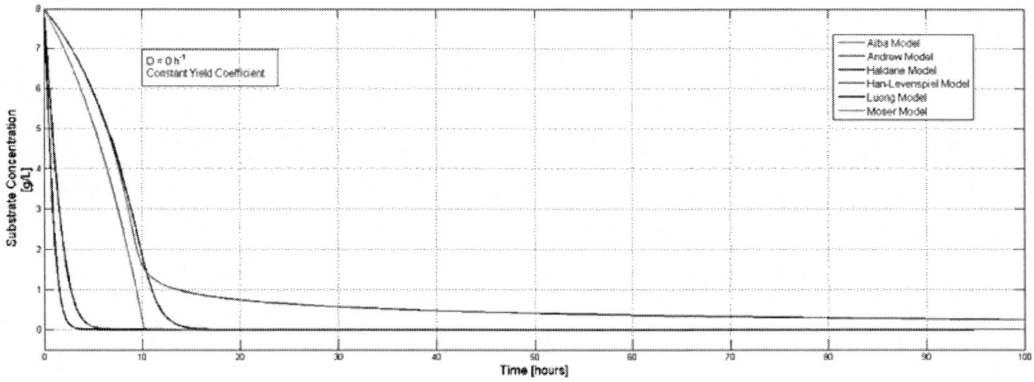

Figure 1. Time series of substrate kinetics under batch operation and constant yield coefficient.

Because of the linear relationship between biomass and substrate concentration, analog behavior is observed for biomass kinetics (Figure 2). Again, Aiba's model shows an equivalent behavior in comparison with Luong's model, as predicted above.

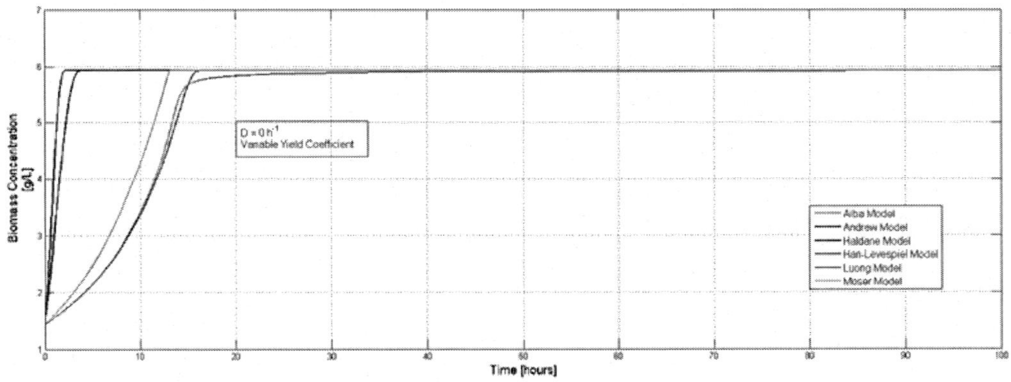

Figure 2. Time series of biomass kinetics under batch operation and constant yield coefficient.

Now, for variable yield coefficient and batch operation, all the considered kinetic model reach the same equilibrium point at *Sss* = 0 g/L and *Xss* = 6.9 g/L; however the transient period is different for almost all the considered kinetic models, the Andrew model reach the equilibrium point at 2 hours, the Luong model at 4 hours, de Moser model at 13 hours and the Haldane model at 16 hours, here the Aiba model shows an equivalent behavior in comparison with the Luong model (see Figures 3 and 4). For continuous operation was selected a dilution rate of D = 0.2 h-1, in order to explore different operation conditions. For variable yield coefficient, the Moser, Han-Levenspiel and the Haldane models reach the same steady state at *Sss* = 6 g/L and *Xss* = 0 g/L, i.e., the washout condition; the Aiba and the Luong models present a similar dynamic behavior reaching a steady state of *Sss* = 2.2 g/L and *Xss* = 1.4 g/L and finally the

Andrew model reached an steady state of Sss = 1.2 g/L and Xss = 1.7 g/L, all of these results are showed in Figures 6 and 7.

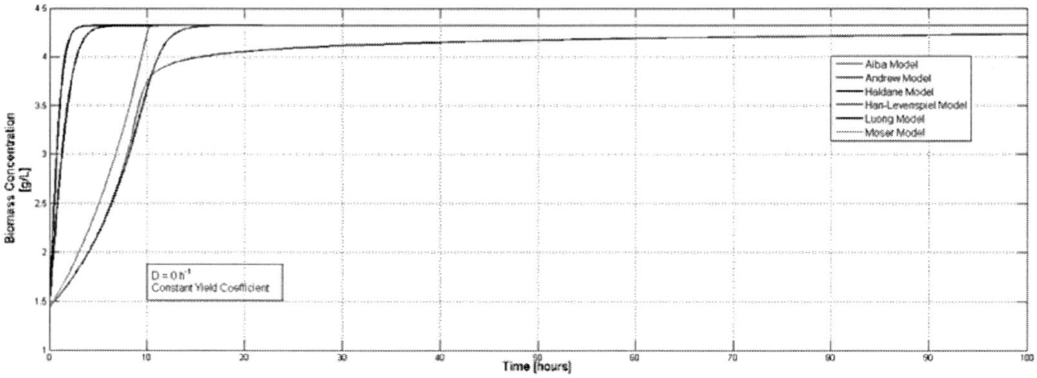

Figure 3. Time series of the set of kinetic models under batch operation for substrate mass concentration and variable yield coefficient.

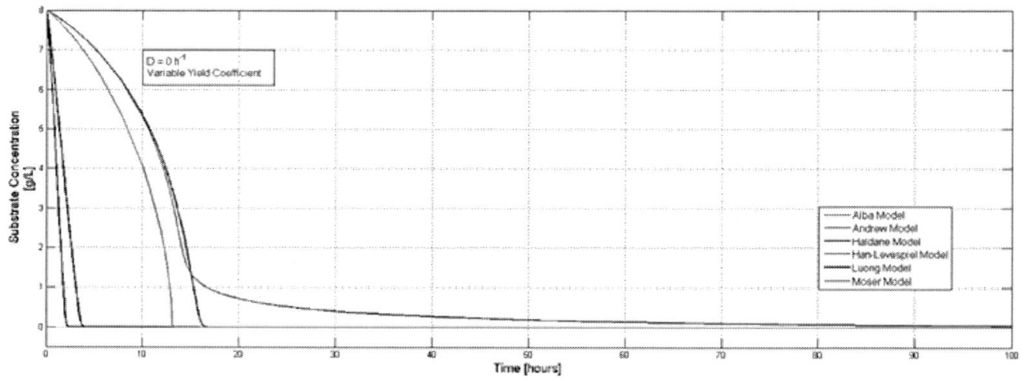

Figure 4. Time series of the set of kinetic models under batch operation for biomass concentration and variable yield coefficient.

Correlation coefficients (R^2) for predicted against experimental data for batch operation are shown (Table 3). Additionally, steady state analysis for continuous operation is performed; most chemostat models reach stable steady state, except the Moser model that cannot reach any steady state. Analysis were performed considering $x_0 = 1.44\ g\ L^{-1}$ and substrate $S_0 = 8.0\ g\ L^{-1}$ as initial conditions of biomass and a substrate, respectively; for continuous operating mode, input concentration of substrate is $S_{in} = 6\ g\ L^{-1}$ and dilution rate is $D = 0.09\ h^{-1}$.

Table 3. Dynamic behavior of the evaluated kinetic models

<table>
<tr><th rowspan="2">Unstructured kinetic model</th><th rowspan="2">R^2</th><th colspan="2">Equilibrium point</th><th colspan="2">Eigenvalues</th><th rowspan="2">Stability</th></tr>
<tr><th>S ($kg\ m^{-3}$)</th><th>X ($kg\ m^{-3}$)</th><th>λ_1</th><th>λ_2</th></tr>
<tr><td>Andrew</td><td>0.7450</td><td>1.102</td><td>0.6897</td><td>-0.2134</td><td rowspan="4">-0.09</td><td rowspan="4">Stable node</td></tr>
<tr><td>Luong</td><td>0.8215</td><td>1.554</td><td>0.6445</td><td>-0.1748</td></tr>
<tr><td>Han-Levenspiel</td><td>0.9393</td><td>1.961</td><td>0.6038</td><td>-0.7028</td></tr>
<tr><td>Haldane</td><td>0.8753</td><td>1.603</td><td>0.6396</td><td>-0.2107</td></tr>
<tr><td>Moser</td><td>0.7478</td><td>-</td><td>-</td><td>-</td><td>-</td><td>Not convergence</td></tr>
<tr><td>Aiba</td><td>0.8412</td><td>1.569</td><td>0.6430</td><td>-0.1706</td><td>-0.09</td><td>Stable node</td></tr>
</table>

MODELS WITH CONSTANT YIELD COEFFICIENT

Aiba's (1982) model is considered as reference in this case. As it was noticed, the Jacobian matrix for any unstructured kinetic model is singular, for batch operating mode; in consequence the equivalence matrix cannot be computed. However, an approach of this matrix was evaluated considering the case of continuous operating mode, with no dilution rate ($D = 0\ h^{-1}$) (Table 4).

In the case of continuous operation, equivalence analysis considering dilution rates of $D = [0.001, 0.01, 0.1]\ h^{-1}$ is shown (Table 4); the equivalence matrix (EM) was computed as expressed in equations (24, 25). Results of the equivalence matrix show that only Luong's model is closed to be equivalent to Aiba's model. Andrews' model is close to be locally equivalent at dilution rate of $D = 0.1\ h^{-1}$; besides the only model closed to be locally equivalent in the washout zone is the Haldane's model; the other kinetic models are not equivalent to Aiba's model.

MODELS WITH VARIABLE YIELD COEFFICIENT

Han-Levenspiel's (1988) model is considered as reference in this case. For the case of batch operating mode Jacobian matrices are not singular, therefore equivalence matrices (EM) were computed. In this case, there are not equivalent models to the Han-Levenspiel's one for all dilution rates (Table 4).

Table 4. Equivalence matrix (*EM*) for chemostat considering constant yield coefficient

	D = 0 (batch operation)	D = 0.001	D = 0.01	D = 0.1	D = 0.2 (washout)
Han-Levenspiel/Aiba	$\begin{bmatrix} 0.008 & 0 \\ 0.168 & 1 \end{bmatrix}$	$\begin{bmatrix} 0.041 & 0 \\ 0.162 & 0.996 \end{bmatrix}$	$\begin{bmatrix} 0.315 & 0 \\ 0.116 & 0.999 \end{bmatrix}$	$\begin{bmatrix} 6.218 & 0 \\ -0.887 & 1 \end{bmatrix}$	$\begin{bmatrix} 1.233 & 0 \\ -0.083 & 0.990 \end{bmatrix}$
Luong/Aiba	$\begin{bmatrix} 1.002 & 0 \\ -0.004 & 1 \end{bmatrix}$	$\begin{bmatrix} 1.002 & 0 \\ -0.004 & 0.992 \end{bmatrix}$	$\begin{bmatrix} 1.003 & 0 \\ -0.005 & 0.999 \end{bmatrix}$	$\begin{bmatrix} 1.052 & 0 \\ -0.008 & 1 \end{bmatrix}$	$\begin{bmatrix} 0.099 & 0 \\ 0.321 & 1.008 \end{bmatrix}$
Andrew/Aiba	$\begin{bmatrix} 1.653 & 0 \\ -0.112 & 1 \end{bmatrix}$	$\begin{bmatrix} 1.652 & 0 \\ -0.110 & 1.005 \end{bmatrix}$	$\begin{bmatrix} 1.641 & 0 \\ -0.109 & 0.999 \end{bmatrix}$	$\begin{bmatrix} 0.970 & 0 \\ 0.005 & 1 \end{bmatrix}$	$\begin{bmatrix} 0.717 & 0 \\ 0.109 & 0.909 \end{bmatrix}$
Haldane/Aiba	$\begin{bmatrix} 0.683 & 0 \\ 0.053 & 0.992 \end{bmatrix}$	$\begin{bmatrix} 0.687 & 0 \\ 0.052 & 0.996 \end{bmatrix}$	$\begin{bmatrix} 0.7332 & 0 \\ 0.0453 & 0.999 \end{bmatrix}$	$\begin{bmatrix} 1.392 & 0 \\ -0.066 & 1 \end{bmatrix}$	$\begin{bmatrix} 1.023 & 0 \\ -0.008 & 0.963 \end{bmatrix}$
Moser/Aiba	-	-	-	-	-

Table 5. Equivalence matrix (*EM*) for chemostat considering variable yield coefficient

	D = 0 (batch operation)	D = 0.001	D = 0.01	D = 0.1	D = 0.2 (washout)
Aiba/Han-Levenspiel	$\begin{bmatrix} -0.003 & -0.001 \\ 0.005 & 0.001 \end{bmatrix}$	$\begin{bmatrix} 0.116 & 0 \\ -3.153 & 0.357 \end{bmatrix}$	$\begin{bmatrix} 1.052 & 0 \\ -2.698 & 0.280 \end{bmatrix}$	$\begin{bmatrix} 1.48 & 0 \\ 0.24 & 0.615 \end{bmatrix}$	$\begin{bmatrix} 0 & -0.131 \\ 0.475 & 0.798 \end{bmatrix}$
Andrew/ Han-Levenspiel	$\begin{bmatrix} 0.007 & 0.001 \\ -2.758 & -0.005 \end{bmatrix}$	$\begin{bmatrix} 0.070 & 0 \\ -3.222 & 0.333 \end{bmatrix}$	$\begin{bmatrix} 0.687 & 0 \\ -3.233 & 0.253 \end{bmatrix}$	$\begin{bmatrix} 1.076 & 0 \\ -0.125 & 0.406 \end{bmatrix}$	$\begin{bmatrix} 0 & -0.430 \\ 0.303 & 0.461 \end{bmatrix}$
Haldane/ Han-Levenspiel	$\begin{bmatrix} 29.0 & 14.596 \\ 0.005 & 0.001 \end{bmatrix}$	$\begin{bmatrix} 14.009 & 5.6 \\ 0.050 & 0 \end{bmatrix}$	$\begin{bmatrix} 7.345 & 3.42 \\ 0.634 & 0 \end{bmatrix}$	$\begin{bmatrix} -31.849 & 0 \\ 16.805 & -1.53 \end{bmatrix}$	$\begin{bmatrix} 1.0 & 0.033 \\ 0 & 0.976 \end{bmatrix}$
Luong/ Han-Levenspiel	$\begin{bmatrix} 0.017 & 0.003 \\ 7.213 & 1.989 \end{bmatrix}$	$\begin{bmatrix} 0.115 & 0 \\ -3.153 & 0.357 \end{bmatrix}$	$\begin{bmatrix} 1.05 & 0 \\ -2.701 & 0.280 \end{bmatrix}$	$\begin{bmatrix} 1.478 & 0 \\ 0.239 & 0.614 \end{bmatrix}$	$\begin{bmatrix} 0 & -0.669 \\ 0.474 & 0.796 \end{bmatrix}$
Moser/ Han-Levenspiel	$\begin{bmatrix} 29.015 & 14.015 \\ 0.004 & 0.009 \end{bmatrix}$	$\begin{bmatrix} 1.342 & 0.56 \\ 0.046 & 0 \end{bmatrix}$	$\begin{bmatrix} 1.199 & 0.171 \\ -0.563 & 0 \end{bmatrix}$	$\begin{bmatrix} -30.555 & 2.203 \\ 22.728 & 0 \end{bmatrix}$	$\begin{bmatrix} 1 & -0.207 \\ 0 & 1.148 \end{bmatrix}$

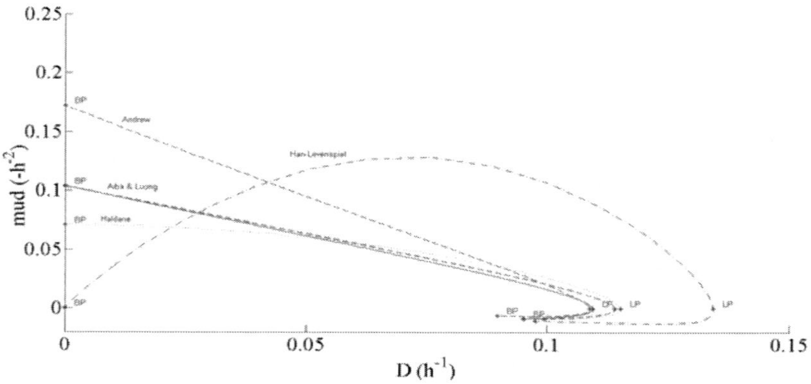

Figure 5. Bifurcation Diagram, derivative of the kinetic models versus Dilution rate (S_{in}=6 g/L; s_0=0.1 g/L; x_0=1.44 g/L).

In order to explain these results, some important factors to determinate trajectory equivalences correspond to the derivative of the considered kinetic models as observed in (15). There is drastically different behavior of the derivatives of the kinetic models at several dilution rates in the corresponding bifurcation diagram (Figure 5).

DYNAMIC ANALYSIS OF THE KINETIC MODELS

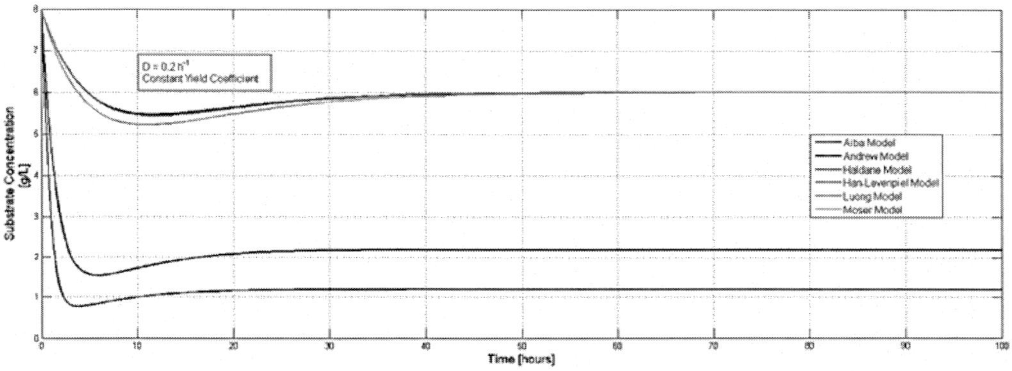

Figure 6. Time series of the set of kinetic models under continuous operation for substrate mass concentration and constant yield coefficient.

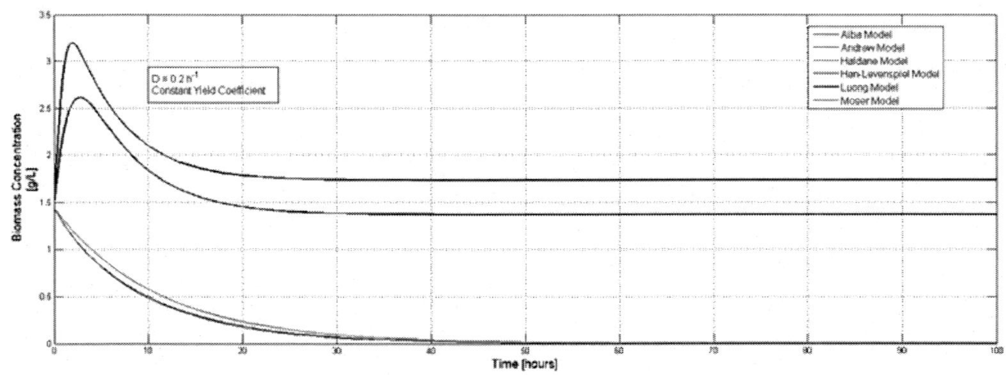

Figure 7. Time series of the set of kinetic models under continuous operation for biomass concentration and constant yield coefficient.

Now Figures 8 and 9 are related with the dynamic behavior of the continuous operation at $D = 0.2\ h^{-1}$ and variable yield coefficient form in the kinetic modeling, as similar as the above case, the bioreactor reach the steady states previously described, despite of some differences of the trajectories in comparison with the constant yield coefficient case.

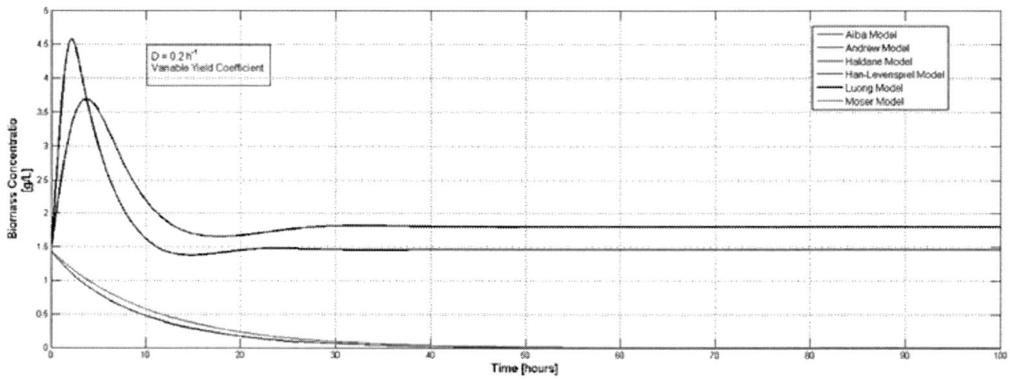

Figure 8. Time series of the set of kinetic models under continuous operation for biomass concentration and variable yield coefficient.

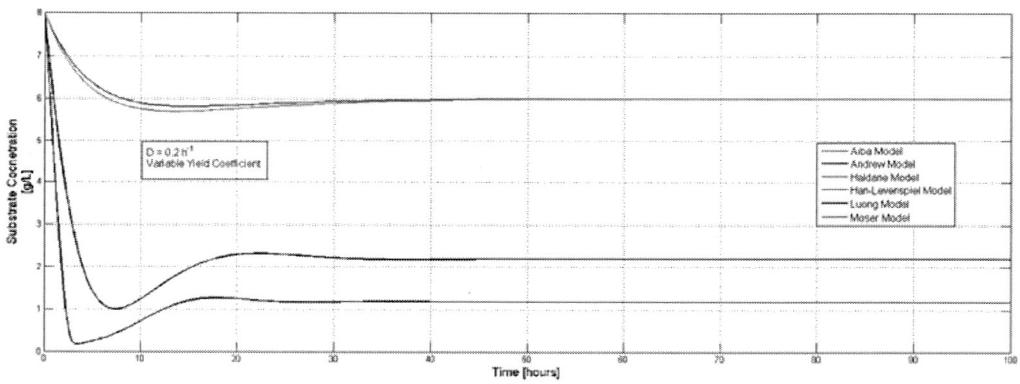

Figure 9. Time series of the set of kinetic models under continuous operation for substrate mass concentration and variable yield coefficient.

Conclusion

Structural indistinguishability is one of the main study cases considered by structural identifiability analysis, where the main goal is to determine the uniqueness of the determination of unknown parameters from a mathematical model given a certain output. In this paper the approach developed in (Evans et al., 2004) for testing the structural identifiability of biochemical systems was extended to one for testing the indistinguishability between different models for a given process.

In this work the dynamical behavior of a chemostat model has been analyzed by numerical bifurcation where the kinetic growth was described by a set or unstructured growth models and under the assumption that the biomass yield is constant, or a function of the substrate concentration. Significant changes in the chemostat behavior were obtained when the Gaussian yield function was used instead of the constant biomass yield value, as there were found differences in the steady state multiplicity intervals, productivity and equilibrium stability. Particularly outstanding are the results from the Han-Levenspiel growth model, as it predicts a maximum dilution rate over the other unstructured models and non-equivalent equilibrium branches. The above results disclose that chemostat model predictions can vary greatly according to the unstructured growth model and the biomass yield. It also shows the necessity of the validation of simulation results with laboratory experiments to choose the most appropriate bioreactor model, thus leading to an improvement of the bioreactor design and control.

REFERENCES

Agarwal, R., Mahanty, B., Venkata Dasu, V. (2009). Modeling the growth of Cellulomonas cellulans NRRL B-4567 under substrate inhibition during cellulase Production. *Chemical and Biochemical Engineering Quarterly* 23 (2): 213-218.

Aguilar Sierra, H., Martinez-Guerra, R., Mata-Machuca, J. (2011). Fault diagnosis via a polynomial observer. *2011 8th International Conference on Electrical Engineering, Computing Science and Automatic Control*, 121-126.

Aguilar-López, R., Maya-Yescas, R. (2008). Inverse dynamics: A problem on transient controllability for industrial plants. *Inverse Problems in Science and Engineering* 16: 811–827.

Aiba, S. (1982). Growth kinetics of photosynthetic microorganisms. Microbial reactions. *Advances in Biochemical Engineering* 23: 85-156.

Bailey, J. E., Ollis, D. F. *Biochemical Engineering Fundamentals*. 2nd ed. McGraw-Hill Science/Engineering/Math; New York, 1986.

Bothast, R. J., Schlicher, M. A. (2005). Biotechnological processes for conversion of corn into ethanol. *Appl Microbiol. Biot.* 67(1): 19-25.

Brown, R. *Biorenewable resources*. Iowa State Press, Blackwell Publishing Co., Iowa, 2003.

Buzzi-Ferraris, G. (1999). Planning of experiments and kinetic analysis. *Catalysis Today* 52: 125-132.

Buzzi-Ferraris, G., Manenti, F. (2009). Kinetic Models Analysis. *Chemical Engineering Science* 64 (5): 1061-1074.

Carere, C. R., Sparling, R., Cicek, N., Levin, D. B. (2008). Third generation biofuels via direct cellulose fermentation. *IJMS* 9(7): 1342-1360.

Counotte, G. H. M., Prins, R. A. (1979). Calculation of Km and Vmax from substrate concentration versus time plot. *Appl. Environ. Microbiol.* 38: 758-760.

Craciun, G., Pantea, C. (2008). Identifiability of chemical reaction networks. *Journal of Mathematical Chemistry* 44 (1): 244-259.

Demirbas, A. (2009). Progress and Recent Trends in Biodiesel Fuels. *Energy Conversion and Management* 50: 14-34.

Dey, P., Saggi, S. K. (2016). An Innovative Approach towards Economic Bio-ethanol Production from Starchy and Ligno-Cellulosic Biomass through Simultaneous Saccharification and Fermentation (SSF). *International Journal of Current Microbiology and Applied Sciences* 5 (5), 870-877.

Dong, Q. L., Ma, W. B. (2013). Qualitative analysis of the chemostat model with variable yield and a time delay. *Journal of Mathematical Chemistry* 51: 1274-1292.

Dufey A. (2006). *Biofuels production, trade and sustainable development: emerging issues*. Environmental economics programme/Sustainable markets group. Paper No.2, International Institute for Environment and Development (IIED).

Evans, N. D., Chappell, M. J., Chapman, M. J., Godfrey, K. R. (2004). Structural indistinguishability between uncontrolled (autonomous) nonlinear analytic systems. *Automatica* 40: 1947 – 1953.

Froment, G. F., Bishoff, K. B., De Wilde, J. *Chemical Reactor Analysis and Design*. 3rd ed. John Wiley & Sons, Singapore, 2011.

Han, K., Levenspiel, O. (1988). Extended Monod kinetics for substrate, product and cell inhibition. *Biotechnol. Bioeng.* 32: 430 – 437.

Hosten, L. H., Froment, G. F. (1976). *Proceedings, IV International Symposium on Chemical Reaction Engineering*, Heidelberg, Apr 6-9.

Hosten, L. H., Froment, G. F. (1976). Non-Bayesian sequential experimental design procedure for optimal discrimination between rival models. In: *Proceedings of Fourth International Symposium on Chemical Reaction Engineering*, Heidelberg.

Hou, Z., Kang, N., Kong, X., Chen, G., Yan, G. (2010). On The Nonequivalence of Lorenz System and Chen System. *Int. J. Bifurcation Chaos* 20, 557.

Kovárová-Kovar, K., Egli, T. (1998). Growth kinetics of suspended microbial cells: from single-substrate-controlled growth to mixed substrate kinetics. *Microbiol. Mol. Biol. Rev.* 62:646-666.

Levenspiel, O. *Chemical Reactor Engineering: An Introduction to the Design of Chemical Reactors*, 2nd ed., Wiley, New York, 1972.

Lo, Y. C., Bai, M. D., Chen, W. M., Chang, J. S. (2008). Cellulosic hydrogen production with a sequencing bacterial hydrolysis and dark fermentation strategy. *Bioresour. Technol.* 99 (17): 8299–8303.

Luong, J. H. T. *Biotechnol. Bioeng.* 1987, 29, 242-248.

Maddipati, P., Atiyeh, H. K., Bellmer, D. D., Huhnke, R. L. (2011). Ethanol production from syngas by *Clostridium* strain P11 using corn steep liquor as a nutrient replacement to yeast extract. *Bioresour. Technol.* 102 (11): 6494–6501.

Maya-Yescas, R., Aguilar-López, R. (2003). Controllability assessment approach for chemical reactors: nonlinear control affine systems. *Chemical Engineering Journal* 92: 69-79.

Méndez, J. M., Femat, R. (2012). Dynamic equivalence in tangent spaces from vector fields of chemical reaction networks. *Chemical Engineering Science* 83: 50-55.

Moo-Young, M., Blanch, H. W.: In: Foundations of Biochemical Engineering (Blanch et al. eds.). Symp. Ser. No. 207, New York: *Amer. Chem. Soc.* 1983.

Moser, A. *Bioprocess Technology. Kinetics and Reactors*. Springer-Verlag, New York, 1981.

Naik, S. N., Goud, V. V., Rout, P. K., Dalai, A. K. (2010). Production of first and second generation biofuels: A comprehensive review. *Renewable and Sustainable Energy Reviews* 14: 578–597.

Okpokwasili, G. C., Nweke, C. O. (2005). Microbial growth and substrate utilization kinetics. *African Journal of Biotechnology* 5 (4): 305-317.

Pandey, A. (2003). Solid-State Fermentation. *Biochemistry Engineering Journal* 13 (2-3): 81-84.

Penfold, W., Norris, D. (1912). The relation of concentration of food supply to generation time bacteria. *J. Hyg.* 12: 527-531.

Pfromm, P. H., Amanor-Boadu, V., Nelson, R., Vadlani, P., Madl, R. (2010). Bio-butanol vs. Bio-Ethanol: A technical and economic assessment for corn and switchgrass fermented by yeast or *Clostridium acetobutylicum*. *Biomass & Bioenergy* 34: 515-524.

Pimentel, D., Patzek, T. W., Gerald, C. (2007). Ethanol production: energy, economic, and environmental losses. *Reviews of Environmental Contamination and Toxicology* 189: 25-41.

Prokop, A. In: *Foundations of Biochemical Engineering* (Blanch et al. eds.), Symp. Ser. No. 207, New York: Amer. Chem. Soc. 1983.

Sandgren, M., Hiberg, J. S. (2005). Structural and Biochemical Studies of GH Family 12 Cellulases: Improved Thermal Stability and Ligand Complexes. *Progress in Biophysics and Molecular Biology* 89 (3): 246-291.

Schnell, S., Chappell, M. J., Evans, N. D., Roussel, M. R. (2006). The mechanism distinguishability problem in biochemical kinetics: The single-enzyme, single-substrate reaction as a case study. *Comptes Rendus Biologies* 329 (1): 51-61.

Sharma, B., Ingalls, R. G., Jones, C. L., Khanchi, A. (2013). Biomass supply chaindesign and analysis: Basis, overview, modeling, challenges, and future. *Renewable and Sustainable Energy Reviews* 24: 608–627.

Smith, J. M. *Chemical Engineering Kinetics*. Mc Graw Hill. 3rd Ed., New York, 1986.

Sterner, R. W., Small, G. E., Hood, J. M. (2012). The conservation of mass. *Nature Education Knowledge* 3: 20-30.

Szederkényi, G., Kristensen, N. R., Hangos, K. M., Jorgensen, S.B. (2001). Nonlinear analysis and control of a continuous fermentation process. *European Symposium on Computer-Aided Process Engineering* 11: 787-792.

Tobajas, M., Garcia-Calvo, E. (2000). Comparison of analysis methods for determination of the kinetic parameters in batch cultures. *World Journal of Microbiology & Biotechnology* 16: 845-851.

Versyck, K. J., Claes, J. E., Van Impe, J. F. (1997). Practical identification of unstructured growth kinetics by application of optimal experimental design. *Biotechnol. Prog.* 13, 524–531.

Zhang, Y., Henson, M. A. (2001). Bifurcation analysis of continuous biochemical reactor models. *Biotechnology Progress* 17: 647-653.

Zhu, L., Huang, X. (2010). Bifurcation in the stable manifold of a chemostat with general polynomial yield. *Mathematical Methods in the Applied Sciences* 33 (3): 119-128.

In: Bioethanol and Beyond
Editor: Michel Brienzo
ISBN: 978-1-53613-478-0

Chapter 5

BRAZILIAN EFFORTS IN TECHNICAL ANALYTICAL STANDARDS APPLIED TO THE QUALITY CONTROL OF BIOETHANOL

Ana Carolina Gandini Panegossi[1], Rodrigo Sequinel[2], Luciana Antoniosi[4], Sandra Imaculada Maintinguer[3,4,*] and Danilo Luiz Flumignan[1,5]

[1]IFSP, São Paulo Federal Institute of Education, Science and Technology, Matão, São Paulo, Brazil
[2]UFPR, Federal University of Paraná, Setor Palotina, Palotina, Paraná, Brazil
[3]IPBEN, Bioenergy Research Institute, UNESP, São Paulo State University, Rio Claro, São Paulo, Brazil
[4]UNIARA, University of Araraquara, Araraquara, São Paulo, Brazil
[5]CEMPEQC, Center for Monitoring and Research of the Quality of Fuels, Biofuels, Crude Oil and Derivatives, Institute of Chemistry, UNESP, São Paulo State University, Araraquara, São Paulo, Brazil

ABSTRACT

The search for renewable energy sources, especially for bioethanol, has attracted a lot of attention from researchers, businesses, and governments, due to economic, geopolitical, and environmental reasons. In the global market, the demand for bioethanol is increasing and being stimulated by public policies aimed at the modernization of the global energy matrix through the use of alternative kinds of fuel derived from petroleum, and which lead to a reduction in CO_2 emissions.

To maintain the leading position in production and marketing, ethanol was formally introduced in the Brazilian energy matrix in 2005, by the federal law no. 11.097, and to

[*] Corresponding Author Email: mainting2008@ipben.unesp.br

ensure it is commercialized in a highly globalized and competitive market, quality and price are decisive. The quality of bioethanol in Brazil is supported by three federal agencies: the Brazilian Technical Standards Association (ABNT), responsible for technical standardization; the National Agency of Petroleum, Natural Gas, and Biofuels (ANP), responsible for the definition of standardized specifications and quality control; and the National Institute of Metrology, Quality, and Technology (Inmetro), responsible for metrology standards. In this paper, the fundamental aspects of bioethanol's quality assurance through standardized specifications, the official technical standards analysis, the analytical methods and the certified reference materials from regulatory authorities have been minutely described. Quality control should be implemented in order to reduce the cost of non-quality products, increase product reliability, reduce the rates of non-conformities, and assuage the technical barriers to commercialization of ethanol worldwide.

Keywords: bioethanol, quality, specifications, standards, analytical methods, certified reference materials

1. Introduction

The growing interest about bioethanol is specially due to its social, economic, and environmental sustainability. Brazil is the pioneer country in using ethanol as a motor fuel and is currently the largest ethanol sugarcane producer in the world. Most of this production is absorbed by the domestic market, but a substantial quantity is prompty traded. Both the domestic and cross-border ethanol have to be certified by national and/or international quality standards in order to guarantee its applicability. Among the several understandings that are possible for the term "quality," the one we shall adopt is "adequacy to use," but, in this paper, the definition according to the notion of non-quality or lack of quality will also be relevant, since, as well, this causes losses – through noxious side-effects – to the engines and to the environment.

In the beginning of the 90s, some researches from the Brazilian Quality and Productivity Program (PBQP) have shown that Brazil had been losing 10% of its GDP (Gross Domestic Product) because of problems related to problems with quality. Improper products and services led to large amounts of wastes, rework, and product substitution during the warranty period, in both local and international market. In order to reduce the economic losses related to the lack of quality, several actions have been implemented by the public and private initiatives and, thus, networks for metrology labs, human resources capacitation programs, capacitation and infrastructure for certifying products and management systems have been consolidated, among others (Toledo et al., 2013).

When it comes to the regulation and quality control of the bioethanol, in 1991, the first Ordinance DNC was created: nº23 of 10/29/1991, from the National Fuel Department (1991). Its technical regulation prescribes the specifications for the anhydrous and hydrated ethanol to be used as fuels.

A product's specification must ensure its quality and, in order to make sure that a product has a satisfactory performance on the application to which it is proposed and on the uniformity of its manufacturing, this product's specifications are stablished. Bioethanol's specification is defined according to the necessary characteristics for a good engine performance, which are evaluated through internationally standardized tests. Besides, its control is made by appropriate analyses and tests that present good precision and reproducibility.

Despite being part of the Brazilian energetical matrix for more than 80 years, bioethanol was only formally inserted in it in 2005, through federal law nº11.097. The regulatory structure of its quality is carried out by three federal organs: the National Agency of Petroleum, Natural Gas, and Biofuels (ANP), responsible for the normalized specifications; the Brazilian Technical Standards Association (ABNT), responsible for official technical analysis norms; and the National Institute of Metrology, Quality, and Technology (Inmetro), responsible for the certified reference materials.

A Work Group, created in 2007, made by specialists from Brazil, the European Community, and the United States (ANP, CEN – Comité Européen de Normalisation, and ASTM – American Society for Testing and Materials, respectively) has gathered efforts to critically analyze and discuss the specifications for bioethanol, such as content, physical characteristics, and contamination levels, factors that determine its quality. These specifications are defined in order to harmonize these aspects worldwide, aiming, thus, to consolidate bioethanol as a global commodity (White Paper, 2007, apud Bonomi, 2010).

Considering that it is essential to adopt regulatory devices that avoid the commercialization of bioethanol, be it as anhydrous or hydrated fuel, aiming to repress fraudulent practices in the market, the technical specifications and obligations towards the quality control that shall be reached by the several economic agents have been recently updated, through ANP resolution nº7, published in 2/10/2011. Parallelly, ABNT has been developing and updating the technical norms setting for bioethanol's analysis in order to keep it as up to date as possible with the present technology, specifications, and regulations. Also, Inmentro has been developing certified reference materials in order to guarantee the quality of the measurement results. These measures aim to soften the existing barriers to trading bioethanol in the international market and to show the importance that Brazil has been giving to this fuel's quality in production, commercialization, and use. Thus, it is justified to show, in this paper, the Brazilian advances in technical norms for the certification and quality control of bioethanol.

2. Specifications

2.1. Bioethanol's Quality and Normalization

According to Flumignan et al. (2012), in the past few years, the biofuels have been aggregated to the Brazilian energetic matrix, constituting a viable alternative for fossil fuels, and being meaningful for energy supply in the future. The participation of the energy source in the energetic matrix depends on a complex arrangement of factors that involve production (feedstock and technology) and sustainability aspects (social, economic, and environmental). Bioethanol awakes big attention from researchers, companies, and from the government because of the pressures regarding prices, environmental worries, and perspectives of depleting nonrenewable fossil fuel sources.

A new configuration of the sugarcane agroindustry system (SAG), which contains the most important of the biofuels - the bioethanol - rises from the deregulation of the sector. With the partial exit of the State from the SAG, the Brazilian economy commercial opening, and the international demand for alcohol as a fuel are the main causes for this situation. In order to maintain and consolidate the governmental support, bioethanol was formally inserted in the Brazilian energetic matrix through law 11.097, published in 1/13/2005, with ANP as the organ responsible by regulating it (Bonomi, 2010).

According to Bonomi (2010), the Brazilian regulatory structure guarantees the possibility of building a regulatory and metrologic mark in order to maintain the highlighting position in the production and commercialization of bioethanol, since this product is regarded as a commodity in a globalized market. This structure is composed by three federal entities:

- ANP, implanted by the Federal Decree nº2.455, of 1/14/1998. It is responsible for defining the specifications for the biofuels and for controlling their quality in all of the steps in their production and in the commercialization chain;
- ABNT, founded in 1940. It is responsible for the technical standardization in Brazil.
- Inmetro, created by federal law nº5.966, of 12/11/1973. Among other competences and attributions, it is responsible for performing the national metrology and quality policies.

Thus, the characterization and quality control of biofuels in general and, in this specific case, of bioethanol, depend substantially on three factors:

- Standardized Specifications, which are a group of characteristics necessary in order to make sure that ethanol fulfils its purpose. The attribution of these characteristics is a responsibility of the regulatory organ ANP;

- Official Technical Analysis Norms, which are the instruments and procedures necessary for evaluating the standardized specifications of bioethanol. ABNT is the organ responsible by this technical standardization;
- Certified Reference Materials, which are the devices used in evaluation, validation, and analytical control of the official technical norms. This is a responsibility of the metrologic organ Inmentro.

Each one of these factors has some fundamental aspects that will be detailed as it follows.

2.2. The Specifications for Bioethanol's Characterization and Quality Control

For Toledo et al. (2013), there are three branches to define the product quality:

- Fitness for Use: this branch suggests that quality is the degree to which the product regards satisfactorily the user's needs during its use;
- Conformance to Requirements: this branch supposes that it is only possible to think about the product's quality if there is a group of previously defined specifications;
- And "the loss, measurable and immeasurable, that a product imposes to society after its embark (after leaving the company), except for the losses that are caused by its intrinsic function," aspect that focuses on the topic as non-quality or as the lack of it. In this branch, the quality is defined as the losses, in monetary units, that a product causes to society after being sold. The losses are restricted to two kinds: the ones caused by the variability in the basic and intrinsic function that the product has, and the ones caused by its noxious side effects.

Flumignan et al. (2012) mention, in their study, that a product's quality may be defined as its suitability for a determined application, i.e., quality as adjustment to use. Therefore, this definition will be prioritized in the present study. However, it is also important to consider that the losses caused by the use of non-renewable biofuels, even if not considered in monetary units, are the main environmental worries that reinforce the use of bioethanol. On the other hand, when it comes to money, the losses caused on the engines through the use of bad quality fuels is one of the main concerns.

According to ABNT NBR ISO 9000:2005, quality control is the part of the quality management that focuses on serving the quality requirements. The quality warranty is the part that focuses on providing confidence that the quality requirements will be served (Associação Brasileira De Normas Técnicas, 2005).

The quality control is made through official or alternative analyses and tests that present good precision and reproducibility. The specifications of a product are stablished in order to guarantee its quality, i.e., the satisfactory performance in the application to which it is applied and the manufacturing uniformity (Flumignan et al. 2012). Therefore, besides ensuring the quality of product, its specifications are also a reference standard for the consuming market (Bonomi, 2010).

For Bonomi (2010), is the specifications of a fuel are defined by the set of characteristics that are necessary for good engine performances, which are evaluated by internationally standardized tests. ANP, like all the other international regulatory organs, does not regulate the fuels' chemical composition, but stablishes the specifications that determine the minimum quality required for the use to which this fuel is destined, through Technical Resolutions and Regulations.

According to Araújo (2011), ANP's attributions, when it comes to ethanol's quality, are, among others:

- Stablishing the specifications;
- Regulating the supply chain;
- Acting conjointly to mitigate and identify problems;
- Monitoring the quality: evaluating the profile of non-conformities of the ethanol that is traded throughout the national territory;
- Working in groups: participating with the GTs' Government in order to reduce technical barriers, as well as taking part in international forums that talk about products' specifications and sustainability;
- International standardizing: accompanying the actions in ASTM's Committee and in the Technical ISO Committees of biofuels, waterway fuels, and sustainability.
- Controlling the quality: increasing the trustworthiness and traceability of the quality analysis of biofuels.

Considering that the proposals for bioethanol's specifications are elaborated by technicians specialized in biofuels, it is possible to notice that they are stablished by the regulatory organ (ANP). ANP works based on the needs and discussions, always looking for a consensus among the agents that are connected to the production and use of bioethanol, and the environmental organs, in order to serve to the consumer's longings. These specifications are periodically reviewed in order to serve, also, to the new engine technologies, the environmental demands, and the need to supply products for society (Bonomi, 2010).

As an example of the continuous refinement of the specifications, there is ANP nº 19, published in 4/16/2015 and currently applied, which follows the Technical Regulation nº2/2015. Its main quality parameters, specifications, and analysis methods are present in Table 2.

Besides the national specifications, other countries also present their own specifications, and it is necessary to answer to them too, in order to trade the bioethanol with these countries. With the objective of facilitating the commerce between countries, a Work Group was created in 2007, composed by specialists from Brazil, the European Community, and the United States, in order to harmonize the bioethanol's specifications (Bonomi, 2010). The specialists from these standardization organizations critically compare bioethanol's specifications, such as content, physical characteristics, and contamination levels, aspects that determine its quality (Flumignan et al. 2012).

The White Paper on Internationally Compatible Biofuels Standard, document elaborated by the trio Brazil, European Community, and United States, has identified that the main specifications for bioethanol that are present in the regulations can be classified in three categories (Table 1): (A) similar characteristics, (B) characteristics with significative differences, and (C) characteristics with fundamental differences (Flumignan et al. 2012). Table 1 brings the classification of the differences in specifications, for ethanol as a fuel, through the characterization of the limit values specified in the resolutions and the classifications of the instrumental analysis techniques.

Below, there are tables with the specifications that are adopted for national ethanol, through Resolution ANP nº19/2015 (ANP, 2015), both for anhydrous fuel ethanol (EAC) and hydrated ethanol (EHC) (Table 2), and the comparison between the specifications that are adopted in Brazil and the ones adopted by the European Community (EN, 2011) and by the United States (ASTM, 2017) (Table 3).

Table 1. Specification differences for Fuel Ethanol

Category (A)	**Category (B)**	**Category (C)**
Aspect	Ethanol Content	Water Content
Color	Total Acidity	
Specific Weight	pH	
Sulfate Content	Evaporation residue	
Sulphur Content		
Copper Content		
Iron Content		
Sodium Content		

Source: (White Paper, 2007, apud Bonomi, 2010, p. 804).

Table 2. Specifications for Anhydrous Ethanol (EAC), Hydrated Fuel Ethanol (EHC) and Premium[1] Hydrated Fuel Ethanol

Characteristic	Unit	Limit			Method	
		EAC	EHC	EHCP[2]	NBR	ASTM
Aspect [3]	-	Clean and impurity-free (LII)			Visual	
Color	-	4	5		Visual	
Total acidity, max. (in milligrams of acetic acid)	mg/L	30			9866 16047	ISO 17315
Conductivity, max. [6]	μS/m	350			10547	ISO 17308
Specific weight at 20°C.[7]	Kg/m³	791.5 max.	805.2 to 811.2	799.7 to 802.8	5992 and 15639	D4052
Alcohol contents [7,8,9]	% weight	99.3 min.	92.5 to 94.6	95.5 to 96.5	5992 15639	-
Hydrogenionic potential (pH)	-	-	6.0 to 8.0		10891	-
Ethanol content, min.[10]	% volume	98.0	94.5	96.3	16041	D5501
Water content, max.[10]	% weight	0.7	7.5	4.5	15531 and 15888	E203 and E1064
Methanol content, max.[10]	% volume	0.5			16041	-
Evaporation residue, max.[12,13]	mg/ 100 ml	5			8644	-
Hydrocarbon content, max.[12]	% volume	3			13993	-
Chloride content, max.[14]	mg/kg	1			10894	D7328 D7319
Sulfate content, max.[15]	mg/kg	4			10894	D7328 D7319
Iron content, max.[15]	mg/kg	5			11331	-
Sodium content, max.[15]	mg/kg	2			10422	-
Copper content, max.[16]	mg/kg	0.07			11331	-
Sulphur content, máx.[17,18]	mg/kg	report			-	D5453 EN15485 EN15486 EN15837

Source: (ANP, 2015).

Notes:

[1] ANP might add additional characteristics and complementary methods, and/or impose new limits to the specifications that are disposed in the Table of the Technical Regulation for the ethanol that is produced from methods or processes that are distinct from the fermentative route, which used the sugarcane broth and/or molasses as feedstock.

[2] On the Quality Certificate, there must be the nomenclature Premium Hydrated Fuel Ethanol, when the results of the analyses are according to the specification limits from EHCP.

[3] The product will only be able to be considered reproved in Aspect if the parameter "waste through evaporation" is not in accordance. Alternatively, the parameter "evaporation residue" can be substituted by the non-volatile material content, according to the norm ABNT NBR 15559: Fuel Ethanol – Determination of the non-volatile material content through evaporation. The acceptable limit is 5mg/100mL.

[4] Orange, after the addition of the pigment specified, according to the Technical Regulation.

[5] It cannot contain the colors orange and blue, which are restricted to EAC and to the aviation gasoline, respectively.

[6] The limit of 300 μS/m have come into operation starting on July 1st, 2015. Until June 30th, 2015, the maximum value was 389 μS/m.

[7] For the case of EHC and EHCP certified after pipeline or waterway transport, and in the case of imported product, the limit for the specific weight and for the alcohol content must be inside the specified values for the distributor, which are constant in the Technical Regulation.

[8] The unit °INPM is equivalent to the unit % weight for the alcohol content.

[9] For calculating the alcohol content, the specific weight value must be considered with only one decimal point. The rounding off that is determined in norm NBR 5891 should be applied in this case.

[10] Mandatory analyses for imported product, as well as in case of doubt about the possibility of contamination, or by request of ANP.

[11] The norm ASTM E1064 will only be able to be used for the Anhydrous Fuel Ethanol.

[12.] Mandatory Analysis for imported product ad, in the case of Fuel Ethanol that comes from pipeline or waterway transport, what does not exempt the economic agent from the responsibility of serving the limit that is predicted in the specification for the cases in which the ethanol is not transported in these ways.

[13] In the case of the Additive Fuel Ethanol, the determination of the characteristic evaporation residue might be conducted according to the procedure 2 in the Norm ABNT NBR 8644.

[14] Mandatory analyses for imported product and in the case of Fuel Ethanol that comes from pipeline or waterway transport, what does not exempt the economic agents that trade the fuel from serving the limit predicted in the specification throughout the chain.

[15] Mandatory analyses for imported product. In the case of national production, monthly analyses are mandatory only for EHC and EHCP, what does not exempt the Fuel Ethanol Supplier and the other chain agents from the responsibility of serving the limit predicted in the specification for the Anhydrous Fuel Ethanol.

[16] Mandatory analyses for imported product. In the case of national production, monthly analyses are mandatory only when the ethanol is produced, stored or transported in equipment or lines that contain metallic alloys made up by copper.

[17] Mandatory analyses for imported product. In the case of national production, trimestral analyses.

[18] The analyses of the Sulphur content will be mandatory, starting on July 1st, 2015.

Table 3. Brazilian, European, and American Specifications for Anhydrous Fuel Ethanol

Characteristic	Resolution ANP N°19*	BS EN 15376	ASTM D4806**
Aspect	Clear and impurity free	Clear and light	Free of suspension-material (clear and light)
Color	Colorless (before adding colorant)	-	-
Total acidity - max.	30 mg/L	0.007% m	56 mg/L
Electric conductivity – max.	300 μS/m	2.5 μS/m	-
Non-volatile material – max.	-	10 mg/100 mL	-
Specific weight at 20°C – max.	791.5 kg/m³	-	-
Alcohol content – min.	99.3 % m		
Ethanol content – min.	98.0% v	98.7% m	92.1% v
Monoalcohol content (C3-C5) – max.	-	2% m	-
Methanol content – max.	0.5% v	1% m	0.5% v
Hydrocarbon content – max.	3% v	-	-
Water content – max.	0.7% v	0.3% m	1% v
Copper content – max.	0.07 mg/kg	0.1 mg/kg	0.1 mg/kg
Sulphur content – max.	report	10 mg/kg	30 mg/kg
Sulfate content – max.	4 mg/kg	3 mg/kg	4 mg/kg
Phosphorus content – max.	-	0.15 mg/L	-
Inorganic chloride content – max.	1 mg/kg	1.5 mg/kg	6.7 mg/kg
Washed gum – max.	-	-	5 mg/100ml
pH	-	-	6.5 to 9.0

Source: own confection.

* Considered specifications for the Denatured Fuel Ethanol.

** California Air Resources Board has approved proper specification and standards for denatured ethanol. The specifications set maximum limits on sulfur, benzene, olefin, and aromatic contents.

2.3. The Brazilian Technical Norms for Bioethanol's Specification

A norm is the document established by a consensus and approved by an acknowledged organ, which provides rules, directives, or minimum characteristics for activities or for one of their results, trying to obtain a great coordination degree in a given context (ABNT, 2012).

When considering exportation and importation, each time there is more exigence of demonstrating conformity of products and their processes to technical requisitions that derive from norms - mainly international ones - and technical regulations. The Brazilian Technical Standards Association – ABNT – is the official organ for emitting the technical norms in Brazil. All the official entities in the world dialogue through the International Organization for Standardization – ISO. In our country, ABNT has been chosen as secretary in the ISO Normalization Commission for biofuels (White Paper, 2007, apud Bonomi, 2010).

The technical norms have their applicability strongly connected to the current technology and to the context of the regulations. Thus they must be updated in order to be compatible, or even to be replaced, with the most current technological context. Therefore, the official standardization organs have technical commissions to keep the set of norms as updated and syntonized as possible with technology, standardizations, and current national regulations.

In this chapter, although there are mentions to technical norms for the characteristics of ethanol and its mixtures (regarding quality), distribution logistics, superior alcohols, industrial-use ethanol, as well as norms for their use in engines and for the fuel alcohol production, the highlight will be the technical norms which refer to quality.

Table 4 brings the technical norms currently working in Brazil when it comes bioethanol quality. The information in the table has fundamental importance for visualizing the historic actuation, in the development of technical norms of the Fuel Alcohol Special Study Commission (CEE-00.001.61) from the Petroleum Sectorial Standardization Organism from the Brazilian Technical Standards Association (ABNT/ONS 34 – Petroleum).

2.3.1. Alternative Analysis Methods

Besides the official analysis methods, there are several alternative methods. Considering the latter, it is possible to highlight the predominance of instrumental methods about the classics (gravimetric and volumetric). Among the instrumental methods to analyze bioethanol, there are three kinds: chromatographic, spectroscopic, and electrochemical. The official analysis methods are recommended by the organs ANP, ASTM, and CEN.

Table 4. Brazilian Technical Norms for Bioethanol Quality

Fuel Alcohol Special Study Commission (CEE-00.001.61) From The Brazilian Technical Standards Association (ABNT/ONS 34 - PETROLEUM)					
Norms	Project	Origin	Title	Scope	Publishing Date
NBR 5992:2016	-	NBR 5992:1980	Determination of the specific weight and the alcohol content - Glass Densimeter Method	This Norm specifies the glass densimeter method for the determination of the specific weight and the alcohol content in ethylic alcohol and in its mixtures with water.	11/25//2016
NBR 8644:2014	-	NBR 8644:2008	Determination of the evaporation residue content	This Norm specifies the method to determine the presence of non-volatile residues (non-washed and washed) in fuel ethanol.	3/28/2014
NBR 9866:2012	-	NBR 9866:2006	Total acidity determination through colorimetric titration	This Norm specifies a method for determining the total acidity in fuel ethanol, expressed in acetic acid weight, in milligrams (mg) by ethanol volume, in liters (L), through volumetric colorimetric titration.	4/5/2012
NBR 10422:2013	-	NBR 10422:2007	Determination of the sodium concentration – flame photometry method	This Norm specifies the flame photometry method to determine the sodium concentration in fuel ethanol.	7/17/2013
NBR 10547:2016	-	NBR 10547:2006	Electric conductivity determination	This Norm specifies a method to determine the electric conductivity in anhydrous and hydrated fuel ethanol that contain up to 3% volume/volume (%v/v) of hydrocarbons.	2/29/2016
NBR 10891:2013	-	NBR 10891:2006	Hydrated fuel ethanol – pH determination – potentiometric method	This Norm specifies the potentiometric method to determine the pH in hydrated fuel ethanol, containing up to 3% volume/volume (%v/v) of hydrocarbons.	3/11/2013

Table 4. (Continued)

Fuel Alcohol Special Study Commission (CEE-00.001.61) From The Brazilian Technical Standards Association (ABNT/ONS 34 - PETROLEUM)					
Norms	Project	Origin	Title	Scope	Publishing Date
NBR 10894:2012	-	NBR 10894:2007	Determination of the chloride and sulfate concentration – ion chromatography method	This Norm describes the method to determine the sulfate and chloride concentration in fuel ethanol through the direct injection in a chromatographer of ions with conductivity detector.	10/9/2012
NBR 11331:2007	-	NBR 11331:1990	Determination of the iron and copper concentration – Atomic Absorption Spectrophotometry Method	This Norm specifies the atomic absorption spectrophotometry method to determine the iron and the copper concentration in ethylic alcohol.	8/27/2007
NBR 13993:2013	-	NBR 13993:2002	Fuel ethanol – hydrocarbon content determination – volumetric method	This Norm specifies the method to determine the hydrocarbon content in anhydrous fuel ethanol and hydrated fuel ethanol.	9/17/2013
NBR 15531:2016	-	NBR 15531:2007	Water content determination – Karl Fischer's volumetric method	This Norm specifies the Karl Fisher's method to determine the water content in ethylic alcohol through volumetry.	7/14/2016
NBR 15559:2014	-	NBR 15559:2008	Determination of the non-volatile material content through evaporation	This Norm specifies the method to determine the non-volatile material content through evaporation in fuel ethanol.	2/19/2014

Fuel Alcohol Special Study Commission (CEE-00.001.61) From The Brazilian Technical Standards Association (ABNT/ONS 34 - PETROLEUM)					
Norms	Project	Origin	Title	Scope	Publishing Date
NBR 15639:2016	-	NBR 15639:2008	Alcohol Content Determination – Electronic Densimetry Method	This Norm specifies the electronic densimetry method to determine the specific weight and the alcohol content of the ethylic alcohol and its mixtures with water.	11/25/2016
NBR 15888:2016	-	NBR 15888:2010	Water Content Determination – Karl Fisher's Coulometric Method	This Norm specifies Karl Fisher's method to determine the water content in ethanol, with maximum content of 2% m/m for anhydrous ethanol, and 8% m/m for the coulometry hydrated ethanol.	6/14/2016
NBR 16041:2015	-	NBR 16041:2012	Methanol and ethanol content determination through gaseous chromatography	This Norm is applied for the quantitative determination of methanol and/or ethanol and fuel ethanol through gaseous chromatography. The presence of hydrocarbons and/or superior acids does not interfere in the determination of methanol and ethanol. This method does not intend to identify or quantify these compounds. This method may be applied for the determination of methanol and ethanol in gasoline.	6/15/2015
NBR 16047:2012	-	-	Acidity determination through potentiometric titration	This Norm specifies a method to determine the total acidity in the ethanol, expressed in acetic acid weight, in milligrams (mg), by ethanol volume, in liters (L), through volumetric titration. If the ethanol's specific weight is known, the total acidity may be expressed in percentage of acetic acid weight (%m/m).	4/5/2012
NBR 16549:2016	-	-	Sulfur concentration determination	This Norm specifies the method to determine the sulfur concentration in fuel ethanol through X-Ray Fluorescence Spectrometry by wave length dispersion.	10/10/2016

The instrumental methods to analyze fuel ethanol have been presenting some changes since the first regulatory norms for checking the quality. This is mainly due to the metrologic evolution, the evolution conducted in the used equipment itself, and the rising of new technical parameters that did not have defined specifications before. The changes may have been motivated by environmental factors, emission of pollutants or chemical species that interfere with the environmental balance or with technical-economic factors, because of the engine evolution or for being oriented by the need of analysis speed, sensitivity, and exactness.

In order to illustrate the instrumental evolution, there are, as examples: the sulfur determination and the chloride determination. The first one was initially made through the gravimetric method, but, later, it started being made through the volumetric method and, further, through ionic chromatography. On the other hand, the chloride determination went from the volumetric and potentiometric chromatographies to the ion chromatography. In order to illustrate the evolution within the methods themselves, it is possible to mention the copper and iron determinations, which went from spectrophotometry in the UV-vis regions to atomic absorption spectrophotometry. And to illustrate the rising of new determinations, there is the hydrocarbon determination in fuel ethanol for importation, distribution, and resale, made through gaseous chromatography (Stradiotto et al. 2010).

2.3.1.1. Separation and Identification Methods: Chromatography and Electrophoresis

The chromatographic methods to analyze bioethanol are, in their most, made to determine the organic species in fuel ethanol, such as alcohols, aldehydes, and ketones. The determination of these chemical species is related to the burnt efficiency, the fuel toxicity and the possible emission of noxious chemical species. Among the alcohols, the main determination is the methanol content, for which all the countries that sell this product stipulate a maximum value. Among the methods, the ones that use gaseous phase chromatography (GC) with detection of flame ionization are the most used ones. The other methods that are used in the determination of organic species are the High-Performance Liquid Chromatography (HPLC) with spectroscopy (UV/Vis), and the electrochemical detectors. The inorganic compounds are usually determined through HPLC or capillary electrophoresis (CE). These compounds include mainly the sulfate and the sulfur, which are metallic elements. They are important because of their relation with the corrosion of the engine's internal metallic parts or even when it comes to the exhaustion, like in the case of sulfur. In Table 5 are presented several chromatographic methods to analyze the fuel ethanol found in the literature.

Table 5. Chromatographic Methods found in the literature

Developed by	Technique	Species
Vilar et al. 2003	Gaseous chromatography coupled with mass spectroscopy	Organic compounds (hydrocarbons, aldehydes, ketones, and alcohols)
Pereira et al. 2006	Capillary electrophoresis technique	Organic and inorganic contaminants (chloride, sulfate, and aldehyde anions)
Munoz et al. 2004	Capillary electrophoresis technique	Inorganic cations (sodium, potassium, calcium, magnesium, and ammonium) and anions (sulfate, chloride, and nitrate)
Saczk et al. 2005	High Performance Liquid Chromatography	Acetaldehyde
Sackz et al. 2006	High Performance Liquid Chromatography with electrochemical detection	Organic compounds (aldehydes and ketones)
Application Note Dionex, s/d	Ion chromatography	Sulfate and chloride

Source: Adapted from (Stradiotto, 2010, p. 814-16).

2.3.1.2. Spectroscopic Methods

The spectroscopic methods to analyze bioethanol are used to determine inorganic species, especially the metallic ions, such as copper, iron, and sodium. Among the methods, it is possible to highlight the ones that use spectroscopy of atomic absorption with flame and with graphite oven. The methods of atomic emissions through induced plasma with weight detectors and the methods involving spectroscopy in the UV-Vis and infrared regions are also important to mention. The nuclear magnetic resonance spectroscopy (NMR) has also been reported for the detection of bioethanol adulteration with methanol. In Table 6 are presented some spectroscopic methods used to analyze fuel ethanol found in the literature.

Table 6. Spectroscopic methods found in the literature

Developed by	Technique	Species
Saint'pierre et al. 2006	Plasma spectroscopy induced with weight detectors	Metals
Virgilio et al. 2015	Inductively coupled plasma tandem mass spectrometry	As, Cr, P, Pb, Si, and V simultaneously
Amais et al. s/d	Spectrophotometry UV-Vis	Fe
Oliveira; Korn, 2006	Spectrophotometry	Sulfate
Saint'Pierre et al. 2008	Atomic absorption	Cd and Pb
Neto et al. 2014	^{1}H NMR	Methanol
Milanez et al. 2016	Near infrared Spectroscopy	Methanol and ethanol

Source: own confection.

2.3.1.3. Electrochemical Methods

The electrochemical methods to analyze bioethanol aim, mainly, to determine inorganic species, especially the metallic ions, such as copper, lead, and zinc, although there are also works towards determining methanol, acetaldehyde, and other organic compounds that involve electrochemical methods. Among the methods, the ones that use anodic redissolution voltammetry with differential pulse and square-wave modalities are predominant, as well as the potentiometric and conductometric methods for non-metallic species, such as chloride and sulfate (Stradiotto et al. 2010). In Table 7 are presented some electrochemical methods to analyze fuel ethanol found in the literature.

Table 7. Electrochemical Methods found in the literature

Developed by	Technique	Species
Godinho et al. 1988	Cathodic redissolution voltammetry with the modality of differential pulse	Total sulfur
Avelar et al. 2007	Conductometric titration	Chloride and total acidity
Oliveira et al. 2004	Anodic redissolution voltammetry, with the modalities of linear sweeping, differential pulse, and square wave	Zn, Cu, Pb and Cd, simultaneously
Rodgher et al. 2006	Square wave voltammetry	Acetaldehyde
Romanini et al. 2009	Linear sweeping and square wave voltammetry	Orange 7 colorant (used as a marker)
Pereira et al. 2013	Cyclic voltammetry	Methanol and ethanol
Abreu et al. 2015	Cyclic voltammetry and multivariate calibration	Methanol

Source: Adapted from (Stradiotto, 2010, p. 819-20).

2.4. Certified Reference Materials (CRMs) for Bioethanol

A reference material (RM) consists in a material or substance that has one or more property values that are sufficiently homogeneous and well stablished in order to be used to calibrate an appliance, to evaluate a measurement method, or to attribute values to materials. A reference material may be a pure substance or a mixture, in the form of gas, liquid, or solid. This material may come along with a certificate (CRM), which is a document emitted by a competent organ that provides one or more specific property values for which the associated uncertainties and the tracking are registered (ABNT, 2012).

The need and the importance of controlling the quality in the chemical measurements, decreasing the costs, and avoiding the analysis duplication give the use of Certified Reference Materials a growing merit, since they are tracked by national and international references and employed in validating and controlling the methods' quality and in calibrating the analytical instruments. Thus, they have national and international relevance in standardization activities, in commercial trades, in proficiency tests, and in labs' credibility (Bonomi, 2010).

However, there is a big variety of chemical substances and different samples that need CRMs. Besides, the high acquisition and importation cost of these materials, the bureaucracy, and the long time spent in transportation and liberation of imported materials by the competent authorities limit the use of CRMs. Thus, the existing number of CRMs is not enough for all the measurements (FRAGA et al. 2009, apud Stradiotto et al. 2010),

Inmetro has been developing these materials so they can be used by several labs in Brazil that act in different areas. The high metrologic level of the measurements made by Inmetro labs comes, mainly, from its participation in key-comparisons coordinated by the International Weights and Measures Committee – CIPM/CCQM. The materials that are prepared by Inmetro follow the requisites stablished by the ISO Guides 30, 31, 32, 34, translated to Portuguese by ABNT, and ISO Guide 35 (Rodrigues et al. 2012).

The development of CRMs for ethanol has come from the need that ethanol producers had to guarantee its quality and to increment the exportation of its products, since the parameters could be measured with trustworthiness and traceability and, thus, accepted by the international commerce. Since 2003, Inmetro has been studying the several parameters related to bioethanol's quality stablished by ANP, which regulate the limits of several parameters that are needed to specify the biofuels according to the Brazilian regulation.

Inmetro has available a reference material certification service. Specifically for bioethanol, there are some CRMs catalogued by Inmetro, however they are unavailable in the moment (2017):

- 8313 – Fuel Ethanol CRM – Total acidity, specific weight, and pH;
- 8314 – Fuel Ethanol CRM – Water content and ethanol content;
- 8648 – Hydrated Ethylic Fuel Alcohol CRM – Total acidity, pH, electrolytical conductivity, specific weight, sulfate, water content, ethanol, and sodium content;
- 8857 – Anhydrous Ethylic Fuel Alcohol CRM – Specific weight and alcohol content;
- 8858 – Hydrated Ethylic Fuel Alcohol CRM – Specific weight and alcohol content;
- 8132 – Hydrated Ethylic Fuel Alcohol CRM – Water content;
- 8315 – Fuel Ethanol CRM – Iron, copper, and sodium.

The parameters that are studied by Inmetro for ethanol's certification and the reasons for their relevance are (Stradiotto et al. 2010):

- Ethanol content and density: important for the product's stability and quality, since the original ethanol content and the evaporation may affect the engine's calorific power and the performance.

- Acidity: in order to avoid the corrosion of the fuel production chain, transport, and use (engines);
- Conductivity: conductivity levels above 350μS/m contribute to increase the corrosion (in 2011, through ANP n. 7, the maximum limit was altered to 350μS/m);
- Water content: the excess of water may increase the conductivity and, thus, the corrosion, besides decreasing the calorific power;
- pH: between 6-8 for hydrated ethanol; its determination also intends to avoid the corrosion of the fuel production chain, transport, and use (engines);
- Copper and iron: these metals, when in contact with gasoline (20% of ethanol, ANP n.40/2013), may cause polymerization (gum, probably) in automotive engines. The ASTM Norm recommends levels below 0.1 mg/kg for the copper content – in Brazil, this level is 0,07 mg/kg tops.
- Chlorides and sulfates: the presence of chlorides (external contamination) and sulfates (deposits in electronic injection systems) may also contribute to increase the corrosion.

Since 2008, Brazil and the United States have stablished a partnership, through National Metrology Institutes – Inmetro and NIST (National Institute of Standards and Technology), for developing a CRM for ethanol, given its visibility and because of environmental questions, since both countries are the biggest biofuel producers in the world. On the other hand, Brazil and the European Union work on the relevant task of comparing and researching analytical methods for the biofuels in order to harmonize them. The main objective of this partnership is to coordinate a proficiency test – PT – with the finality of measuring the different bioethanol parameters. This group will have the participation of countries that are interested in implanting the quality management system in order to trade their products (Rodrigues et al. 2012).

CONCLUSION

Even though Brazil is the biggest ethanol exporter, the access to the international market is still a big obstacle for its commercialization. For that, the Brazilian specifications must be aligned to the specifications from the European Community and the United States. The technical barriers are being discussed by the Study Group, aiming at eliminating them.

The Brazilian standards, despite being constantly reviewed and replaced, not always present test methods that are similar or equivalent to the ones prescribes by the main countries that may represent big commercial partners. They present simple and low-cost methods, which facilitate lab operations, but do not considered more sophisticated methodologies, what restricts their being accepted by importing countries. Besides, there are other gaps that make it difficult to trade ethanol worldwide, but one to be highlighted is still the low availability of specialists acting in the technical commissions.

The certified reference materials may be prepared by harmonizing the specifications and stimulating an effort so that the Brazilian standards present methods that are comparable to the international ones, i.e., consolidating the specifications and standardizations.

The development and production of CRMs shall facilitate the ethanol input in the foreign market, since the Brazilian producers, in order to prove the exactitude of the parameters that are measured in the international commerce, need an RM that may be used as a reference standard for chemical measurements. Ethanol's CRM, which has been being developed in partnership with the United States (Inmetro and NIST, respectively), shall guarantee the quality and the tracking of the parameter measurement results, collaborating with a fairer and more trustworthy commerce.

The quality control should not be considered an expensive system, since, once it is implemented, it leads to increases in the product's trustworthiness, decreases in the non-conformity levels, and the softening of the technical barriers that make it difficult to trade ethanol outside the country.

REFERENCES

Agência Nacional Do Petróleo, Gás Natural E Biocombustíveis. *Resolução ANP nº 07.* Brasília, 2011. [Available on: <http://nxt.anp.gov.br/NXT/gateway.dll/leg/resoluсoes_anp/2011/fevereiro/ranp%207%20-%202011.xml?f=templates$fn=document-frame.htm$3.0$q=$x=$nc=735>. Accessed on: June 24th 2014. [National Agency Of Petroleum, Natural Gas And Biofuels. *ANP Resolution No. 07.* Brasília, 2011.]

Araujo, Rosangela Moreira de. NAP actions on Biofuels Quality. *4th National Biofuels Symposium* – May, 19th-20th, 2011

Associação Brasileira De Normas Técnicas. Available in < http://www.abntcatalogo.com.br/>. Accessed on June 27th 2014. [Brazilian Association of Technical Standards.]

Bastos, Valéria Delgado. *Etanol, alcoolquímica e biorrefinarias*. BNDES setorial, v. 25, p. 5-38, 2007. Available in: <http://www.bndes.gov.br/SiteBNDES/export/sites/default/bndes_pt/Galerias/Arquivos/conhecimento/bnset/set2501.pdf>. Accessed on: June 21th 2014. . [*Ethanol, alcohol chemistry and biorefineries*. Sectoral BNDES]

Bonomi, Antônio. Projeto de Políticas Públicas para o Etanol. In: Cortez, Luís Augusto Barbosa. *Bioetanol de cana-de-açúcar: P&D para produtividade e sustentabilidade.* Blucher, 2010, p. 801-811. [Public Policy Project for Ethanol. In: Cortez, Luís

Augusto Barbosa. *Bioethanol from sugarcane: R & D for productivity and sustainability.*]

Departamento Nacional De Combustíveis. Portaria DNC nº 23. Brasília, 1991. Available in: <http://nxt.anp.gov.br/nxt/gateway.dll/leg/folder_portarias/portarias_dnc/ 1991/ pdnc%2023%20-%201991.xml>. Accessed on June 29th 2014. [National Department of Fuels. DNC nº 23. Brasília, 1991.]

Flumignan, D. L.; Oliveira, J. E. De; Hatanaka, R. R.; Sequinel, R. Avanços brasileiros no desenvolvimento de normas técnicas analíticas para certificação e controle da qualidade de biodiesel. In: Stradiotto, N. R.; Lemos, E. G. S. DE M. Pesquisa e Desenvolvimento em Bioenergia. São Paulo: Editora Unesp, 2012. cap. 5. [*Brazilian advances in the development of technical analytical standards for certification and quality control of biodiesel.* In: Stradiotto, N.R.; Lemos, E. G. S. DE M. *Research and Development in Bioenergy*. São Paulo: Unesp Publishing, 2012. cap. 5]

Instituto Nacional De Metrologia, Qualidade E Tecnologia. Available in: <http://www.inmetro.gov.br>. Accessed on June 1st 2014. [National Institute of Metrology, Quality and Technology.]

Milanez, Artur Yabe; Faveret Filho, Paulo de Sá Campello; Da Rosa, Sergio Eduardo Silveira. Perspectivas para o etanol brasileiro. BNDES Setorial, v. 27, p. 21-38, 2008. Available in: <http://funcex.org.br/material/redemercosul_bibliografia/biblioteca/ ESTUDOS_BRASIL/BRA_161.pdf> Accessed on June 20th 2014. [*Prospects for Brazilian ethanol.* BNDES Sector]

NBR 10422: etanol combustível: determinação da concentração de sódio: método da fotometria de chama. Rio de Janeiro, 2013. [*NBR 10422: fuel ethanol: determination of sodium concentration: flame photometry method.*]

NBR 10547: etanol combustível: determinação da condutividade elétrica. Rio de Janeiro, 2012. [*NBR 10547: fuel ethanol: determination of electric condutivity*]

NBR 10891: etanol hidratado combustível: determinação do pH: método potenciométrico. Rio de Janeiro, 2013. [*NBR 10891: hydrated ethanol fuel: pH determination: potentiometric method.*]

NBR 10894: etanol combustível: determinação da concentração de cloreto e sulfato: método da cromatografia de íons. Rio de Janeiro, 2012. [*NBR 10894: fuel ethanol: chloride and sulphate concentration determination: ion chromatography method.*]

NBR 11331: álcool etílico: determinação da concentração de ferro e cobre: método da espectrofotometria de absorção atômica. Rio de Janeiro, 2007. [*NBR 11331: ethyl alcohol: determination of iron and copper concentration: atomic absorption spectrophotometry method.*]

NBR 13993: etanol combustível: determinação do teor de hidrocarbonetos: método volumétrico. Rio de Janeiro, 2013. [*NBR 13993: fuel ethanol: determination of the hydrocarbon content: volumetric method.*]

NBR 15531: etanol combustível: determinação do teor de água: método volumétrico de Karl Fischer. Rio de Janeiro, 2013. [*NBR 15531:fuel ethanol: determination of water content: Karl Fischer volumetric method.*]

NBR 15559: etanol combustível: determinação do teor de material não volátil por evaporação. Rio de Janeiro, 2014. [*NBR 15559: fuel ethanol: determination of the non-volatile content by evaporation.*]

NBR 15639: álcool etílico e suas misturas com água: determinação da massa especifica e do teor alcoólico: método da densimetria eletrônica. Rio de Janeiro, 2008. [*NBR 15639: ethyl alcohol and its mixtures with water: determination of specific mass and alcoholic strength: electronic densimetry method.*]

NBR 15888: etanol: determinação do teor de água: método coulométrico de Karl Fischer. Rio de Janeiro, 2010. [*NBR 15888: ethanol: determination of water content: coulometric method of Karl Fischer*.]

NBR 16041: etanol combustível: determinação dos teores de metanol e etanol por cromatografia gasosa. Rio de Janeiro, 2012. [*NBR 16041: fuel ethanol: determination of the methanol and ethanol contents by gas chromatography*.]

NBR 16047: etanol combustível: determinação de acidez total por titulação potenciométrica. Rio de Janeiro, 2012. [*NBR 16047: fuel ethanol: determination of total acidity by potentiometric titration*.]

NBR 5992: álcool etílico e suas misturas com água: determinação da massa específica e do teor alcoólico: método do densímetro de vidro. Rio de Janeiro, 2008. [*NBR 5992: ethyl alcohol and mixtures thereof with water: determination of specific mass and alcoholic strength: glass densimeter method*]

NBR 8644: etanol combustível: determinação do teor de resíduo por evaporação. Rio de Janeiro, 2014. [*BR 8644: fuel ethanol: determination of the residue content by evaporation*.]

NBR 9886: etanol combustível: determinação da acidez total por titulação colorimétrica. Rio de Janeiro, 2012. [*BR 9886: fuel ethanol: determination of total acidity by colorimetric titration.*]

NBR ISSO 9000: sistemas de gestão da qualidade: fundamentos e vocabulário. Rio de Janeiro, 2005. [*NBR ISO 9000: quality management systems: fundamentals and vocabulary.*]

Stradiotto, Nelson Ramos; Zanoni, Maria Valnice Boldrin; Fraga, Isabel Cristina Serta; Borges, Paulo Paschoal. Qualidade do Etanol Combustível. In: Cortez, Luís Augusto Barbosa. *Bioetanol de cana-de-açúcar: P&D para produtividade e sustentabilidade*. Blucher, 2010, p. 813-826. [Quality of Fuel Ethanol. In: Cortez, Luís Augusto Barbosa. *Bioethanol from sugarcane: R & D for productivity and sustainability. Blucher*, 2010, p. 813-826.]

Toledo, J. C.; Borrás, M. A. A.; Mergulhão, R. C.; Mendes, G. H. S. *Qualidade: gestão e métodos*. Rio de Janeiro: LTC, 2013. [*Quality: management and methods*.]

In: Bioethanol and Beyond
Editor: Michel Brienzo

ISBN: 978-1-53613-478-0

Chapter 6

Ozonolysis Applied to Lignocellulosic Biomass

Olavo Micali Perrone*[*], *Sidnei Emilio Bordignon*, *Mauricio Boscolo*, *Eleni Gomes and Roberto da Silva
São Paulo State University (UNESP), Institute of Biosciences, Letters and Exact Sciences (IBILCE), São José do Rio Preto, São Paulo, Brazil

Abstract

Lignocellulosic biomass produced by several agroindustrial processes has become the most important source of green carbon able to integrate the development of a mode of production based on the biorefinery concept. The elevated recalcitrance of lignocelluloses demands intensive procedures to deconstruct their polymers and to obtain sugars as the substrate for biochemical conversion using enzymes and microorganisms. Second-generation ethanol is the most studied and advanced product from these biomasses and pretreatment is considered a mandatory step in the development of this technology. Ozonolysis is a chemical pretreatment which has attracted attention due to the high specificity in the delignification and the mild operating conditions. By removing lignin, ozone contributes to higher enzymatic yields in the hydrolysis step. This chapter presents a general contextualization of biomass pretreatment for cellulosic ethanol production focused on ozonolysis, the scientific gains achieved, the chemical reactions involved and the combination of ozonolysis with other pretreatment techniques aimed at reducing costs and increasing yields.

Keywords: ozonolysis, lignin, bioethanol, cellulose, enzyme, pretreatment, hydrolysis, fermentation

[*] Corresponding Author Email: olavo_perrone@hotmail.com.

1. Introduction

In recent years, the number of studies associated with the synthesis of fuels and chemicals from renewable resources has been increasing as a strategy to partially or totally replace compounds obtained from fossil sources. Main reasons for this trend include the need for reducing greenhouse gas emissions, new green energy policies and guaranteeing future energy sources. Interest in the establishment of a bio-based carbon economy (bioeconomy) surges as it is a potentially integrative model of sustainable production and valorization of renewable materials to obtain bioproducts and biomaterials through the biorefinery concept.

A biorefinery consists in the fractionation of biomass from agricultural residues (e.g., sugarcane, corn stalk, corn stover and wheat) and their conversion into energy, fuels (e.g., bioethanol), animal feed, food and/or chemicals. It expects to make full use of the biomass and the energy contained in those residues. These lignocellulosic biomasses are cheap, available in large quantities and their energy sources are considered cleaner-burning due do them not increasing the concentration of CO_2 in the atmosphere. Nowadays, ethanol represents the most important fuel obtained from lignocellulosic material worldwide.

Since the 70s, ethanol has been produced on a large scale from the alcoholic fermentation of sucrose sugars (glucose and fructose) by the yeast *Saccharomyces cerevisiae* ($C_6H_{12}O_6 \rightarrow 2\ C_2H_5OH + 2\ CO_2$), and widely marketed as a liquid fuel or as a gasoline additive in countries such as Brazil and the United States of America. This tendency is becoming global due to the increase of new green energy policies that will further increase the demand for ethanol (Perrone et al. 2016). In this sense, the use of lignocellulosic material for cellulosic ethanol production can increase the total ethanol production without requiring additional plantation areas.

Different biomasses have been extensively studied since the great potential of sugars from these agricultural leftovers can be fermented and converted into new compounds with added-value, such as cellulosic ethanol, was discovered (Perrone et al. 2017). Therefore, a wide variety of feedstocks can be used for the production of cellulosic ethanol, for example, corn stover (Li et al. 2015), sugarcane straw (Barros et al. 2013), sugarcane bagasse (Perrone et al. 2016; Barros et al. 2013), Japanese cedar sawdust (Miura et al. 2012; Sugimoto et al. 2009), corn straw (Shi et al. 2015), wheat straw (Baig et al. 2015), rye straw (García-Cubero et al. 2012), waste paper (Kojima and Yoon 2008), and cotton stalks (Kaur et al. 2012).

Generally, the lignocellulosic material is comprised of around 40% cellulose, 25% hemicellulose, 25% lignin, and 10% other components (e.g., extractives, ashes). The high amount of carbohydrates, represented by the cellulose and hemicellulose portions, makes this material promising for biofuel production, since cellulose and hemicellulose can be chemically or enzymatically hydrolyzed to fermentable sugars.

The enzymatic route has attracted attention and it has been extensively studied in the last two decades because of its mild operating conditions, including the possibility of being carried out at moderate temperatures (50-60ºC), added to lower reactor corrosion, lower energy consumption and lower toxicity (Al-Jibouri et al. 2015). A preliminary projection shows that cellulosic ethanol production by the enzymatic route could achieve around 3 billion liters per year (Moretti et al. 2014).

However, in order to reach cellulosic ethanol using enzymatic conversion, there are four mandatory steps: (i) pretreatment of the lignocellulosic material; (ii) enzymatic hydrolysis; (iii) alcoholic fermentation of sugars; and (iv) distillation. The last two steps are already well known in the traditional process of production of ethanol (first generation) and most of this know-how can be applied to the production of cellulosic ethanol (second generation). On the other hand, there have been intensive efforts in research and development for the first two steps to overcome the natural recalcitrance of the lignocellulosic material by using combined techniques with physical, chemical and biological approaches.

The enzymatic hydrolysis consists of breaking down sugar polymers into their monomers. This is usually done using hydrolytic enzymatic extracts, composed of several classes of enzymes that work in a synergistic way to achieve the complex deconstruction of cellulose and hemicellulose chains (Al-Jibouri et al. 2015). However, the enzyme accessibility to the substrate is quite limited because the presence of lignin, crystalline cellulose and crosslinks between hemicellulose and lignin make the lignocellulosic material highly recalcitrant (Perrone et al. 2016).

The arrangement of the lignocellulosic material affords protection to the plant and it is an effective barrier to prevent microorganism attack. For that reason, processing these materials for industrial purposes is challenging because the direct enzymatic action has a very low conversion rate, making the process economically unfeasible. When there has been no pretreatment, the enzymatic yield of some lignocellulosic material is limited to around 12.5% for sugarcane bagasse, 27.5% for sugarcane straw (Barros et al. 2013) and 17.5% for corn straw (Shi et al. 2015).

Therefore, the pretreatment step is an important preparation of the substrate for enzymatic action. According to Mosier et al. (2005), an ideal pretreatment must be able to change the lignocellulosic structure without damaging the carbohydrates, thus allowing an increase in the efficiency of the enzymatic hydrolysis. Additionally, it must not form byproducts that inhibit the enzymes and/or microorganisms and it must also be a cost-effective process. It has been estimated that the pretreatment step represents up to 20% of the total cellulosic ethanol production cost (Barros et al. 2013).

Several pretreatments have been developed in the last 40 years aiming to promote these effects. Ozonolysis is one of the most promising pretreatments that has shown great potential for scaling-up. Ozone reaction with lignocelluloses occurs at room temperature and at atmospheric pressure, it is an eco-friendly process with no wastewater being generated and, also, ozone is easily produced and decomposed into O_2 and H_2O.

Currently, the main criticism of using ozonolysis is its elevated cost, however, ozone already has some large scale applications, including water purification (drinking water, industrial wastewaters, groundwater remediation, wastewater reuse, recreational water), air purification (gas emissions from industry, air conditioning systems in cars), pulp bleaching (chlorine-free paper production), organic synthesis, food processing (food preservation), medical and esthetic techniques (e.g., ozonotherapy). So, expanding the variety of such applications of ozone tends to reduce its production cost and increase its utilization.

The application of ozonolysis to lignocellulosic biomass pretreatment remains at the research level and it has shown that monitoring some relevant factors (moisture, particle size, etc.) during the gas application can optimize the process and reduce the quantity of ozone needed. In addition, the combination of ozonolysis with other pretreatment techniques could save energy and reduce operating costs by combining the strengths of each one (Barros et al. 2013).

This chapter comprises: ozone pretreatment characteristics; the chemical modifications caused in each one of the main components of the lignocellulosic material; the parameters that influence the efficiency of the pretreatment and consequently increase the enzymatic conversion; analytical methods that demonstrate structural modification, and some possibilities of other pretreatment combinations with ozonolysis.

2. Ozonolysis Effects

Ozone gas is easily produced from the reaction between an oxygen free radical and a molecule of O_2 ($O^{\bullet} + O_2 \rightarrow O_3$). The synthesis of oxygen free radicals is possible through the breakdown of the strong bond between the oxygen atoms. This bond breaking requires a high energy level which results in the formation of the oxygen free radicals ($O2 \rightarrow O\bullet + O\bullet$). Several methods for breaking these bonds can be used, including: (i) Photochemistry (UV-radiation); (ii) Corona (silent electric discharge), and (iii) Electrochemistry (electrolysis of aqueous solutions) (Da Silva and Jardim 2006). Since ozone spontaneously degrades back to oxygen atoms, it cannot be stored and it needs to be produced on site.

The Corona discharge has shown to be the most appropriate method for the generation of large amounts of ozone at low cost (Guzel-Seydim et al. 2004). In this method, two electrodes – one with low tension and the other with high tension – are separated by a dielectric ceramic and a narrow discharge gap is formed (Guzel-Seydim et al. 2004). After getting around 6-7 ev of kinetic energy, the electrons cause dissociation of the oxygen molecule and the ozone molecule is formed (Guzel-Seydim et al. 2004).

Ozone (O_3) is a powerful oxidizing agent (E° = +2.07 V) stable under acidic conditions. Molecular resonance shows the positive terminal as the electrolytic site, which is more reactive than the nucleophilic site. During the ozonolysis, the electro-deficient site of the ozone molecule reacts with the richest regions of electrons in the lignocellulosic material, which are concentrated on the aromatic rings and C=C double bonds.

Lignin is one of the macro-components of the lignocellulosic material and its main role is to reinforce the cellulose wall structure, giving stability and resistance to the cell (Aro et al. 2005). Lignin is a complex and insoluble polymer composed of phenylpropanoid units formed by three major monomeric precursors: coniferyl alcohol, synapyl alcohol and p-coumaryl alcohol (Amen-Chen et al. 2001). The units are linked by carbon-carbon or ether linkages and they form a cross-linking network within the cell wall (Aro et al. 2005). In this way, the lignin C = C double bonds may belong to both the phenyl group (the aromatic ring) or the side chain (the propyl group). This large number of double bonds makes lignin the preferred component to be attacked by the ozone. And so ozonolysis is a selective pretreatment for lignocellulosic substrates focused on their delignification effect (Bule et al. 2013; García-Cubero et al. 2009).

According to Gierer (1986), ozone can attack the lignin structure in four different ways: (i) electrophilic substitution; (ii) 1,3-dipolar cycloaddition; (iii) 1,1-cycloaddition in alkenes of the side chain or opening ring via 1,3-dipolar cycloaddition; (iv) insertion into carbon-hydrogen bonds in alcohol, aldehyde and ether type structures.

Reaction (i) occurs when the positively charged oxygen terminal ozone electrophilically attacks the aromatic ring, causing ring hydroxylation or dealkylation.

Reaction (ii) can be considered the most important because it promotes the opening of the aromatic ring. It consists of 1,3-dipolar cycloaddition that produces a radical ("initial ozonide"), which can recombine generating a second radical ("final ozonide") that is cleaved into the final ozonolysis products.

Reaction (iii) is a 1,1-cycloaddition that can occur both in the products from aromatic ring opening or in the conjugated double bonds belonging to the side chain, and it involves 1,1-cycloaddition σ - and π -complexes leading to epoxide formation.

Finally, reaction (iv) can occur by the introduction of an ozone molecule into the C-H bonds of functional groups (alcohols, aldehydes, ethers), which are present mainly in the lignin side chain (Gierer 1986).

All reactions of ozone with lignin release soluble compounds with lower molecular weight, mainly organic acids (Zhuang et al. 2015; Barrera-Martínez et al. 2016; Kaur et al. 2012).

Ozone is unstable in alkaline conditions and it degrades easily and rapidly leading to the formation of hydroxyl radicals ($O_3 + OH^- \rightarrow HO_2^- + O_2$), ($O_3 + HO_2^- \rightarrow {}^{\bullet}OH + O^{\bullet}{}_2^- + O_2$), which are an even more powerful oxidant than ozone (E° = +3.06 V) (Gunten 2003).

The hydroxyl radical has electrophilic properties and, similar to ozone, it preferentially attacks regions with high electron density in the lignocellulosic material (aromatic and ring-conjugated structures) but, additionally, it could also attack the aliphatic structure of both carbohydrates and lignin (Gierer 1997).

Carbohydrates constitute about two thirds of all lignocellulosic material represented by cellulose and hemicellulose polymers. Cellulose is a linear polysaccharide formed by glucose units linked by β (1→4) glycosidic bonds and it is the most abundant polymer in nature (Aro et al. 2005). Hemicellulose is the second most abundant polysaccharide in nature, composed of several different sugar units (Aro et al. 2005) linked in a branched chain of pentoses (xylose, rhamnose and arabinose), hexoses (glucose, mannose and galactose) and uronic acids (4-O-methylglucuronic acid and galacturonic acid) (Ramos 2003).

The attack of the hydroxyl radical on the carbohydrate chain occurs mainly in the linear chain through glycosidic bond β (1→4) disruption, consequently promoting the opening of the cyclic carbohydrate structure (Peshev et al. 2013). Briefly, the radical attack on glucose starts by abstracting a proton and forming a carbon radical, which is oxidized to form an anion superoxide. Consequently, the breakdown of the glycosidic bond induces the formation of a carbonyl and a carbanion that will later be deprotonated leading to the opening of the ring (Henriksson et al. 2000). The radical can also attack the carbon near the glycosidic bond. In this case, a direct scission of the bond occurs (Peshev et al. 2013).

The products of ozone degradation ($^{\bullet}OH$, $O^{\bullet}{}_2^-$) increase the delignification by promoting multiple attacks on the aromatic ring of lignin by an addition to the π-electron system, resulting in a short-lived charge-transfer adduct. This adduct decays to produce isomeric hydroxycyclohexadienyl radicals that undergoes oxidation, resulting in hydroxylation, aryl alkyl ether cleavage or cleavage of conjugated double bonds. Additionally, these radicals may eliminate hydroxide anions or water, leading to coupling side chain oxidation or C_α-C_β cleavage. (Gierer 1997). The insertion of hydroxyl groups in aromatic rings also improves the solubility and the degradation rate of the lignin by hydroxylation reactions (Gierer 1997).

Different from hydroxyl, superoxides do not degrade lignin or carbohydrates directly because their oxidative properties are weak. Superoxides need to combine with centre radicals, which are released after the attack of hydroxyl on the aromatic ring or the side chain. At this point, peroxide anions and cyclic 4-membered peroxidic intermediate (dioxetan) are formed, resulting in cleavage of a C-C bond, followed by the opening of an aromatic ring, or a scission of a conjugated double bond (Gierer 1997).

Ozone degradation products ($^{\bullet}OH$, $O^{\bullet}{}_2^-$) in alkaline conditions can intensify the treatment and high delignification can be achieved using less ozone gas, contributing to a cost-effective operation. On the other hand, the operational parameters should be monitored to avoid carbohydrate degradation. Although there are more attacks through radicals, a direct reaction can also occur by a direct ionic attack of ozone at the glycosidic linkage (Olkkonen et al. 2000). However, the direct reaction between ozone and carbohydrates is 10^6 times slower than the reaction between ozone and lignin (Travaini et al. 2016). In that sense, carbohydrate degradation via direct action of ozone is generally considered irrelevant.

3. Ozonolysis Treatment in Biomass

The aim of using ozonolysis on lignocellulosic biomass is to prepare the substrate for enzymatic attack by promoting its delignification and making it more susceptible to enzymes, consequently increasing the hydrolysis yield. The chemical preference of ozone for lignin makes this process quite specific and efficient for a variety of biomasses. As a matter of fact, some parameters during ozonolysis pretreatment have a major influence on lignin degradation, for example, the moisture of the material, the amount of ozone and material characteristics including particle size, surface area and lignin content.

The moisture of the material during the pretreatment has been studied a lot. The amount of water present has a direct influence on the ozone reactivity and must be taken into account during ozonolysis treatment (García-Cubero et al. 2009). Generally, materials with lower moisture do not undergo efficient delignification due to mass transfer limitations. Consequently, the access of ozone to lignin is reduced (Baig et al. 2015). On the other hand, excessive moisture is also undesirable because it forms a kind of protection around the surface of the material, inhibiting ozone attack, decreasing ozone efficiency and, consequently, increasing the processing cost (Kaur et al. 2012). With high moisture substrates, the delignification rate decreases as a consequence of the amount of free water in the material. According to Li et al. (2015), besides the water molecules chemically bonded to the material, there is a portion of water (free water) not linked to this material. This unlinked water surrounds the surface with a thick layer limiting the access of ozone to the sample (Ueda et al. 2014).

By adjusting to the ideal moisture content in the material before the pretreatment, the amount of free water can be reduced, thus allowing higher efficiency in the delignification process (Miura et al. 2012). The optimum moisture content differs for each type of lignocellulosic material (Baig et al. 2015) and it depends mainly on the hydrophilic or hydrophobic characteristics as well as its particle size.

It has been reported that 40% of moisture in sugarcane bagasse can improve the removal of acid insoluble lignin 2.7 times more than when using 80% of moisture during the ozonolysis (Travaini et al. 2013). Similarly, Ueda et al. (2014) achieved better results for the removal of lignin using moisture between 20 and 40% in the treatment of spent culture media from shiitake production, whereas no change in the content of lignin was detected using 80% of moisture for this substrate. Also, best conditions for the ozonolysis of different residues included moisture between 30% and 40% for Japanese cedar (Miura et al. 2012), 50% for corn straw (Shi, Xiang and Li 2015), and 45% or 60% for maize stover, depending on the particle size used (Li et al. 2015). According to the authors, smaller particle sizes required less ozone to produce the same effects on the delignification of pretreated maize stover with the same moisture content (Li et al. 2015).

This response may be explained especially because smaller particles have a higher surface area with more reaction sites exposed to ozone, making it more accessible in comparison to larger particle sizes (Baig et al. 2015). In addition, a greater surface area also allows more links to water molecules. Thus, an increase in the moisture content may be needed to achieve an effective ozonolysis for more milled substrates (Li et al. 2015).

As a consequence of lignin depolymerization (called delignification), ozonolysis increases the specific surface area of the pretreated material resulting from the generation of a porous structure by the removal of lignin (Barros et al. 2013; Kojima and Yoon 2008; Ben'ko et al. 2013). The removal of lignin leads to the formation of pores on the surface and, thus, there is an improvement in the accessibility of enzymes to this damaged structure (Shi et al. 2015). To illustrate, Barros et al. (2013) reported an increase of 7 times in the surface area of sugarcane bagasse (from 2.4 to 18.0 m^2 g^{-1}), and 23 times in the surface area of sugarcane straw (from 1.5 to 36.2 m^2 g^{-1}) after 60 minutes ozonolysis.

3.1. Lignin

Lignin is the most modified constituent of the lignocellulosic material by ozone action. There are different aspects of lignin removal over the different lignocellulosic materials. However, there is a consensus that the rate of lignin removal is faster at the beginning of ozonolysis and it tends to decrease during the reaction due to the effects of chemical saturation reactions.

Perrone et al. (2017) reached approximately 50% delignification with 60 minutes of ozonolysis on sugarcane bagasse but then the delignification rate kept stable up to 180 minutes with only a slight variation in lignin content from 60 min onwards. The same phenomenon was observed by García-Cubero et al. (2009) using rye straw, achieving approximately 44% delignification in 150 minutes and minor variations after that. For wheat straw, similar results were observed and the maximal delignification (86%) was reached in 60 minutes using high ozone flux in a pretreatment with a total time of 120 minutes.

The highest rates of delignification at the early stage of the process may be explained by the availability of the substrate and the easier access of ozone to the lignin (Baig et al. 2015). The degradation of lignin releases several fragments of low molecular mass which can block the access of ozone to the deeper cavities of the fiber. Another factor that must be considered is the reduction of moisture of the material due to ozone/oxygen flow during the pretreatment (Al-Jibouri et al. 2015). As discussed in the previous section, lack of moisture in the material does not allow effective action of the ozone.

Some authors have studied separately the effects of ozone on acid insoluble lignin (AIL) and acid soluble lignin (ASL), and they observed that AIL is more affected by ozonolysis than ASL (Li et al. 2015). For example, Travaini et al. (2013) observed a decrease of AIL in all pretreatments using ozone in sugarcane bagasse, while, on the other hand, a slight increase in ASL was detected. This could be related to the conversion of AIL into ASL by the insertion of hydrophilic functional groups after the ozone has attacked lignin radicals.

The lignin content is generally quantified using standard methods for lignocellulosic material, as published by the National Renewable Energy Laboratory (NREL) through the Laboratory Analytical Procedures (LAP) (Sluiter et al., 2012). However, some electronic techniques are suitable to give a complete and precise elucidation about the structural changes occurring in the material after the pretreatment. These techniques include infrared spectroscopy (FTIR), scanning electron microscopy (SEM), transmission electronic microscopy (TEM), ultraviolet photomicrographs (UV) and nuclear magnetic resonance (NMR), which are described in more detail in the following sections.

3.2. Scanning Electron Microscopy (SEM)

Electron microscopy differs from optical microscopy by using electrons as a source of illumination instead of light. For this reason, higher resolution images can be obtained. In this technique, an electron beam scans the surface of the sample and provides information about the topography and composition of the sample.

Generally, the SEM shows a compact, smooth and highly ordered surface of untreated lignocellulosic materials (Perrone et al. 2016; Barros et al. 2013; Perrone et al. 2017). On the other hand, pretreated materials exhibit opened surfaces with rough, porous and fluffy characteristics, which are attributed to the multiple destructive effects of the pretreatments on the cell wall (Perrone et al. 2016).

Additionally, SEM images have revealed the appearance of nanoscopic fibrous morphology which may be originated from the primary cell wall of the material. Also, these nanoscopic fibers were more visible in sugarcane bagasse than in sugarcane straw (Barros et al. 2013).

Perrone et al. (2017) demonstrated, by using SEM images, that the intensification of the pretreatment with ozone releases fragments from the surface of the material. Thus, this deconstruction could be directly associated with the increase of the surface area and, later, this facilitates the enzymatic action.

3.3. Transmission Electron Microscopy (TEM)

Transmission electron microscopy (TEM) is another technique that produces images of the material through an electron microscope. However, unlike SEM, in TEM the electron beam passes through a thin sample and interacts with it. The sample is arranged between the electron source and a shield, where the magnified image is formed by the impact of the transmitted and diffracted electrons.

An image is generated by a two-dimensional projection of the sample in a light or dark field by the diffraction of electrons. This technique allows a clear analysis of the primary (PW) and secondary cell wall (SW), as well as the middle lamella (ML). In addition, differences in electron density can provide important information about changes in the material structure promoted by the pretreatment. For example, Perrone et al. (2017) reported effects such as the reduction of the electron density and ruptures of the secondary cell wall and medial lamella as the time of ozonolysis increased from 60 minutes of pretreatment.

The analysis of TEM images is an important tool that contributes to a general understanding of the main effects of each pretreatment strategy and conditions on the material, in addition to other methods of biomass characterization. Specifically for ozonolysis, TEM shows that lignin is intensively modified by ozone, remembering that the main amount of lignin is contained in the secondary cell wall of lignocelluloses (Perrone et al. 2017).

3.4. Ultraviolet Photomicrographs (UV)

Aromatic structures absorb UV light at 280 nm and produce dark images in the presence of lignin. Kojima and Yoon (2008) produced ultraviolet photomicrographs to evaluate chemical modifications in waste paper delignification using ozonolysis. The untreated material showed whole cellular walls that produce a dark image throughout the area of the fiber. After 30 min of ozonolysis, the distribution of lignin in the cell walls was altered in waste paper samples. According to the authors, the removal of aromatic structures by ozone can be associated with the decrease in the absorbance at 280 nm in the outer layer of the cell wall, as well as the middle lamella of the material (Kojima and Yoon 2008).

3.5. Nuclear Magnetic Resonance (NMR)

NMR is an analytical technique by which a nucleus absorbs the electromagnetic radiation of a specific frequency in the presence of a strong magnetic field. It allows the determination of physical and chemical properties of molecules, giving detailed information about the structure of the sample.

Bule et al. (2013) applied NMR to demonstrate the effects of ozonolysis on lignin units by analyzing NMR in Solid-state (13C CP/MAS NMR), and in Solution-state (1H 13C NMR) spectroscopies of isolated lignin. Solid-state NMR has the advantage of giving information about the material modifications without the need for degradation or isolation of its components.

The comparison between untreated and treated materials indicates some evident signal reduction related to lignin. Authors reported a main signal decrease in the regions at 115-165 ppm belonging to the aromatic structures, at 153 ppm referring to aryl units, at 137 ppm referring to aromatic carbon, and at 177 ppm referring to the carbonyl group. Another important information provided by this technique was the change in the signal at 60 ppm, which is attributed to methoxyl groups, suggesting the break-up of ester-linked structures during ozone pretreatment (Bule et al. 2013).

The study of Solution-state 1H NMR spectra and 13C NMR analysis demonstrates a clear difference in the distribution of the chemical shift values of aromatic protons, confirming that ozone has a prominent effect on the aromatic component of lignin. Furthermore, Solid-state analysis also demonstrated a reduction in the aromatic structure by the decrease of intensity in the regions from 152 to 143 ppm related to G-units, and from 120 to 103 ppm related to S-units. However, the authors reported the biggest difference between 168-173 ppm when compared with the control material, which represents the region of acetate groups (-OAc) released during the isolation procedure from acid functional groups bonded to lignin (Bule et al. 2013). As a consequence of aromatic ring opening, there is an augmentation of the signal at 168-173 ppm indicating an increase in the number of carboxyl groups in the material (Bule et al. 2013).

3.6. Fourier Transform Infrared Spectroscopy (FTIR)

FTIR spectroscopy is a non-destructive and rapid technique often used in the identification of functional groups. It measures the specific vibration frequency of chemical bonds in a molecule and allows its identification according to the intensity and position of each band in the infrared spectrum.

Aromatic rings have specific vibrations and the presence of bands in regions at 1426, 1510, and 1600 cm^{-1} are widely recognized as the typical vibrations of aromatic groups only present in lignin. The pretreatment of lignocellulosic material by ozonolysis induces its delignification, which can be seen using infrared spectroscopy. Several authors have observed reduction or even extinction of those typical bands corresponding to aromatic skeletal vibrations and, sometimes, the reduction of a band at 1240 cm^{-1} (ascribed to the C-O bond in the guaiacyl ring), and at 833 cm^{-1} (ascribed to C-H bond in p-hydroxyphenyl), moieties stretching out of plane in aromatic rings (Ridenti et al. 2013; Perrone et al. 2016; Brienzo et al. 2017).

3.7. Thermogravimetric Analysis (TGA)

Thermogravimetric analysis (TGA) is a destructive technique that monitors the sample mass variation as a function of temperature or time under controlled temperature and atmosphere conditions. By this technique, it is possible to observe the thermal degradation of each polymer (cellulose, hemicellulose and lignin) separately because each one has a specific degradation profile. These differences are based on the types of intermolecular and intramolecular forces and their polymeric structures. For example, cellulose has a narrower range of degradation, while lignin begins to degrade thermally at low temperatures and has a wide range of degradation (Perrone et al. 2016).

Generally, treated materials show very close temperatures of degradation for each component and the separation of peaks is difficult. Treated materials exhibit greater thermal stability over the untreated material. This might be explained due to the increase of the crystallinity index, as a consequence of the increase in the proportion of crystalline cellulose after the removal of amorphous content by ozone (Perrone et al. 2016).

3.8. Carbohydrates

As previously demonstrated, the pretreatment by ozonolysis preferentially attacks high electron density sites of the biomass and promotes an efficient delignification of lignocelluloses. Nevertheless, ozone can also react with carbohydrates, albeit to a lesser extent (Miura et al. 2012).

Travaini et al. (2013) pretreated sugarcane bagasse by ozonolysis under severe conditions (ozone concentration 3.44 ± 0.04% (v/v); moisture 40 ± 0.16% (w/w); ozonolysis time 120 ± 0.02 min; using 0.21 ± 0.05 g of O3 per gram of bagasse) and they observed that the pretreatment slightly affected the carbohydrates as well as lignin, reaching recovery rates of 92.5% for cellulose and 98.7% for xylan.

Binder et al. (1980) detected some degradation of polysaccharides by ozonolysis in wheat straw when more than 60% of the lignin was removed. Similar results were found with sugarcane bagasse (Miura et al. 2012). Perrone et al. (2017) reported a decrease in the glucose yield from hydrolysis after achieving 58% of delignification of sugarcane bagasse.

Xylose and mannose production by hydrolysis were decreased in Japanese cedar biomass with more than 66% of lignin removal (García-Cubero et al. 2009). Using less severe conditions for pretreatment, the action of ozone on carbohydrates tends to diminish and the losses can be considered insignificant. Furthermore, some authors have considered that carbohydrate degradation is more related to the removal of xylan-linked lignin (Li et al. 2015).

After ozonolysis, the main change related to carbohydrates is the increase in the degree of crystallinity of the material. Cellulose presents crystalline and amorphous regions, while hemicellulose and lignin are always amorphous. Crystalline cellulose is formed by chains bound by hydrogen bonds in an organized way. This organization in the bonds renders crystalline cellulose persistent and insoluble, hindering the penetration of water and enzymes. On the other hand, amorphous cellulose is less organized and more soluble, and its hydrolysis is facilitated (Moon et al. 2011).

The crystallinity index of the material is estimated by X-ray diffraction (XRD) and it represents the proportion between crystalline and amorphous portions in the sample. In that sense, the amorphous portion includes hemicellulose and lignin content, which are mainly modified by the pretreatments. Different calculation methods are used to determine the crystallinity index (Terinte et al. 2011). Generally, lignocellulosic biomass presents three peaks in the XRD patterns for cellulose at 2θ at 16, 22, and 35°.

In ozonolysis, the increase in the crystallinity index may be strongly associated with the delignification process or even with the removal of both lignin and hemicellulose in combined treatments (Perrone et al. 2016; Shi et al. 2015). Also, the amorphous portion of the cellulose is probably more susceptible to ozone attack than the crystalline regions (Perrone et al. 2017).

Initially, the crystallinity index was directly associated to biomass digestibility through enzymatic hydrolysis. From this point of view, an increase in the crystallinity means more recalcitrance and, consequently, more difficulties for enzymes to overcome in polymer deconstruction. Nevertheless, pretreated biomasses show better conversion yields in hydrolysis even those presenting a higher crystallinity index. The crystallinity index is just one parameter that refers to the amount of crystalline structures in a portion of the sample, but many others factors are involved in the process of enzymatic conversion, such as accessibility of the cellulose superficial area, cellulose availability, porosity and particle size of material, etc. (Sun et al. 2016; Pereira et al. 2016). Thus, the crystallinity index could be better understood as the potential for removal of non-crystalline components of biomass, instead of an estimate of biomass digestibility.

4. Combined Treatments

The development of pretreatment strategies for lignocellulosic material addressed to second generation ethanol is mostly based on recovering the high content of cellulose as the source of glucose, which is the substrate of interest for C6 fermentation. In fact, the challenge is to find a way to effectively fractionate biomass because hemicellulose and lignin must be separated from the cellulose, which becomes more accessible and available for enzymatic conversion. There is no pretreatment able to remove both fractions in one step because, basically, hemicellulose and lignin have different chemical properties. Some traditional techniques are effective to solubilize hemicellulose, such as hydrothermal processing, acid hydrolysis, steam explosion, etc. However, these pretreatments only have minor effects on the lignin content. At the same time, as discussed in the previous section, ozone is a strong oxidant and it mainly affects lignin by attacking double C=C bonds and aromatic centers. Consequently, the effects of ozone on hemicellulose are minimal.

Considering these limitations, it is possible to combine pretreatments to take advantage of the positive effects of different techniques, aiming to modify the structure of the material, improving its enzymatic conversion, and, if possible, reducing the energy cost of the pretreatment step and, consequently decreasing the cost of the final product.

The next section will discuss the use of ozonolysis combined with one or more other pretreatments to improve the efficiency of this initial step of breaking down lignocellulosic biomass.

4.1. Ozonolysis and Alkaline Washing

The ozonolysis of lignocellulosic biomass can be considered an acidic technique due to the release of weak acids that reduces the pH rapidly, even if carried out in an aqueous medium (Lee et al. 2010; Puri 1983). However, some studies have performed ozonolysis in alkaline conditions by adding an alkali into the reaction (Binder et al. 1980; Kojima and Yoon 2008; Barrera-Martínez et al. 2016). The use of ozonolysis under alkaline condition can minimize the decrease of the pH in the reaction by forcing the degradation of ozone into hydroxyl radicals. Another use is to wash the pretreated material with alkaline solution after ozonolysis aiming to solubilize and remove undesirable products from lignin and xylan degradation. Binder et al. (1980) tested the addition of 10 mL of alkaline solution in the pretreatments to maintain the initial pH between 8-9 and observed faster ozone consumption in alkaline media due to ozone degradation into hydroxyl radicals.

A similar study observed a decrease in the removal of soluble acid lignin after moistening the material with NaOH (20%) (García-Cubero et al. 2009). Additionally, the authors reported an increase in cellulose degradation in rye straw and wheat straw in alkaline conditions, as well as a degradation of more than 50% of the hemicellulose. Therefore, the decomposition of ozone into hydroxyl radicals seems to decrease the rate of delignification and increase the degradation of carbohydrates. Further study on the use of ozonolysis in an alkaline medium is needed and this should include the use of milder operational conditions for the ozonolysis. Considering that ozone decomposes faster in the presence of alkalis, it could result in lower energy expenditure for the pretreatment.

Alkaline washing for the removal of residual compounds of ozonolysis is usually done prior to the enzymatic hydrolysis. Kojima and Yoon (2008) observed an increase in the enzymatic yield after soaking the material in NaOH 0.5% (v/v) solution for 1 h. Similarly, Perrone et al. (2017) obtained a 5-fold higher hydrolysis yield for an alkaline washed sample in comparison to the unwashed sugarcane bagasse pretreated using ozonolysis.

Other authors (Baig et al., 2015) proposed an alkaline washing step before the pretreatment of wheat straw because the presence of wax and proteomic coatings on the material surface could obstruct the access of ozone. They found that soaking the material into NaOH 1% (v/v) solution for 12 h, followed by washing procedures until neutral pH was reached and then ozonolysis, the delignification rate increases 7 times in comparison to the control sample (without an alkaline wash). This shows that some extractive components of the biomass may prevent ozone interaction with lignin, consequently reducing pretreatment efficiency and requiring higher amounts of gas for achieving desirable delignification rates.

4.2. Ozonolysis and Washing Ozonolysis

Lignin fragments with low molecular mass also become an obstacle for ozone action during the pretreatment. In this way, Al-Jibouri et al. (2015) propose an intermittent strategy that includes 30 minutes of ozonolysis pretreatment, one washing step to remove byproducts released from lignin depolymerization (such as carboxylic acids) and then another 30 minutes of ozonolysis. By this procedure, the authors saved 75% in ozone consumption to achieve the same efficiency in the pretreatment, which represents an option for optimizing the energy needs of the process (Al-Jibouri et al. 2015).

4.3. Ozone, Ultrasound and Alkaline Washing

The use of ultrasonic irradiation in liquids leads to the formation of microbubbles under high temperature and pressure, generating a phenomenon known as acoustic cavitation. If ultrasound is performed in alkaline conditions, the penetration of the solvent is favored and, consequently, the efficiency of the pretreatment increases due to more intensive hemicellulose removal (Velmurugan and Muthukumar 2012).

In general, the use of ultrasound as a unique pretreatment is unfeasible for lignocellulosic biomass due to its low efficiency and high energy costs. For this reason, Perrone et al. (2016) associated ozonolysis and ultrasound under alkaline conditions and obtained greater removal of hemicellulose. The ultrasound-assisted step increased the hydrolysis yield by 19% in comparison to the control (without ultrasound).

4.4. Autohydrolysis and Ozonolysis

Autohydrolysis or liquid hot water (LHW) pretreatment is one of the most commonly used techniques to increase the digestibility of cellulose by solubilizing hemicellulose content with water under high pressure and temperature without chemicals. It was initially used in the pulp paper industry (Bobleter and Pape 1968), and, more recently, for biofuels (Weil et al. 1998; Kim et al. 2011; Ko et al. 2015). The high capacity to remove significant amounts of hemicellulose came from the breaking down of heterocyclic esters linkages, which release acetic and uronic acids responsible for a fast drop in the pH inside the reactor, easily solubilizing xylan. In contrast, lignin is minimally affected. For example, Lee et al. (2009) reached up to 83% removal of hemicellulose but only 4% delignification using severe conditions (170°C, 1 h) of autohydrolysis for coastal Bermuda grass.

Combining autohydrolysis with ozonolysis efficiently removes lignin and also efficiently solubilizes hemicellulose, producing a cellulose-rich substrate ready for enzymatic action.

The enzymatic conversion of cellulose from coastal Bermuda grass using only ozonolysis (60 minutes) was around 53%. This conversion was increased to 61% by adding 11 minutes of autohydrolysis to the process. Also, the authors reported an approximate 60% yield in the enzymatic hydrolysis when combining 30 minutes of ozonolysis and 6 minutes of autohydrolysis, which means a 50% reduction in processing time (Lee et al. 2010).

4.5. Ozonolysis and Milling

Physical treatments, such as milling, have the advantages of being environmentally friendly (i.e., they do not use chemical solvents) and efficient in reducing the particle size and decreasing the degree of crystallinity of the cellulose (Shi et al. 2015; Miura et al. 2012). However, a high amount of energy is generally required to achieve good results (Miura et al. 2012).

A combination of milling and ozonolysis reduces the crystallinity and the amount of lignin, respectively, which are two of the major bottlenecks for achieving efficient enzymatic hydrolysis.

This combination of pretreatments can be performed in two ways: (i) ozonolysis followed by milling, or (ii) milling followed by ozonolysis. The first technique focuses on reducing lignin by ozonolysis, making the material less recalcitrant. Then, the ozonated material is milled for greater efficiency and lower energy expenditure (Barros et al. 2013). Shi et al. (2015), for example, reduced the milling time by 8 times using ozonated biomass in comparison to untreated biomass and achieved an equivalent enzymatic yield.

The second way applies milling first aiming to reduce the particle size and consequently increase the surface area of the material. Ozone efficiency is improved due to the greater contact surface of the substrate, increasing delignification with less gas consumption. This means that the energy consumption of the ozonolysis is significantly lower. For example, Barros et al. (2013) found that using milling/ozonolysis has 6 times less energy consumption than ozonolysis/milling.

CONCLUSION

Ozonolysis represents one option of pretreatment of lignocellulosic biomass with a specific focus on lignin. Ozone gas easily interacts with all fractions of the material but preferentially attacks the linkages with higher electron density in lignin, leaving the carbohydrates for biofuel production using, for example, enzymatic hydrolysis and alcoholic fermentation to obtain ethanol. The main advantages of ozonolysis are the easy generation of ozone gas, no chemicals needed, no effluents generated and no pressure or high temperature required for the process, making it an eco-friendly and potentially cost-effective pretreatment. Further improvements are needed to optimize the use of ozone with agricultural residues for green chemistry products, including second-generation ethanol. Critical points in ozonolysis include the control of humidity during the process, generation of organic acids as potential inhibitors for enzymes and fermenting microorganisms, low capacity of fractionating hemicellulose, and reactor development for scaling up. There is a great need for research into and development of emerging technologies in order to reduce production costs and improve applications in the wider concept of the biorefinery.

REFERENCES

Al-Jibouri, Ali Kamel H., Ginette, Turcotte., Jiangning, Wu. & Chil-Hung, Cheng. (2015). "Ozone Pretreatment of Humid Wheat Straw for Biofuel Production." *Energy Science & Engineering*, *3* (6), 541–48.

Amen-Chen, Carlos., Hooshang, Pakdel. & Christian, Roy. (2001). "Production of Monomeric Phenols by Thermochemical Conversion of Biomass: A Review." *Bioresource Technology*, *79* (3), 277–99.

Aro, Nina., Tiina, Pakula. & Merja, Penttilä. (2005). "Transcriptional Regulation of Plant Cell Wall Degradation by Filamentous Fungi." *FEMS Microbiology Reviews*, *29* (4), 719–39.

Baig, K. S., Wu, J., Turcotte, G. & Doan, H. D. (2015). "Novel Ozonation Technique to Delignify Wheat Straw for Biofuel Production." *Energy & Environment*, *26* (3), 303–18.

Barrera-Martínez, Iliana., Nydia, Guzmán., Erika, Peña., Tania, Vázquez., Ricardo, Cerón-Camacho., Jorge, Folch., José, Amador Honorato Salazar. & Jorge, Aburto. (2016). "Ozonolysis of Alkaline Lignin and Sugarcane Bagasse: Structural Changes and Their Effect on Saccharification." *Biomass and Bioenergy*, *94*, 167–72.

Barros, Rodrigo da Rocha Olivieri de., Raquel, de Sousa Paredes., Takashi, Endo., Elba, Pinto Da Silva Bon. & Seung, Hwan Lee. (2013). "Association of Wet Disk Milling and Ozonolysis as Pretreatment for Enzymatic Saccharification of Sugarcane Bagasse and Straw." *Bioresource Technology*, *136*. Elsevier Ltd, 288–94.

Ben'ko, E. M., Manisova, O. R. & Lunin, V. V. (2013). "Effect of Ozonation on the Reactivity of Lignocellulose Substrates in Enzymatic Hydrolyses to Sugars." *Russian Journal of Physical Chemistry A*, *87* (7), 1108–13.

Binder, Arthur., Luciano, Pelloni. & Armin, Fiechter. (1980). "Delignification of Straw with Ozone to Enhance Biodegradability." *European Journal of Applied Microbiology and Biotecnology*, *11* (1), 1–5.

Brienzo, Michel., Simphiwe, Fikizolo., Yuda, Benjamin., Luvuyo, Tyhoda. & Johann, Görgens. (2017). "Influence of Pretreatment Severity on Structural Changes, Lignin Content and Enzymatic Hydrolysis of Sugarcane Bagasse Samples." *Renewable Energy*, *104*, 271–80.

Bule, Mahesh V., Allan, H. Gao., Bill, Hiscox. & Shulin, Chen. (2013). "Structural Modification of Lignin and Characterization of Pretreated Wheat Straw by Ozonation." *Journal of Agricultural and Food Chemistry*, *61* (16), 3916–25.

Da Silva, Leonardo M. & Wilson, F. Jardim. (2006). "Trends and Strategies of Ozone Application in Environmental Problems." *Quimica Nova*, *29* (2), 310–17.

García-Cubero, Maria Teresa., Luis, G. Palacín., Gerardo, González-Benito., Silvia, Bolado., Susana, Lucas. & Mónica, Coca. (2012). "An Analysis of Lignin Removal in a Fixed Bed Reactor by Reaction of Cereal Straws with Ozone." *Bioresource Technology*, *107*, 229–34.

García-Cubero, Teresa., Gerardo, González-Benito., Irune, Indacoechea., Mónica, Coca. & Silvia, Bolado. (2009). "Effect of Ozonolysis Pretreatment on Enzymatic Digestibility of Sugarcane Bagasse." *Agricultural Engineering International: CIGR Journal*, *16* (1). Elsevier Ltd, 151–56.

Gierer, J. (1997). "Formation and Involvement of Superoxide (O2.-/HO2·) and Hydroxyl (OH·) Radicals in TCF Bleaching Processes: A Review." *Holzforschung*, *51* (1), 34–46..

Gierer, J. (1986). "Chemistry of Delignification - Part 2: Reactions of Lignins during Bleaching." *Wood Science and Technology*, *20* (1), 1–33.

Gunten, Urs Von. (2003). "Ozonation of Drinking Water: Part I. Oxidation Kinetics and Product Formation." *Water Research*, *37* (7), 1443–67.

Guzel-Seydim, Zeynep B., Annel, K. Greene. & Seydim, A. C. (2004). "Use of Ozone in the Food Industry." *LWT - Food Science and Technology*, *37* (4), 453–60.

Henriksson, Gunnar., Gunnar, Johansson. & Göran, Pettersson. (2000). "A Critical Review of Cellobiose Dehydrogenases." *Journal of Biotechnology*, *78* (2), 93–113.

Kaur, Ujjal., Harinder, Singh Oberoi., Vinod, Kumar Bhargav., Ratna, Sharma-Shivappa. & Sandeep, Singh Dhaliwal. (2012). "Ethanol Production from Alkali- and Ozone-Treated Cotton Stalks Using Thermotolerant Pichia Kudriavzevii HOP-1." *Industrial Crops and Products*, *37* (1). Elsevier B.V., 219–26.

Kojima, Yasuo. & Seung, Lak Yoon. (2008). "Improved Enzymatic Hydrolysis of Waste Paper by Ozone Pretreatment." *Journal of Material Cycles and Waste Management*, *10* (2), 134–39.

Lee, Jung Myoung., Hasan, Jameel. & Richard, A. Venditti. (2010). "Effect of Ozone and Autohydrolysis Pretreatments on Enzymatic Digestibility of Coastal Bermuda Grass." *Bio Resources*, *5* (2), 1084–1101.

Lee, Jung Myoung., Jian, Shi., Richard, A. Venditti. & Hasan, Jameel. (2009). "Autohydrolysis Pretreatment of Coastal Bermuda Grass for Increased Enzyme Hydrolysis." *Bioresource Technology*, *100* (24). Elsevier Ltd, 6434–41..

Li, Cheng., Li, Wang., Zhengxing, Chen., Yongfu, Li., Ren, Wang., Xiaohu, Luo., Guolin, Cai., Yanan, Li., Qiusheng, Yu. & Jian, Lu. (2015). "Ozonolysis Pretreatment of Maize Stover: The Interactive Effect of Sample Particle Size and Moisture on Ozonolysis Process." *Bioresource Technology*, *183*. Elsevier Ltd, 240–47.

Miura, Toyokazu., Seung-Hwan, Lee., Seiichi, Inoue. & Takashi, Endo. (2012). "Combined Pretreatment Using Ozonolysis and Wet-Disk Milling to Improve Enzymatic Saccharification of Japanese Cedar." *Bioresource Technology*, *126*. Elsevier Ltd, 182–86.

Moon, Robert J., Ashlie, Martini., John, Nairn., Simonsen, J. & Jeff, Youngblood. (2011). Cellulose Nanomaterials Review: Structure, Properties and Nanocomposites. *Chemical Society Reviews*., Vol. *40*.

Moretti, Marcia Maria de Souza., Daniela, Alonso Bocchini-Martins., Christiane, da Costa Carreira Nunes., Maria, Arévalo Villena., Olavo, Micali Perrone., Roberto, da Silva., Maurício, Boscolo. & Eleni, Gomes. (2014). "Pretreatment of Sugarcane Bagasse with Microwaves Irradiation and Its Effects on the Structure and on Enzymatic Hydrolysis." *Applied Energy*, *122*. Elsevier Ltd, 189–95.

Mosier, Nathan., Charles, Wyman., Bruce, Dale., Richard, Elander., Lee, Y. Y., Mark, Holtzapple. & Michael, Ladisch. (2005). "Features of Promising Technologies for Pretreatment of Lignocellulosic Biomass." *Bioresource Technology*, *96* (6), 673–86.

Olkkonen, C., Tylli, H., Forsskåhl, I., Fuhrmann, A., Hausalo, T., Tamminen, T., Hortling, B. & Janson, J. (2000). "Degradation of Model Compounds for Cellulose and Ligno-Cellulosic Pulp during Ozonation in Aqueous Solution." *Holzforschung*, *54* (4), 397–406.

Pereira, Josiani de Cassia., Rodolfo, Travaini., Natalia, Paganini Marques., Silvia, Bolado-Rodríguez. & Daniela, Alonso Bocchini Martins. (2016). "Saccharification of Ozonated Sugarcane Bagasse Using Enzymes from Myceliophthora Thermophila JCP 1-4 for Sugars Release and Ethanol Production." *Bioresource Technology*, *204*. Elsevier Ltd, 122–29.

Perrone, Olavo Micali., Felippe, Mariano Colombari., Jessika, Souza Rossi., Marcia, Maria Souza Moretti., Sidnei Emilio, Bordignon., Christiane da Costa Carreira, Nunes., Eleni, Gomes., Mauricio, Boscolo. & Roberto, Da-Silva. (2016). "Ozonolysis Combined with Ultrasound as a Pretreatment of Sugarcane Bagasse: Effect on the Enzymatic Saccharification and the Physical and Chemical Characteristics of the Substrate." *Bioresource Technology*, *218* (October), 69–76.

Perrone, Olavo Micali., Jessika, Souza Rossi., Moretti, De Souza Maria Marcia., Costa, Carreira., Sidnei Emilio, Bordignon., Eleni, Gomes., Roberto, Da-silva. & Mauricio, Boscolo. (2017). "Influence of Ozonolysis Time during Sugarcane Pretreatment: Effects on the Fiber and Enzymatic Saccharification." *Bioresource Technology*, *224*. Elsevier Ltd, 733–37.

Perrone, Olavo Micali., Jessika Souza, Rossi., Marcia, Maria de Souza Moretti., Christiane, da Costa Carreira Nunes., Sidnei, Emilio Bordignon., Eleni, Gomes., Roberto, Da-Silva. & Mauricio, Boscolo. (2017). "Influence of Ozonolysis Time during Sugarcane Pretreatment: Effects on the Fiber and Enzymatic Saccharification." *Bioresource Technology*, *224* (January), 733–37.

Peshev, Darin., Rudy, Vergauwen., Andrea, Moglia., Éva, Hideg. & Wim, Van Den Ende. (2013). "Towards Understanding Vacuolar Antioxidant Mechanisms: A Role for Fructans?" *Journal of Experimental Botany*, *64* (4), 1025–38.

Puri, Vinod P. (1983). "Ozone Pretreatment to Increase Digestibility of Lignocellulose." *Biotechnology Letters*, *5* (11), 773–76.

Ramos, Luiz Pereira. (2003). "The chemistry involved in the steam treatment of lignocellulosic materials." *Quimica Nova*, *26* (6), 863-871.

Ridenti, M. A., Oliveira, C. & Amorim, J. (2013). "Decomposition of Lignin from Sugar Cane Bagasse during Ozonation Process Monitored by Optical and Mass Spectrometries." *Physical Chemistry B*, *117*, 3110-3119.

Shi, Feng., Heji, Xiang. & Yongfu, Li. (2015). "Combined Pretreatment Using Ozonolysis and Ball Milling to Improve Enzymatic Saccharification of Corn Straw." *Bioresource Technology*, *179*. Elsevier Ltd, 444–51.

Sluiter, A., Hames, B., Ruiz, R., Scarlata, C., Sluiter, J., Templeton, D. & Crocker, D. (2012). "Determination of Structural Carbohydrates and Lignin in Biomass: Laboratory Analytical Procedure (LAP), Technical Report NREL/TP-510-42618." *National Renewable Energy Laboratory*, *42618* (Issue Date: 4/25/2008), 1–17.

Sugimoto, Tomoko., Kengo, Magara., Shuji, Hosoya., Satoshi, Oosawa., Takafumi, Shimoda. & Kozo, Nishibori. (2009). "Ozone Pretreatment of Lignocellulosic Materials for Ethanol Production: Improvement of Enzymatic Susceptibility of Softwood." *Holzforschung*, *63* (5), 537–43.

Sun, Shaoni., Shaolong, Sun., Xuefei, Cao. & Runcang, Sun. (2016). "The Role of Pretreatment in Improving the Enzymatic Hydrolysis of Lignocellulosic Materials." *Bioresource Technology*, *199*. Elsevier Ltd, 49–58.

Terinte, Nicoleta., Roger, Ibbett. & Kurt Christian, Schuster. (2011). "Overview on Native Cellulose and Microcrystalline Cellulose I Structure Studied By X-Ray Diffraction (Waxd): Comparison Between Measurement Techniques." *Lenzinger Berichte*, *89*, 118–31.

Travaini, Rodolfo., Judit, Martín-Juárez., Ana, Lorenzo-Hernando. & Silvia, Bolado-Rodríguez. (2016). "Ozonolysis: An Advantageous Pretreatment for Lignocellulosic Biomass Revisited." *Bioresource Technology*, *199*. Elsevier Ltd, 2–12.

Travaini, Rodolfo., Marian Derly Morales, Otero., Mónica, Coca., Roberto, Da-Silva. & Silvia, Bolado. (2013). "Sugarcane Bagasse Ozonolysis Pretreatment: Effect on Enzymatic Digestibility and Inhibitory Compound Formation." *Bioresource Technology*, *133*. Elsevier Ltd, 332–39.

Ueda, Chisato., Yuya, Takashima., Futoshi, Ishiguri., Kazuya, Iizuka., Nobuo, Yoshizawa. & Shinso, Yokota. (2014). "Ozone Oxidation Pretreatment for Enzymatic Saccharification of Spent Culture Media after Lentinula Edodes Cultivation." *Journal of Wood Science*, *61* (1), 65–69.

Velmurugan, Rajendran. & Karuppan, Muthukumar. (2012). "Ultrasound-Assisted Alkaline Pretreatment of Sugarcane Bagasse for Fermentable Sugar Production: Optimization through Response Surface Methodology." *Bioresource Technology*, *112*. Elsevier Ltd, 293–99.

Zhuang, Xinshu., Qiang, Yu., Zhenhong, Yuan., Xiaoying, Kong. & Wei, Qi. (2015). "Effect of Hydrothermal Pretreatment of Sugarcane Bagasse on Enzymatic Digestibility." *Journal of Chemical Technology and Biotechnology*, *90* (8), 1515–20.

In: Bioethanol and Beyond
Editor: Michel Brienzo
ISBN: 978-1-53613-478-0

Chapter 7

BIOETHANOL FROM MICROALGAE: TECHNICAL ASPECTS TO GUIDE TECHNOLOGICAL DEVELOPMENT

Eduardo Bittencourt Sydney[1,*], Ana Maria Velez Escallon[2], Alessandra Cristine Novak Sydney[1], Juliana Vitoria Messias Bittencourt[1], Júlio Cesar de Carvalho[3] and Carlos Ricardo Soccol[3]

[1]Federal University of Technology - Paraná (UTFPR), Ponta Grossa, PR, Brazil
[2]Federal University of Technology - Paraná (UTFPR), Toledo, PR, Brazil
[3]Federal University of Paraná (UFPR), Curitiba, PR, Brazil

ABSTRACT

The production of bioproducts from microalgal biomass is gaining importance since it allows the use of non-arable land, wastewater reuse, mitigation of carbon dioxide emissions and the partial introduction of solar energy in industrial processes, resulting in environmentally friendly processes. Many microalgae and cyanobacteria can accumulate significant amounts of lipids, proteins and carbohydrates within cells, that can be extracted and transformed into value added products. Ideally, in a biorefinery context, where industrial processes are integrated, every microalgal component can be exploited. The carbohydrates of the microalgal biomass can be directed to bioethanol production, a source of energy to be used within or outside the industrial complex. Bioethanol from microalgae and cyanobacteria, known as 3rd generation bioethanol, has been gaining attention lately but can be classified as incipient. Despite facing economical drawbacks due to limited biomass production and carbohydrate content, microalgal bioethanol technologies are under development. For the production of third generation bioethanol,

[*] Corresponding Author Email: eduardosydney@utfpr.edu.br.

the selection of microalgal species with the ability to accumulate high concentrations of carbohydrates instead of lipids as an energy reserve compound, is of great importance. Carbohydrate accumulations up to 55% (dry basis) are reported in the literature. Production of 3rd generation bioethanol consists of microalgal biomass cultivation (open ponds, closed photobioreactors and hybrid systems) biomass harvest and concentration (flocculation, flotation, sedimentation, filtration and centrifugation). The microalgal biomass must be hydrolyzed to fermentable sugars so that it can serve as a substrate for fermentation to bioethanol. The hydrolysis of microalgae produces a carbohydrate solution rich in simple sugars which can be easily fermented. Considering the best yields described in the literature we can predict that approximately 1.7 tonnes of dried microalgal biomass are necessary to produce 1m^3 of ethanol. Such yield is economically inviable, thus the technology needs to be improved. This chapter focuses on the technological aspects and bottlenecks of 3G bioethanol production from microalgae, followed by considerations on carbon dioxide assimilation/balance and integration of 3G bioethanol with other industrial activities.

Keywords: biofuels, third-generation bioethanol, microalgae, biorefinery

1. Introduction

The recent advances in biotechnology and in worldwide sustainable policies have increased the interest in microalgal technologies. Microalgae are photoautotrophic microorganisms that depend on light, carbon dioxide and nutrients such as nitrogen, phosphorus and potassium to produce biomass, characteristics that give them flexibility to be used strategically in industrial systems. Their capacity to reach high biomass productivities, mitigate carbon dioxide from industrial exhaust, use of solar energy, and the non-reliance to arable land and potable water puts microalgae in an advantageous position on the field of bio and circular economy where biological systems play an important role in the production of new and improved bioproducts.

The first person to culture algae in a laboratory under controlled conditions (artificial light and synthetic culture medium) was Andrei S. Faminstsyn in 1868 (Preisig and Andersen 2005). Since then, our understanding of the physiology, biochemistry and metabolism of microalgaehas led to significant advances. Especially after the 1940s, when large scale cultivation of microalgae began (Borowitzka 2016), different species were evaluated for the production of bioproducts and carbon fixation capacity. In recent years the use of wastewater and industrial effluent gases gained importance because of associated cost reductions of >50% (Slade and Bauen 2013) environmental benefits (bioremediation of wastewater and gases) and the production of biochemicals with commercial interest (Figure 1 – The reuse of liquid and gas effluents from industrial and domestic sources for the culture of microalgae promotes simultaneous bioremediation of wastes, process economics and production of biochemicals of commercial importance).

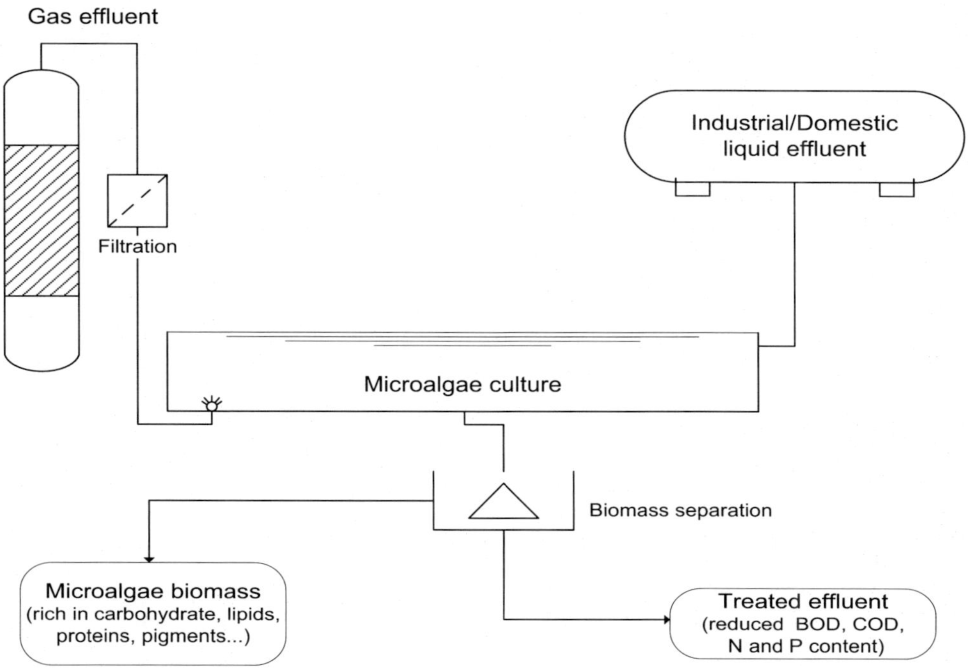

Figure 1. The reuse of liquid and gas effluents from industrial and domestic sources for the culture of microalgae promotes simultaneous bioremediation of wastes, process economics and production of biochemicals of commercial importance.

1.1. Biomass Production and Wastewater Valorization

The higher productivity of biomass production of microalgae and cyanobacteria in comparison to other photoautotrophic organisms (Tan and Lee, 2016) represents one of the great advantages of microalgal technology. Algae can produce 2-10 fold more biomass per unit of land area than the best terrestrial systems (Weyer et al., 2010). In addition, they can be produced any time of the year (rather than seasonally) under controlled conditions, in photobioreactors with different sizes and geometries, as well as in locations where agricultural activity is limited or impossible.

Photoautotrophic microalgal cultures typically reach a maximum of 4 grams per liter of biomass due to the auto-shadowing effect (Sydney et al., 2016): the increase in cell concentration impairs the incidence of solar rays, decreasing photosynthesis. The major consequence is the necessity to optimize reactor geometry to reach maximum sunlight incidence in the culture. Thus, open and closed photobioreactors generally work with a greatly reduced water depth of 20cm maximum (Kumar et al., 2015), resulting in a large area of cultivation (especially in open ponds). Frequent recovery of biomass from the culture to maintain exponential growth is the main strategy in avoiding the shadowing effect.

Because microalgae don't depend on organic carbon to produce biomass, free or low carbon content domestic and industrial wastewater can be used as a source of other nutrients. Wastewater is a low-cost medium for microalgal cultures.

The chemical composition of industrial wastewater varies according to the economic activity and differs from domestic waste composition. Many wastewaters are rich in nitrogen (nitrate, ammonia, urea), phosphorus (phosphates) and micronutrients (essentially minerals), even after going through treatment systems. The presence of these compounds in balanced concentrations, especially nitrogen and phosphorus, makes the wastewater capable of supporting microalgal growth. During microalgal production, these elements are removed from the medium and result in advanced wastewater treatment (Salama et al., 2017).

Several residues have been reported to be suitable for microalgal growth with the benefit of reducing potential pollutant contamination: sugar mill effluent, pulp and paper industry effluent (Polishchuk et al., 2015), fish farm wastewater, coal-fired metal-contaminated wastewater, oil exploration wastewater, pharmaceutical wastewater, textile dye industry effluent, and electroplating industry wastewater. In most cases a pure microalgal culture is used, but in some particular cases cultivation of microalgal consortia with bacteria is beneficial: residue with high inorganic load, in which case microalgal products can increase organic load (e.g., polysaccharides) or when the environmental conditions (temperature, altitude, light incidence) are not the most favorable for microalgal growth (Gonçalves et al., 2017).

Adaptation of microalgae to these wastewaters is a crucial step where the choice of species plays an important role. Industrial wastewater frequently contains toxic organic and inorganic substances (heavy metals, pesticides, antibiotics, surfactants). In many cases, the spectrum of use of microalgal biomass is reduced (food and feed, for example) and this is where the production of biofuels gains importance.

The reuse of wastewater gains further importance when considering that the cultivation of microalgae depends on fertilizer as a source of nitrogen and phosphorus, the production of which results in higher carbon dioxide production than can be mitigated through the microalgal culture. According to Kool, Marinussen, & Blonk (2012), the average carbon footprint for the production of N-fertilizer is 5.66 kg_{CO2eq}/kg_N and 1.36 kg_{CO2eq}/kg_{P2O5}, while microalgae can capture approx. 1.80 $g_{CO2}\ g_{biomass}^{-1}$ (see section 1.2).

1.2. Carbon Fixation in Microalgae and Cyanobacteria

The CO_2 present in atmospheric air (406.05 ppm, May 2017) ("Trends in Atmospheric Carbon Dioxide") is not enough to induce high biomass productivity in microalgae. Therefore, CO_2 supplementation is necessary for the culture medium, further increasing the amount of biomass produced at the end of the process (Colling Klein,

Bonomi, and Maciel Filho 2017). In this scenario, gaseous effluent from the production process that contains CO_2 can be diverted to the microalgal culture for subsequent fixation.

Almost all microalgae and cyanobacteria have a carbon concentrating mechanism (CCM) that increases $CO_{2(aq)}$ concentrations close to the carbon fixing enzyme RuBisCO (Ribulose-1,5-bisphosphate carboxylase/oxygenase). This enzyme acts as oxygenase and carboxylase, depending on the ligand substrate, which means that oxygen and carbon dioxide compete as substrates. RuBisCO has higher affinity for CO_2 than for O_2 in all photosynthetic organisms (van Lun et al. 2014), but one might consider the atmospheric levels of O_2 (21%) and CO_2 (0,035%). O_2 accounts for as much as 30% of the reactions catalyzed by RuBisCO and results in a useless metabolic product (phosphoglycolate) (Moroney and Somanchi 1999). Reengineering RuBisCO has been a primary objective for enhancing photosynthesis, although most of these efforts have met with limited success. The most successful engineering approaches to increase rates of carbon fixation in autotrophs involved modification of neighboring pathways and enzymes (Ducat and Silver 2012).

CCM in unicellular photosynthetic organisms works by concentrating RuBisCO in a specific location (in carboxysomes in cyanobacteria and in the pyrenoid in algae). In addition, these microorganisms have a means of concentrating HCO_3^- in the cytoplasm and can convert the accumulated HCO_3^- to CO_2 rapidly at the location of RuBisCO (reduced diffusing through the membrane) (Moroney and Somanchi 1999). For microalgal cultures it is thus necessary that this CO_2 diffuses in the liquid medium, which occurs 10,000 times slower than in air (Leung, Caramanna, and Maroto-Valer 2014). Moreover, it equilibrates with HCO_3^- on a pH dependent basis where the neutral pH results in relatively low CO_2 availability, with most of it existing as HCO_3^-, which is not a substrate for RuBisCO. Feeding the culture with CO_2 improves carbon uptake and consequently biomass production, but pH control becomes imperative due to medium acidification. One alternative is to associate pH control with CO_2 injection (biomass growth is associated with medium basification).

Apart from the universal mechanism of carbon fixation, species of microalgae and cyanobacteria have peculiarities when it comes to carbon fixation rate, which also depends on culture conditions, mainly the concentration of carbon dioxide in feeding (Cuellar-Bermudez et al. 2015), and reactor design (Kumar et al. 2011). There is no relation between the rate of carbon dioxide captured and species, genus and/or family of microalgae and cyanobacteria; each strain must be assessed individually.

The rate of carbon dioxide capture during biomass production can be estimated by Equation 1, which represents carbon and nitrogen uptake biomass production through photosynthesis. For each gram of biomass produced it is expected that approx. 1.80 g of CO_2 is absorbed.

$$0{,}800\ H_2O + 1{,}000\ CO_2 + 0{,}150\ HNO_3 + 0{,}01\ H_3PO_4 \rightarrow 1{,}000\ CH_{1,78}N_{0,15}O_{0,52}P_{0,01} + 1{,}385\ O_2$$

According to the review paper on CO_2 mitigation through photosynthesis from Ho et al. (2011), most rates described (>90%) are very similar to this estimation (varying from 1.60-1.90 $g_{CO2}\ g_{biomass}^{-1}$). This theoretical yield of carbon fixation may vary according to the strain and/or species, biomass composition (lipid, carbohydrate and protein content) (Sydney et al. 2016), CO_2 content in feeding gas (de Morais and Costa 2007) and culture conditions such as nitrogen source (Dalrymple et al. 2013) and insolation (Laurens et al. 2017), and is directly related to biomass productivity.

The best-case scenario of microalgal biomass productivity is 33-42 g m^{-2} d^{-1}, while the demonstrated yields in closed photobioreactor systems and open ponds are roughly 2-fold lower (Laurens et al. 2017). Considering the realistic scenario, carbon fixation of microalgae could reach 29,7- 37,8 g_{CO2} m^{-2} d^{-1}.

Closed photobioreactors present advantages in comparison to open ponds because it is possible to recirculate the injected CO_2 and increase gas-liquid mass transfer. Fan, Zhang, Zhang, & Chen (2008), for example, reached 23 fold higher CO_2 capture than average using a helical tubular membrane-sparger photobioreactor.

1.3. Valorization of Gaseous Effluents

Recent concerns about greenhouse gas emissions and the necessity to reduce or mitigate carbon dioxide encouraged the development of microalgal based technologies. The objective is to achieve technological viability to associate microalgal systems directly with carbon emitting sources (power plants, industry, etc.), converting emissions into usable bioproducts. The amount of flue gas needed in the microalgal culture depends on its CO_2 concentration, CO_2 uptake rate of specific species and reactor geometry.

Flue gases from fossil-fuel power plants typically have high CO_2 concentrations, ranging from 5-20% in power plants to 99% in a fermentation-based ethanol plants (Laurens et al. 2017). Besides not depending on the high purity of CO_2 gas for growth, microalgae can also use NOx and SOx (frequently presented in biologically significant amounts in flue gases) as nutrients for the production of biomass and value added products. However, it is important to state that flue gases might also be toxic to microalgae (Lara-Gil, Álvarez, and Pacheco 2014) and suppress their metabolism. Many studies on the ability of microalgae to use flue gases from different origins are described in the scientific literature (Abdel-Raouf, N.; Al-Homaidan, A. A.; Ibraheem 2012; Lv et al. 2017; Cuellar-Bermudez et al. 2015; Tan and Lee 2016; Yen et al. 2015; Huang et al. 2016).

The major drawback in using effluent gases in microalgal ponds is the outgassing phenomenon. Outgassing is responsible for the loss of more than half the CO_2 pumped into the culture (Laurens et al. 2017).

1.4. Microalgae for Biofuel Production

Due to the dependence on petroleum for energy production, which impacts economies and the environment, many efforts have been made to develop and implement technologies for renewable energy production. The use of wind, water and sunlight in energy production is of great value in replacing fossil fuels but does not solve the problem of centralized production, since the incidence of sufficient winds and solar rays are exclusive to few regions of the planet and the production of hydroelectric energy depends on the existence of rivers with a medium/large flow of water.

In this context the use of biomass in fuel production has gained much attention in the last decades because it allows the exploitation of local, regional, and national sources. In a bio and circular economy, in which waste and environmental impacts are minimized, microalgae play an interesting role due to their ability to use industrial liquid and gaseous wastes in biomass production. The co-production of biofuels in bioprocesses coupled with industrial facilities (methane, ethanol, biodiesel) is, probably, the most famous example of a circular economy.

Among the biofuels that can be derived from microalgae are: methane, biohydrogen, biodiesel, biooil and ethanol. While methane and biooil can be produced from the integral microalgal biomass (the first by anaerobic digestion and the other by pyrolysis) and hydrogen production is very sensitive to the presence of oxygen (complete anaerobic condition is needed) (Oey, M.; Sawyer, A. L.; Ross, I. L.; Hankamer 2016), the other cited biofuels depend on the accumulation of certain molecules within cells.

The (trans) esterification of microalgal lipids to biodiesel has been largely exploited but never reached industrial scale. On the other hand, some microalgae are known to contain a large amount (>50% of the dry weight) of starch, cellulose and glycogen, which are raw materials for ethanol production (Cuellar-Bermudez et al. 2015). Bioethanol from microalgae, also known as 3G ethanol, is an incipient technology that presents many advantages in relation to 1 and 2G ethanol:

1) The carbohydrates produced by the microalgae are in the form of starch (without lignin), which makes it much easier to convert sugar to alcohol, when compared to the processing of lignocellulosic biomass (Lam and Lee 2015).
2) Short harvesting cycle of algae;
3) Biomass production is not seasonal;

4) Algae can grow in non-ideal or extreme conditions, not competing with conventional agriculture;
5) Can capture CO_2 from flue gases;
6) Promote wastewater bioremediation (removal of nitrogen, phosphate, metals, COD and BOD);

A review of microalgal technology and the production of 3G ethanol will be further presented followed by a discussion on integration of 3G ethanol in the context of a circular economy and the technical drawbacks that must be surpassed to promote the confluence of economic interests with the solid environmental benefits of the referred technology.

2. Biofuels from Microalgae

The use of microalgal biomass for the production of biofuels - such as biodiesel, bioethanol, synthesis gas and hydrogen, has been proposed in the literature (Brennan and Owende 2010; Y. Li et al. 2008; Kirrolia, Bishnoi, and Singh 2013; Jones and Mayfield 2012). In the production of so-called "third-generation bioethanol," the carbohydrates produced by microalgae are used as a substrate for fermentation. Microalgal biomass is an interesting alternative to traditional bioethanol crops because of the lack of inherent disadvantages found in first or second generation bioethanol. The microalgal culture can be carried out in different culture media, without necessarily using potable water and can carry industrial effluents, salt water (seawater) and brackish water in its composition (Phukan et al. 2011). This is advantageous in that microalgal crops do not compete for fresh water intended for irrigation of crops or for human and animal consumption.

The barriers that need to be overcome in order to achieve cost effective and sustainable algae-bioenergy deployment can be categorized as follows (beyond the market and logistics barriers that accompany the deployment of novel technologies) (IEA Bioenergy 2017):

- Biomass productivity, energy, water, nutrient (fertilizer), greenhouse gas (GHG) emissions and land use of any algal operation needs to be sustainable across the entire value chain and data needs to be collected in a consistent and scale-relevant manner to support life-cycle analysis.
- Ecological, genetic and biochemical development of algal species is needed to improve productivity and robustness of species against perturbations such as temperature, seasonality, predation, and competition.

- Physical, chemical, biological, and post-harvest physiological variations of produced algal species need to be researched and understood and integrated with biorefinery operations.
- Integration of co-located inoculation, cultivation, primary harvest, concentration, and preprocessing systems needs to be developed to maximize the economic viability of the process.
- The value of algal biomass, on-site processing or separation of biomass lipids, carbohydrates, and/or proteins need to be maximized at scales compatible with large scale cultivation and farming
- Support of process and operations sustainability is needed to maximize nitrogen recycling, phosphorus, carbon and other nutrients from residual materials remaining after 13 preprocessings and/or residual processings to minimize fresh fertilizer input requirements in upstream cultivation.

Due to their structure, simpler than those of higher plants, microalgae can achieve considerably higher photosynthetic efficiencies than terrestrial plants (Brennan and Owende 2010). Thus, a larger share of the captured solar energy is stored through the accumulation of carbohydrates inside the cell. Considering the potential use of microalgae, the great diversity of species and the different possible conditions of cultivation, knowledge of the physiology and metabolism of these microorganisms become imperative to the development of new industrial processes.

The selection of the appropriate microalgal species for the production of biofuels is an important factor to the success of the productive process as a whole. Some of the desirable characteristics for a microalga to be interesting for the production of biofuels are listed below (Brennan and Owende 2010; Jones and Mayfield 2012).

1) Tolerate shear stresses found in the reactors, especially in closed photobioreactors;
2) Be dominant in relation to lineages of contaminating microorganisms;
3) Large CO_2 absorption capacity;
4) Tolerate large temperature variations resulting from daily and seasonal cycles;
5) Low nutrient requirement;
6) Production potential of high value-added co-products in addition to the desired product;
7) Present a short productive cycle;
8) Have high photosynthetic efficiency;
9) Be capable of auto flocculation: the formation of cell aggregates facilitates the recovery stage of the microalgal biomass.
10) Tolerate and adapt to a variety of environmental conditions, and is also able to produce several different types of biofuels

The microalga *Chlorella vulgaris*, for example, is described in the literature as both for the production of biofuels and as a source of food and pigments (Phukan et al. 2011), is easy to cultivate and exhibits sensitive behavior in relation to different process conditions. C. vulgaris is one among 4 microalgal strains that have reached commercial scale production (others are *Spirulina platensis*, *Dunaliella salina* and *Haematococcus pluvialis*).

Microalgae present several compounds in their cells, such as lipids, carbohydrates, proteins and pigments, in different concentrations. This chemical profile directly reflects the nature of the microorganism (as its species or lineage), the influence of the chosen culture conditions and the stage of growth of the culture.

Microalgae in a nitrogen deprived culture medium direct the flow of carbon to the synthesis of carbohydrates in detriment to the production of proteins (P. Behrens and Kyle 1996). Under optimized cultivation conditions, some microalgal species have been gaining special attention because they present the potential of industrial application for the production of third generation bioethanol.

The versatility of biofuel production from algae may provide answers to both the economic hurdles and the lifecycle challenges faced in renewable energy production. By extracting more than one type of biofuel from algal biomass or an additional co-product, the value of the biomass increases while additionally offsetting environmental impacts. As mentioned above, the combined biorefinery concept can be used to increase ethanol content from algae following extraction of lipids.

2.1. Biodiesel

Algal biodiesel is an algal-derived biofuel with future commercial feasibility. Many species of algae produce large amounts of lipids as storage products, as high as 50–60% of their dry weight. Upon transesterification, these lipids are chemically similar to other oilseed crop derived lipids making algae a very productive potential source of biodiesel.

One of the processes most studied, annual yield estimates for agricultural algal biofuel production can vary between 8,000 l ha^{-1} and 140,000 l ha^{-1} (Savage 2011; Scott et al. 2010). Overall output is of course important in determining the viability of a system, but even if the absolute numbers of LCA outputs vary, they still have a role in identifying the processes in need of optimization. Unlike most agricultural systems, industrial production uses photobioreactors. These photobioreactors vary, but typically have a tubular or large, thin and flat panel design (Shen et al. 2009).

Many factors can be adjusted to improve productivity and reduce costs, including culture media and nutrient sources, flow rates (amount of mixing) and sunlight exposure. Immobilized algal culture systems can be used for growth on a solid surface to prevent shading and improve photosynthesis, which is especially useful for algae that secrete fuel precursors, but early evaluation has deemed this technology too expensive (Georgianna and Mayfield 2012).

In a comparative life cycle assessment of the different growth systems, light-emitting diode (LED)-illuminated photobioreactors produced significantly more biomass (244,668 kg ha^{-1}) than solar-illuminated photobioreactors (8,262 kg ha^{-1}) and open ponds (4,957 kg ha^{-1}). But for the production costs per kilogram of biomass, this order was reversed: the estimated cost of production in open ponds was about $3 per kilogram, five times lower than that achieved with solar-lit photobioreactors and eight times lower than with LED-lit photobioreactors (Amer, Adhikari, and Pellegrino 2011; Georgianna and Mayfield 2012). So despite lower total biomass yield, the reduced input costs can result in a more efficient production system. Irrespective of the organism used, the yield per input cost is the paramount factor for the control of overall economics. Improving algal growth while keeping input costs constant can neutralize many of the efficiency losses associated with the recovery of fuel. Yield is determined by the growth rate, cell density and lipid content of the algae, so optimization of each parameter affects this downstream processing efficiency.

2.2. Biogas

Fermentation of the green alga *Chlamydomonas reinhardtii* was efficient with a production of 587 ml (±8.8 SE) biogas g volatile solids (-1) (VS(-1)), whereas fermentation of *Scenedesmus obliquus* was inefficient with only 287 ml (±10.1 SE) biogas g VS(-1) being produced. Drying as a pretreatment decreased the amount of biogas production to ca. 80%. The methane content of biogas from microalgae was 7-13% higher compared to biogas from maize silage. It was concluded that selected algal species can be good substrates for biogas production and that anaerobic fermentation can seriously be considered as final step in future microalgae-based biorefinery concepts (Mussgnug et al. 2010).

3. Third-Generation Bioethanol from Microalgae

Usage of this first generation feedstock for bioethanol production leads to various discussions about increasing food prices and occupation of agricultural land. These problems are solved partially by using second generation feedstock: lignocellulosic materials such as waste or forest residues. Second generation feedstocks have some advantages over the first generation due to not being used as a food source and having a lesser land requirement. However their harvesting, purification and various pretreatment requirements have made their production quite challenging and not economically sound to date. Algae which are of third generation feedstock for biofuels are an interesting alternative because of their productivity, easy cultivation and convenient harvesting time (John et al. 2011; Singh and Nigam 2014; Daroch, Geng, and Wang 2013). Recently, microalgae have been mostly studied for biodiesel production because of the capacity of some strains to accumulate high content of intracellular lipids.

Many strains of microalgae use carbohydrate biomolecules as an energy stock. For the production of third generation bioethanol, one should select a microalgal species with the ability to produce high concentrations of carbohydrates instead of lipids as an energy reserve compound (Mussatto and Teixeira 2010). Several studies have shown that limiting the amount of nitrogen in the culture medium is one of the main factors that leads to the accumulation of carbohydrates in the microalgae (Dragone et al. 2010; P. Behrens and Kyle 1996). These carbohydrates, especially starch, can be extracted and fermented for the production of bioethanol.

The green algae *Spirogyra* species and *Chlorococum sp.*, for example, have shown to accumulate high levels of polysaccharides both in their complex cell walls and as starch (Harun, Danquah, and Forde 2010; John et al. 2011). Hirano, Ueda, Hirayama, & Ogushi (1997) cultivated several microalgae under non-optimized conditions and identified two species with a high concentration of intracellular starch: *Chlamydomonas reinhardtii* (UTEX 2247) with 45% starch (dry basis) and *Chlorella vulgaris* with 37% starch (dry basis). In this study *C. reinhardtii* and *C. vulgaris* biomass yields were, respectively, 11 and 32 g m^{-2} day^{-1}. Under optimized conditions of nitrogen limitation, (Dragone et al. 2011) achieved 41% starch (dry basis) in the *Chlorella vulgaris* biomass, while (Behrens et al. (1989) reached 55% starch (dry basis) in the same species. A mutant strain for the production of starch is *Chlorella sp.* which can accumulate 70% starch (dry basis) under conditions of suppression of protein metabolism as was described by Doucha & Lívanský (2009).

3.1. Biomass Pretreatment

One of the most important difficulties encountered in the production of bioethanol is the pretreatment of biomass, which is needed to recover the intracell accumulated carbohydrate. Pretreatment results must be balanced against their impact on the cost of the downstream processing steps and the trade-off between operating costs, capital costs, and biomass costs (Mosier et al. 2005).

There are four pretreatment techniques including physical, chemical, physico-chemical and biological pretreatments that are applied to biomass (Vohra et al. 2014). A significant portion of the cost of ethanol production comes from these pretreatment processes (Harmsen and Huijgen 2010). Although there is no technique that can be considered as the best option, research and development efforts are being carried out to reduce cost and improve performance (Sarkar et al. 2012; Özçimen and Inan 2015).

Hydrolysis involves breakdown of the polysaccharides to simple sugars. Three methods are commonly used for hydrolyzing cellulose into glucose (C6 sugar for fermentation) (Lynd et al. 2002; Vohra et al. 2014):

1) Dilute acid hydrolysis (H_2SO_4 < 1%, 215°C and 3 min to achieve satisfactory glucose yields);
2) Concentrated acid (H_2SO_4 30-70%, 40°C, a few hours to achieve > 90% glucose yields); and
3) Enzymatic hydrolysis (cellulase mixture, 50°C several days to reach 75–95% glucose yields).

Generally, sulfuric acid, hydrochloric acid, nitric acid and phosphoric acid are used in these pretreatments. The major drawback in chemical pretreatment is that future fermentation of saccharified sugar is significantly inhibited by the byproducts generated. This is observed in the production of second generation ethanol where the extreme conditions of hydrolysis of biomass produce furfural, which is a toxic component for ethanol yeasts (Wan, Zhou, and Li 2011).

3.1.1. Dilute Acid Hydrolysis

The definition of acid concentration in hydrolysis affects directly the sugar availability for fermentation. Currently, the biomass is pre-treated with dilute sulfuric acid mostly to hydrolyze hemicellulose and facilitate enzymatic hydrolysis (Y. Y. Lee, Iyer, and Torget 1999). Miranda et al. have investigated the effect of acid pretreatments with the concentrations of 0.05-10 N, and have obtained the highest sugar yield under the condition of 2 N acid pretreatment. In their experiments, 2 N to 10 N acid pretreatments, it was reported that a decrease was observed in sugar yields; the best conditions were with 2 N, which obtained 63% of all monosaccharides extracted (Miranda, Passarinho, and Gouveia 2012). Larsson et al. also observed a similar behavior of decrease in ethanol yields with an increasing acid concentration, and the presence of formic acid, which is also a fermentation inhibitor (Larsson et al. 1999). Harun et al. (2011) investigated the best conditions of NaOH finding that the best results are with 0.75% (w/v) NaOH, 120°C and 30 min, the maximum bioethanol yield was 0,26 g ethanol/g algae (Harun et al. 2011) also Saha et al. (2005) found the same optimal concentration of sulfuric acid (Saha et al. 2005).

Thus, dilute acid treatment figures as a possibility and can be considered a cheap and effective pretreatment because microalgae do not contain lignin in their composition, and their hydrolysis requires less extreme conditions than for lignocelullosic biomass (Kim,

Tucker, and Nguyen 2005). Even so, dilute sulfuric acid is described to generate furfural (Miranda, Passarinho, and Gouveia 2012).

3.1.2. Concentrated Acid

Sulfuric (H_2SO_4) and phosphoric (H_3PO_4) acids are widely used since they are relatively cheap and efficient in hydrolyzing lignocellulose, though the latter gives a milder effect and is more benign to the environment (Bensah and Mensah 2013).

Hydrochloric (HCl) acid is more volatile and easier to recover and attacks biomass better than H_2SO_4 (Demirbas 2007); similarly, nitric acid (HNO_3) possesses good cellulose to-sugar conversion rates (Tutt, Kikas, and Olt 2012). However, both acids are expensive compared to sulfuric acid (Bensah and Mensah 2013).

Khan et al. 2017 studied the optimization of pretreatment of *Mycrocystis aeruginosa* with different pretreatments and the highest amount of fermentable sugars was found after biomass hydrolysis with 5 molar H_2SO_4 at 150°C. The lowest amount of carbohydrates resulted from the biomass treated with 10 M H_2SO_4. Stronger acid converts the monomeric sugars to furfural, causing reduction in the final yield of fermentable sugars (Khan et al. 2017).

3.1.3. Enzymatic Hydrolysis

The commercially available hydrolytic enzymes α-amylase and amyloglucosidase were employed to perform liquefaction and saccharification, respectively. Even though bioethanol yield was low compared to the yield obtained using acid pretreatment, α-amylase disrupted the cell wall completely, releasing all carbohydrates without the need for any further pretreatment. In a study performed with *Chlorella vulgari*s comparing the acid treatment to the enzymatic treatment, the results are higher with the first comparable to the enzymatic one (0,233 to 0,178 g ethanol/g algae) (Ho et al. 2013).

Cellulase enzymes consist of a mixture of endoglucanase, exoglucanase and β-glucosidase. While endoglucanase attacks the regions where cellulose fibers have low crystallinity, exoglucanase removes the cellulose units from released chains with the effect of endoglucanase and then degrades the molecule. β-glucosidase hydrolyzes the cellulose units and enables the formation of glucose (Potumarthi, Baadhe, and Bhattacharya 2013).

3.2. Fermentation

The resulting monosaccharides from the hydrolysis step can be used to produce ethanol with *Saccharomyces cerevisiae* or a mixture of acetone, n-butanol and ethanol (ABE fermentation) with *Clostridium* strains (Brethauer and Wyman 2010).

Two strategies for ethanol fermentation from complex carbohydrates are (i) simultaneous saccharification and fermentation, and (ii) separate hydrolysis and fermentation (Özçimen and Inan 2015).

In the first process, (SSF), pretreated algal biomass is hydrolyzed to glucose and subsequently fermented to ethanol in separate units. The main advantages of this process include the low cost of chemicals, short residence time, and simple equipment system, all of which promote its large-scale application (K. Li, Liu, and Liu 2014). The second process, (SHF), is a single process that combines the saccharification and fermentation processes. It has been shown that higher ethanol yields can be obtained using SSF compared with SHF (Canilha et al. 2012).

In Brazil, ethanol production industries have a production capacity that varies from 400 to 2000 m^3 ethanol per day. Typically, high yeast cell concentrations of 8-17% can achieve fermentation times of only 6-12 h with final ethanol concentrations of up to 11% v/v, which corresponds to an average ethanol yield of 91%. Fermentation times of Brazilian distilleries are faster when compared to fermentation in the USA (54–72 h) without yeast cell recycling (Lopes et al. 2016). After each fermentation cycle, the yeast cells are separated, treated with dilute sulfuric acid to kill contaminating bacteria, and then recycled to start a new fermentation process. This sequence can be repeated up to 200 times and minimizes carbon consumption during yeast growth while providing very high ethanol productivity (Basso et al. 2008).

There are several factors which influence the production of bioethanol, the most important being: temperature, pH, fermentation time and agitation rate. The growth rate of the microorganisms is directly affected by the temperature. High temperature, which is unfavorable for cell growth, becomes a stress factor for microorganisms. The ideal temperature range for fermentation is between 20 and 35°C. The pH is important because the permeability of some essential nutrients into the cells is influenced by the concentration of H^+ in the fermentation broth. In fermentation for ethanol production, the optimum pH range of S. *cerevisiae* is 4.0–5.0 (Zabed et al. 2014). Fermentation time affects the growth of microorganisms. Shorter fermentation time causes inefficient fermentation due to the inadequate growth of microorganisms. Otherwise, longer fermentation time produces a toxic effect on microbial growth especially in batch mode due to the high concentration of ethanol in the fermented broth. The fermentation time described in the literature is between 50-60 h (Markou et al. 2013; Harun et al. 2011). The Agitation rate controls the permeability of nutrients from the fermentation broth to inside the cells and removal of ethanol from the cells to the fermentation broth. The common agitation rate for fermentation by yeast cells is 150–200 rpm (Zabed et al. 2014; Ramanan et al. 2010; Mohd Azhar et al. 2017).

There are different groups of microorganisms, including yeast, bacteria, and fungi, which can be used for the fermentation of the pre-treated and saccharified algal biomass for the production of bioethanol. *S. cerevisiae* and *Z. mobilis* have been considered as the main bioethanol fermenting microorganisms (Minh and Hanh 2012).

However, fermentation of mannitol, a sugar present in certain algae is not possible in anaerobic conditions using these well-known microorganisms and requires oxygen during fermentation, which is possible only by *Zymobacter palmae* (Horn, Aasen, and Østgaard 2000). *Z. palmae* is also able to grow on glucose both under aerobic and anaerobic conditions (Okamoto et al. 1993; Horn, Aasen, and Østgaard 2000).

4. Mass Balance and Drawbacks for Bioethanol Production

The Economy plays an important role in the commercial feasibility of microalgal biofuel production. Even though commercial production of microalgae has been established for decades, the total worldwide production of algal biomass is still relatively modest (~1000 tonnes/year) and commercial applications remain primarily for higher value products (Laurens et al. 2017), especially pigments (astaxanthin from *Haematococcus pluvialis* and β-carotene from *Dunaliella salina*) and biomass for human food products (*Spirulina platensis*, *Spirulina maxima*, *Chlorella vulgaris*).

The production of biofuels from algae depends on biomass yield, the scale of production systems, the cost of recovering biomass and products, and cell composition (lipids, carbohydrates, proteins).

Assuming an annual microalgal biomass productivity of 33-42 g m^{-2} d^{-1} (Laurens et al. 2017) and 45% content of intracellular fermentable sugars (starch) (Hirano et al. 1997) that can be directed to bioethanol production, we can assume that 14-19 g_{starch} m^{-2} d^{-1} can be produced. If 95% of these sugars are successfully recovered after acid hydrolysis (Miranda, Passarinho, and Gouveia 2012) and a conversion efficiency of 92% (Walker 2010) of the maximum theoretical starch to ethanol yield (0,567 $g_{ethanol}/g_{starch}$) is achieved, a mere 7-10 $g_{ethanol}$ m^{-2} d^{-1} can be produced. This estimatee was calculated based on yields achieved mostly under laboratory controlled conditions, since these values have not yet been demonstrated in an industrial scale.

From this analysis we can estimate that in practice a maximum of 0.45 grams of ethanol can be produced from 1 gram of microalgae, which is in accordance to the scientific literature (Table 1). This means that approx. 1,7 tonnes of microalgal biomass are needed to produce 1m^3 of ethanol. It is also important to observe that the highest ethanol/microalgae yields were achieved with acid hydrolysis and separate hydrolysis and fermentation.

According to a report from IEA Bioenergy (Laurens et al. 2017), the cost of algal biomass production is between \$1.35/kg and \$1.8/kg, and is expected to reach \$0.54/kg in 2022. It is accepted that cost production of algal biomass has to drop below \$1/kg for fuel applications to be economically viable (PNNL; ANL; NREL 2012). Because the use of fertilizer and, especially, pure CO_2 for microalgal culture represent the major costs for industrial application (Acien, Fernandez, and Molina-Grima 2014), the use of alternative sources of nutrients and carbon is mandatory.

Table 1. Yields of ethanol obtained from microalgae described in the literature according to species, hydrolysis type and fermentation

Microalgal species	Hydrolysis	Fermentation	Yield (g ethanol/ g microalgae)	Reference
Chlorococcum infusionum	Alkaline (NaOH)	Separate hydrolysis and fermentation	0.26	(Harun et al. 2011)
Chlorococcum humicola	Acid (H_2SO_4)	Separate hydrolysis and fermentation	0.48	(Harun et al. 2011)
Chlorococcum sp.	Physical (Supercritical CO_2)	Separate hydrolysis and fermentation	0.38	(Harun, Danquah, and Forde 2010)
Chlorella sp.	Acid ($HCl/MgCl_2$)	Separate hydrolysis and fermentation	0.47	(Zhou et al. 2011)
Chlorella vulgaris	Acid (3% H_2SO_4)	Separate hydrolysis and fermentation with E. coli SJL2526	0.40	(S. Lee et al. 2011)
Chlorella vulgaris	Acid (H_2SO_4)	Separate hydrolysis and fermentation with *Z. mobilis*	0.23	(Ho et al. 2013)
Chlorella vulgaris	Enzymatic (Celulase+amylase)	Simultaneous saccharification and fermentation	0.21	(Ho et al. 2013)
Chlamydomonas reinhardtii	Enzymatic (α-amylase + amyloglocosidase)	Separate hydrolysis and fermentation	0.24	(Choi, Nguyen, and Sim 2010)
Chlamydomonas fasciata	Physical (Ultrasound)	Simultaneous saccharification and fermentation	0.19	(Asada et al. 2012)
Isolated strain	Alkaline (NaOH)	Separate hydrolysis and fermentation	0.43	(Saidane-Bchir et al. 2016)

Other constraints limiting algal biomass yield, biomass harvesting, and biofuel extraction were listed by Sayre (2010): need of fastest-growing, highest-biomass-yielding strains; ability to grow well across a wide range of temperature, light, and environments; increase lipid/carbohydrate accumulation with minimal biomass penalty; control of competing algae, bacteria, viruses, grazers; removal of growth-inhibiting waste products; recycle growth media and nutrients to reduce environmental impact and improve economics of harvesting and extraction systems.

5. G Bioethanol Production in the Biorefinery and the Circular Economy

The production of effluents and residues is inherent to industrial activities. Their reuse as raw materials for the production of new products for chemicals or bio technologies (biorefinery) allows an increase of capital gain in the production chain and a reduction of the polluting potential of that activity.

The current methodologies for nutrient removal from effluents are expensive, complex and energy demanding, but the cultivation of microalgae using them as a culture medium appears as an emergent alternative. This is due to the fact that these photosynthetic organisms require nitrogen and phosphorus for their growth, two of the nutrients most abundant in wastewater. Thus, the cultivation of microalgae (or consortiaof microalgae or between microalgae and bacteria) can be considered a secondary or tertiary treatment of these effluents (Gonçalves, Pires, and Simoes 2017).

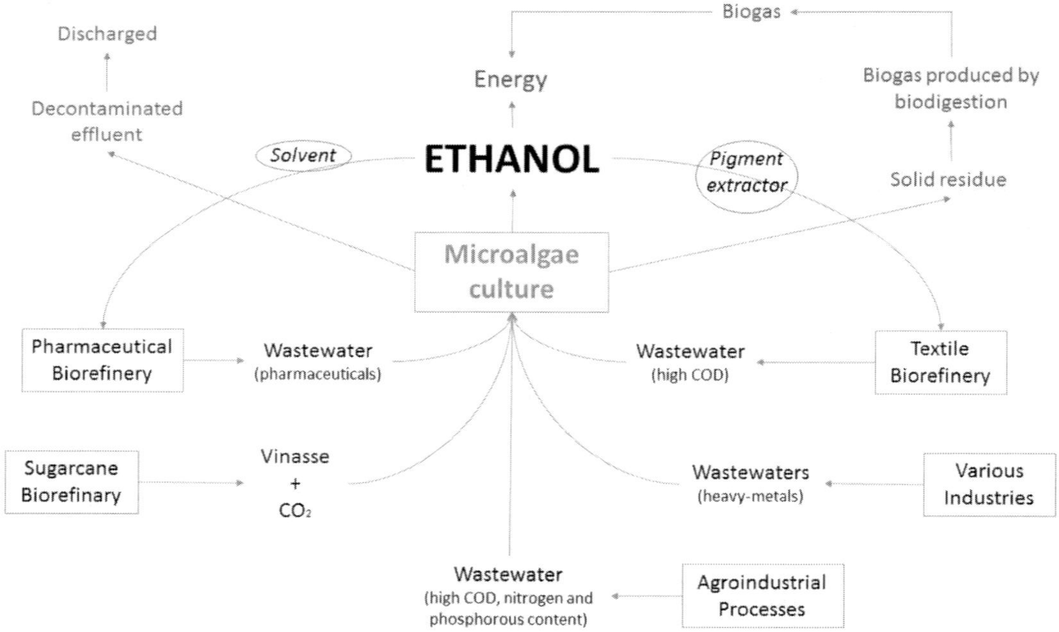

Figure 2. Integrated routes of various microalgal based bioprocesses, emphasizing the common possible end culture media for subsequent ethanol production.

Besides the treatment of liquid effluents, the use of microalgae also allows the capture of CO_2, making possible the simultaneous treatment of two types of effluents. Plants and microalgae also grow by capturing CO_2 and accumulating energy in the form of polysaccharides, which can be used as raw material for the production of alcohol. However, the use of microalgae presents some important advantages (Brasil et al., 2016):

I. High productivity (usually 10 to 100 times greater than land crops);
II. Highly efficient carbon capture;
III. High lipid or starch content, which can be used for biodiesel or ethanol, respectively;
IV. Can be cultivated in seawater, brackish water or even wastewater;
V. Can be grown over non-arable land;
VI. Can be harvested continuously throughout the year.

Despite the many advantages, competitive production of microalgal-based biofuels in the world energy markets depends on the development of integrated routes that promote the valorization of the entire biomass and the reuse of industrial liquid and gas effluent according to the biorefinery and the circular economy concepts, which are summarized in Figure 2.

5.1. Ethanol 1G and Sugar Refineries

The production of ethanol from sugarcane generates enormous amounts of liquid and gas effluents, which can be used for the production of biomass in a biorefinery strategy. Vinasse, a by-product of alcohol distillation produced from sugar and starch crops or cellulosic material, is produced in a rate of 15 liters from eachliter of alcohol. This effluent contains high levels of minerals, COD and BOD and their disposal in soils and rivers can cause environmental problems due to salinization and changes in pH levels (Carrilho, Labuto, and Kamogawa 2016). Many microalgae are able to grow in pure vinasse, with yields of biomass sometimes exceeding the yields obtained by synthetic culture media (Santana et al. 2017). The use of vinasse in a biorefinery strategy, reusing carbon dioxide from fermentation tanks, for the production of microalgal biomass and biofuels are described in patents WO2008083453 (Soccol, Carlos Ricardo; Knoerr, Cristina Elisabete; Costa, 2008), BRPI0705520 (Neto 2008), BRPI0706144 (Neto 2009), BRPI0804115 (Neto 2010), BR132012025497 (Neto 2012) and WO2013056328 (Neto 2013).

Microalgae produce starch molecules as a reservoir of energy, and thus glucose is the monosaccharide obtained in the highest proportion. In vinasse cultures it varied from 46 to 76%. This sugar can be extracted from microalgal cells and added to the fermentation vessels, increasing ethanol productivity per area. The establishment of microalgal cultivation in areas close to sugarcane processing industrial units is essential for process economics due to facilitated logistics and sharing of industrial utilities (electricity, steam, etc) (Colling Klein, Bonomi, and Maciel Filho 2017).

In terms of the quality of treated effluents vinasse acquires milder characteristics after microalgal growth, requiring little or no treatment for final disposal (Santana et al.

2017). *Chlorella sorokiniana*, for example, was described to reduce the COD of a neutral alcohol distillery wastewater by92% in 3 days, concomitant to a 95% decrease in nitrate, 77% in phosphate and 35% in sulfate (Solovchenko et al. 2014). Similarly, Sydney et al. (2016) achieved 98% COD and 89% BOD reduction in vinasse after treatment with microalgae for biodiesel production.

For gas effluents, if half the CO_2 generated in the ethanol fermentation process is directed to microalgal cultivation, greenhouse gases (GHG) are expected to be reduced by 30% (Colling Klein, Bonomi, and Maciel Filho 2017).

The use of vinasse from other raw materials other than sugarcane is also considerably explored. Sugar beet vinasse is described not only to support but increase biomass production of *Spirulina maxima*, and can be used as a supplement in a synthetic medium (Barrocal et al. 2010).

5.2. Pharmaceutical Wastewater

Bioactive pharmaceutical compounds have become a major risk to nature and humans since they were detected in aquatic systems a few decades ago (Xie et al. 2017). These substances include non-steroidal anti-inflammatory drugs, antibiotics, β-blockers, antidepressants and antiepileptics (Petrie, Barden, and Kasprzyk-Hordern 2015).

It is appropriate to consider using this type of residue for the growth of microalgae for the production of alcohol. Even though microalgae accumulate these substances in their biomass, the alcohol generated will not carry these substances with them, enabling their use in the chain either as fuel for the generation of heat within industry, or purified for application in formulations or production processes of such drugs. Many solvents are used for the recovery and extraction of drug products, among which ethanol can be used satisfactorily (Moreira et al. 2009), and in this case, presenting the great advantage of a low production cost.

Some species have demonstrated capacity for growth in these wastes and degradation of important drugs: *Chlamydomonas mexicana*, *Scenedesmus obliquus*, *Chlamydomonas pitschmannii*, *Micractinium reisseri*, *Chlorella pyrenoidosa (Zhang et al. 2012)*, *Scenedesmus quadricauda*, *Chlorella kessleri* (Mackul'ak et al. 2015) and consortia between microalgae and bacteria (Matamoros, Uggetti, García, & Bayona, 2016).

5.3. Dye-Containing Wastewater

The textile industry produces a large amount of effluent, containing dyes and paints as its main constituent. Typically, these effluents contain strong color, high salinity, high pH and high chemical oxygen demand. In a first approach, it is important to check whether the micro-organism is tolerant to the dye to be treated.

Microalgae can remove these dyes and paints through two processes: biosorption or reductive mechanisms (Wang et al. 2016). Depending on the mechanism employed to decrease the dye in the solution, the final biomass may or may not exhibit the molecule or fragments thereof.

There are species of microalgae able to degrade dyes: *Chlorella vulgaris*, *Caulerpa lentillifera*, *Synechocystis sp.*, *Phormidium sp.*, *Lyngbya lagerlerimi*, *Nostoc lincki*, *Oscillatoria rubescens*, *Elkatotrix viridis* and *Volvox aureus (Wang et al. 2016)*, but their content of carbohydrates has not been determined. Methyl Red, Orange II, G-red, Basic Fuchsin (El-Sheekh, Gharieb, and Abou-El-Souod 2009) are examples of industrial dyes that have already been degraded by microalgae.

In a scenario where microalgae are used in industrial dye degradation with 3G ethanol production, the alcohol could be reallocated in the production process as a solvent extractor to obtain dyes from natural products (Santos, Albuquerque, and Meireles 2011; Al-Alwani et al. 2016).

5.4. Metal-Containing Wastewater

The inappropriate disposal of effluents containing heavy metals in the aquatic environment has been a growing concern, and the legislation governing this disposal has come and become stronger over the last few years. Most current technologies are based on physicochemical reactions, mainly precipitation and adsorption on ion-exchange resins. However, these processes have some problems such as low selectivity, intolerance to organic species, low removal when the pollutant is diluted (traces) and the generation of large amounts of secondary effluents that require high costs for treatment (Perales-Vela, Peña-Castro, and Cañizares-Villanueva 2006).

Microalgal cell walls contain negatively charged carbohydrates, which facilitate the binding of positively charged metal ions. In addition to this adsorption on the surface of the cell, other heavy metal removal mechanisms from wastewater involve bioaccumulation, chelating, and internal or surface precipitation (Wang et al. 2016; Suresh Kumar et al. 2015).

As a result, the adsorption of copper and cadmium can be efficiently achieved from the effluent through the growth of microalgae (Wang et al. 2016; Suresh Kumar et al. 2015). Napan et al. (2015) evaluated the impact of heavy metals from flue gas integration with microalgal production. The authors concluded that by applying the same concentration of metals found in this residue for microalgal growth it was possible to increase biomass by 12% and lipid concentration by 61%. However, when tested at higher concentrations, there was inhibition of growth and accumulation of lipids.

Microalgae reported as capable of removing heavy metals include *Chlorella vulgaris*, *Scenedesmus quadricauda*, *Spirulina platensis*, *Spirulina maxima*, *Scenedesmus subspicatus*, *Chlamydomonas reinhardtii*, *Microcystis aeruginosa*, *Porphyridium cruentum*, *Desmodesmus pleiomorphum* among others (Wang et al. 2016; Suresh Kumar et al. 2015). Consortia between different microalgae or between microalgae and bacteria have also been identified as efficient in the removal of heavy metals (Perales-Vela, Peña-Castro, and Cañizares-Villanueva 2006).

The direct use of microalgal biomass produced in heavy metal containing media in the food industry or their by-products and extracts in the pharmaceutical, cosmetic or pharmaceutical industries is contraindicated. Thus, the use of the starch produced by these microalgae for the production of alcohol is a suitable destination, since metals will not reach the final product due to bioethanol distillation in the purification step (the distillation effluent, rich in metals, is still problematic).

5.5. Agroindustrial Wastewater

Agroindustrial effluents may include a wide range of wastes generated from numerous processes. Many of them have high nitrogen content, especially in N-NH_3 form, which is toxic for most microorganisms. Among the N-NH_3 rich agroindustrial effluents, livestock, dairy, slaughterhouse and fish farm wastewater deserve attention because they are produced in large volumes.

Microalgal strains capable of using N-NH3 as a nitrogen source include *Chlorella pyrenoidosa and Chlamydomonas snowiae (Lin et al. 2007), Scenedesmus obliquus* and *C. vulgaris (Ruiz-Marin, Mendoza-Espinosa, and Stephenson 2010).* The co-culture of microalgae and bacteria is also described (Delgadillo-Mirquez et al. 2016) as well as the use of mixed green algae (Su, Mennerich, and Urban 2012).

In any agroindustrial process there are important mechanical processes for grinding, extraction, mixing or processing, which are done by motors. In this context, the use of 3G microalgal ethanol as fuel in one or more of these mechanical processes can be considered.

5.6. Microalgal Biomass in a Biorefinery Concept

The concept of the biorefinery is similar to a traditional petroleum refinery, where biomass is converted into renewable chemicals, fuels and products of commercial interest (Pérez et al., 2017). Microalgae contain a high amount of proteins, lipids and carbohydrates which could be the feedstock for different products (Chew et al. 2017). But to become technically and economic feasible, the major bottleneck in the use of microalgae in a biorefinery approach is the separation of fractions without harming the others (Chew et al. 2017) which must be overcome.

The lipid-free microalgal biomass still contains considerable amounts of carbohydrates. This carbohydrate rich biomass can be added to a fermentation vessel together with simple sugars and/or treated lignocellulosic biomass for ethanol production, directed to an anaerobic pond to generate biogas since microalgae offer good potential for biogas production (Gendy and El-Temtamy 2013). Microalgal proteins, pigments and carbohydrates are described to present important biological activity and can be used by the pharmaceutical industries as immune system boosters (Yen et al. 2013) or in cosmetic formulations (Novak, Sydney, and Soccol 2014).

During biomass processing for the extraction of bioproducts, degrading processes are used (hydrolysis, solvent extraction) that may hinder the recovery of other co-products, especially high valued co-products. The extraction of lipids for biodiesel production, for example, includes biomass hydrolysis (frequently) and solvent extraction, which destroy the biological activity of most microalgal carbohydrates. In this case, the sugar fraction must be used exclusively for bioethanol production. Thus, in a microalgal biorefinery approach where more than 1 product from the biomass is desired, the extraction of high valued/termolable/solvent sensible products must be considered first.

CONCLUSION

Microalgae are versatile microorganisms that can grow very fast using organic and inorganic carbon. The possibility of using wastewater and industrial gases together with bioremediation, CO_2 mitigation and the production of biomass rich in compounds of commercial interest in lands where agriculture is impossible or restricted have made microalgae gain a lot of importance in recent decades. However, very few strains are cultivated in industrial scale and most microalgal products commercially available are of high value. The production of biofuels from microalgae faces not only the common economic constrains for the production of any microalgal product that includes the high cost of fertilizers and carbon dioxide, but also the low biomass concentration and content of carbohydrates (for bioethanol) or lipids (for biodiesel). In the case of bioethanol, 1.6 tonnes of dry biomass are needed to produce $1 m^3$ of ethanol.

Despite the scientific and technological advances in microalgal cultivation and facilities, the reuse of microalgal technology in the concept of the circular economy (where the generation of residue tends to equal zero) and in a biorefinery approach (extraction of many products from the biomass) are expected to be key strategies to overcome economic barriers. In addition, the advance of global warming and the degradation of the environment according to the evolution of humanity must become external factors that will constitute the scenario in which microalgal fuels are economically and technologically viable.

References

Abdel-Raouf, N., Al-Homaidan, A. A., Ibraheem, I. B. M. 2012. "Microalgae and Wastewater Treatment." *Saudi Journal of Biological Sciences* 19: 257–75.

Acien, F. G., J. M. Fernandez, and E. Molina-Grima. 2014. "Economics of Microalgae Biomass Production." In *Biofuels from Algae*, 313–25. Elsevier. doi:10.1016/B978-0-444-59558-4.00014-0.

Al-Alwani, Mahmoud A. M., Norasikin A. Ludin, Abu Bakar Mohamad, Abd. Amir H. Kadhum, Muneer M. Baabbad, and Kamaruzzaman Sopian. 2016. "Optimization of Dye Extraction from Cordyline Fruticosa via Response Surface Methodology to Produce a Natural Sensitizer for Dye-Sensitized Solar Cells." *Results in Physics* 6: 520–29. doi:10.1016/j.rinp.2016.08.013.

Amer, Luke, Birendra Adhikari, and John Pellegrino. 2011. "Technoeconomic Analysis of Five Microalgae-to-Biofuels Processes of Varying Complexity." *Bioresource Technology* 102 (20): 9350–59. doi:https://doi.org/10.1016/j.biortech.2011.08.010.

Asada, Chikako, Keita Doi, Chizuru Sasaki, and Yoshitoshi Nakamura. 2012. "Efficient Extraction of Starch from Microalgae Using Ultrasonic Homogenizer and Its Conversion into Ethanol by Simultaneous Saccharification and Fermentation." *Natural Resources* 3 (4): 175–79. doi:10.4236/nr.2012.34023.

Barrocal, Víctor M., M. Teresa García-Cubero, Gerardo González-Benito, and Mónica Coca. 2010. "Production of Biomass by Spirulina Maxima Using Sugar Beet Vinasse in Growth Media." *New Biotechnology* 27 (6): 851–56. doi:10.1016/j.nbt.2010.07.001.

Basso, Luiz C., Henrique V. De Amorim, Antonio J. De Oliveira, and Mario L. Lopes. 2008. "Yeast Selection for Fuel Ethanol Production in Brazil." *FEMS Yeast Research* 8 (7): 1155–63. doi:10.1111/j.1567-1364.2008.00428.x.

Behrens, Paul, and David Kyle. 1996. "Microalgae as a Source of Fatty Acids." *Journal of Food Lipids* 3 (4). Blackwell Publishing Ltd: 259–72. doi:10.1111/j.1745-4522.1996.tb00073.x.

Behrens, Paul W., Scott E. Bingham, Scot D. Hoeksema, Deborah L. Cohoon, and John C. Cox. 1989. *Studies on the Incorporation of CO_2 into Starch by Chlorella Vulgaris*, 123–30.

Bensah, Edem Cudjoe, and Moses Mensah. 2013. "Chemical Pretreatment Methods for the Production of Cellulosic Ethanol : Technologies and Innovations." *International Journal of Chemical Engineering* 2013: 1–21. doi:10.1155/2013/719607.

Borowitzka, Michel A. 2016. "Algal Physiology and Large-Scale Outdoor Cultures of Microalgae." In *The Physiology of Microalgae*, edited by Michel A. Borowitzka, John Beardall, and John A. Raven, 1st ed., 680. Springer.

Brennan, Liam, and Philip Owende. 2010. "Biofuels from microalgae—A Review of Technologies for Production, Processing, and Extractions of Biofuels and Co-Products." *Renewable and Sustainable Energy Reviews* 14 (2): 557–77. doi: https://doi.org/10.1016/j.rser.2009.10.009.

Brethauer, Simone, and Charles E. Wyman. 2010. "Review: Continuous Hydrolysis and Fermentation for Cellulosic Ethanol Production." *Bioresource Technology* 101 (13). Elsevier Ltd: 4862–74. doi:10.1016/j.biortech.2009.11.009.

Canilha, Larissa, Anuj Kumar Chandel, Thais Suzane, Fernandes Antunes, Wagner Luiz, Maria Grac, Almeida Felipe, and Silvio Silv. 2012. "Bioconversion of Sugarcane Biomass into Ethanol : An Overview about Composition, Pretreatment Methods, Detoxification of Hydrolysates, Enzymatic Saccharification, and Ethanol Fermentation." *Journal of Biomedicine and Biotechnology* 2012: 1–15. doi:10.1155/2012/989572.

Carrilho, E. N. V. M., G. Labuto, and M. Y. Kamogawa. 2016. "Destination of Vinasse, a? Residue from Alcohol Industry." In *Environmental Materials and Waste*, 21–43. Elsevier. doi:10.1016/B978-0-12-803837-6.00002-0.

Chew, Kit Wayne, Jing Ying Yap, Pau Loke Show, Ng Hui Suan, Joon Ching Juan, Tau Chuan Ling, Duu-Jong Lee, and Jo-Shu Chang. 2017. "Microalgae Biorefinery: High Value Products Perspectives." *Bioresource Technology* 229 (April): 53–62. doi:10.1016/j.biortech.2017.01.006.

Choi, Seung Phill, Minh Thu Nguyen, and Sang Jun Sim. 2010. "Enzymatic Pretreatment of Chlamydomonas Reinhardtii Biomass for Ethanol Production." *Bioresource Technology* 101 (14): 5330–36. doi:http://dx.doi.org/10.1016/j.biortech.2010.02.026.

Colling Klein, Bruno, Antonio Bonomi, and Rubens Maciel Filho. 2017. "Integration of Microalgae Production with Industrial Biofuel Facilities: A Critical Review." *Renewable and Sustainable Energy Reviews*, June. doi:10.1016/j.rser.2017.04.063.

Cuellar-Bermudez, Sara P., Jonathan S. Garcia-Perez, Bruce E. Rittmann, and Roberto Parra-Saldivar. 2015. "Photosynthetic Bioenergy Utilizing CO<inf>2</inf>: An Approach on Flue Gases Utilization for Third Generation Biofuels." *Journal of Cleaner Production* 98: 53–65. doi:10.1016/j.jclepro.2014.03.034.

Dalrymple, Omatoyo K., Trina Halfhide, Innocent Udom, Benjamin Gilles, John Wolan, Qiong Zhang, and Sarina Ergas. 2013. "Wastewater Use in Algae Production for Generation of Renewable Resources: A Review and Preliminary Results." *Aquatic Biosystems* 9 (1): 2. doi:10.1186/2046-9063-9-2.

Daroch, Maurycy, Shu Geng, and Guangyi Wang. 2013. "Recent Advances in Liquid Biofuel Production from Algal Feedstocks." *Applied Energy* 102: 1371–81. doi:https://doi.org/10.1016/j.apenergy.2012.07.031.

Delgadillo-Mirquez, Liliana, Filipa Lopes, Behnam Taidi, and Dominique Pareau. 2016. "Nitrogen and Phosphate Removal from Wastewater with a Mixed Microalgae and Bacteria Culture." *Biotechnology Reports* 11 (September): 18–26. doi:10.1016/j.btre.2016.04.003.

Demirbas, A. 2007. "Products from Lignocellulosic Materials via Degradation Processes." *Energy Sources, Part A: Recovery, Utilization, and Environmental Effects* 30 (1). Taylor & Francis: 27–37. doi:10.1080/00908310600626705.

Doucha, J., and K. Lívanský. 2009. "Outdoor Open Thin-Layer Microalgal Photobioreactor: Potential Productivity Outdoor Open Thin-Layer Microalgal Photobioreactor: Potential Productivity." *Journal of Applied Phycology*. doi:10.1007/s10811-008-9336-2.

Dragone, Giuliano, B. D. Fernandes, A. A. Vicente, and J. A. Teixeira. 2010. "Third Generation Biofuels from Microalgae." *Current Research, Technology and Education Topics in Applied Microbiology and Microbial Biotechnology*, 1355–66. doi:10.1016/j.apenergy.2011.03.012.

Dragone, Giuliano, Bruno D. Fernandes, Ana P. Abreu, António A. Vicente, and José A. Teixeira. 2011. "Nutrient Limitation as a Strategy for Increasing Starch Accumulation in Microalgae." *Applied Energy* 88 (10). Elsevier Ltd: 3331–35. doi:10.1016/j.apenergy.2011.03.012.

Ducat, Daniel C., and Pamela A. Silver. 2012. "Improving Carbon Fixation Pathways." *Current Opinion in Chemical Biology* 16 (3–4): 337–44. doi:10.1016/j.cbpa.2012.05.002.

El-Sheekh, Mostafa M., M. M. Gharieb, and G. W. Abou-El-Souod. 2009. "Biodegradation of Dyes by Some Green Algae and Cyanobacteria." *International Biodeterioration & Biodegradation* 63 (6): 699–704. doi:10.1016/j.ibiod.2009.04.010.

Fan, Li Hai, Ya Tao Zhang, Lin Zhang, and Huan Lin Chen. 2008. "Evaluation of a Membrane-Sparged Helical Tubular Photobioreactor for Carbon Dioxide Biofixation by Chlorella Vulgaris." *Journal of Membrane Science* 325 (1): 336–45. doi:10.1016/j.memsci.2008.07.044.

Gendy, Tahani S., and Seham A. El-Temtamy. 2013. "Commercialization Potential Aspects of Microalgae for Biofuel Production: An Overview." *Egyptian Journal of Petroleum* 22 (1): 43–51. doi:10.1016/j.ejpe.2012.07.001.

Georgianna, D. Ryan, and Stephen P. Mayfield. 2012. "Exploiting Diversity and Synthetic Biology for the Production of Algal Biofuels." *Nature* 488 (7411): 329–35. doi:10.1038/nature11479.

Gonçalves, Ana L., Jose C. M. Pires, and Manuel Simoes. 2017. "A Review on the Use of Microalgal Consortia for Wastewater Treatment." *Algal Research* 24 (June): 403–15. doi:10.1016/j.algal.2016.11.008.

Harmsen, P., and W. Huijgen. 2010. *Literature Review of Physical and Chemical Pretreatment Processes for Lignocellulosic Biomass*, no. September: 1–49.

Harun, Razif, Michael K. Danquah, and Gareth M. Forde. 2010. "Microalgal Biomass as a Fermentation Feedstock for Bioethanol Production." *Journal of Chemical Technology & Biotechnology* 85 (2). John Wiley {&} Sons, Ltd.: 199–203. doi:10.1002/jctb.2287.

Harun, Razif, W. S. Y. Jason, Tamara Cherrington, and Michael K. Danquah. 2011. "Exploring Alkaline Pretreatment of Microalgal Biomass for Bioethanol Production." *Applied Energy* 88 (10): 3464–67. doi:http://dx.doi.org/10.1016/j.apenergy.2010.10.048.

Hirano, Atsushi, Ryohei Ueda, Shin Hirayama, and Yasuyuki Ogushi. 1997. "CO_2 Fixation and Ethanol Production with Microalgal Photosynthesis and Intracellular Anaerobic Fermentation." *Energy* 22 (2): 137–42.

Ho, Shih-Hsin, Shu-Wen Huang, Chun-Yen Chen, Tomohisa Hasunuma, Akihiko Kondo, and Jo-Shu Chang. 2013. "Bioethanol Production Using Carbohydrate-Rich Microalgae Biomass as Feedstock." *Bioresource Technology* 135: 191–98. doi:http://dx.doi.org/10.1016/j.biortech.2012.10.015.

Horn, S. J., I. M. Aasen, and K. Østgaard. 2000. "Production of Ethanol from Mannitol by Zymobacter Palmae." *Journal of Industrial Microbiology and Biotechnology* 24 (1): 51–57. doi:10.1038/sj.jim.2900771.

Huang, Guanhua, Feng Chen, Yali Kuang, Huan He, and An Qin. 2016. "Current Techniques of Growing Algae Using Flue Gas from Exhaust Gas Industry: A Review." *Applied Biochemistry and Biotechnology* 178 (6): 1220–38. doi:10.1007/s12010-015-1940-4.

IEA Bioenergy. 2017. *State of Technology Review - Algae Bioenergy* (ISBN: 978-1-910154-30-4) http://www.ieabioenergy.com/wp-content/uploads/2017/01/IEA-Bioenergy-Algae-Report-Update-20170114.pdf.

John, Rojan P., G. S. Anisha, K. Madhavan Nampoothiri, and Ashok Pandey. 2011. "Micro and Macroalgal Biomass: A Renewable Source for Bioethanol." *Bioresource Technology* 102 (1): 186–93. doi:https://doi.org/10.1016/j.biortech.2010.06.139.

Jones, Carla S., and Stephen P. Mayfield. 2012. "Algae Biofuels: Versatility for the Future of Bioenergy." *Current Opinion in Biotechnology* 23 (3). Elsevier Ltd: 346–51. doi:10.1016/j.copbio.2011.10.013.

Khan, Muhammad Imran, Moon Geon Lee, Jin Hyuk Shin, and Jong Deog Kim. 2017. "Pretreatment Optimization of the Biomass of Microcystis Aeruginosa for Efficient Bioethanol Production." *AMB Express*. Springer Berlin Heidelberg. doi:10.1186/s13568-016-0320-y.

Kim, Kyoung Heon, Melvin Tucker, and Quang Nguyen. 2005. *Conversion of Bark-Rich Biomass Mixture into Fermentable Sugar by Two-Stage Dilute Acid-Catalyzed Hydrolysis*. 96: 1249–55. doi:10.1016/j.biortech.2004.10.017.

Kirrolia, Anita, Narsi R. Bishnoi, and Rajesh Singh. 2013. "Microalgae as a Boon for Sustainable Energy Production and Its Future Research {&} Development Aspects." *Renewable and Sustainable Energy Reviews* 20: 642–56. doi:https://doi.org/10.1016/j.rser.2012.12.003.

Kool, A., M. Marinussen, and H. Blonk. 2012. *LCI Data for the Calculation Tool Feedprint for Greenhouse Gas Emissions of Feed Production and Utilization - GHG Emissions of N, P and K Fertilizer Production.*

Kumar, Kanhaiya, Chitralekha Nag Dasgupta, Bikram Nayak, Peter Lindblad, and Debabrata Das. 2011. "Development of Suitable Photobioreactors for CO_2 Sequestration Addressing Global Warming Using Green Algae and Cyanobacteria." *Bioresource Technology* 102 (8): 4945–53. doi:10.1016/j.biortech.2011.01.054.

Kumar, Kanhaiya, Sanjiv K. Mishra, Anupama Shrivastav, Min S. Park, and Ji-Won Yang. 2015. "Recent Trends in the Mass Cultivation of Algae in Raceway Ponds." *Renewable and Sustainable Energy Reviews* 51 (November): 875–85. doi:10.1016/j.rser.2015.06.033.

Lam, Man Kee, and Keat Teong Lee. 2015. "Bioethanol Production from Microalgae." In *Handbook of Marine Microalgae*, 197–208. Elsevier. doi:10.1016/B978-0-12-800776-1.00012-1.

Lara-Gil, Javier A., Mario M. Álvarez, and Adriana Pacheco. 2014. "Toxicity of Flue Gas Components from Cement Plants in Microalgae CO_2 Mitigation Systems." *Journal of Applied Phycology* 26 (1): 357–68. doi:10.1007/s10811-013-0136-y.

Larsson, Simona, Eva Palmqvist, Bärbel Hahn-Hägerdal, Charlotte Tengborg, Kerstin Stenberg, Guido Zacchi, and Nils-Olof Nilvebrant. 1999. "The Generation of Fermentation Inhibitors during Dilute Acid Hydrolysis of Softwood." *Enzyme and Microbial Technology* 24 (3): 151–59. doi:http://dx.doi.org/10.1016/S0141-0229(98)00101-X.

Laurens, Lieve M. L., James D. McMillan, David Baxter, Annette L. Cowie, Jack (John) N. Saddler, Maria Barbosa, Jerry Murphy, et al. 2017. *State of Technology Review – Algae Bioenergy*. doi:10.13140/RG.2.2.11770.90560.

Lee, Soojin, Younghoon Oh, Donghyun Kim, Doyeon Kwon, Choulgyun Lee, and Jinwon Lee. 2011. "Converting Carbohydrates Extracted from Marine Algae into Ethanol Using Various Ethanolic Escherichia Coli Strains." *Applied Biochemistry and Biotechnology* 164 (6): 878–88. doi:10.1007/s12010-011-9181-7.

Lee, Y. Y., Prashant Iyer, and R. W. Torget. 1999. "Dilute-Acid Hydrolysis of Lignocellulosic Biomass." In *Recent Progress in Bioconversion of Lignocellulosics*, edited by G. T. Tsao, A. P. Brainard, H. R. Bungay, N. J. Cao, P. Cen, Z. Chen, J. Du, et al., 93–115. Berlin, Heidelberg: Springer Berlin Heidelberg. doi:10.1007/3-540-49194-5_5.

Leung, Dennis Y. C., Giorgio Caramanna, and M. Mercedes Maroto-Valer. 2014. "An Overview of Current Status of Carbon Dioxide Capture and Storage Technologies."

Renewable and Sustainable Energy Reviews 39 (November): 426–43. doi:10.1016/j.rser.2014.07.093.

Li, Kexun, Shun Liu, and Xianhua Liu. 2014. "An Overview of Algae Bioethanol Production." *International Journal of Energy Research* 38 (8): 965–77. doi:10.1002/er.3164.

Li, Yanqun, Mark Horsman, Bei Wang, Nan Wu, and Christopher Lan. 2008. "Effects of Nitrogen Sources on Cell Growth and Lipid Accumulation of Green Alga Neochloris Oleoabundans." *Applied Microbiology and Biotechnology* 81 (4): 629–36.

Lin, L., G. Y. S. Chan, B. L. Jiang, and C. Y. Lan. 2007. "Use of Ammoniacal Nitrogen Tolerant Microalgae in Landfill Leachate Treatment." *Waste Management* 27 (10): 1376–82. doi:10.1016/j.wasman.2006.09.001.

Lopes, Mario Lucio, Silene Cristina, De Lima Paulillo, Alexandre Godoy, Rudimar Antonio Cherubin, Marcel Salmeron Lorenzi, Fernando Henrique, et al. 2016. "Ethanol Production in Brazil: A Bridge between Science and Industry." *Brazilian Journal of Microbiology* 47. Sociedade Brasileira de Microbiologia: 64–76. doi:10.1016/j.bjm.2016.10.003.

Lun, Michiel van, Jochen S. Hub, David van der Spoel, and Inger Andersson. 2014. "CO_2 and O_2 Distribution in Rubisco Suggests the Small Subunit Functions as a CO_2 Reservoir." *Journal of the American Chemical Society* 136 (8): 3165–71. doi:10.1021/ja411579b.

Lv, Junping, Jia Feng, Qi Liu, and Shulian Xie. 2017. "Microalgal Cultivation in Secondary Effluent: Recent Developments and Future Work." *International Journal of Molecular Sciences* 18 (1): 79. doi:10.3390/ijms18010079.

Lynd, Lee R., Paul J. Weimer, Willem H. Van Zyl, and Isak S. Pretorius. 2002. *Microbial Cellulose Utilization: Fundamentals and Biotechnology* 66 (3): 506–77. doi:10.1128/MMBR.66.3.506.

Mackul'ak, Tomás, Michal Mosný, Roman Grabic, Oksana Golovko, Olga Koba, and Lucia Birosová. 2015. "Fenton-like Reaction: A Possible Way to Efficiently Remove Illicit Drugs and Pharmaceuticals from Wastewater." *Environmental Toxicology and Pharmacology* 39 (2): 483–88. doi:10.1016/j.etap.2014.12.016.

Markou, Giorgos, Irini Angelidaki, Elias Nerantzis, and Dimitris Georgakakis. 2013. *Bioethanol Production by Carbohydrate-Enriched Biomass of*, 3937–50. doi:10.3390/en6083937.

Matamoros, Víctor, Enrica Uggetti, Joan García, and Josep M. Bayona. 2016. "Assessment of the Mechanisms Involved in the Removal of Emerging Contaminants by Microalgae from Wastewater: A Laboratory Scale Study." *Journal of Hazardous Materials* 301 (January): 197–205. doi:10.1016/j.jhazmat.2015.08.050.

Minh, Thi Hong, and Van Hanh. 2012. "Bioethanol Production from Marine Algae Biomass : Prospect and Troubles." *Journal of Vietnamise Environment* 3 (1): 55–29. doi:10.13141/jve.vol3.no1.pp25-29.

Miranda, J. R., P. C. Passarinho, and L. Gouveia. 2012. "Pretreatment Optimization of Scenedesmus Obliquus Microalga for Bioethanol Production." *Bioresource Technology* 104: 342–48. doi:http://dx.doi.org/10.1016/j.biortech.2011.10.059.

Mohd Azhar, Siti Hajar, Rahmath Abdulla, Siti Azmah Jambo, Hartinie Marbawi, Jualang Azlan Gansau, Ainol Azifa Mohd Faik, and Kenneth Francis Rodrigues. 2017. "Yeasts in Sustainable Bioethanol Production: A Review." *Biochemistry and Biophysics Reports* 10 (February): 52–61. doi:10.1016/j.bbrep.2017.03.003.

Morais, Michele Greque de, and Jorge Alberto Vieira Costa. 2007. "Isolation and Selection of Microalgae from Coal Fired Thermoelectric Power Plant for Biofixation of Carbon Dioxide." *Energy Conversion and Management* 48 (7): 2169–73. doi:10.1016/j.enconman.2006.12.011.

Moreira, Tailane Sant'Anna, Yara Peluso Cid, Maria Bernadete Riemma Pierre, Valeria Pereira de Sousa, Arthur Eugen Kummerle, and Carlos Alberto Manssour Fraga. 2009. "Extração E Purificação de Fármacos Anti-Inflamatórios Não Esteroidais Ciclo-Oxigenase-2 Seletivos." ["Extraction and Purification of Non-Steroidal Anti-Inflammatory Drugs Selective Cytoxygenase-2."] *Química Nova* 32 (5): 1324–28. doi:10.1590/S0100-40422009000500040.

Moroney, James V., and Aravind Somanchi. 1999. "No Title." *Plant Physiology* 119 (1): 9–16. doi:http://dx.doi.org/10.1104/pp.119.1.9.

Mosier, Nathan, Charles Wyman, Bruce Dale, Richard Elander, Y. Y. Lee, Mark Holtzapple, and Michael Ladisch. 2005. "Features of Promising Technologies for Treatment of Lignocellulosic Biomass." *Bioresource Technology* 96: 673–86.

Mussatto, Si, and Ja Teixeira. 2010. "Lignocellulose as Raw Material in Fermentation Processes." *Applied Microbiology and Microbial Biotechnology* 2: 897–907. doi:10.1016/j.jrras.2014.02.003.

Mussgnug, J. H., V. Klassen, A. Schlüter, and O. Kruse. 2010. "Microalgae as Substrates for Fermentative Biogas Production in a Combined Biorefinery Concept." *Journal of Biotechnology* 150 (1): 51–56. doi:http://dx.doi.org/10.1016/j.jbiotec.2010.07.030.

Napan, Katerine, Lihong Teng, Jason C. Quinn, and Byard D. Wood. 2015. "Impact of Heavy Metals from Flue Gas Integration with Microalgae Production." *Algal Research* 8 (March): 83–88. doi:10.1016/j.algal.2015.01.003.

Neto, Dolivar Coraucci. 2008. *Integrated technological process for the production of lipids and biodiesel by algae using vinasse and / or CO2 substrates, residues from alcohol distilleries (ethanol)*. BRPI0705520 (A2), issued 2008.

———. 2009. *Technological process for the recycling of vinasse and CO2 from the burning of bagasse in alcohol distilleries and sugar mills through the algal crop.* BRPI0706144 (A2), issued 2009.

———. 2010. *Process for the production of biodiesel and/or fuel oil*. BRPI0804115 (A2), issued 2010.

———. 2012. *Process for the production of biofuels from the pyrolysis of microalgae*. BR20121325497 20121005, issued 2012.

———. 2013. *Carbon fixation method for reducing the turbidity of sugarcane vinasse and thus processes vinasse*. WO2013056328 (A1), issued 2013.

Novak, Alessandra Cristine, Eduardo Bittencourt Sydney, and Carlos Ricardo Soccol. 2014. "Biocosmetics." In *Biotransformation of Waste Biomass into High Value Biochemicals*, 389–411. New York, NY: Springer New York. doi: 10.1007/978-1-4614-8005-1_16.

Oey, M.; Sawyer, A. L., Ross, I. L., Hankamer, B. 2016. "Challenges and Opportunities for Hydrogen Production from Microalgae." *Plant Biotechnol J.* 14 (7): 1487–99.

Okamoto, T., H. Taguchi, K. Nakamura, H. Ikenaga, H. Kuraishi, and K. Yamasato. 1993. "Zymobacter Palmae Gen. Nov., Sp. Nov., a New Ethanol-Fermenting Peritrichous Bacterium Isolated from Palm Sap." *Archives of Microbiology* 160 (5). Germany: 333–37.

Özçimen, Didem, and Benan Inan. 2015. "An Overview of Bioethanol Production from Algae." *Biofuels - Status and Perspective* 2013. doi:10.5772/59305.

Perales-Vela, Hugo Virgilio, Julián Mario Peña-Castro, and Rosa Olivia Cañizares-Villanueva. 2006. "Heavy Metal Detoxification in Eukaryotic Microalgae." *Chemosphere* 64 (1): 1–10. doi:10.1016/j.chemosphere.2005.11.024.

Petrie, Bruce, Ruth Barden, and Barbara Kasprzyk-Hordern. 2015. "A Review on Emerging Contaminants in Wastewaters and the Environment: Current Knowledge, Understudied Areas and Recommendations for Future Monitoring." *Water Research* 72 (April): 3–27. doi:10.1016/j.watres.2014.08.053.

Phukan, Mayur M., Rahul S. Chutia, B. K. Konwar, and R. Kataki. 2011. "Microalgae Chlorella as a Potential Bio-Energy Feedstock." *Applied Energy* 88 (10): 3307–12. doi:https://doi.org/10.1016/j.apenergy.2010.11.026.

PNNL; ANL; NREL. 2012. *Renewable Diesel from Algal Lipids: An Integrated Baseline for Cost, Emissions, and Resource Potential from a Harmonized Model*. Argonne, IL: Argonne National Laboratory; Golden, CO: National Renewable Energy Laboratory; Richland, WA: Pacific Northwest National Laboratory. http://www.nrel.gov/docs/fy12osti/55431.pdf.

Polishchuk, Anna, Dimitar Valev, Marko Tarvainen, Sujata Mishra, Viljami Kinnunen, Taras Antal, Baoru Yang, Jukka Rintala, and Esa Tyystj?rvi. 2015. "Cultivation of Nannochloropsis for Eicosapentaenoic Acid Production in Wastewaters of Pulp and Paper Industry." *Bioresource Technology* 193 (October): 469–76. doi:10.1016/j.biortech.2015.06.135.

Potumarthi, Ravichandra, Rama Raju Baadhe, and Sankar Bhattacharya. 2013. "Fermentable Sugars from Lignocellulosic Biomass: Technical Challenges." In *Biofuel Technologies: Recent Developments*, edited by Vijai Kumar Gupta and Maria

G Tuohy, 3–27. Berlin, Heidelberg: Springer Berlin Heidelberg. doi:10.1007/978-3-642-34519-7_1.

Preisig, Hans R., and Robert A. Andersen. 2005. "Historical Review of Algal Culturing Techniques." In *Algal Culturing Techniques*, edited by Robert A. Andersen, 1st ed., 1–12. Elsevier Academic Press.

Ramanan, Ramakrishnan Nagasundara, Joo Shun Tan, Mohd Shamzi Mohamed, Tau Chuan Ling, Beng Ti Tey, and Arbakariya B. Ariff. 2010. "Optimization of Osmotic Shock Process Variables for Enhancement of the Release of Periplasmic Interferon-α2b from Escherichia Coli Using Response Surface Method." *Process Biochemistry* 45 (2): 196–202. doi:10.1016/j.procbio.2009.09.007.

Ruiz-Marin, Alejandro, Leopoldo G. Mendoza-Espinosa, and Tom Stephenson. 2010. "Growth and Nutrient Removal in Free and Immobilized Green Algae in Batch and Semi-Continuous Cultures Treating Real Wastewater." *Bioresource Technology* 101 (1): 58–64. doi:10.1016/j.biortech.2009.02.076.

Saha, Badal C., Loren B. Iten, Michael A. Cotta, and Y. Victor Wu. 2005. "Dilute Acid Pretreatment, Enzymatic Saccharification and Fermentation of Wheat Straw to Ethanol." *Process Biochemistry* 40 (December): 3693–3700. doi:10.1016/j.procbio.2005.04.006.

Saidane-Bchir, F., A. El Falleh, E. Ghabbarou, and M. Hamdi. 2016. "3rd Generation Bioethanol Production from Microalgae Isolated from Slaughterhouse Wastewater." *Waste and Biomass Valorization* 7 (5): 1041–46. doi:10.1007/s12649-016-9492-6.

Salama, El-Sayed, Mayur B. Kurade, Reda A. I. Abou-Shanab, Marwa M. El-Dalatony, Il-Seung Yang, Booki Min, and Byong-Hun Jeon. 2017. "Recent Progress in Microalgal Biomass Production Coupled with Wastewater Treatment for Biofuel Generation." *Renewable and Sustainable Energy Reviews* 79 (November): 1189–1211. doi:https://doi.org/10.1016/j.rser.2017.05.091.

Santana, Hugo, Carolina R. Cereijo, Valerya C. Teles, Rodrigo C. Nascimento, Maiara S. Fernandes, Patricia Brunale, Raquel C. Campanha, et al. 2017. "Microalgae Cultivation in Sugarcane Vinasse: Selection, Growth and Biochemical Characterization." *Bioresource Technology* 228 (March): 133–40. doi:10.1016/j.biortech.2016.12.075.

Santos, Diego T., Carolina L. C. Albuquerque, and Maria Angela A. Meireles. 2011. "Antioxidant Dye and Pigment Extraction Using a Homemade Pressurized Solvent Extraction System." *Procedia Food Science* 1: 1581–88. doi:10.1016/j.profoo.2011.09.234.

Sarkar, Nibedita, Sumanta Kumar Ghosh, Satarupa Bannerjee, and Kaustav Aikat. 2012. "Bioethanol Production from Agricultural Wastes: An Overview." *Renewable Energy* 37 (1). Elsevier Ltd: 19–27. doi:10.1016/j.renene.2011.06.045.

Savage, Neil. 2011. "Algae: The Scum Solution." *Nature* 474 (7352). Nature Publishing Group, a division of Macmillan Publishers Limited. All Rights Reserved: S15--S16. http://dx.doi.org/10.1038/474S015a.

Sayre, Richard. 2010. "Microalgae: The Potential for Carbon Capture." *Bio Science* 60 (9): 722–27. doi:10.1525/bio.2010.60.9.9.

Scott, Stuart A., Matthew P. Davey, John S. Dennis, Irmtraud Horst, Christopher J. Howe, David J. Lea-Smith, and Alison G. Smith. 2010. "Biodiesel from Algae: Challenges and Prospects." *Current Opinion in Biotechnology* 21 (3). Elsevier Ltd: 277–86. doi:10.1016/j.copbio.2010.03.005.

Shen, Y., W. Yuan, Z. J. Pei, Q. Wu, and E. Mao. 2009. "Microalgae Mass Production Methods." *Transactions of the ASABE* 52 (4): 1275–87. doi:10.13031/2013.27771.

Singh, Anoop, and Poonam Nigam. 2014. "Microbial Biofuels Production." *Microbial Biotechnology*, no. November: 155–68. doi:10.1201/b17587-8.

Slade, R., and A. Bauen. 2013. "Micro-Algae Cultivation for Biofuels: Cost, Energy Balance, Environmental Impacts and Future Prospects." *Biomass & Bioenergy* 53: 29–38. doi:http://dx.doi.org10.1016/j.biombioe.2012.12.019.

Soccol, Carlos Ricardo; Knoerr, Cristina Elisabete; Costa, Jorge Alberto Vieira. 2008. *Process to produce biomass and proteins by microalgae*. WO2008BR00004 20080108, issued 2008.

Solovchenko, Alexei, Sergei Pogosyan, Olga Chivkunova, Irina Selyakh, Larisa Semenova, Elena Voronova, Pavel Scherbakov, et al. 2014. "Phycoremediation of Alcohol Distillery Wastewater with a Novel Chlorella Sorokiniana Strain Cultivated in a Photobioreactor Monitored on-Line via Chlorophyll Fluorescence." *Algal Research* 6 (October): 234–41. doi:10.1016/j.algal.2014.01.002.

Su, Yanyan, Artur Mennerich, and Brigitte Urban. 2012. "Coupled Nutrient Removal and Biomass Production with Mixed Algal Culture: Impact of Biotic and Abiotic Factors." *Bioresource Technology* 118 (August): 469–76. doi:10.1016/j.biortech. 2012.05.093.

Suresh Kumar, K., Hans-Uwe Dahms, Eun-Ji Won, Jae-Seong Lee, and Kyung-Hoon Shin. 2015. "Microalgae? A Promising Tool for Heavy Metal Remediation." *Ecotoxicology and Environmental Safety* 113 (March): 329–52. doi:10.1016/ j.ecoenv.2014.12.019.

Sydney, Eduardo Bittencourt, Carlos José Dalmas Neto, Alessandra Cristine Novak, Adriane Bianchi Pedroni Medeiros, Régis Nouaille, Christian Larroche, and Carlos Ricardo Soccol. 2016. "Bioethanol Wastes: Economic Valorization." In, edited by Carlos Ricardo Soccol, Satinder Kaur Brar, Craig Faulds, and Luiz Pereira Ramos, 255–89. *Green Energy and Technology*. Cham: Springer International Publishing. doi:10.1007/978-3-319-30205-8_11.

Tan, Kenneth Wei Min, and Yuan Kun Lee. 2016. "The Dilemma for Lipid Productivity in Green Microalgae: Importance of Substrate Provision in Improving Oil Yield without Sacrificing Growth." *Biotechnology for Biofuels* 9 (1): 255. doi:10.1186/s13068-016-0671-2.

"Trends in Atmospheric Carbon Dioxide." 2017. *Earth System Research Laboratory*. Accessed June 22. https://www.esrl.noaa.gov/gmd/ccgg/trends/.

Tutt, M., T. Kikas, and J. Olt. 2012. "Influence of Different Pretreatment Methods on Bioethanol Production from Wheat Straw." *Agronomy Research Biosystem Engineering*, no. 1: 269–76.

Vohra, Mustafa, Jagdish Manwar, Rahul Manmode, Satish Padgilwar, and Sanjay Patil. 2014. "Bioethanol Production : Feedstock and Current Technologies." *Journal of Environmental Chemical Engineering*, no. 2: 573–84. doi:10.1016/j.jece.2013.10.013.

Walker, Graeme M. 2010. *Bioethanol: Science and Technology of Fuel Alcohol*. Edited by Graeme M. Walker. 1st ed. bookboon.com.

Wan, Caixia, Yuguang Zhou, and Yebo Li. 2011. "Liquid Hot Water and Alkaline Pretreatment of Soybean Straw for Improving Cellulose Digestibility." *Bioresource Technology* 102 (10): 6254–59. doi:http://dx.doi.org/10.1016/j.biortech.2011.02.075.

Wang, Yue, Shih-Hsin Ho, Chieh-Lun Cheng, Wan-Qian Guo, Dillirani Nagarajan, Nan-Qi Ren, Duu-Jong Lee, and Jo-Shu Chang. 2016. "Perspectives on the Feasibility of Using Microalgae for Industrial Wastewater Treatment." *Bioresource Technology* 222 (December): 485–97. doi:10.1016/j.biortech.2016.09.106.

Weyer, Kristina M., Daniel R. Bush, Al Darzins, and Bryan D. Willson. 2010. "Theoretical Maximum Algal Oil Production." *BioEnergy Research* 3 (2): 204–13. doi:10.1007/s12155-009-9046-x.

Xie, Zhengxin, Guanghua Lu, Zhenhua Yan, Jianchao Liu, Peifang Wang, and Yonghua Wang. 2017. "Bioaccumulation and Trophic Transfer of Pharmaceuticals in Food Webs from a Large Freshwater Lake." *Environmental Pollution* 222 (March): 356–66. doi:10.1016/j.envpol.2016.12.026.

Yen, Hong-Wei, Shih-Hsin Ho, Chun-Yen Chen, and Jo-Shu Chang. 2015. "CO_2, NO X and SO X Removal from Flue Gas via Microalgae Cultivation: A Critical Review." *Biotechnology Journal* 10 (6): 829–39. doi:10.1002/biot.201400707.

Yen, Hong-Wei, I. Chen Hu, Chun-Yen Chen, Shih-Hsin Ho, Duu-Jong Lee, and Jo-Shu Chang. 2013. "Microalgae-Based Biorefinery: From Biofuels to Natural Products." *Bioresource Technology* 135 (May): 166–74. doi:10.1016/j.biortech.2012.10.099.

Zabed, Hossain, Golam Faruq, Jaya Narayan Sahu, Mohd Sofian Azirun, Rosli Hashim, and Amru Nasrulhaq Boyce. 2014. "Bioethanol Production from Fermentable Sugar Juice." *The Scientific World Journal* 2014. doi:10.1155/2014/957102.

Zhang, Wei, Meng Zhang, Kuangfei Lin, Wenfang Sun, Bang Xiong, Meijin Guo, Xinhong Cui, and Rongbing Fu. 2012. "Eco-Toxicological Effect of Carbamazepine on Scenedesmus Obliquus and Chlorella Pyrenoidosa." *Environmental Toxicology and Pharmacology* 33 (2): 344–52. doi:10.1016/j.etap.2011.12.024.

Zhou, Na, Yimin Zhang, Xiaobin Wu, Xiaowu Gong, and Qinhong Wang. 2011. "Hydrolysis of Chlorella Biomass for Fermentable Sugars in the Presence of HCl and MgCl2." *Bioresource Technology* 102 (21): 10158–61. doi:https://doi.org/10.1016/j.biortech.2011.08.051.

In: Bioethanol and Beyond
Editor: Michel Brienzo
ISBN: 978-1-53613-478-0

Chapter 8

PERSPECTIVES FOR THE PRODUCTION OF BIOETHANOL FROM MACROALGAE BIOMASS

Fernando Roberto Paz Cedeno[1a], Eddyn Gabriel Solorzano Chavez[1a], Levi Ezequiel de Oliveira[2], Valéria Cress Gelli[3], Rubens Monti[1b] and Fernando Masarin[1a,*]

[1]Universidade Estadual Paulista (UNESP), Faculdade de Ciências Farmacêuticas, FCF, 14800-903, Departamento de Bioprocessos e Biotecnologia[a] and Alimentos e Nutrição[b], Araraquara, SP, Brazil

[2]Universidade de São Paulo (USP) - Escola de Engenharia de Lorena - Departamento de Engenharia Química, Lorena, SP, Brazil

[3]Instituto de Pesca (IP) - Núcleo de Pesquisa e Desenvolvimento do Litoral Norte - Agência Paulista de Pesquisa Agropecuária- Secretaria de Agricultura e Abastecimento do Estado de São Paulo, Brazil

ABSTRACT

Ethanol is currently produced on large scale from sugarcane, corn (first-generation-ethanol) and lignocellulosic materials (second-generation-ethanol), which results in a competition for space for growing food products. Therefore, future limitations on the availability of land for agriculture may require a shift of focus from land resources to marine resources. It is known that some macroalgae are rich in carbohydrates and devoid of lignin. The structural polysaccharides of the macroalgal cell walls are formed by an external matrix of amorphous mucilage, which can be composed of polysaccharides, such as carrageenan, agar, and starch. In addition, macroalgal biomass also contains other components that provide rigidity to the cell walls, such as cellulose fibers and sodium carbonate. In all the studied species, the main monomeric sugars from the

[*] Correponding Author Email: fmasarin@fcfar.unesp.br.

polysaccharides are glucose and galactose. The low lignin content in the macroalgal cell walls makes them less recalcitrant than terrestrial plants to chemical and enzymatic treatments, which makes polysaccharides more accessible. In addition, macroalgal cultures generally have a productivity between 5.8 and 16.8 $kgm^{-2}year^{-1}$, higher than terrestrial plants which is between 0.5 and 4.4 $kgm^{-2}year^{-1}$. The conversion of macroalgal biomass polysaccharides into fermentable monomeric sugars requires several processing steps. The main methods currently used are chemical and enzymatic, culminating in the hydrolysis of polysaccharides in monosaccharides. Carrageenophyta macroalgae are the best candidates for bioethanol production (third and forth generation-ethanol), since they present substantial amounts of carrageenan and glucans, which are formed by galactose and glucose, sugars that are easily fermentable by yeasts and bacteria.

INTRODUCTION

The use of macroalgae is a promising and economically viable alternative for the manufacture of bioproducts such as biofuels, foods, cosmetics, nutraceuticals, pharmaceuticals, and biofertilizers (Mata, Martins, and Caetano 2010; Zodape et al. 2009; Gelli and Barbieri 2015; Hayashi and Reis 2012). Also, the cultivation of macroalgae species has the advantage of promoting absorption of carbon dioxide through the process of photosynthesis. This capture of carbon dioxide can help to minimize the greenhouse effect (Yanqun Li et al. 2008; Nigam and Singh 2011; Champagne 2007; Brinkhuis et al. 1987). Moreover, in contrast to terrestrial plants, macroalgae produce more oxygen than they consume in their breathing process. In addition, macroalgal cultures show an average productivity of 22 $kgm^{-2}year^{-1}$ while in terrestrial plants it is between 0.5 and 4.4 $kgm^{-2}year^{-1}$ (Brinkhuis et al. 1987; Lewandowski et al. 2003; Wei, Quarterman, and Jin 2013). Therefore, macroalgae biomass can provide many environmental and economic benefits.

Marine macroalgae can be classified into three main groups: brown macroalgae (*Phaeophyceae*), green macroalgae (*Chlorophyceae*), and red macroalgae (*Rhodophyceae*) (Horn, Aasen, and Stgaard 2000; Goh and Lee 2010). The composition of red macroalgae varies from species to species but generally includes cellulose, glucan, galactan, and minerals (measured as ash). Glucans consist of polymers containing glucose as the fundamental unit. Galactans consist entirely of galactose and 3,6-anhydrogalactose. The substitution pattern of the sulfate groups and the amount of 3,6-anhydrogalactose vary among different classes of these compounds, depending on ecological conditions. The cell walls of red macroalgae is made of glucan and two kinds of long-chain structural polysaccharides that are valued for their gel-forming ability, i.e., agar and carrageenan (Meinita et al. 2012). Carrageenan can be classified as lambda (λ), kappa (κ), or iota (i) according to its gel-forming properties and is used mainly for thickening of foods such as yogurt, ice cream, and pudding (Horn, Aasen, and Stgaard 2000; Goh, and Lee 2010). Specie *Kappaphycus alvarezii* belongs to the class of red macroalgae and is mainly cultivated in Indonesia. The cultivation of red macroalgae *K. alvarezii* has been established as an important economic activity in more than 20 countries contributing approximately 6% of the total production of aquatic plant (FAO, 2017). This species is widely used for the production of carrageenan hydrocolloids. Carrageenan represents an annual global market of over US$ 200 million (Hayashi, de Paula, and Chow 2007). Brazil produces small amounts of carrageenan, approximately 10 tons a year, from natural harvests of *Hypnea musciformis* (Wulfen) Lamouroux. In 2015, Brazil produced ~700 tons of *K. alvarezii* (FAO 2017), but this production is not enough to satisfy the national economic demand, which is ~1800 tons a year, demanding importation of 1100 tons of *K. alvarezii* from the Philippines (Hayashi, de Paula, and Chow 2007; Abd-Rahim et al. 2014). The basic chemical composition of *K. alvarezii* includes galactan, glucan, minerals, sulfate groups, and proteins (Meinita et al. 2012; Masarin et al. 2016; Róldan et al. 2017). In addition to the use for the production of carrageenan, macroalgae has been used for bioethanol production, known as third-generation-bioethanol (3G) (Meinita et al. 2012; Khambhaty et al. 2012; Hargreaves et al. 2013). Besides, the residue of carrageenan extraction is a glucan-rich substrate and hydrolyzable with enzymes, serving to produce bioethanol, known as fourth-generation-bioethanol (4G) (Masarin et al. 2016; Roldán et al. 2017).

Taxonomy and Classification of Macroalgae Biomass

Algae are aquatic or terrestrial plants and exist in a wide variety of species living in distinct types of environments. These plants may be autotrophic or heterotrophic. Autotrophic algae are photosynthetic and use sunlight to attach the inorganic carbon of the atmosphere. The assimilated carbon is transformed into more complex organic molecules (such as carbohydrates) and used as food reserve or as a structural constituent of the cell walls of the plant. Heterotrophic algae are able to assimilate small organic molecules that are transformed into more complex compounds like lipids, carbohydrates, and proteins (John et al. 2011).

Generally, the algae are grouped into two categories based on their morphology and size: microalgae and macroalgae. Microalgae are unicellular and photosynthetic organisms. Macroalgae are composed of multiple cells organized in structures similar to roots, stem, and leaves of higher plants. Algae can also be divided into classes: green, brown, and red. The class of brown algae contains only macroalgae (multicellular). However, the class of green and red algae contains both unicellular and multicellular types (John et al. 2011; Yanagisawa et al. 2011; Wei, Quarterman, and Jin 2013).

The marine macroalgae are a group of simple organisms that are non-vascular, flowerless and seedless plants (Avanzo and Fujii 2016). All shows chlorophyll *a* to perform photosynthesis to store energy. Moreover, diverse types of macroalgae contain a large variety of accessory pigments in different structures and quantities, such as chlorophyll *b*, chlorophyll *c*, chlorophyll *d*, phycocyanins, phycoerythrin, fucoxanthin, lutein and beta-carotene. The different combinations and concentrations of accessory pigments are the reason why macroalgae have so many distinct colors. There are about 10,000 species of marine macroalgae, of which 6,500 are red algae (*Rhodophyta*), 2,000 are browns, and 1,500 are greens (about 800 species of *Bryopsidophycea*e, 50 species of *Dasycladophyceae*, 400 *Siphoncladophycea*e, and 250 marine *Ulvophyceae*) (Horn, Aasen, and Stgaard 2000; Goh et al. 2010). Based on the colors, characteristic of each species, marine macroalgae can be divided into the following three phyla:

Brown Macroalgae (*Phaeophyta*)

There are about 1,800 species of brown macroalgae, and most are marine. In general, brown macroalgae are larger and most species are found in colder waters. Virtually, all the biomass worldwide comes from a relatively small number of species of the orders *Laminariales* and *Fucales*. The brown color of these macroalgae results from the dominance of the xanthophyll pigment fucoxanthin, which masks the other pigments, chlorophyll *a* and *c* (there is not Chlorophyll *b*), beta-carotene and other xanthophylls (Horn, Aasen, and Stgaard 2000; Goh et al. 2010). An example of brown macroalgae is the specie *Sargassum sp* (Figure 1).

Figure 1. Brown macroalgae. *Sargassum sp*. Photo credit: Valéria Cress Gelli.

Red Macroalgae (*Rhodohyta*)

Macroalgae databases show that there are about 10,000 species, of which approximately 6,500 are *Rhodophyta*. The vast majority of these algae are marine and are found in the intertidal and subtidal zone, to depths of up to 40 m, and occasionally, 250 m. The red color of these macroalgae results from the pigments phycoerythrin and chlorophyll *c*, which masks other pigments such as chlorophyll *a* (not chlorophyll *b*), beta-carotene, and a some xanthophylls (Horn, Aasen, and Stgaard 2000; Goh et al. 2010). An example of red macroalgae is the specie *K. alvarezii* (Figure 2).

Figure 2. Red macroalgae. *Kappaphycus alvarezii*. Photo credit: Valéria Cress Gelli.

Green Macroalgae (*Chlorophyta*)

Macroalgae databases show that there are about 4,500 species of *Chlorophyta* including approximately 550 species of *Trebouxiophyceae* (mostly subaerial and freshwater), 2,500 *Chlorophyceae* (mostly freshwater), 800 species of *Bryopsidophyceae*, 50 species of *Dasycladophyceae*, 400 *Siphoncladophycea*e, and 250 marine *Ulvophyceae*. *Charophyta* is from fresh water and includes 3,500 species currently allocated to 5 classes.

Figure 3. Green macroalgae. *Caulerpa racemose*. Photo credit: Valéria Cress Gelli.

The green color results from chlorophyll a and b in the same proportions as in the 'higher' plants; beta-carotene (a yellow pigment); and various characteristic xanthophylls (yellowish or brownish pigments) (Horn, Aasen, and Stgaard 2000; Goh et al. 2010). An example of green macroalgae is the species *Caulerpa racemose* (Figure 3).

Cultivation and Production of Macroalgae Biomass

Table 1 briefly summarizes the main techniques of macroalgal cultivation worldwide (Radulovich et al. 2015). The cultivation of macroalgae basically consists in the use of ropes (rows) and nets suspended on the surface or tied at a shallow depth. The propagules are attached to the ropes in densities and sizes ideal for each species. The main objective is to facilitate and enhance the incidence of solar energy on the macroalgae for the photosynthesis process and consequent increase of biomass.

Table 1. Main techniques of macroalgal cultivation worldwide

Techniques of macroalgal cultivation	**Description**
Line cultivation	Seaweeds are attached to ropes of varying lengths (e.g., from 10 m to 50 m or longer) that are placed in a parallel arrangement with varying spacing between them depending on size of the seaweed species at harvest (from 0.3 m to 1.0 m or more), at depths that vary according to:
Off-bottom	Planting close to the bottom near-shore, preferably with a minimum of ca. 0.3 m of water on top at lowest tide. This method is widely used with small and/or frequently harvested species, like *Eucheuma* and *Kappaphycus*.
Submerged hanging line	Planting is mid-water near-shore, submerged several meters at high tides and at the surface or even exposed during low tides.
Floating line (long-line)	Planting at or close to the surface with seaweeds slightly submerged (i.e., not exposed). Excepting the need for anchoring, this method allows cultivation regardless of depth to bottom.
Net cultivation	Seaweed propagules are attached to nets placed at a given water depth, usually floating at the surface or slightly submerged, analogous to line cultivation in relation to depth.
Techniques of macroalgal cultivation	Description
Floating raft cultivation	Planting occurs at the surface, attaching seaweeds to lines or nets with shape given by a floating rigid frame made of bamboo or other material. Submerged raft versions of this technique are used experimentally, particularly in vertical or angled arrangement to increase interception of solar radiation through depth.
Tank or pond culture	Culture in tanks under controlled conditions is important for delicate species for fresh market and/or for intensive production. Seaweeds can be tied (e.g., promoting suspended profuse racemes of *Caulerpa* or *Ulva*) or free-floating with suspension provided by tumbling. Culture for wastewater biofiltration in tanks, ponds or canals used for fish and shrimp aquaculture is practiced to remove excess nutrients from water via a type of seaweed aquaponics.
Other minor or experimental techniques	a) Direct planting on the ocean bottom or placing seaweeds attached to artificial substrate on the seafloor, resembling natural benthic growth like kelp forests and seaweed prairies. b) Free floating rafts: equivalent to line or net cultivation but in spatial arrangements with or without rigid frames that hold shape and do not need anchoring but are left freely floating. This method emulates free-floating seaweed masses, e.g., in the Sargasso Sea and 'green-tides'.

Adapted from Radulovich et al. (2015).

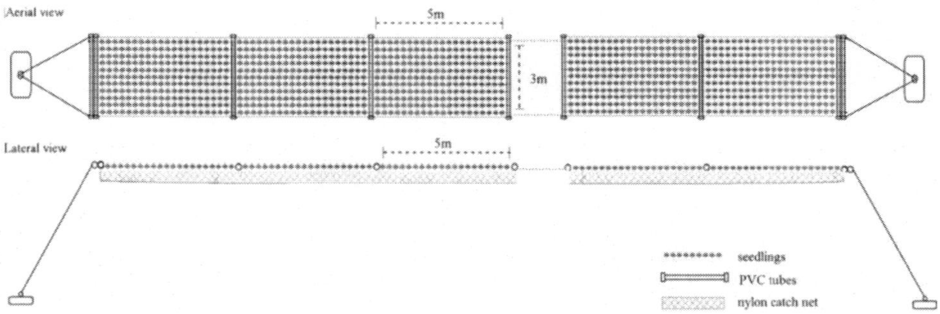

Figure 4. Schematic diagrams of floating raft used in *K. alvarezii* farming at Marambaia Bay. Rio de Janeiro state, Brazil (Góes and Reis 2012).

Figure 4 shows a scheme to demonstrate the method of growing macroalgae by line cultivation (specifically floating line) also known as tie-tie method (Góes and Reis 2012).

According to the latest statistical data available by the FAO (2017), in 2015 61.4% of the cultivated macroalgae species were red. These data also show that the green macroalgae were the most valued, reaching a price of US$ 0.61 each kilogram, that is, 3.6 times the average price reached by the red macroalgae. Table 2 identifies the main species of macroalgae cultivated in the world, their average production, their values and average prices. The total macroalgae produced in more than 30 countries was 27,337,202 tons, generating approximately US$ 4.3 billion. Among the largest producers in the world are China, Indonesia, Philippines, Korea, and Japan. More than 90% of the production is composed by the genera: *Euchema, Kappaphycus, Porphyra, Undaria, Saccharina* (*Laminaria*), and *Gracilaria*.

Table 2. The main cultivated species of macroalgae in the world

Macroalgae	Brown seaweeds	Green seaweeds	Red seaweeds
Family	*Phaeophyceae*	*Chlorophyceae*	*Rhodophyceae*
Species	*Alaria esculenta*	*Enteromorpha clathrata*	*Palmaria palmata*
	Sargassum fusiforme	*Caulerpa spp*	*Kappaphycus alvarezii*
	Macrocystis pyrifera	*Caulerpa racemosa*	*Eucheuma spp*
	Macrocystis spp	*Codium fragile*	*Gelidium spp*
	Saccharina japonica	*Monostroma nitidum*	*Gracilaria spp*
	Nemacystus decipiens		*Asparagopsis spp*
	Saccharina latissima		*Gelidium amansii*
	Laminaria digitata		*Porphyra tenera*
	Undaria pinnatifida		*Porphyra spp*
	Undaria spp		*Eucheuma denticulatum*
	Sargassum spp		*Gracilaria verrucosa*
			Porphyra columbina
			Chondracanthus chamissoi
Quantity (ton)	10.542.822	12.866	16.781.514
Value (US$)	1.463.411	7.874	2.904.283
Price (US$/Kg)	0.13	0.61	0.17

Adapted from FAO (2017).

Figure 5. Cultivation of red macroalgae *K. alvarezii* with tubular networks in Ubatuba-SP. Photo credit: Valéria Cress Gelli.

Figure 6. Cultivation of red macroalgae *K. alvarezii* in Tie-Tie system in Ubatuba-SP. Photo credit: Valéria Cress Gelli.

Brazil has a coastal extension of approximately 8,500 km with high potential for macroalgae production. However, statistical data (FAO 2017) for the year 2015 has shown a still incipient production of approximately 700 tons for the specie *K. alvarezii* and 30 tons for genus *Gracilaria*. It is believed that the macroalgae *K. alvarezii* is a potential for coastal communities, but Brazilian legislation limits its cultivation only to the states of São Paulo and Rio de Janeiro. Brazilian norms are described in the works of Gelli and Barbieri (2015).

Reis et al. (2014) compared the main cultivation techniques used for the production of *K. alvarezii* in Brazil. The results showed that there was not significant difference in the amount of carrageenan in the two systems. It was concluded that the cultivation in nets (Figure 5) was more efficient than the tie-tie system (Figure 6) and retained the macroalgae in the floating rafts system, which prevented them from dispersing in the environment.

ULTRASTRUCTURE AND CELLULAR MORPHOLOGY OF MACROALGAE BIOMASS

Red macroalgae such as *Calliarthron Thalli* contains in its cell walls calcium carbonate. The calcified regions are called "intergeniculum" and the regions that do not contain calcium carbonate are called "geniculum" (Figure 7A-C). The peripheral regions of the geniculum contain cells with thicker cell walls (Mature - Figure 7C and 7E) while in the central region of the geniculum the cells have less thick cell walls (Immature-Figure 7C and 7D) (Martone et al. 2009).

Transmission electron microscopy images in the transverse and longitudinal direction of the *C. thalli* cell walls show deposition of the medial lamella (M), primary cell wall (P) and distinct secondary cell walls (S1) in mature and immature geniculum cells, being remarkable in mature geniculum (Figure 7D-E) (Martone et al. 2009). The regions of the intergeniculum contain the primary and secondary (S_1) cell walls, but not secondary (S_2 and S_3) cell walls (Figure 7F). Secondary (S_2 and S_3) cell walls of red macroalgae have not been described in the literature (Martone et al. 2009). However, in the cells of lignocellulosic materials, it is common in the presence of secondary (S_1, S_2, and S_3) cell walls, show a thicker secondary (S_1+S_2+S_3) cell walls. Thus, most lignin is found in secondary (S_1+S_2+S_3) cell walls (Figure 7F) (H. Chen, 2014). The absence of this secondary cell walls (S_2 and S_3) in red macroalgae may explain the low content of aromatics in its composition (Figure 7F).

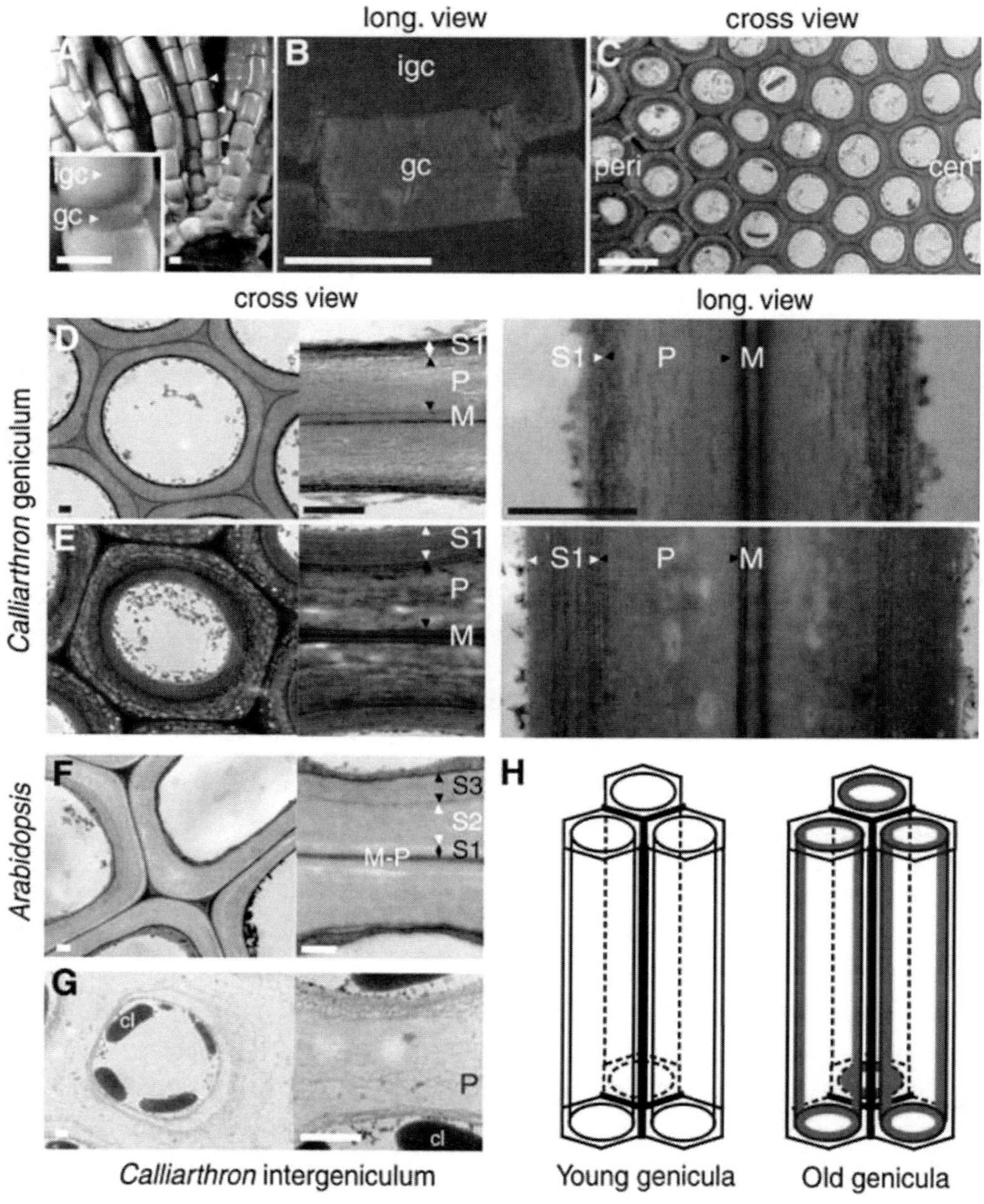

Figure 7. Morphology and Ultrastructure of *Calliarthron Thalli* in which flexible joints, called Genicula, separate calcified segments, called Intergenicula (A) whole aspect, (B) long-section (viewed unstained under ultraviolet light), and (C) cross-section of geniculum. (D) Immature genicular cells have primary cell walls and negligible secondary cell walls. (E) Mature genicular cells develop thick secondary cell walls, resembling (F) fibers from *Arabidopsis* xylem with multiple secondary cell wall layers. (G) Intergenicular cells develop only primary cell walls. (H) Schematic of secondary-wall thickening (red) in genicular cells. Images (C–G) were taken under transmission electron microscopy. Abbreviations are as follows: cen, central genicular cells; cl, chloroplast; gc, genicular cells; igc, intergenicular cells; M, middle lamella; p, primary cell wall; peri, peripheral genicular cells; S1–S3, secondary-cell-wall layers. Scale bar represents 1 mm (A and B), 10 mm (C), and 0.5 mm (D–G). (Martone et al. 2009).

CHEMICAL COMPOSITION OF MACROALGAE BIOMASS

Macroalgae contain carbohydrates, lipids, proteins, ashes, and aromatics (Martone et al. 2009; Yaich et al. 2011; Wei, Quarterman, and Jin 2013; Masarin et. al. 2016; Roldán et al. 2017). The chemical composition of macroalgae can be very variable depending on the classes and species evaluated (Wei, Quarterman, and Jin 2013). Macroalgae do not contain lignin since unlike plants, they do not need rigidity. Thus, the cell walls of macroalgae is structurally flexible. Therefore, process of bioethanol production from macroalgal carbohydrates does not need a complex lignin removal step that is typically necessary to increase digestibility of lignocellulosic materials (Jung et al. 2013). It should be mentioned that brown macroalgae contain different concentrations of polyphenols (Horn, 2000), which are usually referred to as lignin-like materials. The lignin-like materials, which are detected in different macroalgae by the standard methods of biomass characterizations, do not exhibit plant lignin properties (Lewis and Yamamoto 1990; Sluiter et al. 2012)

Carbohydrates

Many types of polymeric carbohydrates are found in macroalgae. The main types of carbohydrates that make up the brown macroalgae are laminarin (linear chain polysaccharide consisting of glucose units linked by β-1-3-type bonds with few branches linked through β-1-6 bonds, mannitol (a type of sugar alcohol) and cellulose (linear chain polysaccharide consisting of glucose units linked by β-1-4-type bonds) (Figure 7). Brown macroalgae may still contain alginate (linear chain polymer consisting of β-D-mannuronic acid and α-D-guluronic acid units linked by type 1-4 linkages) (Figure 7 and Table 3) (Wei, Quarterman, and Jin 2013).

Table 3. Types of carbohydrates and total carbohydrates (w/w%, dry biomass) of some classes and species of macroalgae. Adapted from Wei, Quarterman, and Jin (2013)

Macroalgae	Class	Carbohydrate composition	Total carbohydrates (%)	Reference
Gelidium amanssi	Red	Agar, carrageenan, cellulose	75.2	(Yoon et al. 2010)
Kappaphycus alvarezii		Cellulose, carrageenan	55.4-78.3	(Meinita et al. 2012; Masarin et al. 2016; Roldán et al. 2017)
Laminaria japônica	Brown	Laminarin, mannitol, alginate, fucoidan, cellulose	51.9	(N. J. Kim et al. 2011)
Sargassum fulvellum			39.6	(N. J. Kim et al. 2011)
Ulva lactuca	Green	Starch, cellulose, ulvan	54.3	(N. J. Kim et al. 2011)
Ulva pertusa			65.2	(Jang et al. 2012)

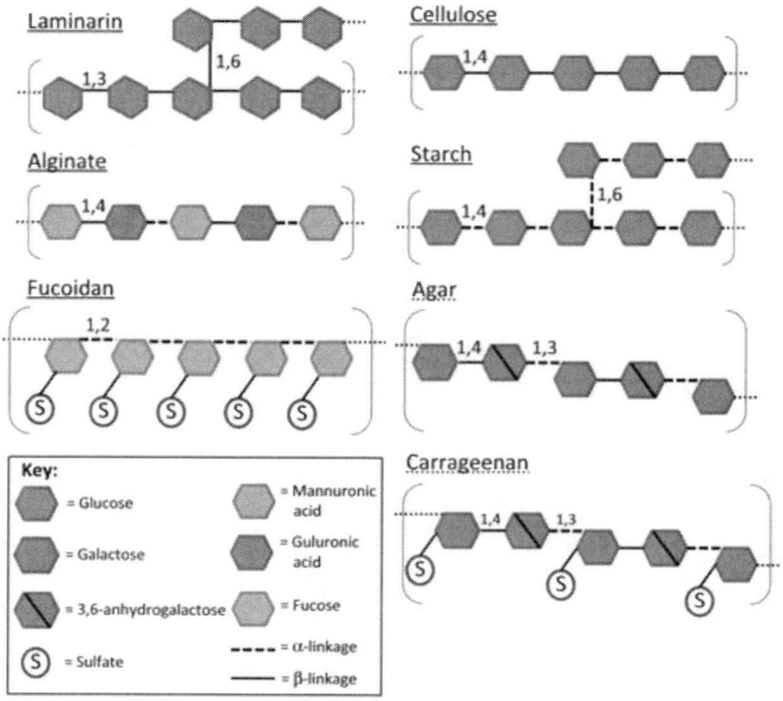

Figure 8. Structural information of the main polysaccharides found in macroalgae (Wei, Quarterman, and Jin 2013).

Green macroalgae contain between 1 and 4% starch (linear and branched polysaccharide consisting of glucose units linked through α-1-4 and α-1-6 linkages) (Figure 8 and table 3). *Ulva* and *Enteromorpha sp.* have water-soluble ulvan and insoluble cellulose (38-52%, w/w, dry mass) in their cell walls (Lahaye and Robic 2007). *Ulvan*, a distinctive feature of green macroalgae, is composed mainly of D-glucuronic acid, D-xylose, L-rhamnose, and sulfate (Wei, Quarterman, and Jin 2013; Lobban and Wynne 1981).

Red macroalgae are different from green and brown algae in the synthesis and metabolism machinery of carbohydrates. Floridean starch is an α-1,4-glucosidic linked glucose homopolymer with α-1,6-branches functioning as carbon and energy reserve in the cells and accounts for up to 80% of the cell volume (Yu et al. 2002). Because floridean starch has the property to weaken gel strength, it is removed as an impurity in the production of agar and carrageenan by using a thermophilic α-amylase treatment (Yu et al. 2002). The major polysaccharide constituents of red macroalgae are galactans such as carrageenan (linear polysaccharide consisting of β-D-galactose-4-sulfate and 3-6-α-D-anhydrogalactose-2-sulfate units connected by type 1-4 links alternating with type 1-3 links) up to 75% (w/w, dry mass). Another important component is agar (linear chain polysaccharide consisting of β-D-galactose and 3-6-α-L-anhydrogalactose units connected by type 1-4 links alternating with type 1-3 links) up to 52%, which are the most commercially important polysaccharides for red macroalgae (McHugh, 2003). The red macroalgae also contains cellulose (Figure 8 and Table 3) (Wei, Quarterman, and Jin 2013).

Table 4. Lipid, protein and ash content (w/w%, dry biomass) of some classes and species of macroalgae. Adapted from Wei, Quarterman, and Jin (2013)

Macroalgae	Class	Lipids (%)	Proteins (%)	Ashes (%)	Reference
Gelidium amanssi	Red	0.6	18.5	5.7	(Yoon et al. 2010)
Gelidium amanssi		1.1	13.1	8.6	(N. J. Kim et al. 2011)
Gelidium amanssi		0.9	12.2	3.3	(Jang et al. 2012)
Kappaphycus alvarezii		0.2-1.3	2.3-3.8	14.6-17.2	(Masarin et al. 2016; Roldán et al. 2017)
Kappaphycus alvarezii		0.6-0.9	2.3-5.4	12.6-19.5	(Meinita et al. 2012)
Laminaria japonica	Brown	1.8	14.8	31.5	(N. J. Kim et al. 2011)
Laminaria japonica		1.5	8.1	30.9	(Jang et al. 2012)
Sargassum fulvellum		1.4	13.0	46.0	(N. J. Kim et al. 2011)
Ulva lactuca	Green	6.2	20.6	18.9	(N. J. Kim et al. 2011)
Ulva pertusa		2.6	7.0	25.2	(Jang et al. 2012)

Proteins, Lipids and Ashes

The protein content in macroalgae is variable depending on the species evaluated. On average the protein content in macroalgae is 12% (w/w, dry biomass) (Table 4) (Yaich et al. 2011; Wei, Quarterman, and Jin 2013). The lipid content is also variable depending on the species evaluated. Lipids can be extracted from biomass and used to produce biodiesel. On average, the lipid content of macroalgae is 5% (w/w, dry biomass) (Table 4) (Ahmad et al. 2011; Yaich et al. 2011; Wei, Quarterman, and Jin 2013). The ash content in macroalgae is also quite variable depending on the species evaluated. Inorganic chemical compounds are common to be found in macroalgae. The ash content is higher in brown macroalgae (average 35%) followed by green macroalgae (average 22%) and red macroalgae (average 6%) (w/w, dry biomass) (Table 4) (Yaich et al. 2011; Wei, Quarterman, and Jin 2013).

Aromatics

Recently the detection of aromatic compounds in macroalgal cell walls has been evaluated. Martone et al. (2009) demonstrated through DFRC method (Derivatization Followed by Reductive Cleavage) that *C. thalli* cell walls contain components derived from radical coupling reactions of hydroxycinnamic alcohols (monolignols) that are precursor components of lignin formation and are used to characterize lignins from terrestrial plants. The DRFC method is considered as a diagnosis of lignin since this method cleaves only the β-aryl-ether-type bonds of acetylated monolignols. The application of the DFRC method in *C. Thalli* showed the presence of acetates derived from p-coumaril, coniferyl and synapyl alcohols (DFRC monomers), typically derived from p-hydroxyphenyl (H), guaiacyl (G) and syringyl (S). The p-coumaril, coniferyl, and synapylacetic acetates are commonly found in terrestrial plant fibers after application of the DFRC method (Martone et al. 2009).

Per gram of tissue, *Calliarthron genicula* release approximately 0.25 mg DFRC monomers (molar ratio H:G:S = 1:1.5:trace), and uncalcified intergenicula release 0.19 mg DFRC monomers (molar ratio H:G:S = 1:2.4:0.1). DFRC monomers are released from gymnosperm (*loblolly pine*) walls at 50 $mg.g^{-1}$ and are essentially 100% G. *Arabidopsis* stems yielded 20 $mg.g^{-1}$ in a 4:1 G:S ratio with very few H units (Martone et al. 2009).

Yaich et al. (2011) quantified the aromatic content in biomass of the *Ulva lactuca* macroalgae by the Klason lignin method (insoluble lignin). The aromatics content was 1.5% (w/w, of dry biomass). Wi et al. (2009) showed the presence of 1.5% of insoluble aromatics and 3.2% of soluble aromatics (w/w, of dry biomass) in biomass of the macroalgae *Gelidium amanssi*. Masarin et al. (2016) showed the presence of insoluble aromatics content of 2.6% (w/w, of dry biomass) from different strains of red macroalgae *K. alvarezii*. However, the mean total lignin contents in gymnosperms and angiosperms are on average 30% and sugar cane 20% (Masarin and Ferraz 2008; Vicentim and Ferraz 2009; Masarin et al., 2011).

Macroalgae for Biofuel Production

Fossil raw material has been the main source in the production of fuels, chemicals, and materials. However, due to debate about feedstock supply and climate change, there has been much effort made to encourage the use of biomass as an alternative feedstock for fuel production. The used feedstock is biomass, an organic material, which can be obtained from various sources. Note, biofuel is any type of fuel generated from biomass. The two most common biofuels are biodiesel and bioethanol which are aimed at replacing, respectively, conventional diesel and gasoline. Biofuels can be classified according their cultivation technology and the feedstock type: first (1G), second (2G) and third generation (3G) (Figure 9) (Nigam and Singh 2011).

Most of the biofuel produced is the 1G bioethanol from sugar biomass by countries such as Brazil, US, and Canada (Chiaramonti 2007). The advantage of using the 1G biofuels is that the growing crop can reduce greenhouse gas levels. Moreover, bioethanol is less toxic, highly biodegradable, and its use produces less airborne pollutants than fossil fuel. Although the growth of feedstock crops for bioethanol production can address environmental issues and it is economically attractive, it has raised doubts about its possible impact on food supply and security. Since the definition of sustainability is the production process being environmentally friendly, economically viable and socially responsible production, the 1G biofuel does not meet the social issue because it is generally in conflict with food security due the need of large extension of arable land for its production.

The second-generation biofuels use agricultural, industrial, forestry, and several wastes that would be discarded. In the case of 2G bioethanol, the use of lignocellulosic materials such as sugarcane bagasse and wood chips has many advantages in terms of resource availability, considering the large amount of lignocellulosic material produced by forestry and agriculture activities. However, the lignocellulosic materials are very recalcitrant to enzymatic hydrolysis of the cellulose, being necessary a previous chemical treatment of the material for conversions above 30% yield (Mendes et al. 2011).

The main drawbacks of 1G and 2G bioethanol can be overcome by the 3G biofuels. The algae are proving themselves as an alternative renewable source for the production of bioethanol, because of its high productivity (Goh and Lee 2010). Moreover, algae biomass is virtually free of lignin, requiring a milder pretreatment process, which is one of the biggest bottleneck of the 2G bioethanol production. Macroalgae do not need of structural biopolymers such as lignin, which are essential for higher plant growth in terrestrial environment. This feature simplifies the bioethanol production by eliminating the chemical pretreatment steps to breakdown these biopolymers into fermentable sugars. Converting seaweed to bioethanol rather than using terrestrial plant resources has some advantages such as no impact on food security, rich in sugar content, not lignin content, and higher mass production (Adams, Gallagher, and Donnison 2009).

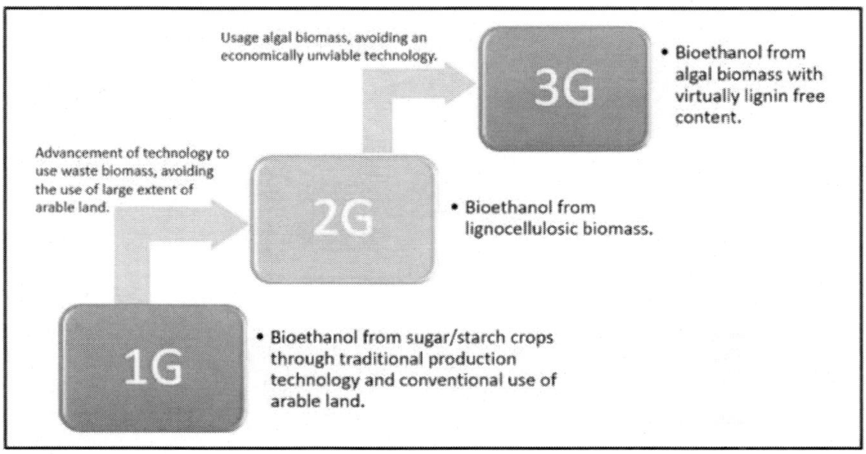

Figure 9. The first, second and third generations biofuels. The advancement on technology allowed the usage of different types of biomass (Nigam and Singh 2011).

The concept of using algae as an energy feedstock dates back to the late 1950s but accelerated due to the world oil crisis in the 1970s (Chen et al. 2010). Over the last three decades, there has been extensive research on algae biofuels, in addition to being renewable, act more efficiently in carbono dioxide sequestration. Moreover, future limitations on terrestrial resources and competition with food production will require a shift from terrestrial-based to marine-based resources.

BIOETHANOL PRODUCTION FROM MACROALGAE BIOMASS

An application for macroalgal biomass is the production of bioethanol. Recent studies have shown that the biomass of *K. alvarezii* can be hydrolyzed to fermentable sugars, which in turn can be transformed into bioethanol through the common process of yeast fermentation (Meinita et al. 2012; Khambhaty et al. 2012; Hargreaves et al. 2013). The route of bioethanol production from macroalgae biomass has been considered as a third generation process (3G) (Meinita et al. 2012; Khambhaty et al. 2012; Hargreaves et al. 2013). The two main methods used to hydrolyze macroalgal biomass are acid and enzymatic hydrolysis.

Acid Hydrolysis

The acid hydrolysis of biomass of different types of macroalgae has been tested in recent years, sulfuric acid being the main used in the macroalgae biomass hydrolysis (Meinita et al. 2012; Khambhaty et al. 2012; Hargreaves et al. 2013).

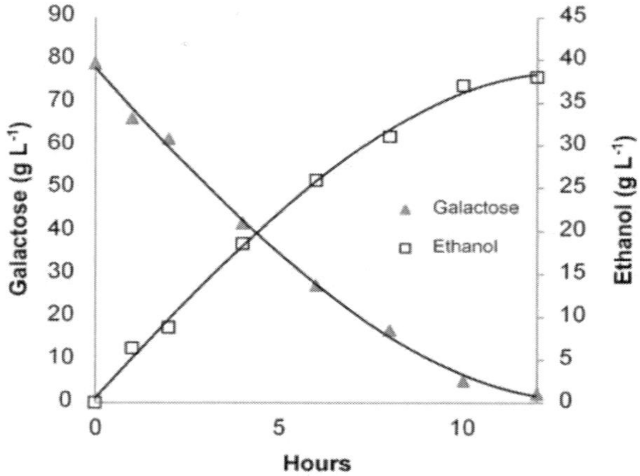

Figure 10. Fermentation of galactose-containing hydrolysate from *K. alvarezii* by *S. cerevisiae* at 30°C with an inoculum size of 7 g.L^{-1} (Hargreaves et al. 2013).

Hargreaves et al. (2013) showed that *K. alvarezii* acid hydrolyzate is rich in galactose (mean concentration of 80 $g.L^{-1}$). However, this hydrolyzate showed to be contaminated with hydroxymethylfurfural (mean concentration of 20 $g.L^{-1}$). Thus, the hydrolyzate of *K. alvarezii* was pretreated with activated charcoal, which eliminated about 90% of hydroxymethylfurfural (remaining an average concentration of 2 $g.L^{-1}$).

Figure 10 shows the alcoholic fermentation profile of the *K. alvarezii* hydrolyzate, previously treated with activated charcoal, with the yeast *Saccharomyces cerevisiae*. It can be seen that after 12h of fermentation, 37 $g.L^{-1}$ of ethanol were obtained reaching a process efficiency of 90.6% (Hargreaves et al. 2013).

Enzymatic Hydrolysis

Enzymatic hydrolysis has also been tested in macroalgae biomass. The enzymatic extracts rich in cellulases activity are the main ones that were tested (Hargreaves et al. 2013; Masarin et al. 2016; Róndan et al. 2017). The advantage of the enzymatic hydrolysis of the macroalgae biomass is the non-formation of hydroxymethylfurfural, since the treatment is performed under mild conditions of acidity and temperature (pH 5 and 50°C) (Hargreaves et al. 2013; Masarin et al. 2016).

Figure 11 shows the simultaneous fermentation and saccharification of a residue from *K. alvarezii* biomass previously acid-treated. Two processes were tested as follows: simultaneous saccharification and co-fermentation (SSCF, Figure 11a) and simultaneous saccharification and fermentation (SSF, Figure 11b) (Hargreaves et al. 2013).

For the SSCF process, a galactose-containing hydrolyzate (63.2 $g.L^{-1}$) was prepared. An macroalgal cellulosic residue and commercial cellulases (Multifect) was then added. The hydrolysis was conducted with a consistency of 18% (w/v), this being considered a process of enzymatic hydrolysis of high consistency (Hargreaves et al. 2013). After 24h of enzymatic hydrolysis, the final hydrolyzate showed a concentration of 65 $g.L^{-1}$ of glucose and 63 $g.L^{-1}$ of galactose (Figure 11). This hydrolyzate was fermented with *S. cerevisiae*, reaching a concentration of ethanol of 65 $g.L^{-1}$ after 46h with the cellulolytic enzymes acting continuously on the cellulose residue (CR) (Figure 11). The results indicated a clear improvement in the assimilation and fermentation of galactose together with glucose (Hargreaves et al. 2013).

The enzymatic hydrolysis with commercial cellulases (Multifect) of the SSF process showed a maximum glucose concentration of 80.8 $g.L^{-1}$ in 24h (Figure 11). The hydrolysis also was conducted with a consistency of 18% (w/v). This process showed 70.3% hydrolysis enzymatic efficiency (Hargreaves et al. 2013). After 24h of enzymatic hydrolysis, began the fermentation using *S. cerevisiae* reaching a concentration of ethanol of 52 g L^{-1} after 46h with the cellulolytic enzymes acting continuously on the CR. This result corresponds to a volumetric productivity of 1.13 $g.L^{-1}.h^{-1}$ and 78.5% fermentation efficiency (Hargreaves et al. 2013).

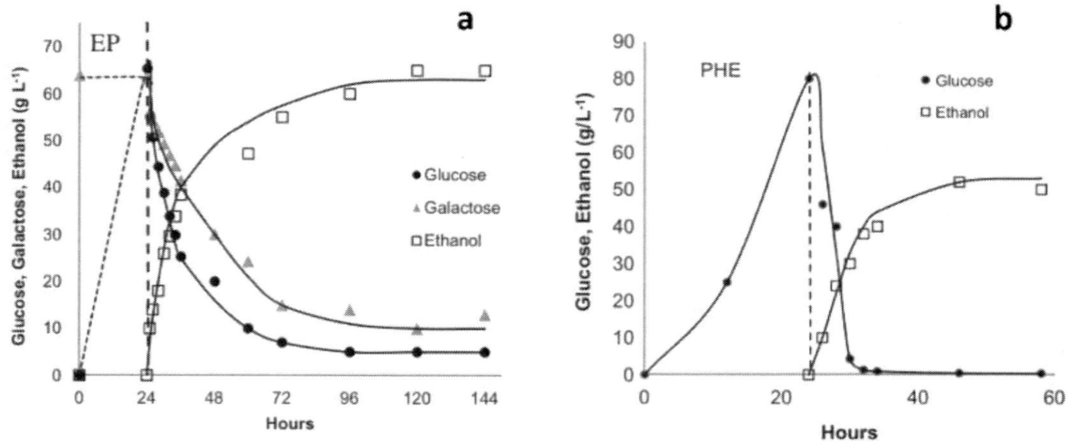

Figure 11. (a) Simultaneous saccharification and co-fermentation (SSCF) from red macroalgae *K. alvarezii* with cellulosic residue (CR) and 45 $FPU.g^{-1}$ for pre-hydrolysis, taking 24 h, and K. *alvarezii* hydrolysate (63.2 $g.L^{-1}$ of galactose). (b) Simultaneous saccharification and fermentation (SSF) with CR and 45 $FPU.g^{-1}$ for pre-hydrolysis, taking 20 h. EP and PH: pre-hydrolysis (50°C, 150 rpm using a commercial cellulases preparation (Multifect), which was supplied by Genencor International), CR with consistency of 18% (w/v). Both fermented with yeast *S. cerevisiae* at 30°C and an inoculum size of 7 $g.L^{1}$ (Hargreaves et al. 2013).

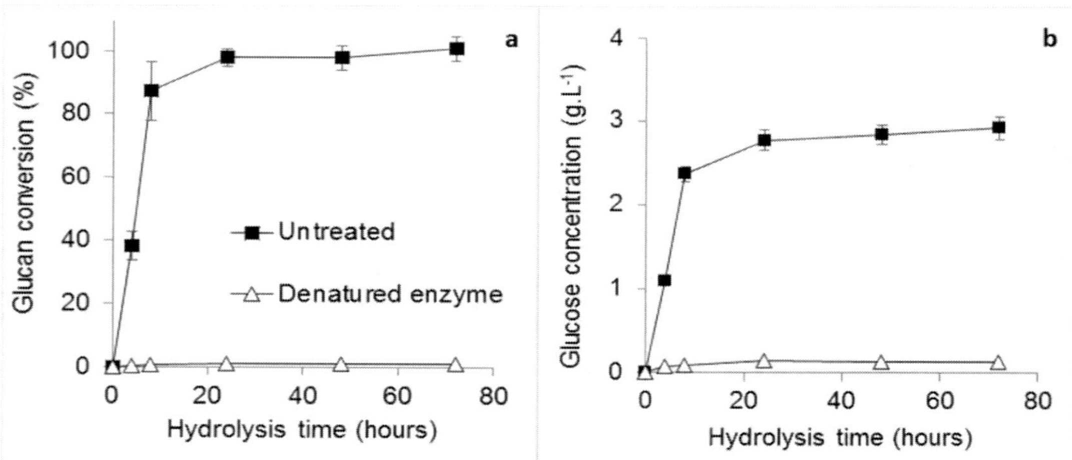

Figure 12. Enzymatic hydrolysis of untreated-biomass of *K. alvarezii*. Enzymatic extract Cellic CTec II (Novozymes). (a) Glucan conversion (w/w%, dry biomass). (b) Glucose concentration ($g.L^{-1}$) (Masarin et al. 2016).

Table 5. Reducing sugars and bioethanol production from different macroalgal biomass

Biomass type	Saccharification/ pretreatment type	Reducing sugar ($g.L^{-1}$)	Microorganism	Ethanol ($g.L^{-1}$)	References
Rhodophyta (Red macroalgae)					
Gelidium amansii	Acid	29.2	*Saccharomyces cerevisiae*	0.83	(Meinita et al. 2013)
Gracilaria tenuistipitata	Acid	26.6	*Saccharomyces cerevisiae*	0.60	(Meinita et al. 2013)
Gracilaria chorda	Acid	23.4	*Saccharomyces cerevisiae*	0.50	(Meinita et al. 2013)
Gelidium amansii	Acid + Enzyme	43.5	*Scheffersomyces stipitis*	20.5	(Ra et al. 2013)
Gelidium elegans	Enzyme	49.0	*Saccharomyces cerevisiae* IAM 4178	18.4	(Yanagisawa et al. 2011)
Gelidium elegans	Acid + Enzyme	124.1	*Saccharomyces cerevisiae* IAM 4178	55.0	(Yanagisawa et al. 2011)
Kappaphycus alvarezii	Acid + Enzyme	-	*Saccharomyces cerevisiae* CBS1782	64.3	(Hargreaves et al. 2013)
Gracilaria corticata	Acid + Bacteria	30.7	*Meyerozyma guilliermondii*	1.7	(Sudhakar et al. 2017)
Gracilaria verrucosa	Enzymes	38.9	*Saccharomyces cerevisiae* HAU	14.9	(Kumar et al. 2013)
Chlorophyte (Green macroalgae)					
Ulva pertusa Kjellman	Acid + Enzyme	59.1	*Saccharomyces cerevisiae* IAM 4178	27.5	(Yanagisawa et al. 2011)
Ulva pertusa Kjellman	Enzyme	78.8	*Saccharomyces cerevisiae* IAM 4178	30.0	(Yanagisawa et al. 2011)
Ulva lactuta	Thermal + Enzyme	22.4	*Clostridium beijerinckii*	0.20	(van der Wal et al. 2013)
Ulva prolifera	H_2O_2 + Enzyme	7.1	*Saccharomyces cerevisiae YSC2 type II*	0.13 (g/g of biomass)	(Yinping Li et al. 2016)
Chaetomorpha linum	Alkali + Enzyme	12.0	*Saccharomyces cerevisiae*	0.093 (g/g of biomass)	(Ben Yahmed et al. 2016)
Phaeophyceae (Brown macroalgae)					
Alaria crassifolia Kjellman	Enzyme	123.0	*Saccharomyces cerevisiae* IAM 4178	34.4	(Yanagisawa et al. 2011)
Alaria crassifolia Kjellman	Acid + Enzyme	66.3	*Saccharomyces cerevisiae* IAM 4178	25.6	(Yanagisawa et al. 2011)
Undaria pinnatifida	Acid	28.8	*Pichia angophorae* KCTC 17574	9.4	(Cho, Kim, and Kim 2013)
Laminaria hyperborea	Acid	~ 40.0	*Pichia angophorae* CBS 5830	5.2	(Horn 2000)
Undaria pinnatifida	Acid + Enzyme	33.1	*Pichia angophorae*	13.0	(H. Kim, Ra, and Kim 2013)
Saccharina japônica	Acid + Enzyme	13.1	*Saccharomyces cerevisiae* DK 410362	6.6	(J. ye Lee et al. 2013)
Laminaria japônica	No treatment	-	*Pichia stipitis* KCTC 7228	2.9	(S. M. Lee and Lee 2012)
Sargassum ilicifolium	Acid + Bacteria	39.7	*Meyerozyma guilliermondii*	2.7	(Sudhakar et al. 2017)

Figure 12 shows the enzymatic hydrolysis of the red macroalgae *K. alvarezii* biomass. The conversion of the glucan fraction was 100%, due to the fact that the material is not recalcitrant (reaching a glucose concentration of 3 $g.L^{-1}$). The hydrolysis was conducted with a consistency of 2% (w/v), being considered a process of enzymatic hydrolysis of low consistency. However, the galactan fraction was not hydrolyzed due to the low galactanase activity found in the commercial extract Cellic CTec II (Masarin et al. 2016; Roldán et al. 2017).

Table 5 summarizes the main types of macroalgae (red, green, and brown) the main processes of saccharification (enzymatic and/or acid), the content of reducing sugars obtained in the hydrolysates, the concentration of ethanol obtained in the fermentation processes and the citation of the source. In general, Table 5 shows that the fermentation of hydrolyzates obtained from acid treatments reached low concentrations of ethanol. In other words, fermentative processes had low efficiency. We can explain these results due to the formation of inhibitors (mainly hydroxymethylfurfural). However, the fermentation of hydrolyzates obtained from acid + enzymatic or only enzymatic treatments showed higher concentrations of ethanol providing high efficiency fermentation processes.

Thus, we can conclude that acid treatment under severe conditions reached the formation of high concentrations of fermentable sugars, but the formation of considerable concentrations of inhibitors (mainly hydroxymethylfurfural) of fermentative processes, which leads to low efficiency of it. In contrast, mild acid and enzymatic treatments reached considerable concentrations of fermentable sugars, without the formation of inhibitors and consequently a better efficiency in fermentative processes.

Production of By-Products from Macroalgae Biomass

One of the main applications of macroalgae biomass is the production of hydrocolloids. The cell walls of red macroalgae is made of glucan and two kinds of long-chain structural polysaccharides that are valued for their gel-forming ability, i.e., agar or carrageenan (Meinita et al. 2012). According to their gel-forming carachteristics, carragenaan can be lambda (λ), kappa (κ), or iota (i). These bioproducts are used in the pharmaceutical, cosmetic, and food industries (Horn et al. 2000; Goh and Lee 2010). Figure 13 shows a basic flowchart of the process of obtaining carrageenan from the red macroalgae *K. alvarezii* (Roldan et al. 2017).

Figure 14 shows the visual appearance of the materials throughout the processing of semi-refined carrageenan production. The material untreated has a darker appearance due to the presence of pigments in the biomass of *K. alvarezii* (Figure 14a). The material treated-KOH showed a clearer appearance (Figure 14b). The semi-refined carrageenan showed a translucent appearance, characteristic of this material, being similar to commercial carrageenan. However, the granulometry of both carrageenan (semi-refined and commercial) is different (Figure 14c and 14e, respectively). The carrageenan-extraction-residue showed a similar appearance to the material treated-KOH-algae (Figure 14d and 14b, respectively) (Masarin et al. 2016).

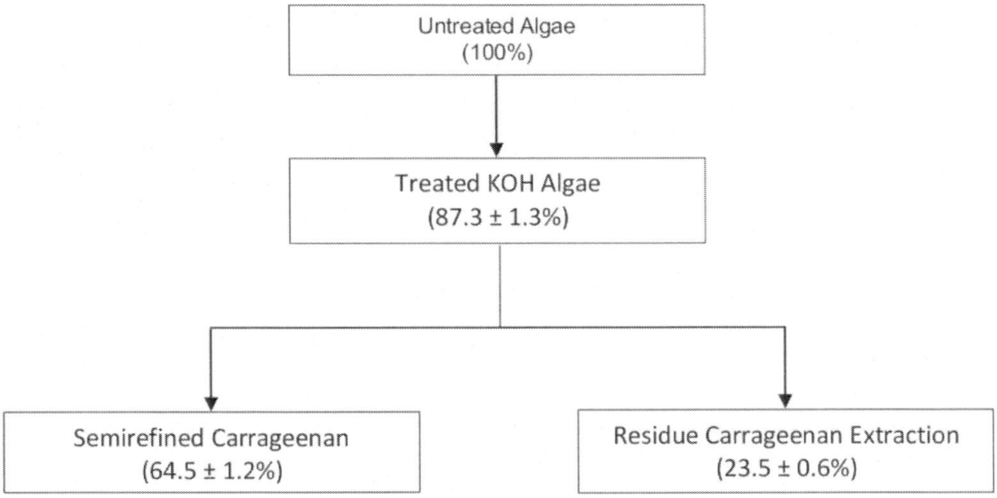

Figure 13. Flowchart and yields (w/w%, dry biomass) of obtaining fractions of Semirefined-Carrageenan and Residue-Carrageenan-Extraction from *K. alvarezii* (Roldán et al. 2017).

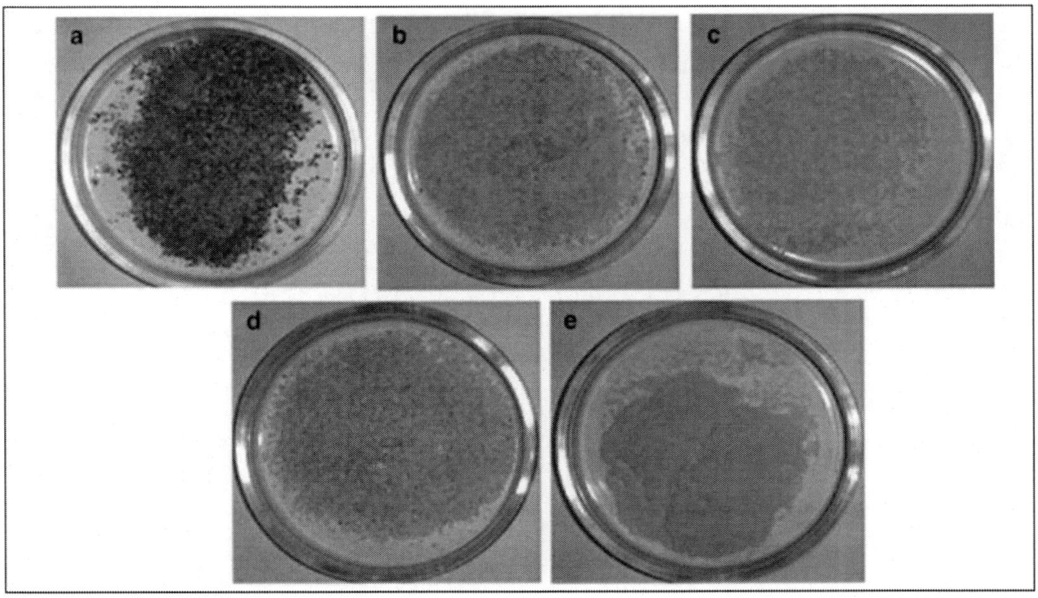

Figure 14. Residual solids of brown strain cultivation for May 2013 after treatment with 6% KOH (w/v) and subsequently extraction of semi-refined carrageenan. (a) Untreated-Algae. (b) Treated-KOH 6% (w/v) Algae. (c) Semirefined-carrageenan. (d) Residue-extraction-carrageenan. (e) Commercial carrageenan (Masarin et al. 2016).

The process of obtaining carrageenan generates a by-product with an overall yield (considering the two process steps, Figure 13) of approximately 20%. The chemical composition of this by-product revealed a glucan and galactan content of 50 and 10%

(w/w, dry biomass), respectively. However, the untreated-biomass (*K. alvarezii* biomass) showed a glucan and galactan content of 13 and 33% (w/w, dry biomass), respectively. In this way, there was an enrichment of the glucan fraction in the by-product of the order of 3.8 times (Masarin et al. 2016; Roldán et al. 2017).

Figure 15 shows the enzymatic hydrolysis of the residue-carrageenan-extraction from *K. alvarezii* biomass. The hydrolysis was conducted with a consistency of 2% (w/v), being considered a process of enzymatic hydrolysis of low consistency. The conversion of the glucan fraction was 100%, due to the lack of recalcitrance of the material. However, the galactan fraction was not hydrolyzed, because of the low galactanase activity found in the commercial extract Cellic CTec II (Masarin et al. 2016; Roldán et al. 2017). When comparing the enzymatic hydrolysis of the *K. alvarezii* untreated-biomass with the residue-carrageenan-extraction from *K. alvarezii* (both using commercial cellulases Cellic CTec II), it is evident that both substrates have no recalcitrance against the hydrolysis of the glucan fraction (reaching a 100% conversion). However, the residue-carrageenan-extraction shows an enrichment of the glucan fraction, which culminates in a concentration of 12 g.L^{-1} of glucose in the hydrolyzate, while the untreated-biomass of *K. alvarezii* produced a hydrolyzate with a glucose concentration of 3 g.L^{-1}, in other words, 4 times more concentrated (Masarin et al. 2016). Future studies related to increase of consistency (consistency values of the order of 20% (w/v)) may raise the glucose concentration from the residue-carrageenan-extraction to the order of 100 g.L^{-1}.

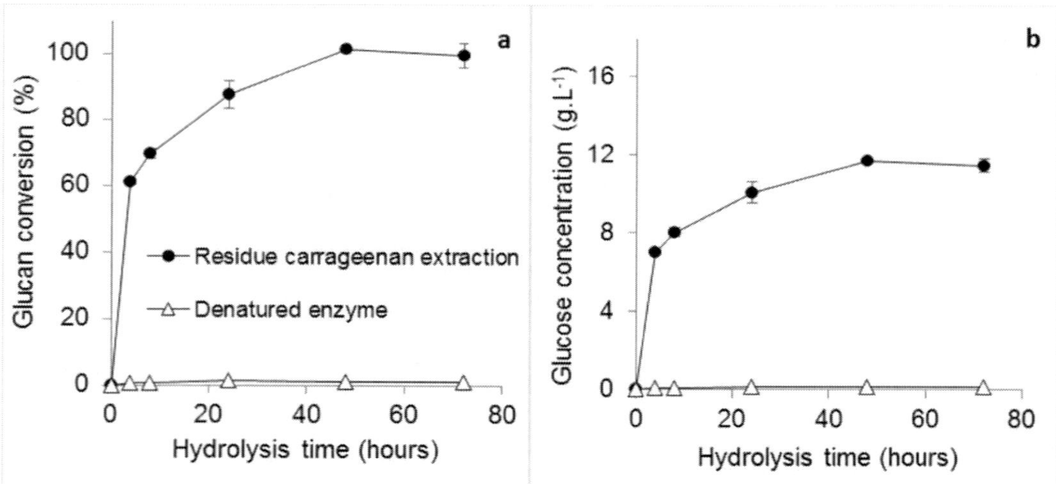

Figure 15. Enzymatic hydrolysis of residue-extraction-carrageenan from *K. alvarezii*. Enzymatic extract Cellic CTec II (Novozymes). (a) Glucan conversion (w/w%, dry biomass). (b) Glucose concentration (g.L^{-1}) (Masarin et al. 2016).

Thus, these by-products from processes of obtaining hydrocolloids from macroalgal biomass, can be converted into fermentable monomeric sugars having potential for the production of products of commercial interest. An example would be the production of fourth-generation-bioethanol (4G) (Masarin et al. 2016; Róndan et al. 2017).

References

Abd-Rahim, Faiqah, Helmi Wasoh, Mohd Rafein Zakaria, Arbakariya Ariff, Rizal Kapri, Nazaruddin Ramli, and Liew Siew-Ling. 2014. "Production of High Yield Sugars from Kappaphycus Alvarezii Using Combined Methods of Chemical and Enzymatic Hydrolysis." *Food Hydrocolloids* 42 (P2). Elsevier Ltd: 309–15. doi:10.1016/j.foodhyd.2014.05.017.

Adams, Jessica M., Joseph A. Gallagher, and Iain S. Donnison. 2009. "Fermentation Study on Saccharina Latissima for Bioethanol Production Considering Variable Pretreatments." *Journal of Applied Phycology* 21 (5): 569–74. doi:10.1007/s10811-008-9384-7.

Ahmad, a. L., N. H Mat Yasin, C. J C Derek, and J. K. Lim. 2011. "Microalgae as a Sustainable Energy Source for Biodiesel Production: A Review." *Renewable and Sustainable Energy Reviews* 15 (1). Elsevier Ltd: 584–93. doi:10.1016/j.rser.2010.09.018.

Avanzo, Neto J., and M. T. Fujii. 2016. *Guia Ilustrado de Identificação E Utilização de Algas Marinhas Bentônicas Do Estado de São Paulo*. [Illustrated Guide on Identification and Use of Benthic Marine Algae from the State of São Paulo.] Rima Edito. Sao Carlos.

Ben Yahmed, Nesrine, Mohamed Amine Jmel, Monia Ben Alaya, Hassib Bouallagui, M. Nejib Marzouki, and Issam Smaali. 2016. "A Biorefinery Concept Using the Green Macroalgae Chaetomorpha Linum for the Coproduction of Bioethanol and Biogas." *Energy Conversion and Management* 119: 257–65. doi:10.1016/j.enconman.2016.04.046.

Brinkhuis, B. H., H. G. Levine, G. G. Schlenk, and S. Tobin. 1987. "Laminaria Cultivation in the Far East and North America." In *Seaweed Cultivation for Renewable Resources*, edited by K. T Bird and P. H. Benson, Elsevier, 107–46. Amsterdam, The Netherlands.

Champagne, Pascale. 2007. "Feasibility of Producing Bio-Ethanol from Waste Residues: A Canadian Perspective. Feasibility of Producing Bio-Ethanol from Waste Residues in Canada." *Resources, Conservation and Recycling*. doi:10.1016/j.resconrec.2006.09.003.

Chen, Hongzhang. 2014. *Biotechnology of Lignocellulose: Theory and Practice. Biotechnology of Lignocellulose: Theory and Practice*. doi:10.1007/978-94-007-6898-7.

Chen, P., M. Min, Y. Chen, L. Wang, Y. Li, Q. Chen, C. Wang, et al. 2010. "Review of Biological and Engineering Aspects of Algae to Fuel Approach." *International Journal of Agricultural and Biological Engineering* 4 (2): 1–30.

Chiaramonti, D. 2007. "Bioethanol: Role and Production Technologies." In *Improvement Crop Plants for Industrial End Uses*, edited by P. Ranalli, 209–51.

Cho, Yukyeong, Hyejin Kim, and Sung Koo Kim. 2013. "Bioethanol Production from Brown Seaweed, Undaria Pinnatifida, Using NaCl Acclimated Yeast." *Bioprocess and Biosystems Engineering* 36 (6): 713–19. doi:10.1007/s00449-013-0895-5.

FAO. 2017. *Fisheries and Aquaculture Information and Statistics Branch*. http://www.fao.org/figis/servlet/TabLandArea?tb_ds=Aquaculture&tb_mode=TABLE&tb_act=SELECT&tb_grp=COUNTRY.

Gelli, Valéria Cress, and Edison Barbieri. 2015. "Cultivation and Utilization of the Kappaphycus Alvarezii Macroalga for Small Maricultores." In *Aquaculture in Brazil: New Perspectives*, edited by P. A. M. Brito and J. R. M. Brito, Pedro & Jo, 641–58. Sao Carlos.

Góes, Henrique Geromel de, and Renata Perpetuo Reis. 2012. "Temporal Variation of the Growth, Carrageenan Yield and Quality of Kappaphycus Alvarezii (Rhodophyta, Gigartinales) Cultivated at Sepetiba Bay, Southeastern Brazilian Coast." *Journal of Applied Phycology* 24 (2): 173–80. doi:10.1007/s10811-011-9665-4.

Goh, Chun Sheng, and Keat Teong Lee. 2010. "A Visionary and Conceptual Macroalgae-Based Third-Generation Bioethanol (TGB) Biorefinery in Sabah, Malaysia as an Underlay for Renewable and Sustainable Development." *Renewable and Sustainable Energy Reviews* 14 (2): 842–48. doi:10.1016/j.rser.2009.10.001.

Goh, Chun Sheng, Kok Tat Tan, Keat Teong Lee, and Subhash Bhatia. 2010. "Bio-Ethanol from Lignocellulose: Status, Perspectives and Challenges in Malaysia." *Bioresource Technology* 101 (13). Elsevier Ltd: 4834–41. doi:10.1016/j.biortech.2009.08.080.

Hargreaves, Paulo Iiboshi, Carolina Araújo Barcelos, Antonio Carlos Augusto da Costa, and Nei Pereira. 2013. "Production of Ethanol 3G from Kappaphycus Alvarezii: Evaluation of Different Process Strategies." *Bioresource Technology* 134. Elsevier Ltd: 257–63. doi:10.1016/j.biortech.2013.02.002.

Hayashi, Leila, Edison José de Paula, and Fungyi Chow. 2007. "Growth Rate and Carrageenan Analyses in Four Strains of Kappaphycus Alvarezii (Rhodophyta, Gigartinales) Farmed in the Subtropical Waters of S{ã}o Paulo State, Brazil." *Journal of Applied Phycology* 19 (5): 393–99. doi:10.1007/s10811-006-9135-6.

Hayashi, Leila, and Renata P. Reis. 2012. "Cultivation of the Red Algae Kappaphycus Alvarezii in Brazil and Its Pharmacological Potential." *Brazilian Journal of Pharmacognosy*. doi:10.1590/S0102-695X2012005000055.

Horn, S J. 2000. *Bioenergy from Brown Seaweeds*. Norwegian University of Science and Technology.

Horn, S J, I M Aasen, and K Stgaard. 2000. "Production of Ethanol from Mannitol by Zymobacter Palmae." *Journal of Industrial Microbiology and Biotechnology* 24 (1): 51–57. doi:10.1038/sj.jim.2900771.

Jang, SS, Y Shirai, M Uchida, and M Wakisaka. 2012. "Production of Mono Sugar from Acid Hydrolysis of Seaweed." *African Journal of Biotechnology* 11 (8): 1953–63. doi:10.4314/ajb.v11i8.

John, Rojan P., G. S. Anisha, K. Madhavan Nampoothiri, and Ashok Pandey. 2011. "Micro and Macroalgal Biomass: A Renewable Source for Bioethanol." *Bioresource Technology* 102 (1). Elsevier Ltd: 186–93. doi:10.1016/j.biortech.2010.06.139.

Jung, Kyung A, Seong-rin Lim, Yoori Kim, and Jong Moon. 2013. "Potentials of Macroalgae as Feedstocks for Biorefinery." *Bioresource Technology* 135. Elsevier Ltd: 182–90. doi:10.1016/j.biortech.2012.10.025.

Khambhaty, Yasmin, Kalpana Mody, Mahesh R. Gandhi, Sreekumaran Thampy, Pratyush Maiti, Harshad Brahmbhatt, Karuppanan Eswaran, and Pushpito K. Ghosh. 2012. "Kappaphycus Alvarezii as a Source of Bioethanol." *Bioresource Technology* 103 (1): 180–85. doi:10.1016/j.biortech.2011.10.015.

Kim, Hyejin, Chae Hun Ra, and Sung Koo Kim. 2013. "Ethanol Production from Seaweed (Undaria Pinnatifida) Using Yeast Acclimated to Specific Sugars." *Biotechnology and Bioprocess Engineering* 18 (3): 533–37. doi:10.1007/s12257-013-0051-8.

Kim, Nag Jong, Hui Li, Kwonsu Jung, Ho Nam Chang, and Pyung Cheon Lee. 2011. "Ethanol Production from Marine Algal Hydrolysates Using Escherichia Coli KO11." *Bioresource Technology* 102 (16): 7466–69. doi:10.1016/j.biortech.2011.04.071.

Kumar, Savindra, Rishi Gupta, Gaurav Kumar, Dinabandhu Sahoo, and Ramesh Chander. 2013. *Bioresource Technology Bioethanol Production from Gracilaria Verrucosa, a Red Alga, in a Biorefinery Approach.* 135: 150–56.

Lahaye, Marc, and Audrey Robic. 2007. "Structure and Functional Properties of Ulvan, a Polysaccharide from Green Seaweeds." *Biomacromolecules* 8 (6): 24–30. doi:10.1021/bm061185q.

Lee, Ji ye, Pan Li, Jieun Lee, Hyun Jin Ryu, and Kyeong Keun Oh. 2013. "Ethanol Production from Saccharina Japonica Using an Optimized Extremely Low Acid Pretreatment Followed by Simultaneous Saccharification and Fermentation." *Bioresource Technology* 127: 119–25. doi:10.1016/j.biortech.2012.09.122.

Lee, Sung Mok, and Jae Hwa Lee. 2012. "Ethanol Fermentation for Main Sugar Components of Brown-Algae Using Various Yeasts." *Journal of Industrial and Engineering Chemistry* 18 (1). The Korean Society of Industrial and Engineering Chemistry: 16–18. doi:10.1016/j.jiec.2011.11.097.

Lewandowski, Iris, Jonathan M O Scurlock, Eva Lindvall, and Myrsini Christou. 2003. "The Development and Current Status of Perennial Rhizomatous Grasses as Energy

Crops in the US and Europe." *Biomass and Bioenergy* 25 (4): 335–61. doi:10.1016/S0961-9534(03)00030-8.

Lewis, Norman G, and Etsuo Yamamoto. 1990. *Lignin: Occurrence, Biogenesis and Biodegradation.*

Li, Yanqun, Mark Horsman, Nan Wu, Christopher Q Lan, and Nathalie Dubois-calero. 2008. "Li Y, Horsman M, Wu N, Lan CQ, Dubois-Calero N. Biofuels from Microalgae. Biotechnology Progress 2008b;24(4):815." *Biotechnology Prog* 24 (1): 815–20. doi:10.1021/bp.070371k.

Li, Yinping, Jiefen Cui, Gaoli Zhang, Zhengkun Liu, Huashi Guan, Hueymin Hwang, Winfred G. Aker, and Peng Wang. 2016. "Optimization Study on the Hydrogen Peroxide Pretreatment and Production of Bioethanol from Seaweed Ulva Prolifera Biomass." *Bioresource Technology* 214: 144–49. doi:10.1016/j.biortech.2016.04.090.

Lobban, C.S., and M.J. Wynne. 1981. *The Biology of Seaweeds*. Edited by Oxford: Blackwell.

Martone, Patrick T., Jos?? M. Estevez, Fachuang Lu, Katia Ruel, Mark W. Denny, Chris Somerville, and John Ralph. 2009. "Discovery of Lignin in Seaweed Reveals Convergent Evolution of Cell-Wall Architecture." *Current Biology* 19 (2): 169–75. doi:10.1016/j.cub.2008.12.031.

Masarin, Fernando, Fernando Roberto Paz Cedeno, Eddyn Gabriel Solorzano Chavez, Levi Ezequiel de Oliveira, Valéria Cress Gelli, and Rubens Monti. 2016. "Chemical Analysis and Biorefinery of Red Algae Kappaphycus Alvarezii for Efficient Production of Glucose from Residue of Carrageenan Extraction Process." *Biotechnology for Biofuels* 9 (1). BioMed Central: 122. doi:10.1186/s13068-016-0535-9.

Masarin, Fernando, and André Ferraz. 2008. "Evaluation of Eucalyptus Grandis Hill Ex Maiden Biopulping with Ceriporiopsis Subvermispora under Non-Aseptic Conditions." *Holzforschung* 62 (1): 1–7. doi:10.1515/HF.2008.001.

Masarin, Fernando, Daniela B Gurpilhares, David Cf Baffa, Márcio Hp Barbosa, Walter Carvalho, André Ferraz, and Adriane Mf Milagres. 2011. "Chemical Composition and Enzymatic Digestibility of Sugarcane Clones Selected for Varied Lignin Content." *Biotechnology for Biofuels* 4 (1): 55. doi:10.1186/1754-6834-4-55.

Mata, Teresa M., António A. Martins, and Nidia S. Caetano. 2010. "Microalgae for Biodiesel Production and Other Applications: A Review." *Renewable and Sustainable Energy Reviews*. doi:10.1016/j.rser.2009.07.020.

McHugh, Dennis J. 2003. *A Guide to the Seaweed Industry.*

Meinita, Maria Dyah Nur, Yong Ki Hong, and Gwi Taek Jeong. 2012. "Comparison of Sulfuric and Hydrochloric Acids as Catalysts in Hydrolysis of Kappaphycus Alvarezii (Cottonii)." *Bioprocess and Biosystems Engineering* 35 (1–2): 123–28. doi:10.1007/s00449-011-0609-9.

Meinita, Maria Dyah Nur, Ji Young Kang, Gwi Taek Jeong, Hyun Min Koo, Sung Min Park, and Yong Ki Hong. 2012. "Bioethanol Production from the Acid Hydrolysate of the Carrageenophyte Kappaphycus Alvarezii (Cottonii)." *Journal of Applied Phycology* 24 (4): 857–62. doi:10.1007/s10811-011-9705-0.

Meinita, Maria Dyah Nur, Bintang Marhaeni, Tjahjo Winanto, Gwi Taek Jeong, Mohammed Nurul Absar Khan, and Yong Ki Hong. 2013. "Comparison of Agarophytes (Gelidium, Gracilaria, and Gracilariopsis) as Potential Resources for Bioethanol Production." *Journal of Applied Phycology* 25 (6): 1957–61. doi:10.1007/s10811-013-0041-4.

Mendes, Fernanda M., Germano Siqueira, Walter Carvalho, Andre Ferraz, and Adriane M F Milagres. 2011. "Enzymatic Hydrolysis of Chemithermomechanically Pretreated Sugarcane Bagasse and Samples with Reduced Initial Lignin Content." *Biotechnology Progress* 27 (2): 395–401. doi:10.1002/btpr.553.

Nigam, Poonam Singh, and Anoop Singh. 2011. "Production of Liquid Biofuels from Renewable Resources." *Progress in Energy and Combustion Science* 37 (1): 52–68. doi:10.1016/j.pecs.2010.01.003.

Ra, Chae Hun, Gwi Taek Jeong, Myung Kyo Shin, and Sung Koo Kim. 2013. "Biotransformation of 5-Hydroxymethylfurfural (HMF) by Scheffersomyces Stipitis during Ethanol Fermentation of Hydrolysate of the Seaweed Gelidium Amansii." *Bioresource Technology* 140. Elsevier Ltd: 421–25. doi:10.1016/j.biortech.2013.04.122.

Radulovich, R., A. Neori, D. Valderrama, C. R. K. Reddy, H. Cronin, and J. Forster. 2015. "Farming of Seaweeds." In *Seaweed Sustainability - Food And NonFood Applications*, edited by B. Tiwari and D. Troy, Elsevier. Amsterdam, The Netherlands.

Reis, Renata Perpetuo, Roberta Rodrigues das Chagas Pereira, and Henrique Geromel de Góes. 2014. "The Efficiency of Tubular Netting Method of Cultivation for Kappaphycus Alvarezii (Rhodophyta, Gigartinales) on the Southeastern Brazilian Coast." *Journal of Applied Phycology* 27 (1): 421–26. doi:10.1007/s10811-014-0330-6.

Roldan, Ismael Ulises Miranda, A. T. Mitsuhara, J. P. M. Desajacomo, L.E. De Oliveira, Valéria Cress Gelli, Rubens Monti, L. V. S. Sacramento, and Fernando Masarin. 2017. "Chemical, Structural and Ultrastructural Analysis of Waste from the Carrageenan and Sugar-Bioethanol Processes for Future Bioenergy Generation." *Biomass and Bioenergy* 127: 233-243. http://dx.doi.org/10.1016/j.biombioe. 2017.10.008.

Sluiter, A, B Hames, R Ruiz, C Scarlata, J Sluiter, D Templeton, and D Crocker Nrel. 2012. *Determination of Structural Carbohydrates and Lignin in Biomass Determination of Structural Carbohydrates and Lignin in Biomass*, no. April 2008.

Sudhakar, M.P., A. Jegatheesan, C. Poonam, K. Perumal, and K. Arunkumar. 2017. "Biosaccharification and Ethanol Production from Spent Seaweed Biomass Using Marine Bacteria and Yeast." *Renewable Energy* 105: 133–39. doi:10.1016/j.renene.2016.12.055.

van der Wal, Hetty, Bram L H M Sperber, Bwee Houweling-Tan, Robert R C Bakker, Willem Brandenburg, and Ana M. Lopez-Contreras. 2013. "Production of Acetone, Butanol, and Ethanol from Biomass of the Green Seaweed Ulva Lactuca." *Bioresource Technology* 128: 431–37. doi:10.1016/j.biortech.2012.10.094.

Vicentim, Marcos Paulo, Robson De Almeida Faria, and André Ferraz. 2009. "High-Yield Kraft Pulping of Eucalyptus Grandis Hill Ex Maiden Biotreated by Ceriporiopsis Subvermispora under Two Different Culture Conditions." *Holzforschung* 63 (4): 408–13. doi:10.1515/HF.2009.067.

Wei, Na, Josh Quarterman, and Yong Su Jin. 2013. "Marine Macroalgae: An Untapped Resource for Producing Fuels and Chemicals." *Trends in Biotechnology* 31 (2). Elsevier Ltd: 70–77. doi:10.1016/j.tibtech.2012.10.009.

Wi, Seung Gon, Hyun Joo Kim, Shobana Arumugam Mahadevan, Duck Joo Yang, and Hyeun Jong Bae. 2009. "The Potential Value of the Seaweed Ceylon Moss (Gelidium Amansii) as an Alternative Bioenergy Resource." *Bioresource Technology* 100 (24). Elsevier Ltd: 6658–60. doi:10.1016/j.biortech.2009.07.017.

Yaich, Hela, Haikel Garna, Souhail Besbes, Michel Paquot, Christophe Blecker, and Hamadi Attia. 2011. "Chemical Composition and Functional Properties of Ulva Lactuca Seaweed Collected in Tunisia." *Food Chemistry* 128 (4). Elsevier Ltd: 895–901. doi:10.1016/j.foodchem.2011.03.114.

Yanagisawa, Mitsunori, Kanami Nakamura, Osamu Ariga, and Kiyohiko Nakasaki. 2011. "Production of High Concentrations of Bioethanol from Seaweeds That Contain Easily Hydrolyzable Polysaccharides." *Process Biochemistry* 46 (11). Elsevier Ltd: 2111–16. doi: 10.1016/j.procbio.2011.08.001.

Yoon, Jeong Jun, Yong Jin Kim, Sang Hyoun Kim, Hyun Jin Ryu, Jung Youn Choi, Gyung Soo Kim, and Myung Kyo Shin. 2010. "Production of Polysaccharides and Corresponding Sugars from Red Seaweed." *Advanced Materials Research* 93–94: 463–66. doi:10.4028/www.scientific.net/AMR.93-94.463.

Yu, Shukun, Andreas Blennow, Maja Bojko1, Finn Madsen, Carl Erik Olsen, and Søren B. Engelsen. 2002. "Physico-Chemical Characterization of Floridean Starch of Red Algae." *Starch - Starke* 54 (2): 66–74.

Zodape, S. T., S. Mukherjee, M. P. Reddy, and D. R. Chaudhary. 2009. "Effect of Kappaphycus Alvarezii (Doty) Doty Ex Silva. Extract on Grain Quality, Yield and Some Yield Components of Wheat (*Triticum Aestivum L.*)." *International Journal of Plant Production* 3 (2): 97–101.

In: Bioethanol and Beyond
Editor: Michel Brienzo
ISBN: 978-1-53613-478-0

Chapter 9

Cellulose Enzymatic Hydrolysis and Its Technological Applications

Marcos Paulo Vicentim *

Directorate of Metrology Applied to Life Sciences (Dimav),
National Institute of Metrology, Quality and Technology (Inmetro),
Xerém, Duque de Caxias, RJ, Brazil

Abstract

Cellulose is the main polymeric component of the plant cell wall, the most abundant polysaccharide on Earth, and an important renewable resource. Its is a polymer consisting of D-anhydroglucopyranose linked by α-1,4-glycosidic bonds forming linear polymeric chains of over 10,000 glucose residues. The importance of cellulose hydrolysis for conversion of plant biomass into fuels and chemicals is widely recognized. For this purpose, thermal, physical, chemical and/or enzymatic processes may be applied. The enzymatic hydrolysis using cellulases presents as the main benefit the higher conversion yield of cellulose to glucose. However, structural features of cellulose impacts on rates of enzymatic hydrolysis. Thus, searching for the best enzyme-producing organisms, for enzymatic complexes in the ideal proportions and for phisical and chemical systems for proper cellulose conversion is a challenge to improve technological processes. Industrial production of cellulases began in the middle of the 1980s. These enzymes are applied in industrial sectors such as food, textile, pulp and paper and laundry. One of the most emerging applications for the cellulolytic enzyme complex is for biomasses hydrolysis. Glucose can be used to obtain a huge range of products, ranging from biofuels to polymers, among which ethanol is one of the most interesting molecules nowadays. This chapter reviews cellulose structure, its enzymatic hydrolysis and the technological applications of cellulases and hydrolyzed cellulose.

Keywords: cellulose, cellulose hydrolysis, cellulases, glycoside hydrolases

---

* Corresponding Author Email: marcospaulov@yahoo.com.br.

Introduction

Cellulose is the main polymeric component of the plant cell wall, the most abundant polysaccharide on Earth, and an important renewable resource (Fengel and Wegener, 1983). It contains both highly crystalline regions where individual chains are linked to each other and less-ordered amorphous regions. Although chemically simple, the intermolecular bonding pattern can result in a very complex morphology (Baldrian, 2008). Its production and application to human activities highly contributes to the global carbon cycle (Falkowski et al., 2000).

The importance of cellulose hydrolysis for conversion of plant biomass into fuels and chemicals is widely recognized (Lynd et al., 1991, 1999; Himmel et al., 1999; Brienzo et al., 2014). For this purpose, thermal, physical, chemical and/or enzymatic processes may be applied. Enzymatic methods, although slower than the other ones, have several advantages due to their greater specificity and action under milder conditions. These two factors practically eliminate the formation of side products (degradation products), which consequently results in higher glucose yield and facilitates subsequent biological processes such as alcoholic fermentation. In fact, most of the side products are toxic and can inhibit microrganisms and enzymes.

Enzymatic hydrolysis of cellulose is performed by cellulases. However, structural features of cellulose like crystallinity index, degree of polymerization, and accessible area impact on the enzymatic hydrolysis rate and yield (Zhang and Lynd, 2004; Brienzo et al., 2015). In order to overcome these obstacles and improve the yield of products, the search for new physicochemical and mainly enzymatic processes is necessary. Thus, searching for the best enzyme-producing organisms and for enzymatic complexes in the ideal proportions for proper cellulose conversion is a challenge to improve technological process. This chapter reviews cellulose structure, its enzymatic hydrolysis and technological applications of cellulases and hydrolyzed cellulose.

Cellulose

Photosynthetic higher plants and algae are by far the most important way of cellulose production. But there are also certain bacteria, marine invertebrates, fungi, slime molds and amoebae that produce cellulose by nonphotosynthetic ways (Coughlan, 1985; Jarvis, 2003; Lynd et al., 2002; Tomme et al., 1995).

Cellulose is a condensed polymer consisting of D-anhydroglucopyranose linked by α-1,4-glycosidic bonds forming linear polymeric chains of over 10,000 glucose residues. Anhydrocellobiose (or simply cellobiose) is the repeating unit of cellulose and these molecules are rotated 180° relative to their neighbors (Figure 1). This rotation makes cellulose highly symmetrical since each side of the chain has an equal number of hydroxyl groups. Coupling of adjacent cellulose molecules by hydrogen bonds and van der Waal's forces results in a parallel alignment and a crystalline structure. The extensive hydrogen bonds of interchain (2 per anhydroglucopyranose) and intrachain (2~3 per anhydroglucopyranose) produces straight, stable supramolecular fibers of great tensile strength (Gardner and Blackwell, 1974a,b; Nevell and Zeronian, 1985; Krassig, 1993; Hon, 1994).

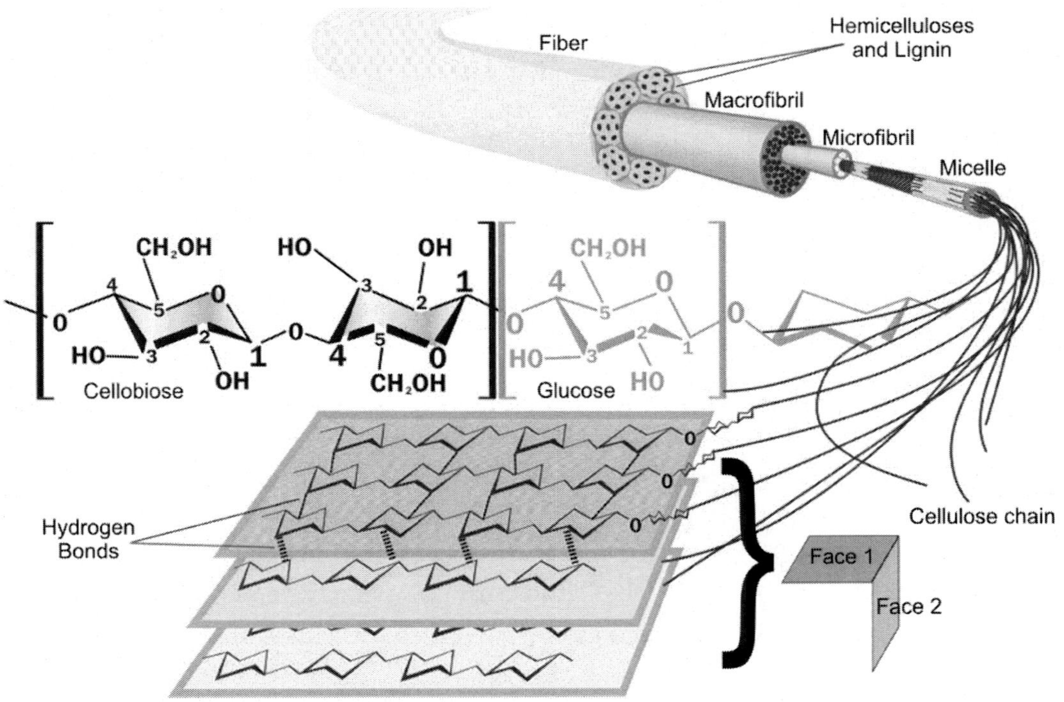

Figure 1. Structure of cellulose featuring the repeating β-1,4-linked anhydrocellobiose units, the crystaline structure and the lignocellulose organization in fibrils.

Cellulose is formed by layers of glucopyranose rings arranged in a specific plane with successive layers. These layers stack up on the top of each other to form a three-dimensional structure. This arrangement makes cellulose surface has distinct "faces," which in turn interact differently with the environment and consequently with the cellulases. The six carbons in the glucopyranose ring and internal β-glucosidic bonds represent one plane (face 1), whereas the edges of the rings build up face 2, which is perpendicular to face 1 (Figure 1). The repeating unit in face 1 is cellobiose, being 1.04 nm long and 0.54 nm wide. About 100 cellulose glucans are aggregated into an elementary fibril, which presents crystalline width of 4–5 nm (O'Sullivan, 1997). Bunches of 7–30 nm thick elementary fibrils are embedded in a matrix of hemicelluloses, representing a microfibril. The lignification of the plant cell wall occurs later during the synthesis of natural fibers, so lignin is located primarily on the exterior of microfibrils (Figure 1). Lignin bonds covalently to hemicelluloses (Klein and Snodgrass, 1993).

CELLULASES

Cellulases are a complex of enzymes, mainly produced by fungi, which hydrolyze cellulose with high specificity (Olsson and Hahn-Hägerdal, 1996). These enzymes act synergistically releasing glucose which has great industrial interest because it can be converted into a series of products, being ethanol one of the main ones currently.

Cellulases and hemicellulases employ hydrolytic reactions to cleave cellulose and hemicellulose, respectively. Almost all of them relies on either: 1) a "retaining" mechanism that yields products of the same anomeric configuration after breaking a glycosidic bond with a "double-displacement" hydrolysis, or 2) an "inverting" mechanism, which yields products of the opposite anomeric configuration after breaking a glycosidic bond with a "single nucleophilic-displacement" hydrolysis, both involving two acidic amino acid residues (Glu or Asp) as a proton donor or general acid and as a nucleophile or base (Vocadlo et al., 2008). "Retaining" hydrolases might also act as glycosyl transferase (Sweeney and Xu, 2012).

Classification of Cellulases

Cellulases are classified according to a range of characteristics. First, according to the cleavage site on cellulose, they are classified as endo-1,4-β-D-glucanases (or simply endoglucanases), exo-1,4-β-D-cellobiohydrolases (or simply cellobiohydrolases or exoglucanases) and 1,4-β-D-glicosidases (or simply β-glicosidases), (Tanimura et al., 2013).

Endoglucanases (EGs) cleave glycosidic linkages mainly at amorphous portions of cellulose chains creating new reducing ends; cellobiohydrolases (CBHs) act at the reducing ends leading to the formation mainly of glucose dimers (cellobiose); and β-glucosidases (BGs) hydrolyze cellobiose to glucose (Olsson et al., 2004). Figure 2 schematically shows the synergistic action of cellulases.

Second, cellulases are classified according to the presence of the carbohydrate-binding module (CBM) within their molecules. Some carbohydrate-degrading enzymes have a CBM independent of the catalytic site, which binds to substrates stabilizing the enzymatic reaction. For example, cellulases that have a cellulose-binding domain (CBD) on a CBM stably binds to cellulose, so the enzyme molecules can continuously keep moving to the next cleavage site after each reaction. In comparison, cellulases that do not have a CBD detach from the cellulose after every cleaving reaction (Watanabe and Tokuda, 2010). Thus, cellulases that have a CBD are assumed to hydrolyze cellulose more efficiently.

Third, cellulases are classified according to their primary structure. For instance, glycoside hydrolases (GH) are classified into families based on amino acid sequence similarities (Henrissat et al., 1989). Currently, 130 GH families are registered in this classification scheme, and cellulases can be found in families GH1, 3, 5, 6, 7, 8, 9, 10, 12, 19, 26, 30, 44, 45, 48, 51, 61, 74, 116, and 124 (CAZy, 2017).

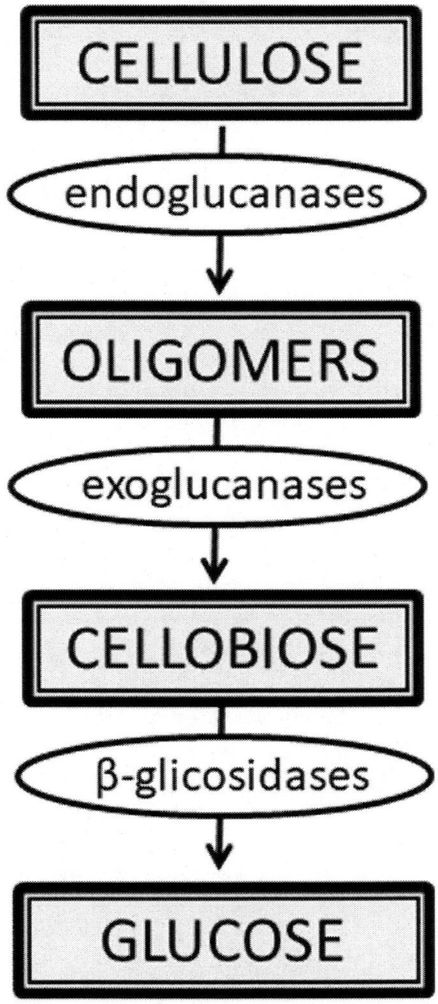

Figure 2. Synergistic action of cellulases.

Many researches focused on cellulases produced by the genus *Trichoderma* mostly due to the high levels of cellulases secreted. *T. reesei* cellulase mixture consists of many catalytically active proteins. At least two cellobiohydrolases (CBH1–2), five endoglucanases (EG1–5), β-glucosidases, and hemicellulases have been identified by 2D electrophoresis (Vinzant et al., 2001). CBH1, CBH2, and EG2 are the three main components of the *T. reesei* cellulase system, representing 60 ± 5%, 20 ± 6%, and 12 ± 3% of total cellulase protein, respectively (Goyal et al., 1991; Gritzali and Brown, 1978; Knowles et al., 1987; Kyriacou et al., 1987; Nidetzky and Claeyssens, 1994).

Table 1. Glycoside hydrolase families in different taxa

GH	Cellulase type	Archaea	Eukaryote			
			Eubacteria	Fungus	Plant	Animal (invertebrate)
1	β-Glucosidases	+	+	+	+	+
3	β-Glucosidases	+	+	+	+	+
5	Endoglucanases, β-Glucosidases	+	+	+	+	+
6	Endoglucanases, cellobiohydrolases	-	+	+	-	-
7	Endoglucanases, cellobiohydrolases	-	-	+	+	+
8	Endoglucanases	+	+	-	-	-
9	Endoglucanases, cellobiohydrolases, β-Glucosidases	+	+	+	+	+
10	Endoglucanases	-	+	-	-	-
12	Endoglucanases	+	+	+	+	-
19	Endoglucanases	+	+	-	-	-
26	Endoglucanases	-	+	-	-	-
30	β-Glucosidases	+	+	+	+	+
44	Endoglucanases	-	+	-	-	-
45	Endoglucanases	-	+	+	-	+
48	Endoglucanases	-	+	+	-	-
51	Endoglucanases	+	+	+	+	-
61	Endoglucanases	-	-	+	+	-
74	Endoglucanases	-	+	+	-	-
116	β-Glucosidases	+	+	+	+	+
124	Endoglucanases	-	+	-	-	-

An important step towards applying *T. reesei* cellulases industrially was the development of efficient strain mutagenesis and screening procedures in the 1970s. In the following two decades, the titer of extracellular protein produced by the original strain QM6a could be increased by up to 20-fold through mutagenesis programs at Natick and Rutgers University. The latter culminated in the isolation of strain RUT-C30 (where "RUT" stands for Rutgers). Although the gold standard for cellulase production in industry was reported to be higher than 100 g/L, this strain still is the prototype cellulase hyperproducer available to the public domain with titers of extracellular protein reaching 30 g/L on the cellulase inducing substrate lactose (Bischof et al., 2016).

Other fungi genus and even other taxa (archaea, eubacteria, plant and invertebrate) are nowadays known to produce cellulases. Table 1 presents the GH classification scheme and which of them are produced by the different taxa.

Endo-1,4-Β-Glucanases

Degradation of amorphous regions of cellulose can be carried out by endoglucanases (EC 3.2.1.4). EGs hydrolyze internal glycosidic bonds in cellulose randomically generating new cellulose chain ends for subsequent CBH action. Few EGs can act "processively" on crystalline cellulose (Wilson, 2008; Li et al., 2010). There is a significant synergism between EGs and CBHs, and their co-presence and cooperation are determinant for highly efficient enzymatic systems for the conversion of industrial biomass.

Widely distributed among several organisms, different EGs have catalytic cores belonging to more than ten GH families, of which GH5, 7, 9, 12, 45 and 48 are representative. Typical cellulolytic fungi secrete EGs at a level of ~ 20% (wt) (Herpoël-Gimbert et al., 2008; Sipos et al., 2010; Chundawat et al., 2010). Also known as EG-I, II, III and V, respectively, GH7, 5, 12 and 45 EGs are the most common in the natural mixtures of fungal cellulase. Most fungi and cellulolytic bacteria produce numerous EGs. Although they all act on the same cellulose substrate, they do so through different mechanisms ("inverting" for GH6, 9, 45 and 48 EGs; "retaining" for GH5, 7, 12 EGs). This variety of EGs may be related to side-activities on hemicellulose in degrading complex lignocellulose (Vlasenko et al., 2010), or synergism between processive and conventional EGs (Wilson, 2008).

The active sites of most EGs are cleft- or groove-shaped, within which a cellodextrin or a cellulose segment can be bound and actuated by EGs. In addition to the catalytic core, EGs may possess CBMs or other domains. CBMs may direct host EGs but it is not a prerequisite for EGs action (Sweeney and Xu, 2012).

Cellobiohydrolases

Degradation of crystalline cellulose is carried out mainly by CBHs, thus these enzymes are indispensable for industrial enzymatic lignocellulose degradation. Archetypical CBHs are found in GH6 and 7, as well as 48, families. GH7 CBH is found in all known cellulolytic fungi (based on secretome or genome information). GH6 CBH is also found in many cellulolytic fungi. Among the secreted proteins and enzymes of cellulolytic fungi, up to 70% (wt) may be CBHs (Herpoël-Gimbert et al., 2008; Sipos et al., 2010; Chundawat et al., 2010). Also known as CBH-I (EC 3.2.1.-), GH7 CBH has specificity towards the reducing end of a cellulose chain. In contrast, GH6 CBH, also known as CBH-II (EC 3.2.1.91), can be specific towards the non-reducing end of a cellulose chain. Such "opposing" specificities render GH7 and 6 CBHs highly synergistic and cooperative in degrading their common substrate (Sweeney and Xu, 2012).

CBHs catalytic core features tunnel-like active sites, a topology that equips CBHs with the ability to hydrolyze cellulose "processively": it threads into the end of a cellulose chain through its active site, cleaves off a cellobiosyl unit, glides down the chain, and starts the next hydrolysis step (Vocadlo et al., 2008; Liu et al., 2011). A CBM may assist the catalytic core with processivity (Beckham et al., 2010). Such processive reactions, plus the insolubility of the cellulose substrate, make CBHs kinetics deviant from the Michaelis-Menten model, and show significant fractal and "local jamming" effect (Xu and Hanshu, 2007; Igarashi et al., 2011; Warden, 2011). Processive CBHs movement can be obstructed by kinks or other impediments on the cellulose surface; and as such it has been suggested that k(off) values may be a major factor in CBHs efficiency (Kurašin and Väljamäe, 2011; Praestgaard et al., 2011).

GH7 CBH-I may have approximately ten anhydro-Glc-binding subsites within its active tunnel, in which a cellulose segment or cellodextrin is bound and activated via H-bonding and π-stacking with key amino acid residues. In addition to the catalytic core, many CBHs also have CBMs, which is believed to be of key importance in CBH's action on crystalline cellulose (Sweeney and Xu, 2012).

β-Glucosidases

Degradation of cellobiose, as well as other cellodextrins, is carried out by BGs or cellobiose hydrolase (EC 3.2.1.21). Unlike CBHs and EGs, BGs, in general, are not modular (lacking distinct CBMs), and have pocket-shaped active sites to act on the non-reducing Glc unit from cellobiose or cellodextrin (Langston et al., 2006). BGs belong to the GH1, 3, and 9 families, with GH1 and 3 BGs being archetypical (Eyzaguirre et al., 2005). Unlike the majority of biomass degrading enzymes, the activity of BGs, which acts upon soluble rather than insoluble substrate, can be studied using traditional kinetic models (Jeoh et al., 2005).

Many cellulolytic fungi produce one or more BGs at levels of about 1% of total secreted proteins, significantly lower than that of CBHs and EGs (Herpoël-Gimbert et al., 2008; Sipos et al., 2010; Chundawat et al., 2010). However, BGs plays a key role in the efficiency of an enzymatic lignocellulose-degrading system, because its action on cellobiose mitigates product inhibition on CBHs and EGs. For industrial biomass conversion targeting high feedstock loads, supplementing BGs to common microbial cellulolytic enzyme preparations can be imperative, because of high cellobiose level during the enzymatic conversion (Kristensen et al., 2009).

GH1 BG tend to be more resilient to Glc (product) inhibition, as well as more active on different di- or oligosaccharides than GH3 BG do. Thus having GH1 BG might enable a cellulolytic enzyme system to be more potent in degrading complex lignocellulose (Sweeney and Xu, 2012).

LPMOs

LPMOs are copper-dependent lytic polysaccharide monooxygenases that cleavage cellulose chains with oxidation of various carbons (C-1, C-4 and C-6). LPMOs were originally classified as a glycoside hydrolase (GH61) based on very weak endoglucanase activity in one family member. They are now reclassified in the auxiliary activity (AA) category of CAZy. Because a significant literature is associated with the old name GH61, CAZy recommends to describe these enzymes as "AA9 (formerly GH61)" (CAZy, 2017). Based on sequence similarity, CAZy (2017) classifies LPMOs in four families (AA9, AA10, AA11 and AA13).

Many cellulolytic fungi encode and express LPMOs (Herpoël-Gimbert et al., 2008; Sipos et al., 2010; Chundawat et al., 2010) and recently RNA-Seq analysis of the termite *Coptotermesgestroi* revealed the occurrence of the LPMO-AA10 (*Cg*AA10-2) (Cairo et al., 2016).

This enzyme release small amounts of soluble cellooligosaccharides from cellulosic substrates (Karlsson et al., 2001). Due to its unusual characteristics, it took more than a decade after the initial discovery of this group of enzymes to begin to understand their function.

The reaction mechanism of LPMOs is still unclear, but several plausible scenarios have been suggested. LPMOs show high sequence and modular diversity and are known, so far, to cleave insoluble substrates such as cellulose, chitin and starch, as well as hemicelluloses such as beta-glucan, xyloglucan and xylan. All LPMOs share a catalytic histidine brace motif to bind copper, but differ strongly when it comes to the nature and arrangement of residues on the substrate-binding surface (Vaaje et al., 2017).

Addition of LPMOs to *Trichoderma* cellulase cocktails greatly enhance the lignocellulose degradation, lowering the required enzyme concentration for substrate breakdown by a factor of two (Sweeney and Xu, 2012; Vaaje et al., 2017). In this context, enzymes currently classified as auxiliary activity 9 (AA9/formerly GH61) and 10 (AA10/formerly CBM 33) or LPMO have emerged as cellulase boosting enzymes. AA9 and AA10 are the new paradigm for deconstruction of lignocellulosic biomass by enhancing the activity and decreasing the loading of classical enzymes to the reaction. Consequently, they represent a promising resource for development of industrial enzyme cocktails for biomass processing and reducing costs of the hydrolysis step in the second-generation ethanol production chain (Correa et al., 2016).

HYDROLYSIS OF LIGNOCELLULOSIC MATERIALS

Natural cellulose molecules occur in elementary fibrils closely associated with hemicellulose and other structural polysaccharides as well as lignin (Figure 1).

Lignocellulosic materials typically contain 35% to 50% (w/w) of cellulose, 20% to 35% (w/w) of hemicelluloses, and 5% to 30% (w/w) of lignin (Chang et al., 1981; Klein and Snodgrass, 1993; Lynd et al., 2002; Mansfield et al., 1999). The enzymatic hydrolysis of cellulose in a biomass is complicated due to low cellulase accessibility and even to the necessity of other types of enzymes, like hemicellulases (such as xylanases and mannanases) and ligninases (such as laccases, lignin peroxidases and manganese peroxidases), for example (Zhang and Lynd, 2004). Lignin has adsorbent properties that reduce cellulose hydrolysis (Bernardez et al., 1993; Ooshima et al., 1990; Sutcliffe and Saddler, 1986). Lignin removal and/or redistribution is important to increase cellulose hydrolysis rates (Chernoglazov et al., 1988; Converse, 1993; Lynd et al., 2002; Brienzo et al., 2017). Even so, residual lignin impacts cellulases to reach cellulose chain (Eriksson et al., 2002; Mansfield et al., 1999). Cellulose crystallinity index also impacts on cellulase accessibility (Zhang and Lynd, 2004).

Cellulases must bind to the surface of substrate particles in order to start hydrolysis of insoluble cellulose. The 3D structure and microstructure of these particles in combination with the size and shape of cellulases determine whether β-glucosidic bonds are accessible to enzymatic attack. Cellulosic particles have both external and internal surfaces. Generally, internal surface area of cellulose is 1–2 orders higher than the external one (Chang et al., 1981).

External surface area is related to shape and particle size. The smaller the particle size the higher the cellulase adsorption and cellulose reactivity (Kim et al., 1992; Mandels et al., 1971). Also, there are linear correlations between the initial hydrolysis rate of pretreated biomass and the pore size accessible to a molecule with a diameter of 51 Å, similar to the size of *T. reesei* cellulase (Grethlein 1985; Wallace et al., 2016).

Hemicellulose is a group of complex polysaccharides made by different sugar units and linked by glycosidic bonds. Degradation of hemicellulose, which not only "makes way" for the action of cellulases but also converts hemicellulose into valuable saccharides, is primarily performed by a series of interdependent and synergistic hemicellulases.

Generally, hemicelluloses include β-glucan, xylan, xyloglucan, arabinoxylan, mannan, galactomannan, arabinan, galactan, polygalacturonan, etc., which are targets of β-glucanase, xylanase, xyloglucanase, arabinase, mannanases, galactanase, polygalacturonase, glucuronidase, acetyl xylan esterase, and other enzymes (van den Brink and Vries, 2011; Scheller and Ulvskov, 2010). Among hemicellulases, glycoside hydrolases (belonging to about 29 GH families) hydrolyze glycosidic bonds, carbohydrate esterases (belonging to about 9 carbohydrate esterases (CE) families) hydrolyze ester bonds, polysaccharide lyases (belonging to about 5 polysaccharide lyases (PL) families) cleave glycosidic bonds. *Endo*-hemicellulases cleave internal/backbone glycosidic bonds, whereas other glycosidases remove mainly the chain's substituents or side chains. Cellulolytic organisms produce many hemicellulases along with cellulases for effective lignocellulose degradation (Sweeney and Xu, 2012).

Different plants have different hemicelluloses: acetylated (galacto)glucomannan (as well as arabinoglucuronoxylan), glucuronoxylan, and arabinoxylan are major hemicellulose in softwood, hardwood, and grass, respectively (Scheller and Ulvskov, 2010). Hence different hemicellulase combinations are needed for different biomass feedstocks in industrial biomass conversion. Synergism of hemicellulases is found both amongst hemicellulases themselves and between hemicellulases and cellulases (Kumar et al., 2008; Banerjee et al., 2010; Gao et al., 2011; Couturier et al., 2011; Gottschalk et al., 2010; Kumar and Wyman, 2009).

Lignin is a highly heterogeneous and recalcitrant aromatic polymer (consisted of various syringyl, guaiacyl, or other hydroxyphenyl units) entangled with hemicellulose or cellulose and inactive/inhibitory to (hemi)cellulases. Lignin degradation is imperative for industrial enzymatic biomass-conversion because it not only increases (hemi)cellulose accessibility for (hemi)cellulase, but also diminishes (hemi)cellulase inactivation caused by lignin adsorption.

Lignin peroxidase (LiP/EC 1.11.1.14), Manganese peroxidase (MnP/EC 1.11.1.13), and versatile peroxidase (VP/EC 1.11.1.16) are extracellular fungal heme peroxidases belonging to LO2 (lignin oxidases) family with high potency to oxidatively degrade lignin. Upon interaction with H_2O_2, these enzymes form highly reactive Fe(V) or Fe(IV)-oxo species, which abstract electrons from lignin causing oxidation or radicalization either directly or via Mn(III) species. Laccase (EC 1.10.3.2) is a multi-copper oxidase belonging to LO1 family secreted by numerous lignocellulolytic fungi. This enzyme can directly oxidize phenolic parts of lignin, or indirectly oxidize non-phenolic lignin parts with the aid of suitable redox-active mediator.

Aryl-alcohol oxidase (EC 1.1.3.7), glyoxal oxidase (EC 1.1.3.-), and various carbohydrate oxidases (EC 1.1.3.4, 9, 10) are also involved in natural lignocellulose degradation. These enzymes, belonging to lignin degrading auxiliary families (LDA1–6), can generate hydrogen peroxide (H_2O_2) from oxygenO_2, with concomitant oxidation of aromatic alcohol, glyoxal, or reducing carbohydrates (Sweeney and Xu, 2012). The generated H_2O_2 may support lignin-degrading peroxidases or power Fenton reaction, which comprises a non-enzymatic mechanism that depolymerizes cellulose, hemicellulose and lignin.

Cellobiose dehydrogenase (CDH/EC 1.1.99.18) is produced by many lignocellulotic fungi. The flavoheme enzyme (belonging to LO3 family) may dehydrogenate or oxidize cellobiose or other cellodextrins to the corresponding aldonolactones, with concomitant quinone reduction to phenol or O_2 reduction to H_2O_2. Until recently, the role of CDH in enzymatic degradation of lignocellulose was largely considered only from the Fenton reaction perspective (Zamocky et al., 2006) or mitigation of cellulases' product inhibition by cellobiose (Bey et al., 2011). However, the recent discovery of the CDH stimulation of

GH61 enzyme activity may shed new light upon its function (Sweeney et al., 2014; Langston et al., 2011; Phillips et al., 2011).

Hydrolysis of Cellulose

Cellulose hydrolysis starts with physical disruption of insoluble cellulose (also called “amorphogenesis”), which is comprised by physical changes of cellulose (swelling, segmentation, or destratification) that enhance enzymatic hydrolysis and render crystalline cellulose more accessible to cellulases (Zhang and Lynd, 2004).

When cellulase system acts *in vitro* on insoluble cellulosic substrates, three processes occur simultaneously: 1) chemical and physical changes in solid-phase cellulose; 2) primary hydrolysis, which represents the release of soluble intermediates from cellulose surface; and 3) secondary hydrolysis, which represents the hydrolysis of soluble intermediates to lower molecular weight intermediates, and ultimately to glucose. Chemical changes in residual cellulose are observed by the decreasing of degree of polymerization (DP) and the increasing of chain ends concentration. Endoglucanases increase the concentration of chain ends and significantly decreases DP by attacking interior (mostly amorphous) portions of cellulose molecules. Exoglucanases highly decrease DP and only occasionally decrease the concentration of chain ends. Thus, endoglucanase activity is primarily responsible for chemical changes in solid-phase cellulose but plays a minor role in solubilization, while exoglucanase activity is primarily responsible for solubilization but plays a minor role in changing the chemical properties of solid cellulose. Physical changes in solid cellulose are observed as changes in accessible surface area due to geometrical changes resulting from the consumption or enlargement of accessible surface of cellulose due to progressive solubilization. Other cellulolitc systems make additional surface area available by mechanisms other than hydrolysis per se (Zhang and Lynd, 2004).

Many factors, like lignin content, crystallinity of cellulose and particle size, limit cellulases access to cellulose (Hendriks and Zeeman, 2009). So, microorganisms count on some chemical mechanisms of cellulose degradation at least to start providing cellulose accessibility to enzymes. H_2O_2 produced by some carbohydrate oxidases is a very powerful oxidant, but at low concentrations its reaction kinetics is very slow; however, the addition of ferrous salt dramatically increases the oxidative strength of H_2O_2. The ferrous ions (Fe^{2+}) catalyze the decomposition of H_2O_2 to hydroxyl radicals ($^{\cdot}OH$) at pH 3–5, as given in Eq. (1). This system comprises the Fenton reaction (Kirk et al., 1991).

$$Fe^{2+} + H_2O_2 \rightarrow Fe^{3+} + {}^{\cdot}OH + OH^- \qquad (1)$$

The highly reactive electrophilic $^{\cdot}OH$ react with organic substrates, such as lignin, hemicelluloses and cellulose, by abstracting a hydrogen atom, resulting in formation of reactive organic radicals, leading to depolymerization and changes in fiber surface topography (Jain and Vigneshwaran, 2012; Park et al., 1999). This depolymerization is brought about by cleavages in the amorphous region or the slightly disordered region (Martinez et al., 2009). The ferric ions (Fe^{3+}) produced during the reaction further react with H_2O_2 regenerating Fe^{2+}, thus continuing the process, as shows Eq. (2):

$$Fe^{3+} + H_2O_2 \rightarrow Fe^{2+} + {}^{\cdot}OOH + H^+ \quad (2)$$

The pretreatment of lignocellulosic materials by Fenton reaction for technological applications has the advantages of low energy requirement and mild environmental conditions; on the other hand, it is very time-consuming (Cheng and Timilsina, 2011). Hellström et al. (2014) demonstrated that Fenton pretreated birch kraft pulps are easier to be processed in the final mechanical stage, indicating a potential to lower the energy consumption in this stage. Furthermore, this pretreatment produced disintegrated pulp fibers that to some extent can be referred to as microfibrillated cellulose.

Depolymerization of cellulose by Fenton reaction has received great attention in recent years, mainly because when it is used as lignocellulosic pretreatment it influences positively on the hydrolysis response of cellulases (Jain and Vigneshwaran, 2012) and on higher bioavailability of cellulose for microbial conversion into bioethanol (Kato et al., 2014). Jung et al. (2015) employed Fenton system for pretreatment of rice straw and observed that enzymatic digestibility reached 93.2% of the theoretical glucose yield.

Cotton fibers can be completely decomposed by the Fenton's reagent in 7 days at pH 4.2–4.3, after passing by a primary stage of disintegration into very short fibers (Rättö et al., 1997).

Some fungi have developed specific systems to control the aggressiveness of the Fenton reaction. This is the case of the white rot fungus *Ceriporiopsis subvermispora*. It is known for the high specificity in degrading lignin without intensive damage to cellulose, and for this reason it has been extensively studied for application in biopulping (Vicentim et al., 2009; Ferraz et al., 2008; Vicentim e Ferraz, 2008; Ferraz et al., 2006). *C. subvermispora* produces ceriporioic acid B (1-nonadecene-2,3-dicarboxylic acid), an extracellular metabolite that strongly inhibites $^{\cdot}OH$ production and the depolymerization of cellulose (Rahmawati et al., 2005)

In addition to the Fenton system, another non-enzymatic mechanism worth highlighting and which has received considerable attention recently is the ultrasound treatment of cellulose (Zhang et al., 2013). Ultrasound acts mainly through cavitation, which occurs at a very large number of sites in the reactor simultaneously and generates high temperature and pressure as well as intense shear forces, shock waves and microjets (Iskalieva et al., 2012). The decomposition of water molecules into extremely reactive radicals such as ·HO and ·H by high temperature and pressure due to the cavitation aids in cleaving cellulose, resulting in decreased DP and crystallinity of cellulose and increased surface area (Velmurugan and Muthukumar, 2012). The intense shear forces, shock waves and microjets due to the cavitation also facilitates cleavage of cellulose and enhance mass transfer through streaming within the solution (Bussemaker and Zhang, 2013).

Combined pretreatment processes for cellulose depolymerization have been recently considered as a promising approach to overcome drawbacks demonstrated by various individual pretreatment processes, by increasing efficiency of sugar production, decreasing formation of inhibitors and/or shortening process time (Mood et al., 2013). Given that both Fenton reagent and ultrasound can depolymerize cellulose, combining Fenton reagent with ultrasound may provide a promising alternative pretreatment process for cellulose depolymerization. Zhang et al. (2016) demonstrated by X-ray diffraction analyses of microcrystalline cellulose that Fenton reagent was more efficient in decreasing the crystallinity while ultrasound was more efficient in decreasing DP.

INDUSTRIAL APPLICATIONS

Naturally occurring lignocellulosic biomass, especially in plant cell walls, serves structural and protective roles for plants, and consequently is recalcitrant and resistant to degradation. It is a major challenge to convert or degrade at industrial scale highly complex and heterogeneous lignocellulosic biomass into simple carbohydrates, phenolics, aromatics, and other more transformable substances. Among the numerous physical, chemical, and biological methods under development, the ones relying on enzymes are particularly attractive. Natural lignocellulose degradation and utilization (as part of natural energy transfer and carbon cycle) are carried out by specific enzymes from lignocellulolytic organisms (especially wood-degrading fungi and bacteria). As potential industrial catalysts for biomass conversion, enzymes might provide high specificity, low energy or chemical consumption, or low environment pollution (Sweeney and Xu, 2012).

Cellulases have shown their potential application in various industries. The industrial production of cellulases began in the middle of the 1980s, aiming its application as an additive for animal feed, in order to increase the digestion of rations by ruminants and other herbivorous. Then, these enzymes began to be used by the food industry, aiming the improvement of sensory properties of pastas. In this sector, cellulases also began to be used in the beverage processing, by promoting the clarification of fruit and wine juices and the maintenance of a stable rheology of the final product. Later, the cellulolytic enzymes began to be used in large scale in the following industries: a) textile, in the just implemented biodepolymerization processes (defibrillation of fabrics such as cotton, linen, wool and viscose) and bio-staining (dye softening and fading); b) pulp and paper, for the controlled modification of mechanical properties of the pulp and release of inks from the surface of the recycled fibers; c) and in laundry, in order to increase the luster, the removal of dirt and the softness of the fabrics, besides decreasing the wear of the pieces, noticed by the formation of lint and pellets after successive washes (Jeffries et al., 1993; Jeffries et al., 1995; Kirk and Jeffries, 1996; Tolan and Foody, 1999; Bhat, 2000; Coelho et al., 2008).

In the 1990s, cellulases, together with hemicellulases, accounted for more than 20% of the world market of enzymes (Bhat, 2000). Considering only the Brazilian exports, from 2010 to July 2017 cellulases moved USD 10,049,065, with 2013 and 2014 being the years of highest sales, with US $ 2.4 million in exports each year (Aliceweb, 2017).

Cellulases applications are summarized following according to (agro)industrial area (Kuhad et al., 2011):

Agriculture

Plant pathogen and disease control; generation of plant and fungal protoplasts; enhanced seed germination and improved root system; enhanced plant growth and flowering; improved soil quality; reduced dependence on mineral fertilizers.

Bioconversion

Conversion of cellulosic materials to ethanol, other solvents, organic acids and single cell protein, and lipids; production of energy-rich animal feed; improved nutritional quality of animal feed; improved ruminant performance; improved feed digestion and absorption; preservation of high quality fodder.

Detergents

Cellulase-based detergents; superior cleaning action without damaging fibers; improved color brightness and dirt removal; remove of rough protuberances in cotton fabrics; antiredeposition of ink particles.

Fermentation

Improved malting and mashing; improved pressing and color extraction of grapes; improved aroma of wines; improved primary fermentation and quality of beer; improved viscosity and filterability of wort; improved must clarification in wine production; improved filtration rate and wine stability.

Food

Release of the antioxidants from fruit and vegetable pomace; improvement of yields in starch and protein extraction; improved maceration, pressing, and color extraction of fruits and vegetables; clarification of fruit juices; improved texture and quality of bakery products; improved viscosity fruit purees; improved texture, flavor, aroma, and volatile properties of fruits and vegetables; controlled bitterness of citrus fruits.

Pulp and Paper

Coadditive in pulp bleaching; biomechanical pulping; improved draining; enzymatic deinking; reduced energy requirement; reduced chlorine requirement; improved fiber brightness, strength properties, and pulp freeness and cleanliness; improved drainage in paper mills; production of biodegradable cardboard, paper towels, and sanitary paper.

Textile

Biostoning of jeans; biopolishing of textile fibers; improved fabrics quality; improved absorbance property of fibers; softening of garments; improved stability of cellulosic fabrics; removal of excess dye from fabrics; restoration of colour brightness.

Others

Improved carotenoids extraction; improved oxidation and colour stability of carotenoids; improved olive oil extraction; improved malaxation of olive paste; improved quality of olive oil; reduced risk of biomass waste; production of hybrid molecules; production of designer cellulosomes.

One of the most emerging applications for the cellulolytic complex enzymes nowadays is for biomasses hydrolysis. Cellulose can potentially be totally converted to glucose, by enzymatic action. In subsequent stages, this monosaccharide can be used to obtain a huge range of products, ranging from biofuels to polymers, among which ethanol is one of the most interesting molecules nowadays. These technologies are part of the concept of cellulosic biorefineries, which aim the integral and integrated use of agroindustrial waste generated in a given production chain, in order to add value to it.

Cellulose saccharification can be performed under two main ways: chemically, by using diluted or concentrated acid, or enzymatically, by using cellulases (Olofsson et al., 2008). The use of cellulases in the hydrolysis of cellulose occurs under milder conditions of pressure, temperature and pH than chemical processes, and exhibits high specificity, thus eliminating the formation of toxic substances (furfural, 5-hidroximethylfurfural and lignin derivatives) to the microbial cells that will be employed for fermentation of the hydrolyzed medium. In the enzymatic way, although the cost of biocatalysts production is still high, there are economic benefits in the process, observed in energy savings and metallurgical requirements, since the equipments can be made with less noble materials (Badger, 2002).

Lignocellulosic biofuels are promising alternatives to fossil fuels due to their less polluting nature, renewability and miscibility with conventional transportation fuels. Concerning second-generation ethanol (2G-EtOH) production from lignocellulosic materials, cellulases are inputs that significantly impact the process from the economic point of view, seen that it can represent up to 18% of the operational cost of a plant (Philippidis and Hatzis, 1997; Wingren et al., 2003). So, it is essential to obtain extracts with ideal proportions between the various enzymes of the cellulolytic complex, in particular endoglucanases and β-glucosidases. The ideal formulation can be obtained by the co-cultivation of organisms and lineages that overproduce the main types of cellulases (Wen et al., 2005); by the separate production of the cellulases and subsequent mixture of the extracts in pre-optimized proportions (Kádár et al., 2004; Saha et al., 2005; Xiao et al., 2004; Pan et al., 2004; Aguiar and Menezes, 2002; Öhgren et al., 2006; Imai et al., 2004); or by the controlled incorporation of cellulase genes in host organisms, so that the modified cell excretes the enzymes in ideal proportions (Srivastava et al., 1999). In addition, it is necessary to study unit operations that allow the recovery and recycling of these biocatalysts (Castro and Pereira Jr, 2010).

The production of biofuels from biomass comprises a multistep process consisting of unit operations such as biomass pretreatment, enzymatic hydrolysis, fermentation and product separation. During the past two decades, significant research efforts have gone into the optimization and integration of each of these processes using a wide variety of biomass and biocatalysts (Limayem and Ricke, 2012; Mattam et al., 2016). At least two different organisms are conventionally used for producing cellulolytic enzymes and for ethanol production through fermentation. The search for a single yeast capable of simultaneously producing (hemi)cellulases and efficiently ferment the released sugars into ethanol is a way to decrease unit operations and consequently the costs for 2G-EtOH production.

References

"Aliceweb." http://aliceweb.mdic.gov.br/ (Accessed in Sep 1, 2017).

Aguiar, C. L., and T. J. B. Menezes. C. L. Aguiar, T. J. B Menezes. "Enzymatic conversion of sugarcane bagasse." *Biotechnology Science & Development*, Brasília 5, no. 26 (2002): 52-55.

Badger, P. C. "Ethanol from cellulose: a general review." *Trends in new crops and new uses* 1 (2002): 17-21.

Baldrian, Petr, and Vendula Valášková. "Degradation of cellulose by basidiomycetous fungi." *FEMS microbiology reviews* 32, no. 3 (2008): 501-521.

Banerjee, Goutami, Suzana Car, John S. Scott-Craig, Melissa S. Borrusch, and Jonathan D. Walton. "Rapid optimization of enzyme mixtures for deconstruction of diverse pretreatment/biomass feedstock combinations." *Biotechnology for biofuels* 3, no. 1 (2010): 22.

Beckham, Gregg T., James F. Matthews, Yannick J. Bomble, Lintao Bu, William S. Adney, Michael E. Himmel, Mark R. Nimlos, and Michael F. Crowley. "Identification of amino acids responsible for processivity in a Family 1 carbohydrate-binding module from a fungal cellulase." *The Journal of Physical Chemistry B* 114, no. 3 (2010): 1447-1453.

Bernardez, Timothy D., Kimberly Lyford, David A. Hogsett, and Lee R. Lynd. "Adsorption of *Clostridium thermocellum* cellulases onto pretreated mixed hardwood, Avicel, and lignin." *Biotechnology and bioengineering* 42, no. 7 (1993): 899-907.

Betcher, Meghan A., Jennifer M. Fung, Andrew W. Han, Roberta O'Connor, Romell Seronay, Gisela P. Concepcion, Daniel L. Distel, and Margo G. Haygood. "Microbial distribution and abundance in the digestive system of five shipworm species (Bivalvia: Teredinidae)." *PLoS One* 7, no. 9 (2012): e45309.

Bey, Mathieu, Jean-Guy Berrin, Laetitia Poidevin, and Jean-Claude Sigoillot. "Heterologous expression of Pycnoporus cinnabarinus cellobiose dehydrogenase in Pichia pastoris and involvement in saccharification processes." *Microbial cell factories* 10, no. 1 (2011): 113.

Bhat, M. K. "Cellulases and related enzymes in biotechnology." *Biotechnology advances* 18, no. 5 (2000): 355-383.

Bischof, Robert H., Jonas Ramoni, and Bernhard Seiboth. "Cellulases and beyond: the first 70 years of the enzyme producer Trichoderma reesei." *Microbial cell factories* 15, no. 1 (2016): 106.

Brienzo, Michel, Luvuyo Tyhoda, Benjamin Y., Görgens J. "Relationship between physicochemical properties and enzymatic hydrolysis of sugarcane bagasse varieties for bioethanol production." *New Biotechnology (Print)*, v. 32 (2015): 253-262.

Brienzo, Michel, Solange Ferreira, Marcos P. Vicentim, Wanderley de Souza, and Celso Sant'Anna. "Comparison study on the biomass recalcitrance of different tissue fractions of sugarcane culm." *Bio Energy Research* 7, no. 4 (2014): 1454-1465.

Brienzo, Michel, Simphewe Fikizolo, Benjamin Y., Tyhoda, L., Görgens, J. F. "Influence of pretreatment severity on structural changes, lignin content and enzymatic hydrolysis of sugarcane bagasse samples." *Renewable Energy*, 104 (2017) 271-280.

Bui, T. H. Hanh, and Shing Yip Lee. "Endogenous cellulase production in the leaf litter foraging mangrove crab *Parasesarma erythodactyla*." *Comparative Biochemistry and Physiology Part B: Biochemistry and Molecular Biology* 179 (2015): 27-36.

Bussemaker, Madeleine J., and Dongke Zhang. "Effect of ultrasound on lignocellulosic biomass as a pretreatment for biorefinery and biofuel applications." *Industrial & Engineering Chemistry Research* 52, no. 10 (2013): 3563-3580.

Cairo, J. P. F., F. Mandelli, R., Tramontina, T. M. Alvarez and F. M. Squina. "Functional characterization of a superoxide dismutase and a lytic polysaccharide monooxygenase AA10 from the lower termite *Coptotermes gestroi*." In *Proceedings of the Symposium on Biotechnology for Fuels and Chemicals*, held April, 25-28, 2016, in Baltimore, MD.

Castro, Aline Machado de, and Nei Pereira Jr. "Produção, propriedades e aplicação de celulases na hidrólise de resíduos agroindustriais." ["Production, properties and application of cellulases in the hydrolysis of agroindustrial residues."] *Química Nova* 33, no. 1 (2010): 181-188.

"Cazy." *Carbohydrate-Active Enzymes Database*. http://www.cazy.org/ (Accessed in Aug 29, 2017).

Chang, Martin M., Terry Y. C. Chou, and George T. Tsao. "Structure, pretreatment and hydrolysis of cellulose." In *Bioenergy*, pp. 15-42. Springer, Berlin, Heidelberg, 1981.

Cheng, Jay J., and Govinda R. Timilsina. "Status and barriers of advanced biofuel technologies: a review." *Renewable Energy* 36, no. 12 (2011): 3541-3549.

Chernoglazov, Vladimir M., Olga V. Ermolova, and Anatole A. Klyosov. "Adsorption of high-purity endo-1, 4-β-glucanases from *Trichoderma reesei* on components of lignocellulosic materials: cellulose, lignin, and xylan." *Enzyme and Microbial Technology* 10, no. 8 (1988): 503-507.

Chundawat, Shishir P. S., Mary S. Lipton, Samuel O. Purvine, Nirmal Uppugundla, Dahai Gao, Venkatesh Balan, and Bruce E. Dale. "Proteomics-based compositional analysis of complex cellulase–hemicellulase mixtures." *Journal of proteome research* 10, no. 10 (2011): 4365-4372.

Cleveland, Lemuel Roscoe. "The physiological and symbiotic relationships between the intestinal protozoa of termites and their host, with special reference to *Reticulitermes flavipes* Kollar." *The Biological Bulletin* 46, no. 5 (1924): 203-227.

Coelho, Maria A. Z., Andrea M. Salgado, and Bernardo D. Ribeiro. *Tecnologia enzimática*. [Enzymatic technology.] Editora EPUB, 2008.

Converse, A. O. "Substrate factors limiting enzymatic hydrolysis." *Biotechnology in Agriculture* (1993): 93-93.

Corrêa, Thamy Lívia Ribeiro, Leandro Vieira Dos Santos, and Gonçalo Amarante Guimarães Pereira. "AA9 and AA10: from enigmatic to essential enzymes." *Applied microbiology and biotechnology* 100, no. 1 (2016): 9-16.

Coughlan, Michael P. "The properties of fungal and bacterial cellulases with comment on their production and application." *Biotechnology and genetic engineering reviews* 3, no. 1 (1985): 39-110.

Couturier, Marie, Mireille Haon, Pedro M. Coutinho, Bernard Henrissat, Laurence Lesage-Meessen, and Jean-Guy Berrin. "Podospora anserina hemicellulases potentiate the Trichoderma reesei secretome for saccharification of lignocellulosic biomass." *Applied and environmental microbiology* 77, no. 1 (2011): 237-246.

Cragg, Simon M., Gregg T. Beckham, Neil C. Bruce, Timothy D. H. Bugg, Daniel L. Distel, Paul Dupree, Amaia Green Etxabe et al. "Lignocellulose degradation mechanisms across the Tree of Life." *Current Opinion in Chemical Biology* 29 (2015): 108-119.

Eriksson, Torny, Johan Börjesson, and Folke Tjerneld. "Mechanism of surfactant effect in enzymatic hydrolysis of lignocellulose." *Enzyme and Microbial Technology* 31, no. 3 (2002): 353-364.

Eyzaguirre, Jaime, Mauricio Hidalgo, and Andrés Leschot. *23 β-Glucosidases from Filamentous Fungi: Properties, Structure, and Applications*. (2005).

Falkowski, Paul, R. J. Scholes, E. E. A. Boyle, Josep Canadell, D. Canfield, J. Elser, Nicolas Gruber et al. "The global carbon cycle: a test of our knowledge of earth as a system." *Science* 290, no. 5490 (2000): 291-296.

Fengel, Dietrich, and Gerd Wegener. "Wood: chemistry, ultrastructure, reactions." *Walter de Gruyter* 613 (1984): 1960-82.

Ferraz, André, Anderson Guerra, Régis Mendonça, Fernando Masarin, Marcos P. Vicentim, André Aguiar, and Paulo C. Pavan. "Technological advances and mechanistic basis for fungal biopulping." *Enzyme and Microbial Technology* 43, no. 2 (2008): 178-185.

Ferraz, André, Fernando Masarin, Marcos P. Vicentim, Paulo C. Pavan. "Aplicações da biodegradação de madeira por Basidiomycetes na produção de celulose e papel." ["Applications of biodegradation of wood by Basidiomycetes in the production of pulp and paper."] In: Mariath, Jorge E. A., Rinaldo P. Santos. (Org.). *Os Avanços da Botânica no início do século XXI*. 1ed. Porto Alegre - RS: Imagine - Design Editorial, 2006, v., p. 244-247.

Gao, Dahai, Nirmal Uppugundla, Shishir P. S. Chundawat, Xiurong Yu, Spencer Hermanson, Krishne Gowda, Phillip Brumm, David Mead, Venkatesh Balan, and Bruce E. Dale. "Hemicellulases and auxiliary enzymes for improved conversion of lignocellulosic biomass to monosaccharides." *Biotechnology for biofuels* 4, no. 1 (2011): 5.

Gardner, K. H., and J. Blackwell. "The hydrogen bonding in native cellulose." *Biochimica et Biophysica Acta (BBA)-General Subjects* 343, no. 1 (1974): 232-237.

Gardner, K. H., and J. Blackwell. "The structure of native cellulose." *Biopolymers* 13, no. 10 (1974): 1975-2001.

Gottschalk, Leda Maria Fortes, Raul Alves Oliveira, and Elba Pinto da Silva Bon. "Cellulases, xylanases, β-glucosidase and ferulic acid esterase produced by Trichoderma and Aspergillus act synergistically in the hydrolysis of sugarcane bagasse." *Biochemical Engineering Journal* 51, no. 1 (2010): 72-78.

Goyal, Anil, Bijan Ghosh, and Douglas Eveleigh. "Characteristics of fungal cellulases." *Bioresource Technology* 36, no. 1 (1991): 37-50.

Grethlein, Hans E. "The effect of pore size distribution on the rate of enzymatic hydrolysis of cellulosic substrates." *Nature Biotechnology* 3, no. 2 (1985): 155-160.

Gritzalli, Mikelina, and Ross D. Brown J. R. *The cellulase system of Trichoderma: relationships between purified extracellular enzymes from induced or cellulose-grown cells*. 1979.

Hellström, Pia, Anette Heijnesson-Hultén, Magnus Paulsson, Helena Håkansson, and Ulf Germgård. "The effect of Fenton chemistry on the properties of microfibrillated cellulose." *Cellulose* 21, no. 3 (2014): 1489-1503.

Hendriks, A. T. W. M., and G. Zeeman. "Pretreatments to enhance the digestibility of lignocellulosic biomass." *Bioresource technology* 100, no. 1 (2009): 10-18.

Henrissat, B., M. Claeyssens, P. Tomme, L. Lemesle, and J. P. Mornon. "Cellulase families revealed by hydrophobic cluster analysis." *Gene* 81, no. 1 (1989): 83-95.

Herpoël-Gimbert, Isabelle, Antoine Margeot, Alain Dolla, Gwénaël Jan, Daniel Mollé, Sabrina Lignon, Hughes Mathis, Jean-Claude Sigoillot, Frédéric Monot, and Marcel Asther. "Comparative secretome analyses of two Trichoderma reesei RUT-C30 and CL847 hypersecretory strains." *Biotechnology for biofuels* 1, no. 1 (2008): 18.

Himmel, Michael E., Mark F. Ruth, and Charles E. Wyman. "Cellulase for commodity products from cellulosic biomass." *Current opinion in biotechnology* 10, no. 4 (1999): 358-364.

Hon, David N. S. "Cellulose: a random walk along its historical path." *Cellulose* 1, no. 1 (1994): 1-25.

Igarashi, Kiyohiko, Takayuki Uchihashi, Anu Koivula, Masahisa Wada, Satoshi Kimura, Tetsuaki Okamoto, Merja Penttilä, Toshio Ando, and Masahiro Samejima. "Traffic jams reduce hydrolytic efficiency of cellulase on cellulose surface." *Science* 333, no. 6047 (2011): 1279-1282.

Imai, Masanao, Kohei Ikari, and Isao Suzuki. "High-performance hydrolysis of cellulose using mixed cellulase species and ultrasonication pretreatment." *Biochemical Engineering Journal* 17, no. 2 (2004): 79-83.

Iskalieva, Asylzat, Bob Mbouyem Yimmou, Parag R. Gogate, Miklos Horvath, Peter G. Horvath, and Levente Csoka. "Cavitation assisted delignification of wheat straw: a review." *Ultrasonics sonochemistry* 19, no. 5 (2012): 984-993.

Jain, Prateek, and Nadanathangam Vigneshwaran. "Effect of Fenton's pretreatment on cotton cellulosic substrates to enhance its enzymatic hydrolysis response." *Bioresource technology* 103, no. 1 (2012): 219-226.

Jarvis, Mike. "Chemistry: cellulose stacks up." *Nature* 426, no. 6967 (2003): 611-612.

Jeffries, T. W., M. S. Sykes, K. Rutledge-Cropsey, J. H. Klungness, and S. Abubakr. "Enhanced removal of toners from office waste papers by microbial cellulases." In *Biotechnology in the Pulp and Paper Industry. Proceedings of 6 th International Conference on Biotechnology in Pulp and Paper Industry,* Facultas-Univ. Verl., Vienna, Austria, pp. 141-144. 1995.

Jeffries, T., J. H. Klungness, M. S. Sykes, and K. Rutledge-Cropsey. "Recycling Synposium–TAPPI Notes." *Recycling Synposium-TAPPI Notes* (1993).

Jeoh, Tina, John O. Baker, Mursheda K. Ali, Michael E. Himmel, and William S. Adney. "β-d-Glucosidase reaction kinetics from isothermal titration microcalorimetry." *Analytical biochemistry* 347, no. 2 (2005): 244-253.

Jung, Young Hoon, Hyun Kyung Kim, Hyun Min Park, Yong-Cheol Park, Kyungmoon Park, Jin-Ho Seo, and Kyoung Heon Kim. "Mimicking the Fenton reaction-induced wood decay by fungi for pretreatment of lignocellulose." *Bioresource technology* 179 (2015): 467-472.

Kádár, Zs, Zs Szengyel, and K. Réczey. "Simultaneous saccharification and fermentation (SSF) of industrial wastes for the production of ethanol." *Industrial Crops and Products* 20, no. 1 (2004): 103-110.

Karlsson, Johan, Markku Saloheimo, Matti Siika-aho, Maija Tenkanen, Merja Penttilä, and Folke Tjerneld. "Homologous expression and characterization of Cel61A (EG IV) of Trichoderma reesei." *The FEBS Journal* 268, no. 24 (2001): 6498-6507.

Kato, Dawn M., Noelia Elía, Michael Flythe, and Bert C. Lynn. "Pretreatment of lignocellulosic biomass using Fenton chemistry." *Bioresource technology* 162 (2014): 273-278.

Kern, Marcelo, John E. McGeehan, Simon D. Streeter, Richard N. A. Martin, Katrin Besser, Luisa Elias, Will Eborall et al. "Structural characterization of a unique marine animal family 7 cellobiohydrolase suggests a mechanism of cellulase salt tolerance." *Proceedings of the National Academy of Sciences* 110, no. 25 (2013): 10189-10194.

Kim, Dong Won, Tae Seung Kim, Young Kyu Jeong, and Jae Kuk Lee. "Adsorption kinetics and behaviors of cellulase components on microcrystalline cellulose." *Journal of fermentation and bioengineering* 73, no. 6 (1992): 461-466.

King, Andrew J., Simon M. Cragg, Yi Li, Jo Dymond, Matthew J. Guille, Dianna J. Bowles, Neil C. Bruce, Ian A. Graham, and Simon J. McQueen-Mason. "Molecular insight into lignocellulose digestion by a marine isopod in the absence of gut microbes." *Proceedings of the National Academy of Sciences* 107, no. 12 (2010): 5345-5350.

Kirk, T. Kent, and Thomas W. Jeffries. "Roles for microbial enzymes in pulp and paper processing." 1996. In: Jeffries, T. W., & Viikari, L. (Eds.). *Enzymes for pulp and paper processing*. American Chemical Society.

Kirk, T. Kent, Rebecca Ibach, Michael D. Mozuch, Anthony H. Conner, and L. Highley. "Characteristics of cotton cellulose depolymerized by a brown-rot fungus, by acid, or by chemical oxidants." *Holzforschung-International Journal of the Biology, Chemistry, Physics and Technology of Wood* 45, no. 4 (1991): 239-244.

Klein, G. L., and W. R. Snodgrass. "Cellulose." *Encyclopedia of food science, food technology and nutrition.* London: Academic Press. p (1993): 758-767.

Knowles, Jonathan, Päivi Lehtovaara, and Tuula Teeri. "Cellulase families and their genes." *Trends in Biotechnology* 5, no. 9 (1987): 255-261.

Krassig, Hans. "Structure of cellulose and its relation to properties of cellulose fibers." *Cellulose and its derivatives: chemistry, biochemistry and applications*/editors, J. F. Kennedy et al. (1985).

Kristensen, Jan B., Claus Felby, and Henning Jørgensen. "Yield-determining factors in high-solids enzymatic hydrolysis of lignocellulose." *Biotechnology for Biofuels* 2, no. 1 (2009): 11.

Kuhad, Ramesh Chander, Rishi Gupta, and Ajay Singh. "Microbial cellulases and their industrial applications." *Enzyme research* 2011 (2011).

Kumar, Rajeev, and Charles E. Wyman. "Effect of xylanase supplementation of cellulase on digestion of corn stover solids prepared by leading pretreatment technologies." *Bioresource Technology* 100, no. 18 (2009): 4203-4213.

Kumar, Rajeev, Sompal Singh, and Om V. Singh. "Bioconversion of lignocellulosic biomass: biochemical and molecular perspectives." *Journal of industrial microbiology & biotechnology* 35, no. 5 (2008): 377-391.

Kurašin, Mihhail, and Priit Väljamäe. "Processivity of cellobiohydrolases is limited by the substrate." *Journal of Biological Chemistry* 286, no. 1 (2011): 169-177.

Kyriacou, Andreas, C. Roger MacKenzie, and Ronald J. Neufeld. "Detection and characterization of the specific and nonspecific endoglucanases of Trichoderma reesei: evidence demonstrating endoglucanase activity by cellobiohydrolase II." *Enzyme and microbial technology* 9, no. 1 (1987): 25-32.

Langston, James A., Tarana Shaghasi, Eric Abbate, Feng Xu, Elena Vlasenko, and Matt D. Sweeney. "Oxidoreductive cellulose depolymerization by the enzymes cellobiose dehydrogenase and glycoside hydrolase 61." *Applied and environmental microbiology* 77, no. 19 (2011): 7007-7015.

Langston, James, Neil Sheehy, and Feng Xu. "Substrate specificity of Aspergillus oryzae family 3 β-glucosidase." *Biochimica et Biophysica Acta (BBA)-Proteins and Proteomics* 1764, no. 5 (2006): 972-978.

Li, Yongchao, Diana C. Irwin, and David B. Wilson. "Increased crystalline cellulose activity via combinations of amino acid changes in the family 9 catalytic domain and family 3c cellulose binding module of Thermobifida fusca Cel9A." *Applied and environmental microbiology* 76, no. 8 (2010): 2582-2588.

Limayem, Alya, and Steven C. Ricke. "Lignocellulosic biomass for bioethanol production: current perspectives, potential issues and future prospects."*Progress in Energy and Combustion Science* 38, no. 4 (2012): 449-467.

Liu, Yu-San, John O. Baker, Yining Zeng, Michael E. Himmel, Thomas Haas, and Shi-You Ding. "Cellobiohydrolase hydrolyzes crystalline cellulose on hydrophobic faces." *Journal of Biological Chemistry* 286, no. 13 (2011): 11195-11201.

Lynd, Lee R., Janet H. Cushman, Roberta J. Nichols, and Charles E. Wyman. "Fuel ethanol from cellulosic biomass." *Science* 251, no. 4999 (1991): 1318-1323.

Lynd, Lee R., Paul J. Weimer, Willem H. Van Zyl, and Isak S. Pretorius. "Microbial cellulose utilization: fundamentals and biotechnology." *Microbiology and molecular biology reviews* 66, no. 3 (2002): 506-577.

Lynel, L. R., C. E. Wyman, and T. U. Gerngross. "Biocommodity Engineering Biotechnol." *Prog* 15, no. 5 (1999): 777-793.

Mandels, Mary, John Kostick, and Richard Parizek. "The use of adsorbed cellulase in the continuous conversion of cellulose to glucose." In *Journal of Polymer Science: Polymer Symposia*, vol. 36, no. 1, pp. 445-459. Wiley Subscription Services, Inc., A Wiley Company, 1971.

Mansfield, Shawn D., Caitriona Mooney, and John N. Saddler. "Substrate and enzyme characteristics that limit cellulose hydrolysis." *Biotechnology progress* 15, no. 5 (1999): 804-816.

Martin, Michael M., C. G. Jones, and E. A. Bernays. "The evolution of cellulose digestion in insects." *Philosophical Transactions: Biological Sciences* (1991): 281-288.

Martinez, Diego, Jean Challacombe, Ingo Morgenstern, David Hibbett, Monika Schmoll, Christian P. Kubicek, Patricia Ferreira et al. "Genome, transcriptome, and secretome analysis of wood decay fungus Postia placenta supports unique mechanisms of lignocellulose conversion." *Proceedings of the National Academy of Sciences* 106, no. 6 (2009): 1954-1959.

Mattam, Anu Jose, Arindam Kuila, Niranjan Suralikerimath, Nettem Choudary, Peddy VC Rao, and Harshad Ravindra Velankar. "Cellulolytic enzyme expression and simultaneous conversion of lignocellulosic sugars into ethanol and xylitol by a new Candida tropicalis strain." *Biotechnology for biofuels* 9, no. 1 (2016): 157.

Mood, Sohrab Haghighi, Amir Hossein Golfeshan, Meisam Tabatabaei, Gholamreza Salehi Jouzani, Gholam Hassan Najafi, Mehdi Gholami, and Mehdi Ardjmand. "Lignocellulosic biomass to bioethanol, a comprehensive review with a focus on pretreatment." *Renewable and Sustainable Energy Reviews* 27 (2013): 77-93.

Nevell, T. P., and S. H. Zeronian. "Cellulose chemistry fundamentals." In: *Cellulose chemistry and its applications*, pp. 15-29. Ellis Horwood Limited, 1985.

Nidetzky, Bernd, Walter Zachariae, Günther Gercken, Marianne Hayn, and Walter Steiner. "Hydrolysis of cellooligosaccharides by Trichoderma reesei cellobiohydrolases: Experimental data and kinetic modeling." *Enzyme and microbial technology* 16, no. 1 (1994): 43-52.

Nozaki, Mana, Chiemi Miura, Yuzuru Tozawa, and Takeshi Miura. "The contribution of endogenous cellulase to the cellulose digestion in the gut of earthworm (Pheretima hilgendorfi: Megascolecidae)." *Soil Biology and Biochemistry* 41, no. 4 (2009): 762-769.

Öhgren, Karin, Andreas Rudolf, Mats Galbe, and Guido Zacchi. "Fuel ethanol production from steam-pretreated corn stover using SSF at higher dry matter content." *Biomass and Bioenergy* 30, no. 10 (2006): 863-869.

Olofsson, Kim, Magnus Bertilsson, and Gunnar Lidén. "A short review on SSF–an interesting process option for ethanol production from lignocellulosic feedstocks." *Biotechnology for biofuels* 1, no. 1 (2008): 7.

Olsson, Lisbeth, and Bärbel Hahn-Hägerdal. "Fermentation of lignocellulosic hydrolysates for ethanol production." *Enzyme and Microbial technology* 18, no. 5 (1996): 312-331.

Olsson, Lisbeth, Henning Jørgensen, Kristian Bertel Rømer Krogh, and Christophe Roca. "Bioethanol production from lignocellulosic material." In *Polysaccharides*, pp. 957-993. Marcel Dekker Incorporated, 2004.

Ooshima, Hiroshi, Douglas S. Burns, and Alvin O. Converse. "Adsorption of cellulase from *Trichoderma reesei* on cellulose and lignacious residue in wood pretreated by dilute sulfuric acid with explosive decompression." *Biotechnology and Bioengineering* 36, no. 5 (1990): 446-452.

O'sullivan, Antoinette C. "Cellulose: the structure slowly unravels." *Cellulose* 4, no. 3 (1997): 173-207.

Pan, Xuejun, Xiao Zhang, David J. Gregg, and John N. Saddler. "Enhanced enzymatic hydrolysis of steam-exploded Douglas fir wood by alkali-oxygen post-treatment." In *Proceedings of the Twenty-Fifth Symposium on Biotechnology for Fuels and Chemicals.* Held May 4–7, 2003, in Breckenridge, CO, pp. 1103-1114. Humana Press, 2004.

Park, JonathanáS B., PauláM Wood, BruceáC Gilbert, and AdrianáC Whitwood. "EPR Evidence for hydroxyl-and substrate-derived radicals in Fe (II)-oxalate/hydrogen peroxide reactions. The importance of the reduction of Fe (III)-oxalate by oxygen-conjugated radicals to regenerate Fe (II) in reactions of carbohydrates and model compounds."*Journal of the Chemical Society, Perkin Transactions 2* 5 (1999): 923-932.

Philippidis, George P., and Christos Hatzis. "Biochemical Engineering Analysis of Critical Process Factors in the Biomass-to-Ethanol Technology." *Biotechnology progress* 13, no. 3 (1997): 222-231.

Phillips, Christopher M., William T. Beeson IV, Jamie H. Cate, and Michael A. Marletta. "Cellobiose dehydrogenase and a copper-dependent polysaccharide monooxygenase potentiate cellulose degradation by Neurospora crassa." *ACS chemical biology* 6, no. 12 (2011): 1399-1406.

Praestgaard, Eigil, Jens Elmerdahl, Leigh Murphy, Søren Nymand, K. C. McFarland, Kim Borch, and Peter Westh. "A kinetic model for the burst phase of processive cellulases." *The FEBS journal* 278, no. 9 (2011): 1547-1560.

Rahmawati, Noor, Yasunori Ohashi, Takahito Watanabe, Yoichi Honda, and Takashi Watanabe. "Ceriporic acid B, an extracellular metabolite of *Ceriporiopsis subvermispora*, suppresses the depolymerization of cellulose by the Fenton reaction." *Biomacromolecules* 6, no. 5 (2005): 2851-2856.

Rättö, M., A. C. Ritschkoff, and L. Viikari. "The effect of oxidative pretreatment on cellulose degradation by Poria placenta and Trichoderma reesei cellulases." *Applied microbiology and biotechnology* 48, no. 1 (1997): 53-57.

Saha, Badal C., Loren B. Iten, Michael A. Cotta, and Y. Victor Wu. "Dilute acid pretreatment, enzymatic saccharification and fermentation of wheat straw to ethanol." *Process Biochemistry* 40, no. 12 (2005): 3693-3700.

Scheller, Henrik Vibe, and Peter Ulvskov. "Hemicelluloses." *Annual review of plant biology* 61 (2010).

Shelomi, Matan, Hirofumi Watanabe, and Gaku Arakawa. "Endogenous cellulase enzymes in the stick insect (Phasmatodea) gut." *Journal of insect physiology* 60 (2014): 25-30.

Simmons, Christopher W., Amitha P. Reddy, Patrik D'haeseleer, Jane Khudyakov, Konstantinos Billis, Amrita Pati, Blake A. Simmons, Steven W. Singer, Michael P. Thelen, and Jean S. VanderGheynst. "Metatranscriptomic analysis of lignocellulolytic microbial communities involved in high-solids decomposition of rice straw." *Biotechnology for biofuels* 7, no. 1 (2014): 495.

Sipos, Bálint, Zsuzsa Benkő, Dóra Dienes, Kati Réczey, Liisa Viikari, and Matti Siika-Aho. "Characterisation of specific activities and hydrolytic properties of cell-wall-degrading enzymes produced by Trichoderma reesei Rut C30 on different carbon sources." *Applied biochemistry and biotechnology* 161, no. 1-8 (2010): 347-364.

Srivastava, K. K., P. K. Verma, and R. Srivastava. "A recombinant cellulolytic *Escherichia coli*: cloning of the cellulase gene and characterization of a bifunctional cellulase." *Biotechnology letters* 21, no. 4 (1999): 293-297.

Sutcliffe, Roger, and John N. Saddler. *The role of lignin in the adsorption of cellulases during enzymatic treatment of lignocellulosic material.* (1986).

Sweeney, Matthew D., and Feng Xu. "Biomass converting enzymes as industrial biocatalysts for fuels and chemicals: recent developments." *Catalysts* 2, no. 2 (2012): 244-263.

Sweeney, Matthew D., Elena Vlasenko, and Eric Abbate. *Methods for increasing hydrolysis of cellulosic material in the presence of cellobiose dehydrogenase*. U.S. Patent 8,809,033, issued August 19, 2014.

Tolan, Jeffrey, and Brian Foody. "Cellulase from submerged fermentation." *Recent progress in bioconversion of lignocellulosics* (1999): 41-67.

Tomme, P., R. A. J. Warren, and N. R. Gilkes. "Cellulose hydrolysis by bacteria and fungi." *Advances in microbial physiology* 37 (1995): 1-81.

Tsuji, Akihiko, Keiko Tominaga, Nami Nishiyama, and Keizo Yuasa. "Comprehensive enzymatic analysis of the cellulolytic system in digestive fluid of the Sea Hare *Aplysia kurodai*. Efficient glucose release from sea lettuce by synergistic action of 45 kDa endoglucanase and 210 kDa ß-glucosidase." *PloS One* 8, no. 6 (2013): e65418.

Vaaje-Kolstad, Gustav, Zarah Forsberg, Jennifer S. M. Loose, Bastien Bissaro, and Vincent G. H. Eijsink. "Structural diversity of lytic polysaccharide mono-oxygenases." *Current opinion in structural biology* 44 (2017): 67-76.

van den Brink, Joost, and Ronald P. de Vries. "Fungal enzyme sets for plant polysaccharide degradation." *Applied microbiology and biotechnology* 91, no. 6 (2011): 1477.

Velmurugan, Rajendran, and Karuppan Muthukumar. "Ultrasound-assisted alkaline pretreatment of sugarcane bagasse for fermentable sugar production: optimization through response surface methodology." *Bioresource technology* 112 (2012): 293-299.

Vicentim, Marcos Paulo, and Andre Ferraz. "Alkaline-sulfite chemithermomechanical pulping of Eucalyptus grandis biotreated by Ceriporiopsis subvermispora under varied culture conditions." *Holzforschung* 62, no. 4 (2008): 375-382.

Vicentim, Marcos Paulo, Robson de Almeida Faria, and Andre Ferraz. "High-yield kraft pulping of Eucalyptus grandis Hill ex Maiden biotreated by Ceriporiopsis subvermispora under two different culture conditions." *Holzforschung* 63, no. 4 (2009): 408-413.

Vinzant, T. B., W. S. Adney, S. R. Decker, J. O. Baker, M. T. Kinter, N. E. Sherman, J. W. Fox, and M. E. Himmel. "Fingerprinting Trichoderma reesei hydrolases in a commercial cellulase preparation." *Applied biochemistry and biotechnology* 91, no. 1 (2001): 99-107.

Vlasenko, E., M. Schülein, J. Cherry, and F. Xu. "Substrate specificity of family 5, 6, 7, 9, 12, and 45 endoglucanases." *Bioresource Technology* 101, no. 7 (2010): 2405-2411.

Vocadlo, David J., and Gideon J. Davies. "Mechanistic insights into glycosidase chemistry." *Current opinion in chemical biology* 12, no. 5 (2008): 539-555.

Wallace, Joshua, Brienzo, Michel, García-Aparicio, M. P., Görgens, J. F. "Lignin enrichment and enzyme deactivation as the root cause of enzymatic hydrolysis slowdown of steam pretreated sugarcane bagasse." *New Biotechnology (Print)*, v. 33 (2016): 361-371.

Warden, Andrew C., Bryce A. Little, and Victoria S. Haritos. "A cellular automaton model of crystalline cellulose hydrolysis by cellulases." *Biotechnology for biofuels* 4, no. 1 (2011): 39.

Watanabe, Hirofumi, and Gaku Tokuda. "Cellulolytic systems in insects." *Annual review of entomology* 55 (2010): 609-632.

Watanabe, Hirofumi, Hiroaki Noda, Gaku Tokuda, and Nathan Lo. "A cellulase gene of termite origin." *Nature* 394, no. 6691 (1998): 330.

Wen, Zhiyou, Wei Liao, and Shulin Chen. "Production of cellulase/β-glucosidase by the mixed fungi culture *Trichoderma reesei* and *Aspergillus phoenicis* on dairy manure." *Process Biochemistry* 40, no. 9 (2005): 3087-3094.

Wilson, David B. "Three microbial strategies for plant cell wall degradation." *Annals of the New York Academy of Sciences* 1125, no. 1 (2008): 289-297.

Wingren, Anders, Mats Galbe, and Guido Zacchi. "Techno-economic evaluation of producing ethanol from softwood: Comparison of SSF and SHF and identification of bottlenecks." *Biotechnology progress* 19, no. 4 (2003): 1109-1117.

Xiao, Zhizhuang, Xiao Zhang, David J. Gregg, and John N. Saddler. "Effects of sugar inhibition on cellulases and β-glucosidase during enzymatic hydrolysis of softwood substrates." In *Proceedings of the Twenty-Fifth Symposium on Biotechnology for Fuels and Chemicals.* Held May 4–7, 2003, in Breckenridge, CO, pp. 1115-1126. Humana Press, 2004.

Xu, Feng, and Hanshu Ding. "A new kinetic model for heterogeneous (or spatially confined) enzymatic catalysis: contributions from the fractal and jamming (overcrowding) effects." *Applied Catalysis A: General* 317, no. 1 (2007): 70-81.

Zamocky, Marcel, R. Ludwig, C. Peterbauer, B. M. Hallberg, Christina Divne, P. Nicholls, and D. Haltrich. "Cellobiose dehydrogenase-a flavocytochrome from wood-degrading, phytopathogenic and saprotropic fungi." *Current protein and peptide science* 7, no. 3 (2006): 255-280.

Zhang, Mei-Fang, Yuan-Hang Qin, Jia-Yu Ma, Li Yang, Zai-Kun Wu, Tie-Lin Wang, Wei-Guo Wang, and Cun-Wen Wang. "Depolymerization of microcrystalline cellulose by the combination of ultrasound and Fenton reagent." *Ultrasonics sonochemistry* 31 (2016): 404-408.

Zhang, Qinghua, Maud Benoit, Karine De Oliveira Vigier, Joël Barrault, Gwenaëlle Jégou, Michel Philippe, and François Jérôme. "Pretreatment of microcrystalline cellulose by ultrasounds: effect of particle size in the heterogeneously-catalyzed hydrolysis of cellulose to glucose." *Green Chemistry* 15, no. 4 (2013): 963-969.

Zhang, Yi-Heng Percival, and Lee R. Lynd. "Toward an aggregated understanding of enzymatic hydrolysis of cellulose: noncomplexed cellulase systems." *Biotechnology and bioengineering* 88, no. 7 (2004): 797-824.

In: Bioethanol and Beyond
Editor: Michel Brienzo
ISBN: 978-1-53613-478-0

Chapter 10

CELLULASES FROM INVERTEBRATE ANIMALS

Youssef Bacila Sade**, *Daniela Toma de Moraes*, *Paulo José Beltrão and Marcos Paulo Vicentim
Life Sciences Metrology, National Institute of Metrology, Quality and Technology - INMETRO, Duque de Caxias, Rio de Janeiro, Brazil

ABSTRACT

Most publications concerning the search for novel enzymes and proteins that can improve lignocellulose degradation, focus on bacteria and fungi due to their great diversity and their presence in different environments and also the easy cultivation of such organisms for enzyme production in large scale. The microflora of invertebrate animals is a rich source of lignocellulose-degrading organisms and has been the subject of many works. It was also considered that animal endogenous enzymes did not play an important role in the lignocellulose degradation. Since the recent discovery of endogenous animal cellulases in termites and nematodes, many animals have been the subject of research aiming the discovery of new enzymes. Animal cellulases can be considerably different in structure from their bacterial and fungal counterparts and have the potential to offer more efficient possibilities for lignocellulose degradation. The search for endogenous enzymes in animals allows the discovery of new enzymes, which in turn can be modified and improved aiming to achieve greater efficiency and, consequently, the reduction of bioethanol production costs. Besides arthropods and nematodes, mollusks are emerging as promising sources of such enzymes. In chapter, it will be discussed the recent discoveries concerning cellulases from invertebrates and their microflora, industrial applications and the mechanisms of cellulose enzymatic hydrolysis.

Keywords: achatina, teredinidae, glycoside hydrolases, biofuels

* Corresponding Author Email: ybsade@inmetro.gov.br.

INTRODUCTION

Many types of research focused on cellulases produced by the genus *Trichoderma* mostly due to the high levels of cellulases secreted. *T. reesei* cellulase mixture consists of many catalytically active proteins. At least two cellobiohydrolases (CBH1-2), five endoglucanases (EG1–5), β-glucosidases, and hemicellulases have been identified by 2D electrophoresis (Vinzant et al., 2001). CBH1, CBH2, and EG2 are the three main components of the *T. reesei* cellulase system, representing 60 ± 5%, 20 ± 6%, and 12 ± 3% of total cellulase protein, respectively (Goyal et al., 1991; Gritzali and Brown, 1978; Knowles et al., 1987; Kyriacou et al., 1987; Nidetzky and Claeyssens, 1994). Cellulolytic fungi and bacteria have complex cellulase systems to use cellulose fibrils as a nutritional source (Tomme et al., 1995). In plants, cellulases hydrolyze their cell walls at various developmental stages, such as in bean abscission, fruit ripening and in pedicel abscission (Bennett et al., 1986; Tucker et al., 1987; Tucker and Milligan, 1991; Lashbrook et al., 1994; Milligan and Gasser, 1995; del Campillo and Bennett, 1996; Yung et al., 1999; Trainotti et al., 1999a; Trainotti et al., 1999b; Brummell et al., 1994).

When animals were not recognized to produce endogenous cellulases, symbiotic protozoa or bacteria used to explain cellulose digestion in invertebrates and herbivorous cattle. This theory used to be supported by the relevant Cleveland's (1924) work, who demonstrated that the termite *Reticulitermes flavipes* lost its viability after removing the protozoa fauna from the hindgut. Since then, symbiotic cellulase production became the most favored explanation for cellulose digestion in higher animals (Watanabe and Tokuda, 2001).

The activity of possible endogenous cellulases was firstly investigated in 74 higher animals, including several vertebrates (Yokoe and Yasumasu, 1964). The first endogenous cellulose demonstration was performed in the Japanese subterranean termite *Leucotermes* (*Reticulitermes*) *speratus* (Yokoe, 1964), which cellulases synergistically break down native cellulose like fungal cellulase system (Yamaoka and Nagatani, 1975; Reese et al., 1950). Another investigation employing the arboreal higher termite *Nasutitermes walkeri* and the mound-building termite *Coptotermes lacteus* showed that the intestinal fauna and flora play no role (case of *N. walkeri*) or a relatively small role (*C. lacteus*) in cellulase production by these species. The removal of intestinal bacteria and protozoa (by tetracycline treatments) did not affect cellulase activities in the guts of either species (except for a 20% decrease in *C. lacteus* hindgut) (O'Brien, 1979).

Cellulase activity in insects was then not only observed in xylophagous species (like termites) and wood-eating cockroaches but also in many other species with different food habits, which do not have symbiotic relationships with cellulolytic protozoa or microorganisms, like domestic omnivorous cockroaches (Martin, 1983).

In 1998 a novel gene encoding cellulases (GHF9; β-1-endoglucanases) was identified in *Reticulitermes speratus* (Arthropoda, Insecta) (Watanabe et al., 1998). At the same time, four genes of endoglucanases from family GH5 were identified in the plant nematodes *Globodera rostochiensis* and *Heterodera glycines* (Smant et al., 1998). Subsequently, the presence of endogenous β-1,4-endoglucanases belonging to various GH families started to be reported in various insects, crustaceans, mollusks, echinoderms, and nematodes (Sugimura et al., 2003; Nishida et al., 2007; Sakamoto et al., 2007; Imjongjirak et al., 2008; Milatovič and Štrus, 2010; Lo et al., 2003; Davison and Blaxter, 2005).

The origin of cellulase in metazoans can be explained by two alternative hypotheses. The first refers to the horizontal transfer of cellulase genes from symbiotic protozoa. The second hypothesis is the vertical transfer of cellulase genes from ancient organisms that are now extinct. In this second hypothesis, cellulase genes were transmitted through inheritance from generation to generation over a long period of time. According to reports about GH9 (Lo et al., 2003; Davison and Blaxter, 2005), which encodes the most intensively studied cellulases, vertical transference is considered more likely. The amino acid sequence of GH9 is very similar in several organisms, indicating the presence of a shared common cellulase ancestor from which the GH9 gene was inherited over one hundred million years. Unfortunately, evolutionary evidence of other GHs is not available, and information about their primary structures remains fragmented (Lo et al., 2003).

CELLULASES IN INVERTEBRATES

Vast researches have been carried aiming at the discovery of invertebrates'cellulolytic potential, concerning both endogenous enzymes and those from their gut microbiota. Among the invertebrates, insects have been, by far, one of the most studied class and involves mainly the Blattodea order, which include both cockroaches and termites. The subject has been extensively reviewed (e.g., Brune, 2014; Watanabe and Tokuda, 2010). Gut metagenomic analysis of a termite from the genus *Nasutitermes* revealed more than 700 GH genes, of which more than a hundred coding enzymes related to carbohydrate enzymes, mainly endoglucanases from GH5 family and β-glucosidases from GH1 and 3 families (Warnecke et al., 2007). An approach that mixed/combined a metagenomic analysis with a functional screen was used to identify and characterize a bacterial GH5 endoglucanase in *Microcerotermes* sp. (Nimchua et al., 2012).

A metatranscriptome analysis found that endogenous enzymes in the termite *Reticuliterme flavipes* correspond to 16% of its cellulase genes and that these genes were provenient, mainly, from Protists (66%) and Prokaryotes (18%). Cellulases originated from termites included GH1 β-glucosidases, GH9 endoglucanases and numerous GH7, which includes endo-β-1,4-glucanase and cellobiohydrolases (Tartar et al., 2009). Similar results were found for the termite *Coptotermes formosanus*, which presented many GH7 from Protists; its endogenous cellulases were GH1 and 9, while cellulases from symbionts were more abundant from GH3, 5 and 45 families (Zhang et al., 2012). A proteomic analysis found similar repertory for *Coptotermes gestroi* (Cairo et al., 2010).

Cellulases produced by termites that were characterized are mainly GH1 and 9. In *Reticulitermes flavipes*, a GH1 β-glucosidase was found to be produced at salivary glands and foregut, while another isoform was shown to be produced in the whole gut (Scharf et al., 2010). Endogenous GH9 endoglucanases from the genus *Reticulitermes* and *Nasutitermes takasagoensis* were successfully expressed in bacteria and fungus (Hirayama et al., 2010; Zhou et al., 2010). Cellulases from the genus *Coptotermes* were also characterized, and the synergism between a GH9 endoglucanase and a GH1 β-glucosidase was observed with a 9.79 fold increase in relative activities when they are used to hydrolyze amorphous cellulose (Zhang et al., 2009; Cairo et al., 2013). A β-glucosidase GH1 from the termite *Neotermes koshunensis* was intriguingly stimulated by its product, glucose (Uchima et al., 2011). That seems to happen only in enzymes from GH1 family and this mechanism was recently elucidated (Yang et al., 2015).

Coleoptera is another order that has recently received more attention (Hatefi et al., 2017; Liu et al., 2015; Willis et al., 20 recently 11). Studies of the microbiota present in the gut of different species from Coleoptera order showed a large presence of the Phylum Proteobacteria. Other prominent phyla included Firmicutes and Bacteroidetes (Scully et al., 2013, El-Sayed and Ibrahim, 2015, Sheng et al., 2016,). A highly cellulolytic strain of *Pseudomonas* sp. was isolated from the hindgut of the scarab beetle *Holotrichia parallela.* This strain presented a thermostable endoglucanase activity, even higher than those observed in microorganisms used in industry (Sheng et al., 2012). Endogenous enzymes are also subject of research in these animals, and transcriptomic analysis showed the presence of putative endogenous cellulases from families GH5 and 45 in *Mesosa myops* (Liu et al., 2015). These are believed to be the main endogenous cellulases in Coleoptera (Pauchet et al., 2010), but GH9 endoglucanase acting against carboxymethyl cellulose (CMC) in alkaline conditions was described in *Tribolium castaneum* (Willis et al., 2011).

Other arthropods, such as crustaceans, have also been shown to have cellulase activity in their guts. Some work has been carried with marine isopods from the genus *Limnoria*, and it presented the identification, by transcriptomic analysis, of GH5, 7 and 9, and further characterization of GH7, which were the most abundant ones (King et al., 2010; Kern et al., 2013). *Limnoria* is described as having no cellulolytic activity associated to microbiota, reinforcing the production of endogenous cellulolytic enzymes. Another isopod, from the terrestrial *Porcelio* genus, was also shown to possess endogenous GH9 enzymes by cDNA analysis (Kostanjšek et al., 2010).

There is also some literature on the subject concerning the Crustacean Decapoda order. The works included different genus of land crabs, mangrove crabs and crayfish. In all of them, GH9 was identified endogenously (Allardyce and Linton, 2008; Bui and Lee,

2015; Linton et al., 2006; Linton and Shirley, 2011). The work involved DNA screening techniques and the use of degenerate primers to seek similar enzymes in other animals from the Decapoda order. There was also a purified and characterized GH9 cellulase form Krill (Tsuji et al., 2012).

GH5 and 45 sequences were identified in a few studies with a different genus of plant-parasitic nematodes (Rybarczyk-Mydłowska et al., 2012; Wu et al., 2016). An interesting discussion in this context concerns the origins of such genes. Although there is a lot of evidence showing the widespread similarities among some animal cellulases, some GH45 enzymes from nematodes have shown similarities to those from fungi, including an intron position, which suggests that, at least in this case, horizontal gene transfer might explain the origins of some GH45 enzymes in these animals (Palomares-Rius et al., 2014).

Cellulases have also been described in oligochaetes from the Clitellata class (Nozaki et al., 2009; Park et al., 2012). These studies used the expressed gene sequence to clone and characterize different enzymes form the most widespread endogenous GH family among animals (Tsuji et al., 2012). Park et al. (2012) highlighted the possibility that cellulases in earthworms might play an additional immunologic role, due to their intense expression on the epidermis.

Despite the fact that the first studies on animal cellulases on the early 20th century had been performed on escargot (*Helix pomatia*) land snail (Watanabe and Tokuda, 2010), only more recently mollusks (Mollusca) arose as a more studied group for the prospection of cellulases. In the past years, works showed cellulase activity in different aquatic and terrestrial mollusks. Afterwards, this subject will be better discussed regarding two species that have drawn the researcher's attention for unexplored biotechnological potential.

CELLULASES IN *ACHATINA FULICA*

Mollusks are a potentially valuable source for the research of new cellulases. Different research groups identified cellulases in the brackish water clam *Corbicula japonica* (Sakamoto and Toyohara, 2009), *Ampullaria crossean* (Yin et al., 2011) and in the golden apple snail *Pomacea canaliculata* (Yang et al., 2017). Two cellulolytic enzymes were identified in the gastric juice of the African snail *Achatina fulica* using chromatographic and biochemical methods, however without obtaining the sequence of cellulase-producing genes in these animals (Maeda et al., 1996; Teng et al., 2010). In recent studies, the role of its microbiota in cellulase degradation has been addressed, however this animal is still poorly explored as a source of enzymes, especially the endogenous ones, with activity over cellulose and lignocellulosic substrates (Cardoso et al., 2012a; Pawar et al., 2015).

Figure 1. The snail *Achatina fulica* (A) and its digestive system with the following parts denoted: salivary gland (SG), crop (C), esophagus (O), stomach (S), intestine (I), digestive gland (DG) and rectum (R). This figure was adapted from Cardoso et al. (2012).

The snail *Achatina fulica* (Bowdich, 1822), also known as the giant African snail (Figure 1), is a terrestrial mollusk of the Pulmonata and Gastropoda's order class. The Achatinidae family is composed by more than 200 species, and it is believed that this family is derived from a common ancestor from southwestern Africa. Currently, species from Achatinidae are found throughout the sub-Saharan Africa, in great concentration in the central regions of the continent. Species of the genus *Achatina* are concentrated in places with higher moisture and are not found at altitudes higher than 1500 m. In addition, several species are found in higher densities in regions of modified forests, and the *Achatina fulica* snail is found predominantly on the edges of forests in East Africa. These data show the adaptability of this species in tropical regions modified by the action of men (Barker, 2002).

It is believed that *Achatina fulica* is one of the main invasive mollusks, being considered the most widely dispersed terrestrial snail (Thiengo et al., 2007). It is also included in the list of the 100 main invasive species maintained by the International Union for Conservation of Nature (IUCN) in the Global Invasive Species Database

(GISD) (Pagad et al. 2015). Its dispersion began from East Africa to the islands of the Indian Ocean between the eighteenth and nineteenth centuries, heading eastward during the early twentieth century, reaching Japan and Southeast Asia. In the 1940s, the species was introduced to Hawaii and to several islands in the Pacific Ocean. By the 1970s, it was taken to the United States, where its dispersion was controlled. Recently the number of individuals of this species increased, especially in the Florida state. At the end of the 1980s, the species was carried to the Caribbean and South American islands, having been described in Colombia, Ecuador, Peru, Brazil and Argentina (Fontanilla et al., 2014).

The first research article to describe cellulase activity in *A. fulica* snail was published in 1963 and compared the hydrolase activity present in the digestive juice of this species with the digestive activity from the escargot *Helix pomatia*. The hydrolytic activity was similar in both species, and the digestive juice of *A. fulica* showed higher activity on CMC, lichenan and laminarin. The activity of digestive juice on several oligosaccharides was also evaluated, and higher activities were found with the substrates gentiobiose, cellobiose and lactose. The β-glucosidase activity was found to be lower than in *H. pomatia*. Additionally, the digestive juice of *A. fulica* showed *in vitro* lytic activity against *Micrococcus lysodeikticus*, *Bacillus subtilis* and *Saccharomyces cerevisiae* (Thirlwell et al., 1963). A cysteine-rich antimicrobial peptide that is active against bacteria and fungus was recently identified in the mucus of *A. fulica* (Zhong et al., 2013).

Afterwards, a β-mannosidase was isolated from the snail viscera by extraction of the total amount of proteins with acetone followed by the fractionation with ammonium sulfate. The purification step was performed by hydroxyapatite column chromatography followed by ion exchange on diethylaminoethyl cellulose (DEAE-C) resin. At this stage, the β-acetylglucosaminidase activity was also detected in the protein extract. The purified and concentrated β-mannosidase presented higher activity at the pH range of 4.0 to 5.0 and the activity was reduced by more than 80% after incubation at 55°C for 5 minutes. The speciality of the purified enzyme for β-1,4-type mannosidic bonds has also been demonstrated (Sugahara et al., 1972).

The β-glucosidase activity was also identified in another species of the Achatinidae family, the *Achatina achatina* snail. The hydrolytic activity on 4-nitrophenyl β-D-glucopyranoside (pNPG) was detected in fractions of intestinal contents submitted to gel filtration chromatography. The molecular weight of the identified enzyme was approximately 41 kDa, the enzyme formed dimers, tetramers and octamines. The degree of oligomerization was related to the nutritional state of the animals (Umezurike, 1976a). The monomer was also identified in the protein extracts of the digestive gland, gut, stomach and intestine, previously washed out of its contents. The optimum pH range (5.0 to 5.6) was determined and kinetic studies showed that the monomer had a higher affinity for the substrate and that oligomerization favored the thermal stability of the enzyme (Umezurike, 1976b).

The β-glucuronidase, N-acetylglucosaminidase, galactosidase, β-glucosidase and fucosidase activities were detected in the the *Achatina balteata's digestuve juice* species. Two enzymes with β-glucosidase activity were purified and higher activities were observed against β-D-fucosides. The purified and characterized enzymes had a molecular weight of 110 and 300 kDa (2 subunits of 190 and 170 kDa). The activity profile at different pHs was similar, with the optimum range between 5.2 and 5.6, and the optimum temperature determined was close to 50°C for both (Colas 1978, 1980).

About 20 years after the initial studies on enzymes in the *Achatina* genus, a new study described the identification of a cellobiohydrolase in the intestinal contents of *Achatina fulica*. The proteins of the intestinal contents were extracted and subjected to gel filtration chromatography and fractions with endoglucanase, cellobiohydrolase and β-glucosidase activities were detected. A 23 kDa cellobiohydrolase enzyme was then purified and characterized. In addition to the activity towards microcrystalline cellulose Avicel®, this enzyme was also active over salicylin, evidencing its β-glucosidase activity. Analysis of the cellopentaose's enzymatic hydrolysis by thin layer chromatography showed the production of cellobiose and cellotriose, without the presence of glucose and cellotetraose. The enzyme amino acid composition further analyzed was rich in aspartic acid, asparagine, alanine and threonine. A peptide of 20 amino acids was elucidated by N-terminal sequencing (Maeda et al., 1996).

The first endo-β-1,4-glucanase from *Achatina fulica* digestive juice was identified and characterized in 2010. In this work, zymography assays by native electrophoresis in gels containing CMC revealed the presence of three cellulolytic enzymes in the digestive juice of the snail with molecular weights around 20, 35 and 60 kDa. The lower molecular weight enzyme was purified by three chromatographic steps, and mass spectrometry analysis revealed the sequence of a 16-amino acid peptide and the determined molecular weight was 22 kDa. This purified enzyme did not show activity against other substrates besides CMC, which was degraded with higher efficiency in acidic pHs (2.5 to 4.5) and at temperatures between 45 and 50°C (Teng et al., 2010).

Although many of the studies described above suggest the endogenous origin of the enzymes identified, no molecular evidence was presented to confirm this hypothesis. Recently, several papers have linked the microbiota presence in the snail's digestive system with the digestion of cellulose. Sequencing of the 16S fragment of ribosomal RNA extracted from microorganisms found in the digestive juice of the crop, showed the predominant presence of the phylum Proteobacteria. In the intestinal contents, the predominance of Bacteroidetes was altered when animals were subjected to a diet composed of sugarcane bagasse, with an increase in the proportion of both Proteobacteria and Firmicutes (Cardoso et al., 2012b). Later, the composition of the microbiota was detailed in the different segments of the digestive tract, also by 16S rRNA sequencing. The Gammaproteobacteria class was predominant in the esophagus, gut, intestine and rectum, while in the stomach was found the lowest bacterial diversity, with an abundance of Tenericutes. This phylum was abundantly found in all studied segments, and in 4 of them, it was the second most represented phylum. It is interesting to note that in the first paper this phylum had not been detected, which may indicate the influence of geographic location and diet over the composition of the snail's microbiota (Pawar et al., 2012).

Metagenomic analysis of the microbiota present in the snail's digestive fluid confirmed the predominance of Gammaproteobacteria in this particular region of the digestive tract. In addition to stress-related genes and detoxification processes, the attention was drawn to the large presence of genes related to carbohydrate metabolism. Genes from 53 GH families and 11 carbohydrate-binding modules were identified, most of them from Gammaproteobacteria. The larger number of GH genes were mainly coding for β-glucosidases (GH1 and 3), xylanases (GH8 and 10), as well as fucosidases (GH29) and galacturanases (GH28). In addition, genes corresponding to the endoglucanases from GH9 and 6 families were also identified. This metagenomic analysis found genes were involved in all steps of the lignocelluloses degradation, reaffirming the preponderant role of microorganisms in the digestion of biomass in *A. fulica* (Cardoso et al., 2012a).

Structural and biological characterization of one of these GHs has been performed recently. It is a GH8 endoglucanase of the bacteria *Raoultella ornithinolytica*, and its gene sequence was obtained from the metagenome of the snail's digestive fluid described above. This enzyme showed activity on substrates such as CMC, β-glucan and lichenan, with high catalytic activity in a wide temperature range of 30 to 55°C, and maximum activity at pHs near 6.5. This GH8 endoglucanase showed a preference for β-1,4-type glycosidic bonds in polysaccharides containing at least 5 glucose units, and its structure, the first of a cellulolytic enzyme obtained from *A. fulica* snail, revealed the similarity of this enzyme with bacterial cellulose synthesized enzymes described in several species of bacteria (Scapin et al., 2017).

The isolation and identification of cellulolytic bacteria were carried in two works and confirmed the relevance of Proteobacteria from the snail's digestive tract in cellulose degradation. Microorganisms of the genus *Klebsiella* and *Enterobacter* were found in a greater ratio in intestinal isolates with cellulolytic activity in *A. fulica* specimens collected in India. An isolate of *Citrobacter freundii* species showed the highest activity towards CMC and Avicel®, and this species had already been identified as one of the most abundant in the esophagus and intestine of the snail (Pawar et al., 2012; Pawar et al., 2015).

In snails collected in Brazil, the genus Aeromonas was the most representative of cellulolytic isolates from the crop. Cellulolytic isolates from intestine and rectum showed a high abundance of Actinobacteria, with the genus *Cellulosimicrobium* being the most predominant in intestinal isolates. It is interesting to note that Actinobacteria was not identified as the most representative in metagenomic and rRNA 16S analysis. Nevertheless, this phylum may have critical relevance in the cellulose degradation that takes place at the snail's intestine, considering its remarkable presence in isolates as well as its high level of degradation considering different substrates (Pinheiro et al., 2015). Genomic analysis of *Streptomyces* sp. isolated from the snail showed genes coding for endoglucanases (GH5, 6 and 12), cellobiohydrolases (GH6 and 48) and β-glucosidases (GH1 and 3). These activities were identified in culture media of the bacteria grown with cellulosic substrates available, confirming the biotechnological potential of this strain isolated from *A. fulica* (Pinheiro et al., 2017).

CELLULASES IN TEREDINIDAE

There are many animals in nature able of boring holes into wood and use it as shelter and source of food. Among them, featured insects (termites, cockroaches, grasshoppers) and cattle. Considering marine invertebrates, mollusks and crustaceans are known to have species with this ability, and among mollusks, bivalves of the family Teredinidae are the most specialized in boring and wood-feeding (Turner, 1966) (Figure 2).

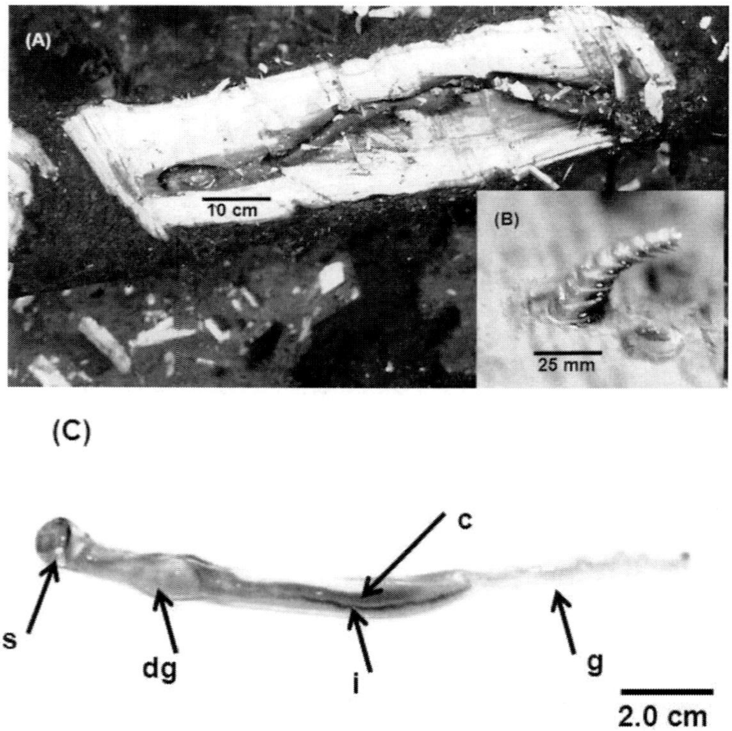

Figure 2. (A) Wood log in a mangrove forest (São Paulo State, Brazil) with a shipworm of *Neoteredo reynei* species boring into it. Notice that the log is still alive and attached to the tree. (B) Pallets of *Bankia* species closing the burrow excavated by the animal. Pallets are the only organ in contact with the external environment. (C) Morphology of a *Bankia* species, showing the organs related to the wood digestion. Wood is particulate with the aim of the shell (s) and the particles are swallowed and conducted to the caecum (c) where they are stored; enzymes of the digestive glands (dg) and the gills (g) are acting to digest the wood. In the intestine (i), wood particles are also digested and the fecal pellets are formed.

The Teredinidae, popularly known as shipworms, live in marine and brackish environments. They are being known since Antiquity in the writings of Ovidio, Pliny and Theophrastus (Distel, 2003). These ancient reports, and many others that followed them, always referred to the extremely destructive hability of these animals and, therefore, to the economic losses caused by the damage to the ships and other wood structures exposed to the marine environment (Distel, 2003). Because of these damages and prejudicial economic impact, they have been popularly called "termites of the sea", in association with these mollusks.

From the 20th century and further, studies with shipworms focused on the elucidation of the mechanisms of wood digestion by these animals as a form to avoid the financial losses that they cause. The wood digestion is related to the capability to break the cell wall of the plants, releasing the lignin and accessing the cellulose. The shipworms make the cellulose accessible in the cell wall with the mechanic grinding of wood by the shell (Dean, 1978).

Cellulolytic activity in some parts of the digestive system of the shipworms was reported for the first time by Harrington (1921). In this study, an extract of the digestive gland would be able of release glucose in an enzymatic hydrolysis. Nevertheless, this extract would not be able to complete the reduction of pure cellulose (like Avicel®) into glucose. The author suggested that there are not cellulolytic enzymes in adequate concentration. Moreover, it also could be related to the absence of one of the enzymes from the cellulolytic complex, e.g., endo-glucanase, cellobiohydrolase or β-glucosidase.

Subsequent studies suggested the presence of cellulolytic enzymes in many parts of the digestive tract of shipworms, mainly in the digestive glands, wood-storing caecum (appendix) and intestine, due to the measurement of enzymatic activity in these organs (Dore and Miller, 1923; Boynton and Miller, 1927; Deschamps, 1953; Greenfield and Lane, 1953; Liu and Walden, 1970; Crosby and Reid, 1971). However, at this point, it is not clear the origin of the enzymes. The cellulolytic activity of the organs could be provided by microorganisms (exogenous) or could be from the shipworm itself (endogenous). Moreover, it could be a combination of both endogenous and exogenous (Boynton and Miller, 1927; Greenfield and Lane, 1953; Liu and Walden, 1970; Crosby and Reid, 1971).

Enzymes found in the digestive system of the Western Atlantic shipworm *Bankia gouldi* showed a pattern of activity very similar to those of *Aspergillus niger*, a fungus that grows well in soluble celluloses (Dean, 1978). Cellulolytic bacteria were isolated from the digestive system of 5 species of shipworms and identified as *Cellulomonas* sp. (Rosenberg and Breiter, 1972). The bacterial activity of *Cellulomonas* sp. would be a major contribution to the wood degradation (Rosenberg and Cutter, 1973).

Although it seems promising, the discovery of these microorganisms in the digestive system of shipworms did not advance through the years until cellulolytic bacteria were found in the gills of several species of shipworms (Popham and Dickson, 1973;

Waterbury et al., 1983). From this discovery, many works investigated the capability of these bacteria to produce cellulases and their role in the wood digestion by the shipworms (Waterbury et al., 1983; Greene et al., 1988; Distel, 2003; Xu and Distel, 2004; Betcher et al., 2012; O'Connor et al., 2014; Cragg et al., 2015).

At first, the fact that symbiont bacteria were found in limited structures, the bacteriocytes, inside the gill tissue, far away from the digestive system, causes certain discomfort. In the past few years, questions like how the cellulolytic enzymes will be transported to the digestive system and what are their role in the digestive process were asked but still very little is known (Distel, 2003).

Sequences analysis of the 16S rRNA genes from bacteria found in many species of shipworms' gill showed that the isolated cultured bacteria are identical, suggesting that these isolates represent one single species (Distel et al., 1991). Further studies characterized and phylogenetically identified these bacteria as a gamma-subdivision of Proteobacteria (Distel et al., 1991; Distel et al., 2002a), which was named *Teredinibacter turnerae* (Distel et al., 2002a).

The cellulase (endo-beta-1,4-glucanase) and the protease secreted by *T. turnerae* were purified and characterized (Greene et al., 1988; Greene et al., 1989). The endoglucanase showed maximum relative activity in a range of pH from 5.0 to 8.0. The maximum relative activity was obtained about 70°C, however the thermal stability was reached about 40°C. When salt concentration (NaCl) varied, the maximum relative activity observed was about 0.2 and 0.4 M of NaCl (Greene et al., 1988,). The protease showed an optimal pH range from 6.5 to 9.0 and as optimal temperature 40°C. It was suggested that this protease may be related to increasing hydrolysis of wood in the shipworms or may be involved in the digestion of proteins when the animal is filter feeding (Greene et al., 1989). This protease could be tolerant to high concentrations of oxidants and stable at high concentrations of salts, heavy metals, pH higher than 11.9 and temperatures higher than 40°C. These unique characteristics may have enhanced contact lenses cleansing industry by using this protease (Greene et al., 1996).

Posteriorly, another type of cellulase was purified and characterized, but from entire specimens of shipworms (Xu and Distel, 2004) and compared to those previously described as secreted by the bacterial symbiont *T. turnerae* (Greene et al., 1988). The cellulase described by Xu and Distel (2004) is an endo-β-1,4-glucanase, with different characteristics of those described as the endoglucanase secreted by *T. turnerae* as molecular weight, optimal temperature, kM for CMC and inhibition by cellobiose and cellotriose (Xu and Distel 2004). These different biochemical characteristics suggest that these enzymes are coded by distinct genes, but it is not possible to establish if these two enzymes are secreted by *T. turnerae* in different moments. Another doubt is whether each of them is secreted by a different symbiont, even if noncultivated microbiota have been found in the digestive system of shipworms (Xu and Distel, 2004). However, it has not been observed a significant amount of microbiota present in the shipworms' digestive system that may possibly confirm this hypothesis (Liu and Walden, 1970; Betcher et al., 2012).

Recent evidence indicated that, in the same specimen, the gill may present different phylotypes, like different lineages of the same symbiont bacteria (Distel et al., 2002b; Luyten et al., 2006), what would explain the fact that bacteria have produced endoglucanases with distinct properties. Such phylotypes, with its own secreted enzymes, could contribute to digest different kinds of wood or to adapt to environmental conditions (Distel et al., 2002b).

Despite the absence of data concerning the digestion of wood in the Teredinidae family, it can suggest that the interaction with the symbiont bacteria is quite different of that which occurs in terrestrial animals like termites (Distel, 2003). Once the enzymes secreted by bacteria are located in the gills, they are never in direct contact with the wood or digestive system. Somehow not yet elucidated, these enzymes can be found in the digestive system (O'Connor et al., 2014). Unlike termites, shipworms do not possess a variety of microbiota in the digestive system (Cragg et al., 2015; Bletcher et al., 2012). It is possible that the lack of abundant microbiota in the digestive system reduces the competition for soluble nutrients between the "shipworm host" and the microorganisms (O'Connor et al., 2014; Cragg et al., 2015).

This strategy of digestion in shipworms might also be an advantage for the discovery of new enzymes of industrial interest. Considering that shipworms do not have a diverse microbiota, they can be considered as models of a simplier cellulolytic system. It might be possible to explore combinations of more efficient enzymes for lignocellulosic digestion instead of a more complex systems like the ones of herbivores and other xyllotrophic organisms. These organisms have a great number of enzymes and microorganisms coexisting in the digestive system and can obscure the search for enzymes capable of synergistic combination for the biofuel industry purposes (O'Connor et al., 2014).

REFERENCES

Allardyce, Benjamin J; Stuart, M Linton. "Purification and characterisation of endo-β-1,4-glucanase and laminarinase enzymes from the gecarcinid land crab Gecarcoidea natalis and the aquatic crayfish Cherax destructor." *Journal of Experimental Biology*, 211, no. 14 (2008), 2275-2287.

Barker, Gary Michael. ed. Molluscs as crop pests. *CABI*, 2002.

Bennett, Alan B; Rolf, E Christoffersen. "Synthesis and processing of cellulase from ripening avocado fruit." *Plant physiology*, 81, no. 3 (1986), 830-835.

Betcher, Meghan A; Jennifer, M Fung; Andrew, W Han; Roberta, O'Connor; Romell, Seronay; Gisela, P Concepcion; Daniel, L Distel; Margo, G Haygood. "Microbial distribution and abundance in the digestive system of five shipworm species (Bivalvia: Teredinidae)." *PLoS One*, 7, no. 9 (2012), e45309.

Boynton, Lyman C; Robert, C Miller. "The occurrence of a cellulase in the ship-worm." *Journal of Biological Chemistry*, 75, no. 2 (1927), 613-618.

Brummell, David A; Coralie, C Lashbrook; Alan, B Bennett. *Plant endo-1, 4-β-D-glucanases: structure, properties, and physiological function*. 1994.

Brune, Andreas. "Symbiotic digestion of lignocellulose in termite guts." *Nature Reviews Microbiology*, 12, no. 3 (2014), 168-180.

Bui, TH Hanh; Shing, Yip Lee. "Endogenous cellulase production in the leaf litter foraging mangrove crab Parasesarma erythodactyla." *Comparative Biochemistry and Physiology Part B: Biochemistry and Molecular Biology*, 179 (2015), 27-36.

Cairo, João Paulo L Franco; Flávia, C Leonardo; Thabata, M Alvarez; Daniela, A Ribeiro; Fernanda, Büchli; Ana, M Costa-Leonardo; Marcelo, F Carazzolle; Fernando, F Costa; Adriana, F Paes Leme; Gonçalo, AG Pereira. "Functional Characterization and Target Discovery of Glycoside Hydrolases from the Digestome of the Lower Termite Coptotermes Gestroi." *Biotechnology for biofuels*, 4, no. 1 (2011), 50.

Cairo, João Paulo L Franco; Leandro, C Oliveira; Cristiane, A Uchima; Thabata, M Alvarez; Ana, Paula da S Citadini; Júnio, Cota, Flávia Costa Leonardo; Ana, M Costa-Leonardo; Marcelo, F Carazzolle; Fernando, F Costa. "Deciphering the Synergism of Endogenous Glycoside Hydrolase Families 1 and 9 from Coptotermes Gestroi." *Insect biochemistry and molecular biology*, 43, no. 10 (2013), 970-81.

Cardoso, Alexander M; Janaína, JV Cavalcante; Maurício, E Cantão; Claudia, E Thompson; Roberto, B Flatschart; Arnaldo, Glogauer; Sandra, MN Scapin; et al. "Metagenomic analysis of the microbiota from the crop of an invasive snail reveals a rich reservoir of novel genes." *PLoS One*, 7, no. 11 (2012), e48505.

Cardoso, Alexander M; Janaína, JV Cavalcante; Ricardo, P Vieira; Joyce, L Lima; Maria, Angela B Grieco; Maysa, M Clementino; Ana, Tereza R Vasconcelos; et al. "Gut bacterial communities in the giant land snail *Achatina fulica* and their modification by sugarcane-based diet." *PloS one*, 7, no. 3 (2012), e33440.

Cleveland, Lemuel Roscoe. "The physiological and symbiotic relationships between the intestinal protozoa of termites and their host, with special reference to Reticulitermes flavipes Kollar." *The Biological Bulletin*, 46, no. 5 (1924), 203-227.

Colas, Bernard. "Kinetic studies on β-fucosidases of Achatina balteata."*Biochimica et Biophysica Acta (BBA)-Enzymology*, 613, no. 2 (1980), 448-458.

Colas, Bernard. "Some physicochemical and structural properties of two β-fucosidases from Achatina balteata." *Biochimica et Biophysica Acta (BBA)-Enzymology*, 527, no. 1 (1978), 150-158.

Cragg, Simon M; Gregg, T Beckham; Neil, C Bruce; Timothy, DH Bugg; Daniel, L Distel; Paul, Dupree; Amaia, Green Etxabe; et al. "Lignocellulose degradation mechanisms across the Tree of Life." *Current Opinion in Chemical Biology*, 29 (2015), 108-119.

Crosby, Neil D; Robert, GB Reid. "Relationships between food, phylogeny, and cellulose digestion in the Bivalvia." *Canadian journal of zoology*, 49, no. 5 (1971), 617-622.

Cutter, Joanne M; Fred, A Rosenberg. "The role of cellulolytic bacteria in the digestve processes of the shipworm." In *Biodeterioration of materials. Proceedings of the 2nd international biodeterioration symposium.* Lunteren, the Netherlands, 13th-18th September, 1971, pp. 42-51. Applied science, 1972.

Davison, Angus; Mark, Blaxter. "Ancient origin of glycosyl hydrolase family 9 cellulase genes." *Molecular Biology and Evolution*, 22, no. 5 (2005), 1273-1284.

Dean, Robert, C. "Mechanisms of wood digestion in the shipworm Bankia gouldi Bartsch: enzyme degradation of celluloses, hemicelluloses, and wood cell walls." *The Biological Bulletin*, 155, no. 2 (1978), 297-316.

del Campillo, Elena; Alan, B Bennett. "Pedicel breakstrength and cellulase gene expression during tomato flower abscission." *Plant Physiology*, 111, no. 3 (1996), 813-820.

Deschamps, Paul. "Recherche de la cellulase chez Teredo navalis au moyen de la chromatographie de partage." *Bull. Soc. Zool. Fr*, 78 (1953), 174-177.

Distel, DL; DeLong, EF; Waterbury, JB. "Phylogenetic characterization and in situ localization of the bacterial symbiont of shipworms (Teredinidae: Bivalvia) by using 16S rRNA sequence analysis and oligodeoxynucleotide probe hybridization." *Applied and environmental microbiology*, 57, no. 8 (1991), 2376-2382.

Distel, Daniel L. "The biology of marine wood boring bivalves and their bacterial endosymbionts." In: *American Chemical Society Symposium Series*, 845 (2003), 253-271.

Distel, Daniel L; David, J Beaudoin; Wendy, Morrill. "Coexistence of multiple proteobacterial endosymbionts in the gills of the wood-boring bivalve Lyrodus pedicellatus (Bivalvia: Teredinidae)." *Applied and environmental microbiology*, 68, no. 12 (2002a), 6292-6299.

Distel, Daniel L; Wendy, Morrill; Noelle, MacLaren-Toussaint; Dianna, Franks; John, Waterbury. "Teredinibacter turnerae gen. nov., sp. nov., a dinitrogen-fixing, cellulolytic, endosymbiotic gamma-proteobacterium isolated from the gills of wood-boring molluscs (Bivalvia: Teredinidae)." *International Journal of Systematic and Evolutionary Microbiology*, 52, no. 6 (2002b), 2261-2269.

Dore, Walter H; Robert, C Miller. "The digestion of wood by Teredo navalis." *Univ. Calif. Publ. Zool*, 22 (1923), 383-400.

El-Sayed, Wael S; Reda, A Ibrahim. "Diversity and Phylogenetic Analysis of Endosymbiotic Bacteria of the Date Palm Root Borer Oryctes Agamemnon (Coleoptera: Scarabaeidae)." *BMC microbiology*, 15, no. 1 (2015), 88.

Fontanilla, Ian Kendrich C; Inna, Mikaella P Sta Maria; James, Rainier M Garcia; Hemant, Ghate; Fred, Naggs; Christopher, M Wade. "Restricted genetic variation in populations of Achatina (Lissachatina) fulica outside of East Africa and the Indian Ocean Islands points to the Indian Ocean Islands as the earliest known common source." *PloS one*, 9, no. 9 (2014), e105151.

Goyal, Anil; Bijan, Ghosh; Douglas, Eveleigh. "Characteristics of fungal cellulases." *Bioresource Technology*, 36, no. 1 (1991), 37-50.

Greene, Richard V; Harold, L Griffin; Michael, A Cotta. "Utility of alkaline protease from marine shipworm bacterium in industrial cleansing applications." *Biotechnology letters*, 18, no. 7 (1996), 759-764.

Greene, Richard V; Harold, L Griffin; Shelby, N Freer. "Purification and characterization of an extracellular endoglucanase from the marine shipworm bacterium." *Archives of biochemistry and biophysics*, 267, no. 1 (1988), 334-341.

Greene, Richard V; Michael, A Cotta; Harold, L Griffin. "A novel, symbiotic bacterium isolated from marine shipworm secretes proteolytic activity." *Current Microbiology*, 19, no. 6 (1989), 353-356.

Greenfield, Leonard J; Charles, E Lane. "Cellulose digestion in Teredo." *Journal of Biological Chemistry*, 204, no. 2 (1953), 669-672.

Gritzali, Mikelina; Ross, D Brown. Jr. *The cellulase system of Trichoderma: relationships between purified extracellular enzymes from induced or cellulose-grown cells.* 1979.

Guo, Rui; Ming, Ding; Si-Liang, Zhang; Gen-jun, Xu; Fu-kun, Zhao. "Molecular cloning and characterization of two novel cellulase genes from the mollusc Ampullaria crossean." *Journal of Comparative Physiology B*, 178, no. 2 (2008), 209-215.

Harington, Charles Robert. "A note on the physiology of the ship-worm (Teredo norvegica)." *Biochemical Journal*, 15, no. 6 (1921), 736.

Hatefi, Atousa; Ali, Makhdoumi; Ahmad, Asoodeh; Omid, Mirshamsi. "Characterization of a bi-functional cellulase produced by a gut bacterial resident of Rosaceae branch borer beetle, Osphranteria coerulescens (Coleoptera: Cerambycidae)." *International Journal of Biological Macromolecules*, 103 (2017), 158-164.

Hirayama, Kayoko; Hirofumi, Watanabe; Gaku, Tokuda; Katsuhiko, Kitamoto; Manabu, Arioka. "Purification and Characterization of Termite Endogenous Β-1, 4-Endoglucanases Produced in Aspergillus Oryzae." *Bioscience, biotechnology, and biochemistry*, 74, no. 8 (2010), 1680-86.

Imjongjirak, Chanprapa; Piti, Amparyup; Siriporn, Sittipraneed. "Cloning, genomic organization and expression of two glycosyl hydrolase family 10 (GHF10) genes from golden apple snail (Pomacea canaliculata) Full Length Research Paper." *DNA Sequence*, 19, no. 3 (2008), 224-236.

Kern, Marcelo; John, E McGeehan; Simon, D Streeter; Richard, NA Martin; Katrin, Besser; Luisa, Elias; Will, Eborall; et al. "Structural characterization of a unique marine animal family 7 cellobiohydrolase suggests a mechanism of cellulase salt tolerance." *Proceedings of the National Academy of Sciences*, 110, no. 25 (2013), 10189-10194.

King, Andrew J; Simon, M Cragg; Yi, Li; Jo, Dymond; Matthew, J Guille; Dianna, J Bowles; Neil, C Bruce; Ian, A Graham; Simon, J McQueen-Mason. "Molecular insight into lignocellulose digestion by a marine isopod in the absence of gut microbes." *Proceedings of the National Academy of Sciences*, 107, no. 12 (2010), 5345-5350.

Knowles, Jonathan; Päivi, Lehtovaara; Tuula, Teeri. "Cellulase families and their genes." *Trends in Biotechnology*, 5, no. 9 (1987), 255-261.

Kyriacou, Andreas; Roger MacKenzie, C; Ronald, J Neufeld. "Detection and characterization of the specific and nonspecific endoglucanases of Trichoderma reesei: evidence demonstrating endoglucanase activity by cellobiohydrolase II." *Enzyme and microbial technology*, 9, no. 1 (1987), 25-32.

Lashbrook, Coralie C; Carmen, Gonzalez-Bosch; Alan, B Bennett. "Two divergent endo-beta-1, 4-glucanase genes exhibit overlapping expression in ripening fruit and abscising flowers." *The Plant Cell*, 6, no. 10 (1994), 1485-1493.

Linton, Stuart M; Alicia, J Shirley. "Isozymes from the herbivorous gecarcinid land crab, Gecarcoidea natalis that possess both lichenase and endo-β-1, 4-glucanase activity." *Comparative Biochemistry and Physiology Part B: Biochemistry and Molecular Biology*, 160, no. 1 (2011), 44-53.

Linton, Stuart M; Peter, Greenaway; David, W Towle. "Endogenous production of endo-β-1, 4-glucanase by decapod crustaceans." *Journal of Comparative Physiology B*, 176, no. 4 (2006), 339-348.

Liu, DL; Walden, CC. "Enzymes of glucose metabolism in the caecum of the marine borer Bankia setacea." *Journal of the Fisheries Board of Canada*, 27, no. 6 (1970), 1141-1146.

Liu, Jie; Keqing, Song; Huajing, Teng; Bin, Zhang; Wenzhu, Li; Huaijun, Xue; Xingke, Yang. "Endogenous cellulolytic enzyme systems in the longhorn beetle Mesosa myops (Insecta: Coleoptera) studied by transcriptomic analysis." *Acta biochimica et biophysica Sinica*, 47, no. 9 (2015), 741-748.

Lo, Nathan; Hirofumi, Watanabe; Masahiro, Sugimura. "Evidence for the presence of a cellulase gene in the last common ancestor of bilaterian animals." *Proceedings of the Royal Society of London B: Biological Sciences*, 270, no. Suppl 1 (2003), S69-S72.

Luyten, Yvette A; Janelle, R. Thompson, Wendy Morrill, Martin F. Polz, and Daniel L. Distel. "Extensive variation in intracellular symbiont community composition among members of a single population of the wood-boring bivalve Lyrodus pedicellatus (Bivalvia: Teredinidae)." *Applied and environmental microbiology*, 72, no. 1 (2006), 412-417.

Lynel, LR; Wyman, CE; Gerngross, TU. “Biocommodity Engineering Biotechnol.” *Prog*, 15, no. 5 (1999), 777-793.

Maeda, Iori; Yasuyuki, Shimohigashi; Hiroshi, Kihara; Motonori, Ohno. “Purification and characterization of a cellulase from the giant snail *Achatina fulica*.” *Bioscience, biotechnology, and biochemistry*, 60, no. 1 (1996), 122-124.

Mandels, Mary; John, Kostick; Richard, Parizek. “The use of adsorbed cellulase in the continuous conversion of cellulose to glucose.” In *Journal of Polymer Science: Polymer Symposia*, vol. 36, no. 1, pp. 445-459. Wiley Subscription Services, Inc., A Wiley Company, 1971.

Martin, Michael M. “Cellulose digestion in insects.” *Comparative Biochemistry and Physiology Part A: Physiology*, 75, no. 3 (1983), 313-324.

Martin, Michael M; Jones, CG; Bernays, EA. “The evolution of cellulose digestion in insects.” *Philosophical Transactions: Biological Sciences*, (1991), 281-288.

Milatovič, Maša; Jasna, Štrus. “Endogenous origin of endo-β-1, 4-glucanase in common woodlouse Porcellio scaber (Crustacea, Isopoda).” *Journal of Comparative Physiology B*, 180, no. 8 (2010), 1143-1153.

Milligan, Stephen B; Charles, S Gasser. “Nature and regulation of pistil-expressed genes in tomato.” *Plant molecular biology*, 28, no. 4 (1995), 691-711.

Nidetzky, Bernd; Walter, Zachariae; Günther, Gercken; Marianne, Hayn; Walter, Steiner. “Hydrolysis of cellooligosaccharides by Trichoderma reesei cellobiohydrolases: Experimental data and kinetic modeling.” *Enzyme and microbial technology*, 16, no. 1 (1994), 43-52.

Nimchua, Thidarat; Taksawan, Thongaram; Tanaporn, Uengwetwanit; Somchai, Pongpattanakitshote; Lily, Eurwilaichitr. “Metagenomic analysis of novel lignocellulose-degrading enzymes from higher termite guts inhabiting microbes.” *J Microbiol Biotechnol*, 22, no. 4 (2012), 462-469.

Nishida, Yukiko; Ken-ichi, Suzuki; Yuya, Kumagai; Hiroyuki, Tanaka; Akira, Inoue; Takao, Ojima. “Isolation and primary structure of a cellulase from the Japanese sea urchin Strongylocentrotus nudus.” *Biochimie*, 89, no. 8 (2007), 1002-1011.

Nozaki, Mana; Chiemi, Miura; Yuzuru, Tozawa; Takeshi, Miura. “The contribution of endogenous cellulase to the cellulose digestion in the gut of earthworm (Pheretima hilgendorfi: Megascolecidae).” *Soil Biology and Biochemistry*, 41, no. 4 (2009), 762-769.

O’Connor, Roberta M; Jennifer, M Fung; Koty, H Sharp; Jack, S Benner; Colleen, McClung; Shelley, Cushing; Elizabeth, R Lamkin; et al. “Gill bacteria enable a novel digestive strategy in a wood-feeding mollusk.” *Proceedings of the National Academy of Sciences*, 111, no. 47 (2014), E5096-E5104.

O’Brien, GW; Veivers, PC; Sandra, E McEwen; Slaytor, M; O’Brien, RW. “The origin and distribution of cellulase in the termites, Nasutitermes exitiosus and Coptotermes lacteus.” *Insect Biochemistry*, 9, no. 6 (1979), 619-625.

O'sullivan, Antoinette C. "Cellulose: the structure slowly unravels." *Cellulose*, 4, no. 3 (1997), 173-207.

Pagad, Shyama; Piero, Genovesi; Lucilla, Carnevali; Riccardo, Scalera; Mick, Clout. *IUCN SSC Invasive Species Specialist Group: invasive alien species information management supporting practitioners, policy makers and decision takers*. (2015).

Palomares-Rius, Juan E; Yuuri, Hirooka; Isheng, J Tsai; Hayato, Masuya; Akina, Hino; Natsumi, Kanzaki; John, T Jones; Taisei, Kikuchi. "Distribution and evolution of glycoside hydrolase family 45 cellulases in nematodes and fungi." *BMC evolutionary biology*, 14, no. 1 (2014), 69.

Park, In Yong; Ju, Roung Cha; Suk-Mi, Ok; Chuog, Shin; Jin-Se, Kim; Hee-Jin, Kwak; Yun-Sang, Yu; et al. "A new earthworm cellulase and its possible role in the innate immunity." *Developmental & Comparative Immunology*, 67 (2017), 476-480.

Pauchet, Yannick; Paul, Wilkinson; Ritika, Chauhan. "Diversity of Beetle Genes Encoding Novel Plant Cell Wall Degrading Enzymes." *PloS One*, 5, no. 12 (2010), e15635.

Pawar, Kiran D; Mudasir, A Dar; Bharati, P Rajput; Girish, J. Kulkarni. "Enrichment and identification of cellulolytic bacteria from the gastrointestinal tract of giant African snail, *Achatina fulica*." *Applied biochemistry and biotechnology*, 175, no. 4 (2015), 1971-1980.

Pawar, Kiran D; Sunil, Banskar; Shailendra, D Rane; Shakti, S Charan; Girish, J Kulkarni; Shailesh, S Sawant; Hemant, V Ghate; Milind, S Patole; Yogesh, S Shouche. "Bacterial diversity in different regions of gastrointestinal tract of Giant African snail (*Achatina fulica*)." *Microbiologyopen*, 1, no. 4 (2012), 415-426.

Pinheiro, Guilherme L; Allan, C de Azevedo-Martins; Rodolpho, M Albano; Wanderley, de Souza; Susana, Frases. "Comprehensive analysis of the cellulolytic system reveals its potential for deconstruction of lignocellulosic biomass in a novel Streptomyces sp." *Applied microbiology and biotechnology*, 101, no. 1 (2017), 301-319.

Pinheiro, Guilherme L; Raquel, F Correa; Raquel, S Cunha; Alexander, M Cardoso; Catia, Chaia; Maysa, M Clementino; Eloi, S Garcia; Wanderley, De Souza; Susana, Frasés. "Isolation of aerobic cultivable cellulolytic bacteria from different regions of the gastrointestinal tract of giant land snail *Achatina fulica*." *Frontiers in microbiology*, 6 (2015).

Popham, JD; Dickson, MR. "Bacterial associations in the teredo Bankia australis (Lamellibranchia: Mollusca)." *Marine Biology*, 19, no. 4 (1973), 338-340.

Rahman, Mohammad M; Akira, Inoue; Takao, Ojima. "Characterization of a GHF45 cellulase, AkEG21, from the common sea hare *Aplysia kurodai*." *Frontiers in chemistry*, 2 (2014).

Reese, Elwyn T; Ralph, GH Siu; Hillel, S Levinson. "The biological degradation of soluble cellulose derivatives and its relationship to the mechanism of cellulose hydrolysis." *Journal of Bacteriology*, 59, no. 4 (1950), 485.

Rybarczyk-Mydłowska, Katarzyna; Hazel, Ruvimbo Maboreke; Hanny, van Megen; Sven, van den Elsen; Paul, Mooyman; Geert, Smant; Jaap, Bakker; Johannes, Helder. "Rather than by direct acquisition via lateral gene transfer, GHF5 cellulases were passed on from early Pratylenchidae to root-knot and cyst nematodes." *BMC evolutionary biology*, 12, no. 1 (2012), 221.

Sakamoto, Kentaro; Haruhiko, Toyohara. "A comparative study of cellulase and hemicellulase activities of brackish water clam Corbicula japonica with those of other marine Veneroida bivalves." *Journal of Experimental Biology*, 212, no. 17 (2009), 2812-2818.

Sakamoto, Kentaro; Ken, Touhata; Michiaki, Yamashita; Akihide, Kasai; Haruhiko, Toyohara. "Cellulose digestion by common Japanese freshwater clam *Corbicula japonica*." *Fisheries science*, 73, no. 3 (2007), 675-683.

Scapin, Sandra Mara Naressi; Flavio, Henrique Moreira Souza; Leticia, Maria Zanphorlin; Thamyres, Silva de Almeida; Youssef Bacila, Sade; Alexander, Machado Cardoso; Guilherme, Luiz Pinheiro; Mario, Tyago Murakami. "Structure and function of a novel GH8 endoglucanase from the bacterial cellulose synthase complex of *Raoultella ornithinolytica*." *PloS one*, 12, no. 4 (2017), e0176550.

Scharf, Michael E; Elena, S Kovaleva; Sanjay, Jadhao; James, H Campbell; George, W Buchman; Drion, G Boucias. "Functional and Translational Analyses of a Beta-Glucosidase Gene (Glycosyl Hydrolase Family 1) Isolated from the Gut of the Lower Termite *Reticulitermes Flavipes*." *Insect biochemistry and molecular biology*, 40, no. 8 (2010), 611-20.

Scully, Erin D; Scott, M Geib; Kelli, Hoover; Ming, Tien; Susannah, G Tringe; Kerrie, W Barry; Tijana, Glavina del Rio; Mansi, Chovatia; Joshua, R Herr; John, E Carlson. "Metagenomic Profiling Reveals Lignocellulose Degrading System in a Microbial Community Associated with a Wood-Feeding Beetle." *PLoS One*, 8, no. 9 (2013), e73827.

Shelomi, Matan; Hirofumi, Watanabe; Gaku, Arakawa. "Endogenous cellulase enzymes in the stick insect (Phasmatodea) gut." *Journal of insect physiology*, 60 (2014), 25-30.

Sheng, Ping; Jiangli, Huang; Zhihong, Zhang; Dongsheng, Wang; Xiaojuan, Tian; Jiannan, Ding. "Construction and Characterization of a Cellulolytic Consortium Enriched from the Hindgut of *Holotrichia Parallela* Larvae." *International journal of molecular sciences*, 17, no. 10 (2016), 1646.

Sheng, Ping; Shengwei, Huang; Qi, Wang; Ailing, Wang; Hongyu, Zhang. "Isolation, Screening, and Optimization of the Fermentation Conditions of Highly Cellulolytic Bacteria from the Hindgut of Holotrichia Parallela Larvae (Coleoptera: Scarabaeidae)." *Applied biochemistry and biotechnology*, 167, no. 2 (2012), 270-84.

Simmons, Christopher W; Amitha, P Reddy; Patrik, D'haeseleer; Jane, Khudyakov; Konstantinos, Billis; Amrita, Pati; Blake, A. Simmons, Steven W. Singer, Michael P.

Thelen, and Jean S. VanderGheynst. "Metatranscriptomic analysis of lignocellulolytic microbial communities involved in high-solids decomposition of rice straw." *Biotechnology for biofuels*, 7, no. 1 (2014), 495.

Smant, Geert; Jack, PWG Stokkermans; Yitang, Yan; Jan, M De Boer; Thomas, J Baum; Xiaohong, Wang; Richard, S Hussey; et al. "Endogenous cellulases in animals: isolation of β-1, 4-endoglucanase genes from two species of plant-parasitic cyst nematodes." *Proceedings of the National Academy of Sciences*, 95, no. 9 (1998), 4906-4911.

Srivastava, KK; Verma, PK; Srivastava, R. "A recombinant cellulolytic Escherichia coli: cloning of the cellulase gene and characterization of a bifunctional cellulase." *Biotechnology letters*, 21, no. 4 (1999), 293-297.

Sugahara, Kazuyuki; Tadayoshi, Okumura; Ikuo, Yamashina. "Purification of β-mannosidase from a snail, *Achatina fulica*, and its action on glycopeptides." *Biochimica et Biophysica Acta (BBA)-Enzymology*, 268, no. 2 (1972), 488-496.

Sugimura, Masahiro; Hirofumi, Watanabe; Nathan, Lo; Hitoshi, Saito. "Purification, characterization, cDNA cloning and nucleotide sequencing of a cellulase from the yellow-spotted longicorn beetle, *Psacothea hilaris*." *The FEBS Journal*, 270, no. 16 (2003), 3455-3460.

Tartar, Aurélien; Marsha, M Wheeler; Xuguo, Zhou; Monique, R Coy; Drion, G Boucias; Michael, E Scharf. "Parallel metatranscriptome analyses of host and symbiont gene expression in the gut of the termite *Reticulitermes flavipes*." *Biotechnology for biofuels*, 2, no. 1 (2009), 25.

Teng, Yigang; Qiuyu, Yin; Ming, Ding; Fukun, Zhao. "Purification and characterization of a novel endo-β-1, 4-glucanase, AfEG22, from the giant snail, *Achatina fulica frussac*." *Acta Biochim Biophys Sin*, 42, no. 10 (2010), 729-734.

Thiengo, Silvana C; Fábio, André Faraco; Norma, C Salgado; Robert, H Cowie; Monica, A Fernandez. "Rapid spread of an invasive snail in South America: the giant African snail, *Achatina fulica*, in Brasil." *Biological Invasions*, 9, no. 6 (2007), 693-702.

Thirlwell, MP; Strasdine, GA; Whitaker, DR. "A comparison of hydrolase and lytic activities of the digestive juices of the giant African snail, *Achatina fulica*, and *Helix pomatia*." *Canadian journal of biochemistry and physiology*, 41, no. 7 (1963), 1603-1610.

Tomme, P; Warren, RAJ; Gilkes, NR. "Cellulose hydrolysis by bacteria and fungi." *Advances in microbial physiology*, 37 (1995), 1-81.

Trainotti, L; Ferrarese, L; Dalla Vecchia, F; Rascio, N; Casadoro, G. "Two different endo-β-1, 4-glucanases contribute to the softening of the strawberry fruits." *Journal of plant physiology*, 154, no. 3 (1999), 355-362.

Trainotti, Livio; Silvia, Spolaore; Anna, Pavanello; Barbara, Baldan; Giorgio, Casadoro. "A novel E-type endo-β-1, 4-glucanase with a putative cellulose-binding domain is highly expressed in ripening strawberry fruits." *Plant molecular biology*, 40, no. 2 (1999), 323-332.

Tsuji, Akihiko; Keiko, Tominaga; Nami, Nishiyama; Keizo, Yuasa. "Comprehensive enzymatic analysis of the cellulolytic system in digestive fluid of the Sea Hare Aplysia kurodai. Efficient glucose release from sea lettuce by synergistic action of 45 kDa endoglucanase and 210 kDa ß-glucosidase." *PloS one*, 8, no. 6 (2013), e65418.

Tsuji, Akihiko; Shiori, Sato; Ayumi, Kondo; Keiko, Tominaga; Keizo, Yuasa. "Purification and characterization of cellulase from North Pacific krill (Euphausia pacifica). Analysis of cleavage specificity of the enzyme." *Comparative Biochemistry and Physiology Part B: Biochemistry and Molecular Biology*, 163, no. 3 (2012), 324-333.

Tucker, Mark L; Stephen, B Milligan. "Sequence analysis and comparison of avocado fruit and bean abscission cellulases." *Plant physiology*, 95, no. 3 (1991), 928-933.

Tucker, Mark L; Mary, L Durbin; Michael, T Clegg; Lowell, N Lewis. "Avocado cellulase: nucleotide sequence of a putative full-length cDNA clone and evidence for a small gene family." *Plant molecular biology*, 9, no. 3 (1987), 197-203.

Turner, Ruth D. A Survey and Illustrated Catalogue of the Teredinidae. *Museum of Comparative Zoology*, 1966.

Uchima, Cristiane Akemi; Gaku, Tokuda; Hirofumi, Watanabe; Katsuhiko, Kitamoto; Manabu, Arioka. "Heterologous Expression and Characterization of a Glucose-Stimulated Β-Glucosidase from the Termite *Neotermes koshunensis* in *Aspergillus oryzae*." *Applied microbiology and biotechnology*, 89, no. 6 (2011), 1761-71.

Umezurike, Gabriel M. "The β-glucosidase in the gut contents of the snail Achatina achatina." *Biochemical Journal*, 157, no. 2 (1976), 381-387.

Umezurike, Gabriel M. "Metabolic regulation of β-glucosidase in the gut content of the snail *Achatina achatina*." *Biochimica et Biophysica Acta (BBA)-Enzymology*, 438, no. 1 (1976), 333-338.

Vinzant, TB; Adney, WS; Decker, SR; Baker, JO; Kinter, MT; Sherman, NE; Fox, JW; Himmel, ME. "Fingerprinting *Trichoderma reesei* hydrolases in a commercial cellulase preparation." *Applied biochemistry and biotechnology*, 91, no. 1 (2001), 99-107.

Warnecke, Falk; Peter, Luginbühl; Natalia, Ivanova; Majid, Ghassemian; Toby, H Richardson; Justin, T Stege; Michelle, Cayouette; et al. "Metagenomic and functional analysis of hindgut microbiota of a wood-feeding higher termite." *Nature*, 450, no. 7169 (2007), 560.

Watanabe, Hirofumi; Hiroaki, Noda; Gaku, Tokuda; Nathan, Lo. "A cellulase gene of termite origin." *Nature*, 394, no. 6691 (1998), 330.

Watanabe, H; Tokuda, G. "Animal cellulases." *Cellular and Molecular Life Sciences CMLS*, 58, no. 9 (2001), 1167-1178.

Waterbury, John B; Bradford Calloway, C; Ruth, D Turner. "A cellulolytic nitrogen-fixing bacterium cultured from the gland of Deshayes in shipworms (Bivalvia: Teredinidae)." *Science*, 221, no. 4618 (1983), 1401-1403.

Willis, Jonathan D; Brenda, Oppert; Cris, Oppert; William, E Klingeman; Juan, L Jurat-Fuentes. "Identification, cloning, and expression of a GHF9 cellulase from Tribolium castaneum (Coleoptera: Tenebrionidae)." *Journal of insect physiology*, 57, no. 2 (2011), 300-306.

Wu, Guan-Long; Tzu-Hao, Kuo; Tung-Tsuan, Tsay; Isheng, J Tsai; Peichen, J. Chen. "Glycoside hydrolase (GH) 45 and 5 candidate cellulases in *Aphelenchoides besseyi* isolated from Bird's-nest fern." *PloS one*, 11, no. 7 (2016), e0158663.

Xu, Pei-Ning; Distel, DL. "Purification and characterization of an endo-1, 4-β-D glucanase from the cellulolytic system of the wood-boring marine mollusk *Lyrodus pedicellatus* (Bivalvia: Teredinidae)." *Marine Biology*, 144, no. 5 (2004), 947-953.

Yamaoka, I; Nagatani, Y. "Cellulose digestion system in the termite, Reticulitermes speratus (Kolbe), 1: Producing sites and physiological significance of two kinds of cellulase in the worker." *Zoological Magazine*, (1975).

Yang, Lei; Tian-yin, Cheng; Fei-yan, Zhao. "Comparative profiling of hepatopancreas transcriptomes in satiated and starving Pomacea canaliculata." *BMC genetics*, 18, no. 1 (2017), 18.

Yang, Yang; Xinxin, Zhang; Qiang, Yin; Wei, Fang; Zemin, Fang; Xiaotang, Wang; Xuecheng, Zhang; Yazhong, Xiao. "A Mechanism of Glucose Tolerance and Stimulation of Gh1 B-Glucosidases." *Scientific reports*, 5 (2015).

Yin, Qiuyu; Yigang, Teng; Yanhong, Li; Ming, Ding; Fukun, Zhao. "Expression and characterization of full-length Ampullaria crossean endoglucanase EG65s and their two functional modules." *Bioscience, biotechnology, and biochemistry*, 75, no. 2 (2011), 240-246.

Yokoe, Y. "Cellulase activity in the termite, Leucotermes speratus, with new evidence in support of a cellulase produced by the termite itself." *Sci. Papers Coll. Gen. Educ. Univ. Tokyo*, 14 (1964), 115-120.

Yokoe, Y; Yasumasu, I. "The distribution of cellulase in invertebrates." *Comparative Biochemistry and Physiology*, 13, no. 4 (1964), 323-338.

Yung, Mei-Hing; Robert, Schaffer; Jo, Putterill. "Identification of genes expressed during earlyArabidopsiscarpel development by mRNA differential display: characterisation ofATCEL2, a novel endo-1, 4-β-D-glucanase gene." *The Plant Journal*, 17, no. 2 (1999), 203-208.

Zhang, Dunhua; Alan, R Lax; Ashok, K Raina; John, M Bland. "Differential Cellulolytic Activity of Native-Form and C-Terminal Tagged-Form Cellulase Derived from Coptotermes Formosanus and Expressed in *E. Coli*." *Insect biochemistry and molecular biology*, 39, no. 8 (2009), 516-22.

Zhang, Dunhua, AR; Lax, B Henrissat; Coutinho, P., Katiya, N; William, C. Nierman; Natalie, Fedorova. "Carbohydrate-active enzymes revealed in *Coptotermes formosanus* (Isoptera: Rhinotermitidae) transcriptome." *Insect molecular biology*, 21, no. 2 (2012), 235-245.

Zhong, Jian; Wenhong, Wang; Xiaomei, Yang; Xiuwen, Yan; Rui, Liu. "A novel cysteine-rich antimicrobial peptide from the mucus of the snail of *Achatina fulica*." *Peptides*, 39 (2013), 1-5.

Zhou, Xuguo; Elena, S Kovaleva; Dancia, Wu-Scharf; James, H Campbell; George, W Buchman; Drion, G Boucias; Michael, E Scharf. "Production and Characterization of a Recombinant Beta-1, 4-Endoglucanase (Glycohydrolase Family 9) from the Termite *Reticulitermes Flavipes*." *Archives of insect biochemistry and physiology*, 74, no. 3 (2010), 147-62.

In: Bioethanol and Beyond
Editor: Michel Brienzo

ISBN: 978-1-53613-478-0

Chapter 11

ENZYMES FOR THE BIOCHEMICAL ROUTE OF SECOND-GENERATION ETHANOL: PRODUCTION BY SOLID-STATE CULTIVATION AS A FEASIBLE AND SUSTAINABLE ALTERNATIVE

Fernanda Perpétua Casciatori[1,*]
and Priscila Aparecida Casciatori-Frassatto[2]
[1]Department of Chemical Engineering, Federal University of São Carlos, São Carlos, São Paulo State, Brazil
[2]Department of Chemistry and Environmental Sciences, São Paulo State University, São José do Rio Preto, São Paulo State, Brazil

ABSTRACT

Considering the main four steps for second-generation ethanol (E2G) production from a vegetal biomass, namely (1) pretreatment, (2) hydrolysis, (3) fermentation and (4) distillation, hydrolysis may be considered one of the most challenging operations. Although chemical or acidic hydrolysis is possible and somehow efficient, the enzymatic route of vegetal biomasses saccharification is prefered in most of the applications because it presents some advantages like mild conditions of operation, high specificity of the enymes and low generation of toxic residues, either for the environment or for the sucessive steps of the E2G production. However, mainly due to the restricted worldwide marketshare of enzymes, the cost of hydrolytic enzymes (such as cellulases, hemicellulases, ligninases and amylases) and of auxilar enzymes associated can make the enzymatic route of hydrolysis of vegetal biomasses unfeasible. In this context, the solid-state cultivation (SSC), especially of filamentous fungi on agro-industrial by-products (such as bagasses, brans, fruit pulp and peels and remaining on crops leaves) appears as a feasible and sustainable alternative for the supply of enzymes for E2G production chain.

[*] Corresponding Author Email: fernanda.casciatori@ufscar.br.

The main idea defended here is that the enzymes can be produced within the same biorefinery that have E2G as the major product, in which a paralel line can, at one end, be supplied with highly available vegetal biomasses as substrates for microbial cultivation and enzymes synthesis and, at the opposite end, supply the needed enzymes cocktails as an input for enzymatic hydrolysis or saccharification step on the central line of the biorefinery, the E2G production chain. On the above, in the current chapter, an overview of several works on enzymes production (cellulases, hemicellulases and amylases) by SSC using different types of bioreactors (packed-beds and rotating drums) is presented and discussed on the light of the available literature and on the findings of the research group with which the author has contributed since last decade.

Keywords: solid-state cultivation, agro-industrial by-products, enzymatic hydrolysis, second-generation ethanol

INTRODUCTION

Agricultural products like crops and fruits are basic components of human food and animal feed. Moreover, some of them are used for biofuels generation, such as corn (important source of starch and sugars for ethanol fuel production by fermentation in United States), soybean (important source of vegetal oil which may be converted into biodiesel), sorghum sacharinum and sugarcane (outstanding feedstock for ethanol production in Brazil from the rich in sucrose sugarcane juice). Usually, the processing of the crops creates a high amount of industrial by-products, and currently only part of them is used in secondary areas, for instance for animal feed. In Brazil, the major part of the wastes is used for steam generation through combustion or it is left on the fields as compost. For the main Brazilian agricultural products such as sugarcane, soy and corn, the total amount of agroindustrial wastes was roughly estimated to 130 million tons per year in 2005/2006 (Graminha et al. 2008). However, in more recent data, 160 million tons per year were estimated for agroindustrial wastes from sugarcane alone, representing a large increase over the last decade and considering a forecast of 770 million tons of sugarcane produced in Brazil in 2015 (FAOSTAT 2015).

Most part of the wastes consists of straw, bran, husk or bagasse, which means they are fibrous or particulate materials. Although these materials cannot be used directly for food and feed applications, they have a huge potential for other applications due to their caloric and nutritional value, although not immediately accessible. Instead of burning or composting the high amount of low degrading residues, the alternative of adding value to the residues and by-products has been intensively explored in recent years. Moreover, the destination of such vegetal biomasses to second-generation ethanol (E2G) production is emerging and becoming a promise for countries looking for energetic independence in a sustainable way. Assessments for 2030 estimates the potential production of a volume of lignocellulosic ethanol enough to provide around 10^{19} J of energy, equal to 10.5% of current transport fuel demand, assuming that 25% of global forestry and agricultural residues are available for biofuels production (OECD/IEA 2010). According to Kumar et al. (2014), E2G serves as the best alternative for natural resources substituting gasoline and may reduce greenhouse gas emissions by about 86%, which calls attention of developed countries that need to fulfil Kyoto protocols. Therefore, investments are being made by developed economies in promoting the production and use of E2G.

The main four steps for E2G production from vegetal biomasses comprise: (1) pretreatment, either to remove lignin or hemicellulose to liberate cellulose; (2) hydrolysis, which consists on the depolymerization of carbohydrate polymers to produce fermentable sugars; (3) fermentation of hexose and/or pentose sugars to produce ethanol; and (4) distillation for concentration of ethanol (Canilha et al. 2012).

Hydrolysis may be considered the most challenging operation. Although chemical hydrolysis is possible and somehow efficient, the enzymatic route of vegetal biomasses saccharification is prefered in most of the applications. Enzymatic hydrolysis is an ideal approach for degrading cellulose into reducing sugars because mild reaction conditions can be used (pH near 5.0 and temperature around 50ºC); it does not present corrosion problems in the reactors; it results in negligible formation of toxic residues either for the environment or the successive steps of the E2G production, due to the characteristic high specificity of the enymes; and it presents high sugar yields.

A cellulosic enzyme system consists of three major components, namely endoglucanase, exoglucanase and β-glucosidase. The mode of action of each component is like the following sequence:

(1) Endoglucanase (or CMCase) promotes random scission of cellulose chains, yielding glucose and cellooligosaccharides;
(2) Exoglucanase (or Avicelase) promotes an exo-attack on the non-reducing end of cellulose with cellobiose as the primary structure;
(3) β-glucosidase (or cellobiase) promotes the hydrolysis of cellobiose to glucose.

Although cellulase isolation techniques have not been fully developed, the hypothesis above described is now accepted. According to this hypothesis, in a synergistic sequence of events, endoglucanase acts randomly on the cellulose chain, while exoglucanase acts on exposed chain ends by splitting off cellobiose or glucose. Cellobiose is subsequently hydrolysed by β-glucosidase to glucose.

Xylanases are hydrolases comprised in an enzymatic complex responsible by depolymerizing the plant cell wall component xylan, the second most abundant polysaccharide. Xylan is the principal type of hemicellulose, characterized as a linear polymer of xylopyranosyl units linked by glycosidic bonds. In nature, the polysaccharide backbone may be added to glucuronopyranosyl units, acetyl groups, arabinofuranosyl, etc. in variable proportions. The main enzymes involved are endoxylanase and βxylosidase, whose mechanisms of action on xylan are similar to the above described mechanisms of action of endoglucanase and β-glucosidase on cellulose.

However, enzymatic hydrolysis depends on optimized conditions for maximal efficiency (temperature, time, pH, enzyme and substrate concentrations) and suffers from end-product inhibition and biomass structural restraints. Moreover, the worldwide market share of enzymes is very restricted, in a way that the cost of hydrolytic enzymes (such as cellulases, hemicellulases, ligninases and amylases) and of auxiliary enzymes associated can make the enzymatic route of hydrolysis of vegetal biomass unfeasible.

Considerations about the Worldwide Market Share of Enzymes

Increased demand of energy all across the globe because of economic improvement, growing population and urbanization leads to a great interest on the expansion of fuel ethanol market, driving the upward market demand for industrial enzymes. The global market for industrial enzymes reached nearly $ 4.6 and $ 4.9 billion dollars in 2014 and 2015, respectively. This market is expected to increase from nearly $ 5.0 in 2016 to $ 6.3 billion dollars in 2021 at a compound annual growth rate of 4.7% for 2016-2021 (BCC 2017).

Due to all developments and demands, the world market for industrial enzymes is in clear and constant expansion since last decade. The key market players are BASF, DSM, Novozymes, DuPont, Danisco, Enmex, Roche, Lonza Group, Adisseo and Amano. Novozymes is the largest manufacturing company and offers various products for household care, food and beverages, bioenergy, agriculture and feed, and technical and pharmaceuticals applications. The company puts in massive efforts to retain its market share by continuous innovation and research and development. In 2015, the world market for industrial enzymes expanded by 10% to an estimated $ 3.8 billion dollars. Novozymes remained the market leader with a 48% share of the market, followed by DuPont and DSM with estimated 20 and 6%, respectively. Additionally, the commercial value of enzymes is significantly controlled by patents and rights. Novozymes is also the leading company with maximum number of patents obtained on enzymes, which is well correlated to its domination in market share of global enzymes industry as a leader with the biggest market (Kumar et al. 2014).

Enzymes for industrial applications on bioenergy corresponded to 15% of the market share in 2012 (Li et al. 2012), and this demand is expected to increase with the development of the second-generation biofuels derived from cellulosic raw materials (Frost and Sullivan 2011). Major enzymes related to E2G production are the group of the cellulases. As already commented, the cost of cellulases is a significant technical barrier to the conversion of lignocellulosic biomass to fuels associated with commercializing processes (Himmel et al. 2007). Two main strategies may be followed to reduce the cellulases cost: firstly, processes and strains enhancements; secondly, improvement in cellulase specificity and use efficiency (Zhang et al. 2006; Gusakov et al. 2007).

However, the same companies who dominate the enzymes market share are the ones who have devoted themselves on the development of new and improved cellulase preparations by using genetic engineering techniques, aiming to make production of biofuels more cost effective. In this series of studies, Novozymes launched a new enzyme Cellic CTec3 for production of E2G from agricultural wastes and residues. According to Kumar et al. (2014), the new enzyme product has been claimed to be 1.5 times better than the previous Novozymes Cellic CTec2 and five times more dose efficient than competing enzymes. Even so, the almost single supplier continues to be Novozymes, implying that the price of preparations is still on their hands.

On the above, competing with the big enzymes market players seems not to be a possibility and decreasing the prices has also a limit. Hence, an interesting proposal is that a biorefinery become self-sufficient on enzymes provision by producing its own cellulases and hemicellulases (and other enzymes like lipases and amylases) at the same industrial plant in which E2G will be produced as one of the main products.

SOLID-STATE CULTIVATION (SSC)

Solid-state cultivation (SSC) is a general term for a biotechnological process for the cultivation of microorganisms on moist solid substrates without dripping water; however, substrate must possess enough moisture to support growth and metabolism of the microorganism (Pandey 2003). Although droplets of water may be present in between the particles and there may be thin films of water at the particle surface, the inter-particle water phase is discontinuous and most of the inter-particle space is filled by the gas phase (Mitchell et al. 2006).

Couto and Sanromán (2006) presented a review in which it is clear that this process is known and applied for the production of food from ancient times. Examples include the fermentation of rice by *Aspergillus oryzae* to koji production and the cultivation of *Penicillium roquefortii* for cheese production. In Japan, SSC is used commercially to produce industrial enzymes (Suryanarayan 2003). In Brazil, since 80's, several research projects for the value-addition of tropical agroindustrial products and by-products by SSC have been developed. Hence, SSC is another way of increasing the value of agro-industrial by-products (such as bagasses, brans, fruit pulp and peels and remaining on crops leaves), which may be used as substrates that are inoculated with microorganisms which are able to metabolise those substrates, producing specific and very active enzymes or even organic acids, pigments and polymers.

In order to use the agro-industrial wastes as substrates, they may need to be cut, milled or cracked in order to become particles with appropriate size. It may be necessary to add water and nutritional supplements or to cook or pre-treat the substrate to increase the availability of nutrients. Moreover, the substrates use to be sterilized previously to the cultivation of the desired microbial agent, for full elimination of microorganisms formerly present and that could act like contaminants of the SSC processes.

In SSC systems, the substrate (straw, husks or bagasse) forms a porous pack, and air is supplied continuously through the pore space. The air flow fulfils the functions of providing the oxygen required for the metabolic needs of the microorganism and removing the carbon dioxide produced and the heat released by the fungus. Therefore, an ideal substrate for SSC should fulfill the nutritional microbial needs and provide adequate physical structure for air percolation, in order to guarantee efficient transfer of the respiratory gases and facilitate metabolic heat removal (Casciatori et al. 2014). On this, also the moisture content of the substrate have to be considered. According to Asha Poorna and Prema (2007), who studied the production of endoxylanase from *Bacillus pumilus* cultivated in wheat bran, low moisture contents of the substrate reduced swelling capacity, increasing water surface tension and consequently reducing the water activity for the microbial metabolic requirement. In opposition, very high moisture contents reduced inter particle spaces, leading to a decrease in the bed voidage and resulting in less space for microbial growth and deficient gaseous exchange.

Due to the lower water activities typical of SSC processes, filamentous fungi are the most suitable microorganisms for this kind of cultivation. Unlike other microorganisms, fungi typically grow in nature on solid substrates such as pieces of wood, seeds, fruits, roots, parts of dead animals and at other solid supports with low moisture content, in which the water necessary for microbial growth exists in an absorbed state or in complex with the solid matrix (Bhargav et al. 2008). Hence, for filamentous fungi, SSC systems are closer to their natural habitat, which contributes to make their metabolic activities to be increased. In this case, usually a spore-based inocula is prepared with high spores concentration. For that, a cultivation on agar is followed either by suspension of spores in a liquid or by drying and grinding of the solid to produce a powder that can be used as the inoculum for the solid substrate composed by agro-industrial by-products.

The main benefits of this process for enzymes production are two-fold: the produced enzymes are of high economic value and can be used in many different areas such as animal feed, food industry, detergents, textiles or biofuels; the fungi modify the structure of the organic material and, thereby, increase its nutritional value; afterwards, the bio-processed residues can be potentially used, for instance, for animal feeding, reducing the cost for feed stock and also minimizing the problem of residues handling.

SSC offers a number of advantages over the submerged cultivation (SmC) such as the possibility of using agroindustrial wastes in their raw or slightly handled forms, the easy disposition of the fermented residues as animal feed and the lower possibility of contamination by opportunist microbes, once water activity or pH of the substrates can be adjusted to values low enough to be unfavorable for most bacteria. Furthermore, Couto and Sanromán (2006) concluded it has been well established that enzyme titres produced in SSC are much higher than the ones achieved in SmC. Although the reasons for this are not clear, this fact is kept in mind while developing novel SSC bioprocesses.

SSC can be especially interesting for processes in which the cultivated product may be used directly as the enzyme source. Moreover, Brazil and other countries with high agricultural and forestry production have a great potential for the development of this biotechnological process. Hence, SSC appears as a feasible and sustainable alternative for the supply of enzymes for E2G production chain, once the cultivated solid material, containing the enzymes, can be mixed with new vegetal biomasses for the hydrolysis or saccharification, saving costs of enzymes extraction and purification.

The major factors that affect microbial synthesis of enzymes in SSC include the selection of a suitable combination of substrate and microorganism, the appropriate pretreatment of the substrate, the water content and water activity of the solid substrate, the relative humidity, the type and size of the inoculum, the control of temperature by removal of the metabolic heat, the supply of oxygen and removal of carbon dioxide and the duration of the cultivation. Table 1 presents several recent studies of enzymes production by SSC, including cellulases, hemicellulases, lipase and amylases, all useful enzymes for biofuels production, either first or second generations ethanol or biodiesel.

The main idea defended here is that the enzymes can be produced within the same biorefinery that have E2G as the major product. In this industry, a parallel line of production could, at one end, be supplied with highly available vegetal biomasses as substrates for microbial cultivation and enzymes synthesis; at the opposite end, this biotechnological line of production could supply the needed enzymes cocktails (mainly cellulases and xylanases) as inputs for enzymatic hydrolysis or saccharification step on the central line of the biorefinery, the E2G production chain. Moreover, the biorefinery could also produce biodiesel as another main product, together with E2G. In this case, lipolytic enzymes could also be produced by SSC at the same plant, for supply of transesterification step on biodiesel production chain. With this, the biorefinery would become independent of the big companies as enzyme suppliers and could even sell or share the produced cocktails with cooperatives and other small industries, increasing the economic feasibility of biofuels production and commercialization. Furthermore, the same residues can be used for enzymes production and further saccharification.

On the table below, an overview of several works on enzymes production (cellulases, hemicellulases, lipases and amylases) by SSC using different types of bioreactors (packed-beds and rotating drums) is presented and discussed on the light of the available literature and on the findings of the research group with which the author of this chapter has contributed since last decade. It will be clear that the main reason for that SSC is still not a technology well established industrially is the lack of industrial scale operable and commercially available bioreactors, adressing the need of efforts on engineering development of this kind of equipments, aiming to make feasible the enzymes production by SSC on industrial scale.

Table 1. Enzymes produced by SSC on agroindustrial wastes

Microorganism	Substrate	Enzyme	Reference
Aspergillus niger 3T5B8	Wheat bran and mango peel	Cellulase, xylanase and protease	Couri et al. (2000)
Phanerochaete chrysosporium	Corn cob	Lignolytic enzymes	Couto et al. (2001)
Thermoascus aurantiacus	Wheat straw	Cellulases and hemicellulases	Kalogeris et al. (2003)
Thermoascus aurantiacus	Cassava bagasse and rice bran	Xylanase	Palma (2003)
Penicillium simplicissimum	Babassu cake suplemmented with molasses	Lipases	Gutarra et al. (2005)
Thermoascus aurantiacus	Wheat bran, sugarcane bagasse, corn cob and straw	Xylanase and endoglucanase	Da Silva et al. (2005)
Penicillium echinulatum	Sugarcane bagasse and wheat bran	Cellulases and xylanase	Camassola and Dillon (2007)
Aspergillus niger, A. clavatus and A. casiellus	Soybean bran oil-free	Amylase, lipase and cellulase	Paris (2008)
Trichoderma reesei and *Aspergillus niger* MTCC 7956	Wheat bran	Cellulases	Sukumaran et al. (2009)
Aspergillus casiellus	Soybean bran	Invertase	Novaki (2009)
A. fumigatus M.7.1; *Myceliophthora* sp M.7.7	Sugarcane bagasse and wheat bran	Cellulases and xylanase	Moretti (2010)
Bacillus sp. KR-8104	Wheat bran	α-amylase	Derakhti et al. (2012)
Myceliophthora thermophila I-1D3b	Sugarcane bagasse and wheat bran	Cellulases and xylanase	Zanelato et al. (2012); Casciatori, F. (2015)
Rhizopus oligosporus	Wheat bran	α-amylase	Escaramboni (2014)

Bioreactors for Solid-State Cultivation (SSC)

The bioreactor is a key feature in any biotechnological process, and the same is true for SSC, because it is in this equipment where the target bioconversion takes place. According to Mitchell et al. (2006), the bioreactor has two main functions: to hold the substrate bed and provide a barrier against both the release of the microorganism to the surroundings and the contamination of the substrate bed by microorganisms from the surroundings; to control environmental conditions such as temperature and water activity at values which are optimal (or near-optimal) for growth and product formation by the microorganism.

In total, SSC is a multi-phase and multi-component system governed by flow, reaction and heat and mass transfers. For enzymes production, the main task is to provide the optimal conditions to the microbe throughout the bioreactor in order to maximize the enzyme yields.

There are many possible configurations of bioreactors for SSC. The most appropriate configuration will depend on many factors, such as the sensitivity of the microorganism to agitation, the growth rate, the sensitivity of the microorganism or of the desired product to increases in temperature and the aeration requirements. Other considerations are the need of aseptic operation, the pathogenicity of the microorganism, the amount of substrate to be cultivated and the difficulties for loading and unloading the bioreactors.

Usually, the bioreactors for SSC are divided in two categories: packed-beds and moving beds, often rotating drums. For both the cases, heat and mass transfers play a crucial role in determining the performance of the SSC processes.

A typical packed-bed bioreactor (PBB) consists of a column of cylindrical or rectangular cross section vertically oriented and with a perforated base plate on the bottom which supports the bed of substrate (Figure 1a). Air (normally saturated with water) is blown up through the base plate and it percolates through the void spaces in between the particles of the porous media. A water jacket is commonly present for helping on the maintenance of the ideal temperature. PBBs are the natural choice when the microorganism does not tolerate mixing, as it is the case of many filamentous fungi. This is also the configuration usually employed in commercial koji production.

A rotating drum bioreactor (RDB) typically consists of a drum of cylindrical cross section horizontally oriented, partially filled with the bed of substrate and in which air is blown through the headspace (Figure 1b). The whole drum rotates around its central axis to mix the bed. Water may be sprayed over the bed of substrate along the cultivation. RDBs are operationally flexible, allow adequate aeration and mixing of the substrate and provide efficient automatic control alternatives, although they may not suit for shear-stress sensitive filamentous fungi, with applications being limited by possible damage to the inoculum or to the product. Another disadvantage is that the drum usually has to be filled to up to 50% of its capacity, otherwise mixing becomes inefficient.

Concerning bioreactors for industrial scale production of enzymes by SSC, there are some challenges that have to be overcome in order to achieve a maximum enzyme yield (Santiago et al. 2014). The main challenges of SSC are related to the significant spatial dimension of either packed or moving bed bioreactors. Therefore, all quantities of interest such as microbe biomass concentrations, moisture contents, temperatures and enzymes concentrations will vary during the fermentation with the position within the bed, resulting in spatial profiles.

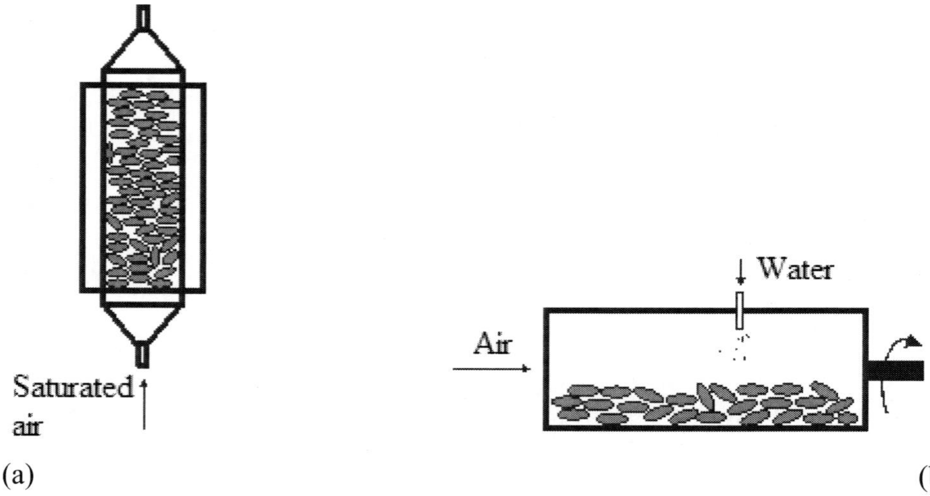

Figure 1. Main bioreactor types for SSC: (a) packed-bed; (b) rotating drum. (Extracted and adapted from Casciatori, F. 2015).

Such non-uniformity becomes especially challenging for the generated heat. It must be quickly removed from the hot spots in order to avoid damage of the fungi or of the produced enzymes. Another very important variable is the water activity, closely related to the metabolic needs of the microorganisms, which is determined by the thermodynamic properties of the flowing air and by the adsorption behavior of the substrate. In addition, agglomeration of the substrate may occur depending on size, shape and stickiness of the particles as well as the thickness of the liquid film on the surface of the substrate. As a result, the inter-particle pore space is reduced, the oxygen supply is impeded and the removal of gaseous reaction products and heat is hampered.

Therefore, thermal management and control of moisture content of the substrate are crucial to attain high-yield processes. One approach to circumvent the problem of profiles formation is to mix the system intermittently or to operate fermentation in a rotating drum. However, the influence of the drum rotation on the mixing of the moist substrate, on the fungal biomass and on the overall process efficiency is not well understood so far.

Fungi Producing Enzymes for Biofuels

The most widely used cellulases and hemicellulases on industrial applications are the ones produced by *Trichoderma* and *Aspergillus* fungal species, which are mesophilic microorganisms. Enzymes produced by them present optimal activities at a temperature range from 40 to 50 °C. At this range of temperatures, complete saccharification of vegetal biomasses requires long reaction times and reactors are susceptible to contamination. To raise the temperature of hydrolysis is a way to increase productivity and avoid contamination risks. However, this alternative requires enzymes that are more thermostable than the available preparations from mesophilic fungi (Berka et al. 2011). Some additional advantages of performing hydrolysis at higher temperatures include enhanced mass transfer coefficients, reduced broth viscosity and the possibility of saving cooling energy after a step of pretreatment of the vegetal biomass using heat (steam explosion, for instance).

In this context, thermophilic fungi species represent a potential source of thermostable enzymes for industrial applications and for biofuels production, once enzymes from thermophilic fungi often tolerate higher temperatures than enzymes from mesophilic species. Gomes et al. (2007) pointed out that higher thermostability of enzymes from thermophilic organisms may be due to factors like hydrophobic packing, enhanced secondary structure propensity, helix dipole stabilization, absence of residues sensitive to oxidation or deamination and increased electrostatic interactions.

According to Singh (2016), *Myceliophthora thermophila* syn. *Sporotrichum thermophile* is a ubiquitous thermophilic fungus with strong ability to degrade all major polysaccharides found in vegetal biomasses during optimal growth at 45 °C due to secrete several hydrolytic enzymes (cellulases, xylanases, phytases, lipases, pectinases) employed in many biotechnological applications at both moderate and high temperatures. Moreover, the native enzymes produced by the fungus present higher activities than their mesophilic counterparts.

Zanelato (2011) tested several strains of thermophilic fungi for endoglucanase and xylanase production. All the strains belonged to the fungi collection of Process Biochemistry and Applied Microbiology Laboratory from Institute of Biosciences, Humanities and Exact Sciences of São Paulo State University (Ibilce/UNESP, São José do Rio Preto-SP, Brazil) and have been isolated from piles of sugarcane bagasse from an ethanol plant located in Olímpia-SP, Brazil. Zanelato (2011) selected the strain *Myceliophthora thermophila* I-1D3b as the best producer among the twelve ones cultivated in flasks scale in a mixture 1:1 (w/w) of sugarcane bagasse and wheat bran with 80% of moisture content (wet basis) at 45ºC for up to 336 h.

The colonies of this strain presented a cotton aspect on agar and on the mixture of sugarcane bagasse and wheat bran, which was pointed out by Zanelato (2011) as an interesting characteristic due to making easier the detection of possible contaminants with different aspects along the cultivations. *Myceliophthora thermophila* I-1D3b presented xylanase activity around 715 U/gds (units of enzyme activity per gram of initial dry solid substrate) after 96 h of cultivation and endoglucanase activity of 730 U/gds after 288 h of cultivation.

Table 2. Endoglucanase and xylanase activities obtained by SSC

Microorganism	Substrate	Endo-glucanase (U/gds*)	Xylanase (U/gds*)	Reference
Trichoderma longibrachiatum	Wheat bran	--	738	Ridder et al. (1999)
Bacillus sp	Wheat bran	--	700	Gessesse and Mano (1999)
Aspergillus niger 3T5B8	Wheat bran + celobiose (0.1%)	--	806	Couri et al. (2000)
Fusarium oxysporum	Corn cob	304	--	Panagiotou et al. (2003)
Funalia trogi IBB 146	Wheat straw	356	--	Kachlishivili et al. (2006)
Sporotrichum thermophile	Beetroot pulp	109	--	Grajek (1987)
Trichoderma koningi F244	Wheat bran	287	--	Li et al. (2004)
Thermoascus aurantiacus	Wheat straw	1572	--	Kalogeris et al. (2003)
Aspergillus niger KK2	Rice straw	129	--	Kang et al. (2004)
Aspergillus niger M11	Corn cob	563	--	Gao et al. (2008)
Thermoascus aurantiacus miehe	Sugarcane bagasse	60	220	Silva et al. (2005)
	Wheat bran	600	1280	
Myceliophthora sp. (V2A2)	Rice straw	31	590	Soni et al. (2008)
Myceliophthora fergusii (T41)		37	885	
Myceliophthora sp. (MYC)		35	900	
Myceliophthora sp. (IMI 389099)	Sugarcane bagasse	7	620	Badhan et al. (2007)
Myceliophthora sp. (IMI 389099)	Wheat bran	27	129	
Myceliophthora thermophila. I-1D3	Sugarcane bagasse + wheat bran	800	900	Zanelato (2011)

* U/gds: Units of enzyme activity per gram of initial dry solid substrate.

Table 2 presents xylanase and endoglucanase activities obtained by SSC from the literature, where it is possible to observe that values obtained by Zanelato (2011) were close to the highest ones reported. For xylanase, the single value very superior to the one found for *M. thermophila* I-1D3b was that presented by Silva et al. (2005) for the cultivation of *Thermoascus aurantiacus* on wheat bran. For endoglucanase, only the result presented by Kalogeris et al. (2003) for the cultivation of the same fungus *Thermoascus aurantiacus* also on wheat bran overcame the activities reported for *M. thermophila* I-1D3b, suggesting the high potential of the fungus selected by Zanelato (2011) to produce enzymes for biofuels by SSC.

Zanelato et al. (2012) optimized temperature of cultivation (40, 45 and 50ºC), substrate composition (sugarcane bagasse [SCB] and wheat bran [WB] weight proportions 1:1, 7:3 and 9:1 w/w, dry basis) and initial moisture content of the substrate (75, 80 and 85% w/w, wet basis) for endoglucanase production by *Myceliophthora thermophila* I-1D3b in SSC tests using polypropylene plastic bags. The authors reported highest enzyme activities (878 U/gds) for substrate composition SCB:WB 7:3 w/w, temperature 50ºC and 80% initial moisture content.

Based on all the above considerations, studies of cellulases and xylanase production have been performed in different bioreactor configurations with the filamentous fungi *Myceliophthora thermophila* I-1D3b cultivated on sugarcane bagasse mixed with wheat bran.

CELLULASES PRODUCTION BY SSC IN PBB

Several studies of enzymes production by microorganisms on SSC have been conducted in PBBs. Some of them, foccused on enzymes involved in biofuels production, are presented on Table 3, where some observations about the processes are summarized.

Besides the high potential of application of PBBs for SSC, such equipments are rarely operated in large scale along the total time of cultivation. For design and development of PBBs, several information about the microorganism, the substrate and the operational conditions are required. Some important variables to be considered are, for instance, morphology and rate of growth of the microorganism, structural, heat and mass transfer properties of the porous media (porosity, bulk density, effective thermal conductivity and water difusivity) and airflow rate and relative humidity.

Zanelato et al. (2012) were the first authors to report the cultivation of the thermophilic fungus *Myceliophthora thermophila* I-1D3b in a simple jacketed PBB made of stainless steel and composed by five modules of 3 inches internal diameter and 10 cm length each module. Their PBB is shown in Figure 2. Through the jacket, water at the same temperature of the air inlet was pumped and flowed along the cultivation.

Table 3. Enzymes produced by SSC in PBBs

Microorganism	Substrate	Enzyme	Observations about the process	Reference
Thermoascus aurantiacus	Sugarcane bagasse	Xylanases	Airflow rate played important effect on enzyme activities.	Milagres et al. (2004)
Penicillium simplicissimum	Babassu cake supplemented with molasses	Lipases	Lipolytic activities in PBB 30% higher than the ones obtained on tray bioreactors.	Gutarra et al. (2005)
Myceliophthora thermophile	Sugarcane bagasse and wheat bran	Cellulases	Similar enzyme activities on PBB and on flask scale.	Zanelato (2011)
Bacillus sp. KR-8104	Wheat bran	α-amilase	Enzyme yield decreased from the bottom to the top of the PBB.	Derakhti et al. (2012)

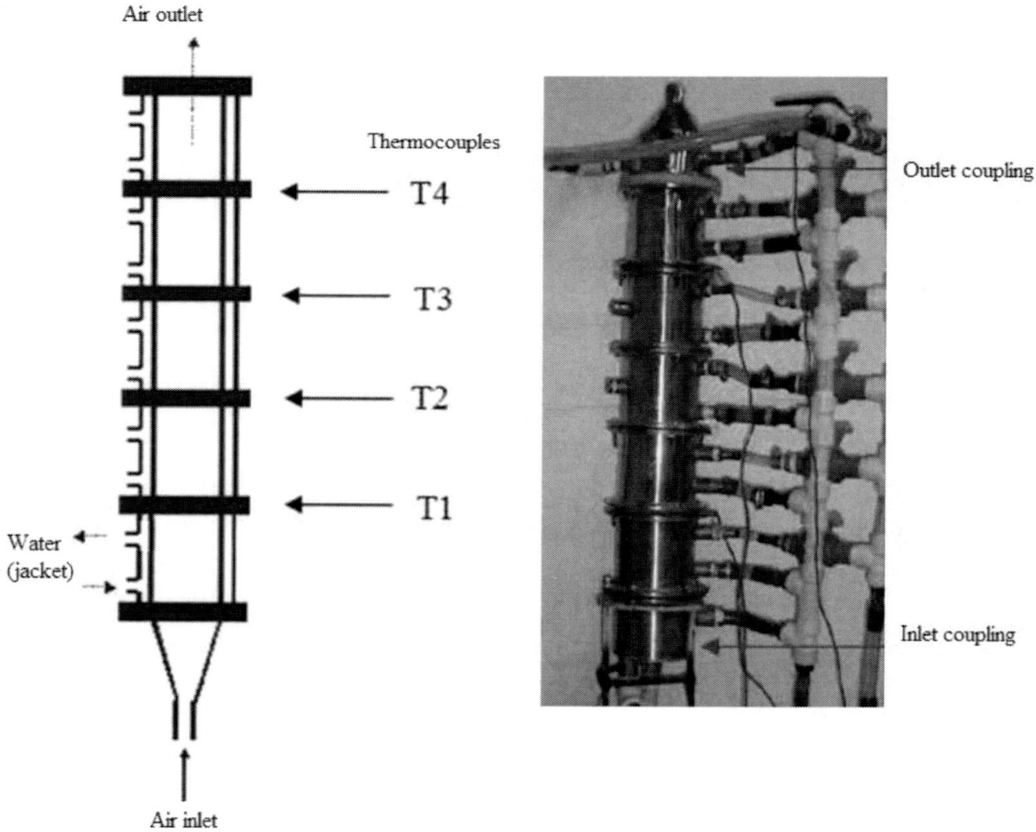

Figure 2. PBB used by Zanelato et al. (2012) for endoglucanase production by *Myceliophthora thermophila* I-1D3b on SSC using SCB:WB 7:3 w/w as substrate. (Extracted and adapted from Zanelato, 2011).

Between consecutive modules, nylon flanges were placed, making possible to insert sheathed T-type thermocouples (1.5mm diameter) with tips positioned at the central axis of the PBB. With this, the authors registered a longitudinal temperature profile. The thermocouples were connected to a data acquisition system (National Instruments, Austin, USA), managed by a LabView version 8.5 routine (National Instruments, Austin, USA) and the signals were recorded in a computer. Two jacketed couplings were used near the air inlet (bottom of the PBB) and outlet (top of the PBB), once the PBB was end-to-end aerated. Authors mentioned the equipment was not sterilized previously to the experiments, as well as any biological filter was not used. Air was provided by a radial compressor, filtered to restrainment of oil and filth and conditioned with respect to temperature and relative humidity (95%). The air inlet temperature (45 and 50°C) and its flow rate (80, 100 and 120L/h) were the controlled variables.

The solid material necessary to pack each individual module was handled in plastic bags, where a nutrient solution containing mineral salts (0.35% $(NH4)_2SO_4$, 0.3% KH_2PO_4, 0.05% $MgSO_4x7H_2O$, 0.05% $CaCl_2$ and 0.1% Tween 20, pH 5.0) and the inoculum of spores suspension (10^7 spores/mL) were added. Afterwards, the solid material was gently accommodated within each module. Longitudinal (axial) profiles of moisture contents and of endoglucanase activities were obtained. Authors report endoglucanase activities similar to the ones obtained in flasks scale and a steep longitudinal gradient of moisture content, increasing from the bottom to the top of the reactor, even though the longitudinal enzyme activity profile was flat for almost the whole bed. Zanelato et al. (2012) also concluded that the air flow rate did not affect enzyme activity and that overheating was not significant, once the maximum temperature registered was about 6°C above the process temperature.

Later on, Casciatori, F. (2015) improved the design of the PBB used by Zanelato et al. (2012) for the cultivation of the same fungus *Myceliophthora thermophila* I-1D3b on sugarcane bagasse and wheat bran (7:3 w/w, dry basis) for cellulase production on two larger scales. The first scale consisted of a jacketed cylindrical column composed by modules equal to the ones previously used by Zanelato et al. (2012), with internal diameter 3 inches (7.62 cm) and 10 cm length each module. However, Casciatori, F. (2015) used up to 15 modules, totalizing up to 1.5 m in height for cellulases production in this thin-PBB configuration. The second scale consisted also of a jacketed cylindrical column composed by modules similar to the ones used for the thin-PBB building, but in this case made of aluminium and with internal diameter 20 cm and 20 cm length each module. For this large-PBB configuration, Casciatori, F. (2015) used up to 7 modules, totalizing up to 1.4 m in height for cellulases production in the large-PBB configuration. PBBs used by Casciatori, F. (2015) are shown in Figure 3.

Like in the PBB of Zanelato et al. (2012), water at the process temperature was continuouly pumped through the jacket. Casciatori, F. (2015) worked with a new kind of flexible tubes that were easier to handle and connect to inlet and outlet of the jacket. Air was provided by a compressor and conditioned at 45°C and 100% relative humidity (saturated) by passing through a 20 cm high jacketed-column filled with glass beads immersed in water.

As the length of the thin-PBB was scaled-up in comparison to the PBB of Zanelato et al. (2012), air flow rates employed by Casciatori, F. (2015) were 120 and 240 L/h (superficial velocities 52.6 and 26.3 m/h, respectively). As already mentioned, air flow rate has to be high enough to supply oxygen to the fungal cells and low enough to avoid undesirable drying of the substrate within the PBB.

(a) (b)

Figure 3. PBBs used by Casciatori, F. (2015) for cellulases production by *Myceliophthora thermophila* I-1D3b on SSC with SCB:WB 7:3 w/w as substrate: (a) thin-PBB (internal diameter 7.62 cm); (b) large-PBB (internal diameter 20 cm). (Extracted from Casciatori, F. 2015).

Table 4. Temperature increases in PBB from literature

Microorganism	Process temperature (ºC)	Temperature increase (ºC)	Reference
Aspergillus niger ATCC 10 864	30	6 – 12	Khanahmadi et al. (2006)
Aspergillus niger	35	9	Mitchell et al. (2010)
Aspergillus niger CFTRI 1105	35	2	Ghildyal et al. (1994)
Kluyveromyces marxianus NRRLY-7571	30	20	Mazutti et al. (2010)
Penicillium italicum IZ 1584	28	4	Rizzatto et al. (2003)
Thermomucor indicae- seudaticae N-31	45	3 – 5	Umsza-Guez (2009)
Myceliophthora thermophila I-1D3b	45 e 50	4 – 5	Zanelato et al. (2012)

Temperature registration was done using T-type thermocouples radially inserted into the PBB, as done by Zanelato et al. (2012). For the thin-PBB, thermocouples were placed at the central axis of the bed; for the large-PBB, some thermocouples were also placed in between the central axis and the near-wall region of the bed. Another improvement of the PBBs operated by Casciatori, F. (2015) was the insertion of a carbon dioxide capacitive probe, for measurement of concentration of this respiratory gas in the outlet air stream along the cultivation, in order to provide information if the fungus was growing well. The probe Carbocap (Vaisala, Vantaa, Finland) was also connected to the data acquisition system (National Instruments Austin, USA), managed by the same LabView version 8.5 routine (National Instruments, Austin, USA) in which thermocouples signals were read and recorded.

Substrate preparation, moisturization, inoculation and packing procedures were like done by Zanelato et al. (2012). However, Casciatori, F. (2015) used to sterilize the modules of the PBBs in autoclave. At the end of the total time of cultivation (96 h), the cultivated materials from each module of the PBBs were transferred to plastic bags for liquid-solid extraction of the extracellular enzymes with water (20 mL of water per gram of initial dry solid substrate packed within each module). Enzymatic activities were based on Ghose (1987) for endoglucanase and filter paper (FPA) activities quantifications and on Ghose and Bisaria (1987) for xylanase activity. The processes yields were measured as enzymatic activities.

Temperature profiles along the processes and final moisture contents of the fermented materials were obtained. In the thin-PBB, there was no overheating of the bed, agreeing with simulations based on heat and mass transfer models proposed and validated by Casciatori et al. (2016). However, in the large-PBB, an expressive radial thermal profile has been observed, with temperature increasing up to 10ºC at bed central axis. Overheating was also observed, even in the thin-PBB, when the substrate was composed by SCB in smaller particle sizes, addressing the effect of bed-porosity on thermal conditions within PBBs of SSC (Casciatori et al. 2014).

Most information in the literature about the dynamics of the fermentation in PBB is related to mesophilic microorganisms with optimal process temperatures between 25 and 35 °C; there is little information on thermophilic microorganisms. Due to the metabolic heat generation by the microorganism along the cultivation, usually is quite difficult to keep the optimum temperature inside the PBB over the whole time of the process. For mesophilic fungi, temperature increases may reach up to 15 °C (Mitchell et al. 2010), while for thermophilic fungi such variation is more discreet, reaching no more than 5 °C, as presented in Table 4.

The removal of metabolic heat is difficult in PBB due to both the low effective thermal conductivity of the substrate (Casciatori et al. 2013) and low air flow rates employed. This drawback is being well studied in the literature since high temperatures may inhibit the microorganism growth and consequently affect the metabolites production and even damage the interesting products, such as enzymes (Mitchell et al. 2010).

However, contrary to trends typically reported in literature for mesophilic fungi, the temperature increase inside the thin-PBB operated by Casciatori, F. (2015) reached maximum 3ºC, allowing to conclude that overheating was not a significant problem for this configuration. Similar results have been reported in the cases of thermophilic fungi cultativation by Rizzatto et al. (2003), Umsza-Guez (2009) and Zanelato et al. (2012) and these observations can be considered as an extra positive aspect of using thermophilic fungi as microbial agents in SSC, besides the advantages already mentioned of higher thermostability of the extracellular enzymes released by those microorganisms.

Mitchell et al. (2010) have simulated the temperature increase in a PBB with *Aspergillus niger* growing in soybean. They showed temperature was predicted to increase along the bed height, coming from 36 °C at the bottom to 45 °C at the top of the PBB with 1 m total length. Gowthaman et al. (1993), who studied the amyloglucosidase production by *Aspergillus niger* CFTRI 1105 with wheat bran as substrate, observed that temperatures near PBB inlet were lower than the ones near the outlet, a trend which was attributed to the intense removal of latent heat by evaporation closer to the non-saturated air inlet.

Casciatori, F. (2015) adopted some strategies in order to minimize moisture gradients along the PBB due to end-to-end aeration. Besides of using saturated air at the inlet at the same temperature of the process, the author included a non-inoculated first module (near air inlet) packed with wet SCB and a last module (near air outlet) packed with dry SCB. The first module served to transfer water to the air in case the gas became undersaturared in between the saturation column and the inlet of the PBB; last module served to absorb condensed water from the top of the PBB when the gas flowed to room temperature. Using these strategies, drying of the substrate did not occur in any cultivation, either in thin or large-PBB.

For the large-PBB, Casciatori, F. (2015) reported non-negligible axial and mainly radial temperature profiles. Maximum temperature was registered near the central position of the large-PBB and reached 54ºC. However, the yields in terms of enzyme acitivities were just a little bit lower than those obtained in thin-PBB, once the most part of the large-PBB was kept at temperatures between 45 and 50ºC. Remember that, in flasks scale, Zanelato et al. (2012) had concluded that there was no significant difference in endoglucanase activities produced by *Myceliophthora thermophila* I-1D3b at temperatures of 45 or 50 °C.

Based on simulations of the processes using heat and water transfer models proposed by Casciatori et al. (2016) for PBBs of SSC, overheating was to be expected during the operation of the large-PBB and not for the thin-PBB, agreeing with the experiments. One interesting observation is that simulation indicates clearly that overheating in the large-PBB is due to the increase of the radial distance in between the central axis of the bed and the wall, in a way that heat removal by the cooling jacket became inefficient.

Table 5 presents a summary of enzymatic activities produced within the PBBs in the SSCs performed by Casciatori, F. (2015), comprising different lenghts, diameters and air flow rates. One may observe that yields of the cultivations in terms of enzymatic activities depends on dimensions of the PBBs and on operational conditions. Average activities of endoglucanase and xylanase reached, respectively, 786 and 2124 U/gds in thin-PBB and 620 and 2228 U/gds in large-PBB. Given the above, the bioreactors have potential for application at pilot plants and good performance in industrial scale may be expected.

Table 5. Summary of enzymatic activities found by Casciatori, F. (2015)

Substrate	Diameter (cm)	Length (m)	Air flow rate (L/h)	Endo-glucanase (U/gds)*	Xylanase (U/gds)*	FPA (U/gds)*
SCB:WB 7:3 (w/w)	7.62	1.0	60	619	1300	16
			120	547	2027	11
			240	786	2124	13
		0.5	30	282	992	9
			60	1043	2395	16
		1.5	360	514	1308	13
	20	1.0	900	482	1267	NA*
		1.4	1500	620	2228	NA*

* NA: not analyzed; U/gds: Units of enzyme activity per gram of initial dry solid substrate.

CELLULASES PRODUCTION BY SSC IN RDB

Grajales-Agudelo (2014) cultivated the same fungus *Myceliophtora thermophila* I-1D3b on sugarcane bagasse and wheat bran (SCB:WB 7:3 w/w, dry basis) for cellulase production using RDBs on two scales.

The first scale was a small-RDB made in aluminum with internal diameter 10 cm and length 20 cm (approximately 1.6 L). Air was supplied through a tube 0.5 cm external diameter and 19 cm length with perforations each 1 cm. Temperature was controlled by a jacket through which water was pumped. The equipment was supported by a metal structure with two rollers that were moved by an electric motor connected to a source of direct current responsible for the variation in the frequency of the drum rotation.

The second scale was a large-RDB made in stainless steel with internal diameter 30 cm and length 60 cm (approximately 42 L). A schematic diagram of the large-RDB apparatus is shown in Figure 4. Air was supplied by a radial compressor and entered into the RDB through a tube 2 cm external diameter and 59 cm length with perforations each 1 cm, item (6) in Figure 4. This configuration makes the air flows preferably radially through the bed of particles, covering a short distance from the tube towards the wall or the bed surface. Similarly to the cultivations of Zanelato et al. (2012) and of Casciatori, F. (2015) in PBBs, air was conditioned at 45°C and 100% relative humidity (saturated) by passing through a 20 cm high jacketed-column filled with glass beads immersed in water at the same temperature of the jacket of the RDBs. The same tube (6) was used for vapor supply into the large-RDB for sterilization previously to the SSC experiments.

Another auxiliar tube (3), to which agricultural sprinklers (4) were linked, has been placed at the headspace for aspersion of nutrient solution, spores suspension, cooling water and water for the extraction of the enzymes at end-time of cultivation. Differently of Zanelato et al. (2012) and Casciatori, F. (2015), Grajales-Agudelo (2014) used a nutrient solution composed by 10 g/L $(NH_4)_2SO_4$, 3 g/L agricultural fertilizer, 2 g/L KCl, 0.5 g/L $MgSO_4x7H_2O$, 0,1 mL/L $CaCl_2$ and 1 g/L Tween 80, pH 5 (Moretti, 2010). Moreover, the author added only 1.5 mL water (instead of 20 mL) per gram of initial dry solid substrate for enzymes extraction, due to the limit in the capacity of the drum.

The RDB was jacketed (8) and temperatures of water inlet (1) and outlet (2) were monitored by T-type thermocouples (5 and 7) connected to a data acquisition system (National Instruments, Austin, USA), managed by a LabView version 8.5 routine (National Instruments, Austin, USA). The RDB had also a drainer (9) for removal of non-absorbed water, a purger (11) for removal of condensed water during the sterilization and a tube for fermentation gases removal (10). The large-RDB was also supported by a metal structure with two rollers that were moved by an electric motor connected to a source of direct current responsible for the variation in the frequency of rotation of the drum.

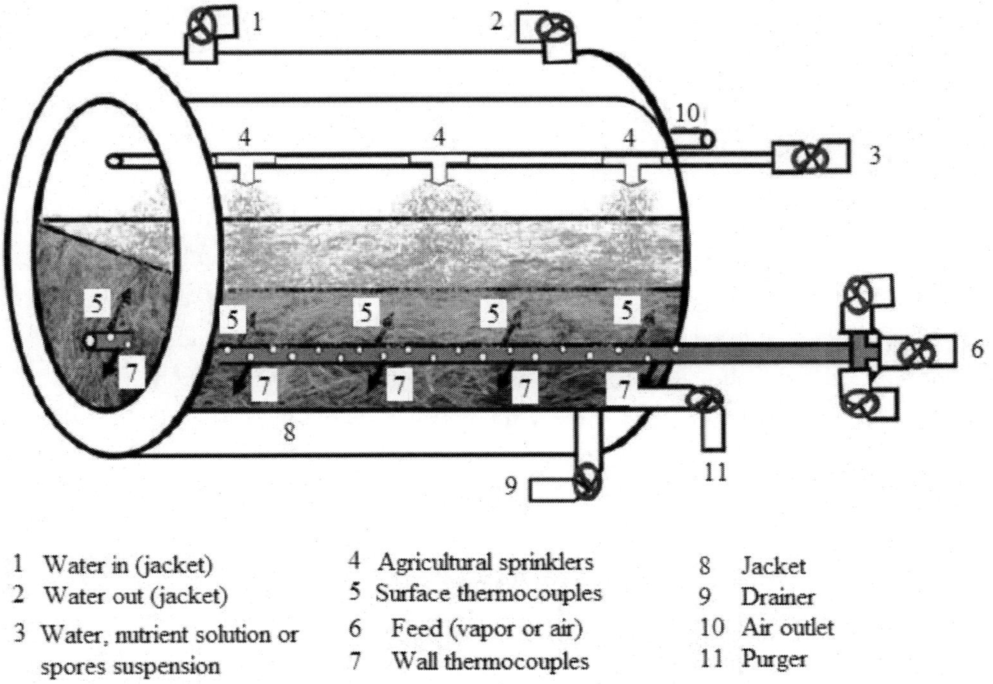

Figure 4. Schematic diagram of the large-RDB apparatus used by Grajales-Agudelo (2014). (Extracted and adapted from Grajales-Agudelo, 2014).

One of the main objectives of Grajales-Agudelo (2014) when using RDBs for cellulases production by SSC was to combine several process steps in the same equipment, aiming to reduce the possibility of contamination by sequential manual practices, reduce labor costs and improve productivity and operactionability of the process. Once the author could perform the steps of sterilization of the substrate and of the equipment itself, aspersion of nutrient solution and of spore suspension into the porous medium, the cultivation itself and finally the sprinkling of distilled water over the bed for the solid-liquid extraction of the enzymes all inside the large-RDB, that objective has been successfully achieved, even the volume of water to extraction had to be much lower than the volume usually employed per gram of dry solid, possibly unfavoring a complete extraction of the secreted enzymes.

Concerning cellulase activities produced in the small and large-RDBs, the author reports endoglucanase activities up to 1.7-fold the ones obtained in flasks scale. In the small-RDB with filling degrees 0.4 and 0.5, the number of rotations of the drum (12 and 15 rotations, respectively, at 18 rpm) did not harm enzyme production; for filling degree 0.6, 17 rotations at 18 rpm every 12 h of cultivation caused a decrease of endoglucanase activities.

Results of sensitivity of fungi to shear stress during SSC, and consequent effects on bioproducts yields, have already been studied in literature. Dillon et al. (2013) studied the production of citric acid by *Aspergillus niger* in a 12 days long SSC using apple bagasse as substrate in a RDB with 12 L capacity. The highest production was found for 1 rotation at 2 rpm every 12 h. Rodriguez-Jasso et al. (2013) studied the production of fucoidanases from *Mucor sp.* using algi as substrate and, in comparison to static cultivation, higher enzyme activities were found for continuous rotation at 10 rpm.

For the large-RDB, Grajales-Agudelo (2014) observed that, only for filling degree 0.4, the effect of rotation (1 rpm, 12 rotations) was negative on enzyme activities; for 0.5 and 0.6 filling degrees (15 and 17 rotations, respectively), enzyme activies were similar when the drum was kept static or rotated every 24 h. Hence, *Myceliophthora thermophila* I-1D3b may be considered moderately tolerant to shear stress and RDBs may be successfully used for cellulase production as well. Industrially, the choice of PBB or RDB will rely on project, buiding and maintenance costs, as well as on operationability of the processes.

FINAL CONSIDERATIONS

Mendes et al. (2017) appointed the cost of enzymes for the enzymatic route of E2G production as a crucial bottleneck to turn this process into a reality. In this context, the authors performed a detailed sutdy on the costs associated to the production of cellulases by SSC, analyzing economical scenarios of enzymes production by SSC in a pilot plant integrated to both a first and a second generation ethanol processes, as it is proposed and defended in the current chapter as well. Authors pointed out the advantagens of integrating E2G to enzymes production by SSC, markedly the possibility of sharing some available utilities such as steam, electrical energy, water, feedstock and some technical personnel.

The case study for the economic evaluation of the enzymes production by SSC was composed by sugarcane bagasse and wheat bran (SCB:WB 7:3 w/w) cultivated by the thermophilic fungus *Myceliophthora thermophila* I-1D3b at 45 °C during 96 h. Enzymatic activities were based on the results of Casciatori, F. (2015). Concerning the real application of the enzymes produced by this fungus in SSC, Casciatori, P. (2015) studied their use for the enzymatic hydrolysis of sugarcane bagasse and promissing results have been reported in terms of fermentable sugars release.

According to the estimations of Mendes et al. (2017), the most important economic indicators showed that the SSC process is economically attractive for the cellulases and hemicellulases production of the case-study due to its easy integration to the main processes. Moreover, the estimated revenue is up to four fold greater than the electricity cogeneration. Economic indicators, such as the internal rate of return (IRR) and payback, were higher than those usually accepted by Brazilian investors in the sucro-energetic sector. Nevertheless, return on the investment (ROI) was under than that recommended by the literature.

However, Mendes et al. (2017) highlights that the sensitivity analysis showed strong influence of the enzymatic activity resulting from the SSC on the economic indicators of the process, being the most important parameter for the project profitability. FPA has to be higher than 12 U/gds, a value that has been reached or even overcame by Casciatori, F. (2015) for some tested operational conditions of the PBBs. Hence, the equipments developed by either Casciatori, F. (2015) or Grajales-Agudelo (2014) are actually promising for the combination of the thermophilic fungus *Myceliophthora thermophila* I-1D3b and the substrate SCB:WB 7:3 w/w or for other combinations that can provide as high as or higher enzymatic activities in terms of endoglucanase, xylanase and mainly FPA.

References

Asha Poorna, Chundakkadu, and Prema, Palpandi. 2007. "Production of cellulase-free endoxylanase from novel alkalophilic thermotolerent Bacillus pumilus by solid-state fermentation and its application in wastepaper recycling." *Bioresource Technology* 98:485–490. Accessed May 10, 2017. doi:10.1016/j.biortech.2006.02.033

Badhan, Ajay K., Chadha, Bhupinder S., Kaur, Jatinder, Saini, Harvinder S. and Bhat, Mahalingeshwara K. 2007. "Production of multiple xylanilytic and cellulolytic enzymes by thermophilic fungus *Myceliophthora sp.* IMI 387099." *Bioresource Technology* 98:504–510. Accessed May 24, 2017. doi:10.1016/j.biortech.2006.02.009.

BCC. 2017. *Global Markets for Enzymes in Industrial Applications*. Accessed May 30. http://www.bccresearch.com/market-research/biotechnology/enzymes-industrial-applications-report-bio030j.html#.

Berka, Randy M., Grigoriev, Igor V., Otillar, Robert, Salamov, Asaf, Grimwood, Jane, Reid, Ian, Ishmael, Nadeeza, John, Tricia, Darmond, Corinne, Moisan, Marie C., Henrissat, Bernard, Coutinho, Pedro M., Lombard, Vincent, Natvig, Donald O., Lindquist, Erika, Schmutz, Jeremy, Lucas, Susan, Harris, Paul, Powlowski, Justin, Bellemare, Annie, Taylor, David, Butler, Gregory, Vries, Ronald P., Allijn, Iris E., Brink, Joost V.D., Ushinsky, Sophia, Storms, Reginald, Powell, Amy J., Paulsen, Ian T., Elbourne, Liam D.H., Baker, Scott E., Magnuson, Jon, Laboissiere, Sylvie, Clutterbuck, John A., Martinez, Diego, Wogulis, Mark, Lopez de Leon, Alfredo, Rey, Michael W., and Tsang, Adrian. 2011. "Comparative genomic analysis of the thermophilic biomass-degrading fungi *Myceliophthora thermophila* and *thielavia terrestris*." *Nature biotechnology* 29:922–927.

Bhargav, Samir, Panda, Bibhu P., Ali, Mohd, and Javed, Saleen. 2008. "Solid-state fermentation: an overview." *Chemical and Biochemical Engineering Quarterly* 22:49–70.

Camassola, Marli, and Dillon, Aldo J.P. 2007. "Production of cellulases and hemicellulases by *Penicillium echinulatum* grown on pretreated sugar cane bagasse and wheat bran in solid-state fermentation." *Journal of Applied Microbiology* 103:2196–2204.

Canilha, Larissa, Chandel, Anuj K., Milessi, Thais S.S., Antunes, Felipe A.F., Freitas, Wagner L.C., Felipe, Maria G.A., and Da Silva, Silvio S. 2012. "Bioconversion of Sugarcane Biomass into Ethanol: An Overview about Composition, Pretreatment Methods, Detoxification of Hydrolysates, Enzymatic Saccharification, and Ethanol Fermentation." *Journal of Biomedicine and Biotechnology* 2012:1–15.

Casciatori, Fernanda P. 2015. *Production of fungal cellulases by solid state fermentation: scale-up of fixed bed bioreactors*. PhD diss., São Paulo State University.

Casciatori, Fernanda P., Laurentino, Carmem L., Lopes, Karen C.M., Souza, André G., and Thoméo, João C. 2013. "Stagnant effective thermal conductivity of agro-industrial residues for solid state fermentation." *International Journal of Food Properties* 16:1578–1593.

Casciatori, Fernanda P., Laurentino, Carmem L., Taboga, Sebastião R., Casciatori, Priscila A., and Thoméo, João C. 2014. "Structural properties of beds packed with agro-industrial solid by-products applicable for solid-state fermentation: Experimental dataand effects on process performance." *Chemical Engineering Journal* 255:214–224.

Casciatori, Fernanda P., Bück, Andreas, Thoméo, João C., and Tsotsas Evangelos. 2016. "Two-phase and two-dimensional model describing heat and water transfer during solid-state fermentation within a packed-bed bioreactor." *Chemical Engineering Journal* 287:103–116

Casciatori, Priscila A. 2015. *Production of cellulolytic enzymes by fungi Trichoderma reesei and Myceliophthora thermophila and application in saccharification of sugarcane bagasse*. MSc diss., São Paulo State University.

Couri, Sonia, Terzi, Selma C., Pinto, Gustavo A.S., Freitas, Suely P., and Costa, Antonio C.A. 2000. "Hydrolytic enzyme production in solid-state fermentation by *Aspergillus niger* 3T5B8." *Process Biochemistry* 36:255–261.

Couto, Susana R., and Sanromán, Ángeles M. 2006. "Application of solid-state fermentation to food industry – A review." *Journal of Food Engineering* 76:291–302. Accessed May 10, 2017. doi:10.1016/j.jfoodeng.2005.05.022.

Couto, Susana R., Rivela, Isabel, and Sanromán, Ángeles M. 2001. "Design of different bioreactor configurations: application to ligninolytic enzyme production in semi-solid-state cultivation." *Journal of Chemical Technology and Biotechnology* 76:78–82.

Da Silva, Roberto, Lago, Ellen S., Merheb, Carolina W., Macchione, Mariana M., Park, Yong K., and Gomes, Eleni. 2005. "Production of xylanase and CMCase on solid state fermentation in different residues by *Thermoascus aurantiacus* Miehe." *Brazilian Journal of Microbiology* 36:235–241.

Derakhti, Sorour, Shojaosadati, Seyed A., Hashemi, Maryam, and Khajeh, Khosro. 2012. "Process parameters study of α-amylase production in a packed-bed bioreactor under solid-state fermentation with possibility of temperature monitoring." *Preparative Biochemistry and Biotechnology* 42:203–216.

Dillon, Gurpreet S., Brar, Satinder K., Kaur, Surinder, and Verm, Mausam. 2013. "Bioproduction and extraction optimization of citric acid from *Aspergillus niger* by rotating drum type solid-state bioreactor." *Industrial crops and products* 41:78– 84.

Escaramboni, Bruna. 2014. *Amilases production by solid state cultivation of Rhizopus microsporus var. oligosporus and its use in obtaining glucose syrup*. MSc diss., São Paulo State University.

FAOSTAT. 2015. *Food and Agriculture Organization of the United Nations: Crops*. Accessed May 30. http://faostat3.fao.org.

Frost and Sullivan. 2011. *Next-generation Biofuels Set to Displace Fossil Fuels in the Transportation Sector*. Accessed June 23. https://www.frost.com/prod/servlet/press-release.pag?docid=226428717.

Gao, Jianmin, Weng, Haibo, Zhu, Daheng, Yuan, Mingxue, Guan, Fangxia, and Xi, Yu. 2008. "Production and characterization of cellulolytic enzymes from the thermoacidophilic fungal *Aspergillus terreus* M11 under solid-state cultivation of corn stover." *Bioresource Technology* 99:7623-7629. Accessed May 24, 2017. doi:10.1016/j.biortech.2008.02.005.

Gessesse, Amare, and Mamo, Gashaw. 1999. "High-level xylanase production by an alkaliphilic *Bacillus sp.* by using solid-state fermentation." *Enzyme and Microbial Technology* 25:68–72.

Ghildyal, N. P., Gowthaman, M. K., Raghava Rao, K. S. M. S., and Karanth, N. G. 1994. "Interaction of transport resistances with biochemical reaction in packed-bed solid-state fermentors: Effect of temperature gradients." *Enzyme and Microbial Technology* 16:253–257.

Ghose, T. K. 1987. "Measurement of cellulase activities." *Pure and Applied Chemistry* 59:257–268.

Ghose, T. K., and Bisaria, Virendra S. 1987. "Measurement of hemicellulase activities – part I: xylanases." *Pure and Applied Chemistry* 59:1739–1751.

Gomes, Eleni, Guez, Marcelo A.U., MARTIN, Natalia, and Da Silva, Roberto. 2007. "Enzimas termoestáveis: fontes, produção e aplicação industrial." ["Thermostable Enzymes: Sources, Production and Industrial Application."] *Química Nova* 30:136–145.

Gowthaman, Marichetti K., Raghava Rao, K. S. M. S., Ghildyal, N. P., and Karanth, Nayakanakatte G. 1993. Gas concentration and temperature gradients in a packed bed solid-state fermentor. *Biotechnology Advances* 11:611–620.

Grajales-Agudelo, Lina M. 2014. *Development of a rotating drum biorreator for the production of cellulolytic enzymes by solid state fermentation.* PhD diss., São Paulo State University.

Grajek, Wlodzimierz. 1987. "Comparative studies on the production of cellulases by thermophilic fungi in submerged and solid state fermentation." *Applied Microbiology and Biotechnology* 26:126–129.

Graminha, Erika B.N., Gonçalves, Aline Z.L., Pirota, Rosangela D.P.B., Balsalobre, Marcos A.A., Da Silva, Roberto, and Gomes, Eleni. 2008. "Enzyme production by solid-state fermentation: application to animal nutrition." *Animal Feed Science Technology* 144:1–22.

Gusakov, Alexander V., Salanovich, Tatyana N., Antonov, Alexey I., Ustinov, Boris B., Okunev, Oleg N., Burlingame, Richard, Emalfarb, Mark, Baez, Marco, and Sinitsyn, Arkady P. 2007. "Design of highly efficient cellulase mixtures for enzymatic hydrolysis of cellulose." *Biotechnology and Bioengineering* 97:1028–1038.

Gutarra, Melissa L.E., Cavalcanti, Elisa D.C., Castilho, Leda R., Freire, Denise M.G., and Sant'anna Jr, Geraldo L. 2005. "Lipase production by solid-state fermentation: cultivation conditions and operation of tray and packed-bed bioreactors." *Applied Biochemistry and Biotechnology* 121:105–116.

Himmel, Michael E., Ding, Shi-You, Johnson, David K., Adney, William S., Nimlos, Mark R., Brady, John W., and Foust, Thomas D. 2007. "Biomass recalcitrance: Engineering plants and enzymes for biofuels production." *Science* 316:982–982.

Kachlishvili, Eva, Penninckx, Michael J., Tsiklauri, Nino, and Elisashvili, Vladimir. 2006. "Effect of nitrogen source on lignocellulolytic enzyme production by white-rot basidiomycetes under solid-state cultivation." *World Journal of Microbiology and Biotechnology* 22:391–397.

Kalogeris, Emanuel, Christakopoulos, Paul, Katapodis, Petros, Alexiou Alexios, Vlachou, Styliani, Kekos, Dimitris, and Macris Basil J. 2003. "Production and characterization of cellulolytic enzymes from the thermophilic fungus *Thermoascus aurantiacus* under solid state cultivation of agricultural wastes." *Process Biochemistry* 38:1099–1104.

Kang, Seong W., Park, Young S., Lee, Jin S., Hong, Suk I., and Kim, Seung W. 2004. "Production of cellulases and hemicellulases by *Aspergillus niger* KK2 from lignocellulosic biomass." *Bioresource Technology* 91:153–156. Accessed May 24, 2017. doi:10.1016/S0960-8524(03)00172-X.

Khanahmadi, Morteza, Roostaazad, Reza, Mitchell, David A., Miranzadeh, Mokhtar, Bozorgmehri, Ramin, and Safekordi, Aliakbar. 2006. "Bed moisture estimation by monitoring of air stream temperature rise in packed-bed solid-state fermentation." *Chemical Engineering Science* 61:5654–5663.

Kumar, Vinod, Singh, Dharmendra, Sangwan, Punesh, and Gill, Prabhjot K. 2014. "Global market scenario of industrial enzymes." In *Industrial Enzymes: Trends,*

Scope and Relevance, edited by Vikas Beniwal, and Anil K. Sharma, 173-196. New York: Nova Science Publishers.

Li, Pei-jun, Jing, De-bing, Zhou, Qi-xing, and Zhang, Chun-gui. 2004. "Optimization of solid fermentation of cellulase from *Trichoderma koningii*." *Journal of Environmental Sciences* 16:816–820.

Li, Shuang, Yang, Xiaofeng, Yang, Shuai, Zhu, Muzi, Wang Xiaoning. 2012. "Technology Prospecting on Enzymes: Application, Marketing and Engineering." *Computational and Structural Biotechnology Journal* 2:1–11.

Mazutti, Marcio A., Zabot, Giovani, Boni, Gabriela, Skovronski, Aline, OLIVEIRA, Débora, Di Luccio, Marco, Rodrigues, Maria I., Treichel, Helen, and Maugeri, Francisco. 2010. "Kinetics of inulinase production by solid-state fermentation in a packed-bed bioreactor." *Food Chemistry* 120:163–173.

Mendes, Fabrício B., Atala, Daniel I.P., and Thoméo, João C. 2017. "Is cellulose production by solid-state fermentation economically attractive for the second generation ethanol production?." *Renewable Energy* 114:525–533.

Milagres, Adriane M.F., Santos, Eder, Piovan, Thais, and Roberto, Inês C. 2004. "Production of xylanase by *Thermoascus aurantiacus* from sugar cane bagasse in an aerated growth fermentor." *Process Biochemistry* 39:1387–1391.

Mitchell, David A., Krieger, Nadia, and Berovic, Marin. 2006. *Solid-state fermentation bioreactors: fundamentals of design and operation, 1-32*. Berlin: Springer-Verlag.

Mitchell, David A., Cunha, Lara E.N., Machado, Alex V.L., Luz Jr., Luis F.L., and Krieger, Nadia. 2010. "A model-based investigation of potential advantages of multi-layer packed beds in solid-state fermentation." *Biochemical Engineering Journal* 48:195–203.

Moretti, Marcia M.S. 2010. *Isolation of thermophilic fungi producing cellulases, xylanases and ferruloyl esterase for bioconversion of sugarcane bagasse into fermentable sugars*. MSc diss., São Paulo State University.

Novaki, Lexandra. 2009. *Production, purification and partial characterization of the invertase obtained by solid state fermentation of soybean with Aspergillus casiellus*. MSc diss., West State University of Paraná.

OECD/IEA. 2010. *Sustainable Production of Second-Generation Biofuels*. Accessed June 23. http://www.iea.org/publications/freepublications/publication/second_generation_biofuels.pdf.

Palma, Márcia B. 2003. *Xylanases production by Thermoascus aurantiacus in solid state cultivation.* PhD diss, Federal University of Santa Catarina

Panagiotou, Gianni, Kekos, Dimitris, Macris, Basil J., and Christakopoulos, Paul. 2003. "Production of cellulolytic and xylanolytic enzymes by *Fusarium oxysporum* grown on corn stover in solid state fermentation." *Industrial Crops and Products* 18:37–45.

Pandey, Ashok. 2003. "Solid-state fermentation." *Biochemical Engineering Journal* 13:81–84.

Paris, Leandro D. 2008. *Production of fungal enzymes by solid state fermentation of organic, transgenic and conventional soybeans*. MSc diss., West State University of Paraná.

Ridder, Elizabeth R., Nokes, Sue E., and Knutson, Barbara L. 1999. "Optimization of solid-state fermentation parameters for the production of xylanase by *Trichoderma longibrachiatum* on wheat bran in a forced aeration system." *American Society of Agricultural and Biological Engineers* 42:1785–1790.

Rizzatto, Márcia L., Da Silva, Roberto, Gomes, Eleni, and Alegre, Ranulfo M. 2003. "Study of the temperature variation in column reactor with different apparent densities in semi-solid fermentation for the production of pectinases." *Proceedings of XIV Sinaferm.*

Rodriguez-Jasso, Rosa M., Mussatto, Solange I., Sepúlveda, Leonardo, Agrasar, Ana T., Pastrana, Lorenzo, Aguilar, Cristóbal N., and Teixeira, José A. 2013. "Fungal fucoidanase production by solid-state fermentation in a rotating drum bioreactor using algal biomass as substrate." *Food and bioproducts processing* 91:587–594.

Santiago, Angela M., Conrado, Líbia S., Mélo, Beatriz C.A., Sousa, Carlo A.B., Oliveira, Paloma L., and Lima, Flávia C.S. 2014. "Solid State Fermentation: Fundamentals and Application." In: *Transport Phenomena and Drying of Solids and Particulate Materials*, edited by João M.P.Q Delgado, and Antonio G. Barbosa de Lima, 117–139. Cham: Springer.

Singh, Bijender. 2016. "*Myceliophthora thermophila syn. Sporotrichum thermophile*: A thermophilic mould of biotechnological potential." *Critical Reviews in Biotechnolohy* 36:59–69.

Soni, Rohit, Nazir, Asiya, Chadha, Bhupinder S., and Saini, Harvinder S. 2008. "Novel sources of fungal cellulases for efficient deinking of composite paper waste." *Bioresources* 3:234–246.

Sukumaran, Rajeev K., Singhania, Reeta R., Mathew, Gincy M., and Pandey, Ashok. 2009. "Cellulase production using biomass feed stock and its application in lignocellulose saccharification for bio-ethanol production." *Renewable Energy* 34:421–424. Accessed May 17, 2017. doi:10.1016/j.renene.2008.05.008.

Suryanarayan, Shrikumar. 2003. "Current industrial practice in solid state fermentations for secondary metabolite production: the Biocon India experience." *Biochemical Engineering Journal* 13:189– 195.

Umsza-Guez, Marcelo A. 2009. *Production of polygalacturonase in solid state fermentation by the fungus Thermomucor indicae-seudaticae N31 in bottle scale and fixed bed bioreactor*. PhD diss., São Paulo State University.

Zanelato, Alex I. 2011. *Production of cellulolytic enzymes by solid state fermentation in fixed bed bioreactors*. MSc diss., São Paulo State University.

Zanelato, Alex I., Shiota, Viviane M., Gomes, Eleni, Da Silva, Roberto, and Thoméo, João C. 2012. "Endoglucanase production with the newly isolated *Myceliophthora sp.* I-1D3b in a packed bed solid state fermentor." *Brazilian Journal of Microbiology* 43:1536–1544.

Zhang, Percival Y.H., Himmel, Michael E., and Mielenz, Jonathan R. 2006. "Outlook for cellulase improvement: screening and selection strategies." *Biotechnology Advances* 24:452–481. Accessed May 10, 2017. doi:10.1016/j.biotechadv.2006.03.003.

In: Bioethanol and Beyond
Editor: Michel Brienzo
ISBN: 978-1-53613-478-0

Chapter 12

THE USES OF BIOETHANOL IN CHEMICAL LOOPING PROCESSES

F. García-Labiano*, E. García-Díez, L. F. de Diego, A. Abad, P. Gayán and J. Adánez
Instituto de Carboquímica (ICB-CSIC), Zaragoza, Spain

ABSTRACT

The concentration of CO_2 in the atmosphere has greatly increased over the last decades, very recently exceeding 400 ppm as a consequence of the anthropogenic CO_2 emissions. The use of new alternatives to reduce these emissions is therefore highly encouraged. BioEnergy and Carbon Capture and Storage (BECCS) is emerging as an interesting option for the removal of CO_2 from the atmosphere, and is seen as one of the best alternatives for meeting the target set out by the United Nations Framework Convention on Climate Change (UNFCCC) in the Paris Agreement: to limit the increase in the average world temperature to 2ºC above pre-industrial levels.

In this context, Chemical Looping (CL) is considered to be one of the most promising CO_2 capture technology due to its low economic and energy costs. The process is based on the use of an oxygen carrier to transfer the oxygen from air to the fuel, preventing the contact between them. The large-scale deployment of CL technologies could present an important advantage for future bioethanol uses.

This chapter describes the current status of bioethanol use in two processes: Chemical Looping Combustion (CLC) for energy production (heat & electricity), and Chemical Looping Reforming (CLR) for syngas/H_2 generation. The chapter includes the basic concepts of the technology with a focus on the use of bioethanol and potential oxygen carriers, and operational experience in continuous units.

Special interest is given to the use of bioethanol in autothermal-CLR processes, where H_2 can be obtained as a final product without the need for external heat input, and with high CO_2 capture efficiencies and producing a pure N_2 stream. In addition, the

* Corresponding Author Email: glabiano@icb.csic.es.

possibility of using low-grade bioethanol in the process implies important savings in energy and economic costs during the bioethanol production process.

Keywords: bioethanol, chemical looping combustion, chemical looping reforming, CO_2 capture, H_2 production, BECCS

1. Introduction

Climate is being affected by the emission of greenhouse gases, mainly carbon dioxide, into the atmosphere. The concentration of CO_2 in the atmosphere has been increasing continuously, evidenced by the fact that the world's most important CO_2 monitoring station at Mauna Loa very recently recorded short term CO_2 concentrations above 400 ppm, the highest levels found on earth in millions of years. According to the 5th Assessment Report of the International Panel on Climate Change (IPCC 2014), anthropogenic CO_2 emissions were responsible for much of these emissions, mainly resulting from the combustion of fossil fuels. Therefore, new solutions are urgently needed to enforce CO_2 emission reduction. In this sense, the Paris Agreement of 2015 set the target for limiting the global average temperature increase to 2ºC above pre-industrial levels and pursuing efforts to limit it to 1.5ºC (UNFCC 2015). This means a massive reduction in the amount of CO_2 emissions is required, from the current 43 Gt $CO_{2,e}$ per year, to reach negative emissions by the end of this century (IPCC 2014).

Among the portfolio of emission reduction strategies (use of renewable and nuclear energies, improvements in efficiency, a shift from coal to natural gas, etc.), carbon capture and storage (CCS) has been identified as an essential technology to meet the required reduction in emissions (Rubin et al. 2012). This option fits mainly with the scenario of large power plants and industrial processes. However, it must be also considered that diffuse emissions derived from transport represent an important part (≈21%) of the global CO_2 emissions (IEA 2013). The use of renewable fuels, such as bioethanol and biodiesel, can be of great utility in this field.

It is imperative that the development of new technologies for energy and/or fuel for transport allows these CO_2 emissions to be reduced. Innovative technologies for CO_2 capture in large power plants have emerged in recent years. Several studies comparing different potential combinations of CCS technologies found chemical looping (CL) to be a highly competitive process, with the highest cost reduction benefit from among all available options (Fan et al. 2012) (Bhave et al. 2017).

In recent years, CL processes have been thoroughly investigated with gaseous and solid fuels (Adánez et al. 2012) (Adánez et al. 2017) (Fan 2010) (Lyngfelt and Mattison 2011), although there is now growing interest in the use of liquid fuels. Most of the experience gained with regard to liquid fuels in CL processes has been related to the use of fossil fuels (Moldenhauer et al. 2014) (Moldenhauer et al. 2017) (García-Díez et al. 2017a) (García-Díez et al. 2017b) (García-Labiano et al. 2017) (Serrano et al. 2017); however, additional advantages can be obtained if the liquid fuel comes from renewable sources. Among the possible options, bioethanol has significant advantages:

- it is the most widely produced bio-fuel in the world and this trend is expected to continue in the near future
- it is readily available in many countries
- it is easily transportable and non-toxic
- given its biomasic origin, its use results in net atmospheric carbon reduction

Furthermore, the integration of bioenergy with CCS, defined as BECCS, is becoming popular, since it could result in greatly reduced CO_2 emissions and a "carbon-negative" power system (Gough and Upham 2011) (Sánchez et al. 2015. In fact, negative emissions are required at significant levels (>1 Gt C per year) to meet the 2DS target established in the Paris Agreement (Gasser et al. 2015). This implies that the use of renewable fuels, such as bioethanol, in CL processes may be a very promising BECCS technology; see Figure 1.

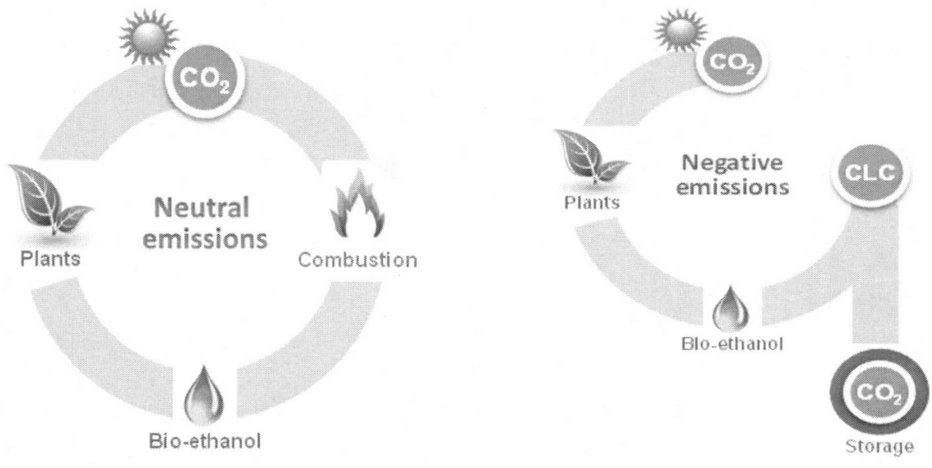

Figure 1. CO_2 balance in the atmosphere without and with BECCS technologies.

2. Ethanol Reactions

Figure 2 shows the gas composition obtained in thermodynamic equilibrium as a function of the amount of oxygen reacted per mol of ethanol. Equilibrium data was obtained with HSC Chemistry 6.1 software (HSC Chemistry 2008), which uses the Gibbs free energy minimization in the system.

Oxygen to ethanol molar ratios above 6 lead to the complete fuel combustion to CO_2 and H_2O through the reaction (R.1).

$$C_2H_5OH_{(l)} + 3\ O_{2(g)} \rightarrow 2\ CO_{2\ (g)} + 3\ H_2O_{(g)}\ \Delta H_c^{\circ}=-1235.5\ kJ/mol \quad (R.1)$$

Otherwise, if the ratio of reacted oxygen ranges between 1 and 6, the reactions that can take place are partial oxidation (R.2), reforming reactions – both steam (R.3) and dry (R.4) – and water gas shift (R.5). The final product is a syngas with a specific composition based on the amount of reacted oxygen.

$$C_2H_5OH_{(l)} + 1/2\ O_2(g) \rightarrow 2\ CO + 3\ H_2 \quad (R.2)$$

$$C_2H_5OH_{(l)} + H_2O \rightarrow 2\ CO + 4\ H_2 \quad (R.3)$$

$$C_2H_5OH_{(l)} + CO_2 \rightarrow 3\ CO + 3\ H_2 \quad (R.4)$$

$$CO + H_2O \leftrightarrow CO_2 + H_2 \quad (R.5)$$

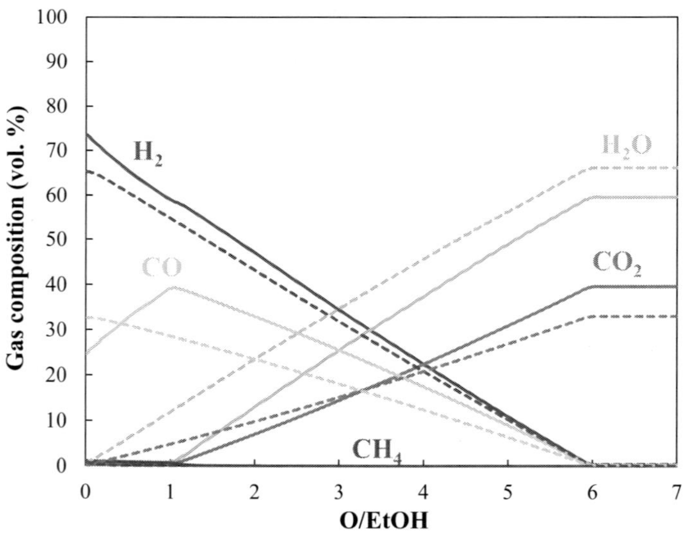

Figure 2. Gas composition at equilibrium during reaction of ethanol and oxygen without H_2O addition (continuous lines) or with H_2O addition (discontinuous lines).

If the oxygen reacted per mol of ethanol is less than 1 and no water is added (continuous lines), there will not be enough oxygen for the partial oxidation, and therefore the decomposition of the fuel will take place, with CH_4 as a possible intermediate compound.

$$C_2H_5OH \rightarrow (CH_4) \rightarrow 3\ H_2 + CO + C \quad (R.6)$$

The addition of a gasifying agent together with the ethanol prevents carbon presence by means of gasification reactions and at the same time reinforces the reforming reactions (R.3) and (R.4); see discontinuous lines in Figure 2.

$$C + H_2O \rightarrow CO + H_2 \quad \text{(R.7)}$$

$$C + CO_2 \rightarrow 2\ CO \quad \text{(R.8)}$$

Therefore, bioethanol can be used to produce a combination of heat and electricity during combustion, or syngas for different uses, including hydrogen production, during partial oxidation or reforming processes. Gaseous oxygen present in air tends to be used in conventional processes. However, if a high purity stream is required, an air separation unit (ASU) becomes necessary to produce the oxygen, which leads to a significant increase in process costs. In contrast, CL processes use lattice oxygen contained in an oxygen carrier for the reactions instead of gaseous oxygen, which provides a relevant advantage since no costly separation steps are necessary.

3. Chemical Looping Combustion

The idea underlying any process for CO_2 capture during combustion is the achievement of a CO_2 stream as pure as possible at the lowest cost. In this sense, chemical looping combustion (CLC) is a two-step combustion process that produces a pure CO_2 stream that is ready for compression, transport and storage. CLC technology is based on the transfer of the oxygen from air to the fuel by means of a solid oxygen carrier, which prevents direct contact between fuel and air. Figure 3 is a schematic diagram of this process for the specific case where bioethanol is used as fuel.

In a first step, the fuel is oxidized to CO_2 and H_2O by a metal oxide (M_xO_y) in the fuel reactor (R.9). The resulting gas primarily contains CO_2 and H_2O. After water condensation and purification, a highly concentrated stream of CO_2 is achieved that is ready for transport and storage. This concept is the main advantage of the CLC process in relation with other CO_2 capture technologies. In this sense, CLC is a combustion process with inherent CO_2 separation, i.e., doing away with the need for CO_2 separation units and with a very low energy penalty.

The reduced metal oxide (M_xO_{y-1}) is sent to the air reactor, where it is oxidized with air in a second step (R.10), and the regenerated material is then ready to start a new cycle. The flue gas obtained at the air reactor outlet contains N_2 and unreacted O_2.

$$C_2H_5OH_{(l)} + 6\ M_xO_y \rightarrow 6\ M_xO_{y-1} + 2\ CO_{2\,(g)} + 3\ H_2O_{(g)} \quad \Delta H_r \quad \text{(R.9)}$$

$$6\ M_xO_{y-1} + 3\ O_2 \rightarrow 6\ M_xO_y \qquad \Delta H_o \qquad (R.10)$$

$$C_2H_5OH_{(l)} + 3\ O_{2(g)} \rightarrow 2\ CO_{2\ (g)} + 3\ H_2O_{(g)} \qquad \Delta H_r + \Delta H_o = \Delta H_c^\circ \qquad (R.11)$$

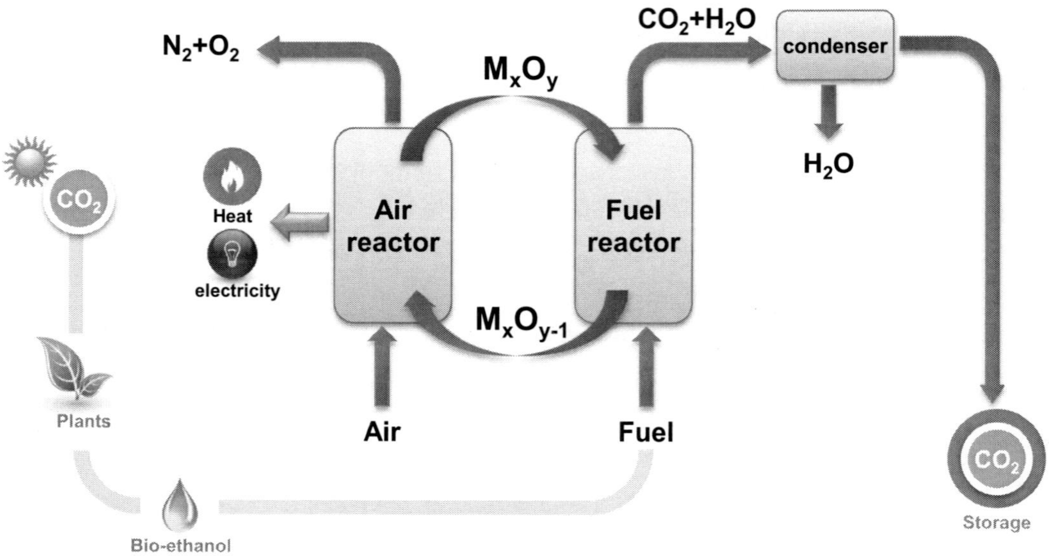

Figure 3. CLC process using bioethanol as fuel.

The net chemical reaction over the two steps (R.11), and therefore the combustion enthalpy, is the same as that of a conventional combustion (R.1) where the fuel is burned in direct contact with oxygen from air. Therefore, the total amount of heat evolved in the CLC process is the same as that of conventional combustion.

In summary, the CLC concept is based on the transport of oxygen from the air to the fuel by means of an oxygen carrier. Consequently, the solids circulation rate between reactors must be high enough to transfer the oxygen necessary for fuel combustion and the heat necessary to maintain the heat balance in the system, if necessary.

3.1. Mass Balances

The key aspect in any CLC process is the supply of oxygen from the air to the fuel by means of an oxygen carrier (Abad et al. 2007). The amount of oxygen supplied by the oxygen carrier to the fuel in a CLC process is evaluated by means of the oxygen carrier-to-fuel ratio, ϕ. This parameter is defined as the ratio between the oxygen available in the oxygen carrier and the oxygen required to stoichiometrically react with the fuel flow.

$$\phi = \frac{F_{MeO}}{b_r \cdot F_{EtOH}} \tag{1}$$

where F_{MeO} is the molar flow rate of metal oxide (M_xO_y) and F_{EtOH} is the EtOH inlet molar flow rate in the FR. b_r is the stoichiometric value for the combustion reaction of ethanol and the oxygen carrier and takes the value of 6 (see R.9). From a practical point of view, $\phi = 1$ means that the oxygen supplied by the oxygen carrier is the exact amount of stoichiometric oxygen needed to fully convert the fuel to CO_2 and H_2O. It should be taken into consideration that ϕ is related to the variation of the solid conversion in the fuel and air reactors, ΔX_S:

$$\Delta X_S = X_{o,AR} - X_{o,FR} = \frac{\eta_c}{\phi} \tag{2}$$

where η_c is combustion efficiency, and $X_{o,AR}$ and $X_{o,FR}$ are the solid oxidation state at the outlet of the air and fuel reactors, respectively.

In addition, the circulation rate required to fulfill the oxygen mass balance in the CLC system depends on the type of oxygen carrier and fuel used, as well as on the conversion variation obtained in the oxygen carrier in the fuel and air reactors, ΔX_S. Expressed as the mass flow of oxygen carrier totally oxidized per MW of fuel, the solids circulation rate is defined by the following equation:

$$\dot{m}_{OC} = \frac{b_r \ M_{MeO}}{x_{MeO} \ \Delta H_c^0 \ \Delta X_S} \tag{3}$$

where M_{MeO} is the molecular weight of the metal oxide, x_{MeO} the mass fraction of metallic oxide in the fully oxidized oxygen carrier, and ΔH_c^0 the standard heat of combustion of the fuel.

Thus, considering the combustion heat of ethanol, 1235.5 kJ/mol, a combustion of 0.81 mol EtOH/s and transport of 4.85 mol O/s is required to obtain 1 MW_{th}. However, this transport will depend on the type of oxygen carrier used. Account should be taken of the fact that oxygen carriers contain metal oxides combined with an inert. Thus, the oxygen transport capacity of the material, R_{OC}, depends on the oxygen transport capability of the redox pair considered, R_O, and the fraction of the active compound in the oxygen carrier, x_{MeO}. Adánez et al. (2012) compiled the values of R_O for the main redox pair considered in the literature.

The solids circulation rate necessary to fulfil the mass balance increases as the oxygen transport capacity of the oxygen-carrier decreases, either owing to a lower R_O value for the redox pair or to a lower fraction of the active compound in the oxygen-carrier, x_{MeO}. Figure 4 shows the circulation rates, $\dot{m}_{oc}$, necessary for the combustion of ethanol for the different redox pairs, as a function of the conversion difference, ΔX_S, obtained during operation. In this case, a 100% metal oxide content was considered for the calculations. Real solids circulation rates will consider the metal oxide content of the oxygen carrier, which usually takes values of between 15 and 30 wt%. It should be remembered that ΔX_S is linked to the ϕ value by means of equation (2).

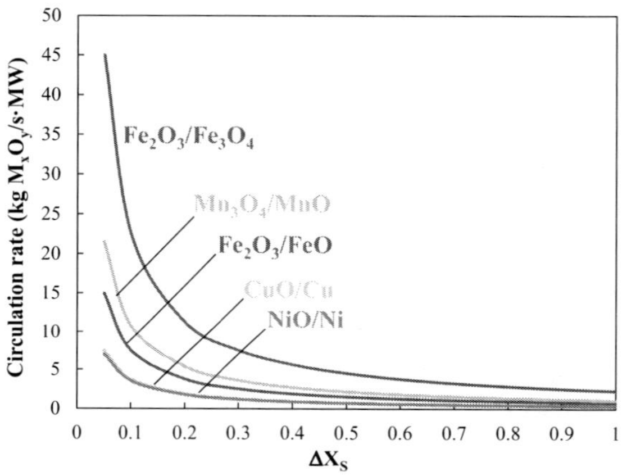

Redox pair	Ro
CuO/Cu	0.20
$CuAl_2O_4/Cu.Al_2O_3$	0.07
NiO/Ni	0.21
$NiAl_2O_4/Ni.Al_2O_3$	0.09
Fe_2O_3 / Fe_3O_4	0.033
Fe_2O_3 / FeO	0.10
$Fe_2O_3.Al_2O_3/FeAl_2O_4$	0.10
Mn_3O_4 /MnO	0.07

Figure 4. Solids circulation rate for different metal oxides in CLC using bioethanol as fuel.

Obviously, those materials with lower oxygen transport capacity need higher solid circulation rates. From a practical point of view, circulation rates should also consider other aspects related with the hydrodynamic behaviour, especially in the riser, and with the heat balance in the whole system. The solids circulation rate in a circulating fluidized bed (CFB) system depends on the operating conditions and configuration of the riser. Abad et al. (2007) selected 16 kg/s MW_{th} as the maximum circulation rate feasible in a CLC plant at atmospheric pressure without increased costs according to commercial experience. Assuming this value, oxygen carriers with oxygen transport capacity values lower than ≈ 0.4% could not be used for CLC because it would not be possible to transfer the required oxygen to fully convert the fuel to CO_2 and H_2O. Thus, materials containing 2 wt% of NiO or CuO may be enough given their high transport capacity. However, higher metal oxide contents are required for Mn_3O_4 (>6 wt%) or Fe_2O_3 (>12 wt%) owing to their lower transport capacity. In contrast, low circulation rates lead to high ΔX_S values, which imply strong chemical stress inside the particles as a consequence of the changes taking place in their internal structure, reducing the lifetime of the materials.

3.2. Heat Balances

As mentioned previously, the overall heat release taking place between the two reactors of a CLC system is the same as that of conventional combustion (R.11). However, the distribution between the fuel and air reactors depends both on the fuel and the oxygen carrier. The standard heat of reaction for the reduction and oxidation reactions of different fuels in relation to the main oxygen carriers considered in literature can be found in the work by Adánez et al. (2012). Figure 5 shows the reduction (ΔH_r) and oxidation (ΔH_o) enthalpies for different redox systems during ethanol combustion in CLC. The heat corresponding to the combustion of 1 mol of ethanol is included as a discontinuous line.

It can be observed that oxidation reaction is always exothermic, with subsequent heat release. However, the reduction reaction can be endothermic or exothermic depending on the redox system considered. This heat distribution can vary even for the same metal-based material depending on the support material used. In fact, the formation of new compounds by reaction between the metal oxide and the support can result in aluminates or titanates formation that affect both oxygen transport capacity and heat distribution.

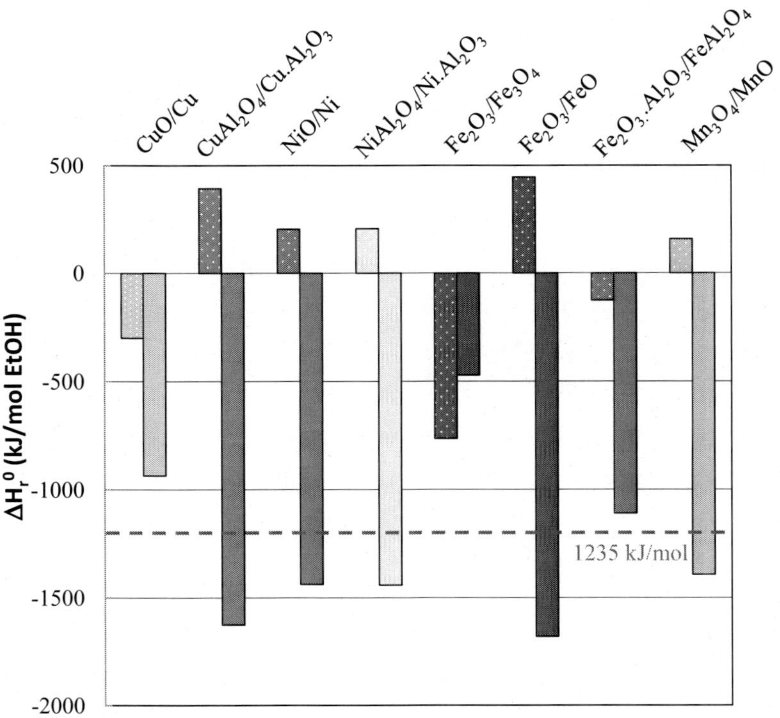

Figure 5. Standard heat for the reduction (left) and oxidation (right) reactions of the main oxygen carriers using ethanol as fuel. Basis: 1 mol ethanol. ΔH_c^o=1235 kJ/mol.

To analyse the effect of heat distribution on a CLC system, the mass and heat balances have been calculated for the different redox pairs considered. Figure 6 shows data regarding the temperature in the air reactor and the heat extracted from the system as a function of the solids circulation rate (or ΔX_S) for different oxygen carriers. Three different redox pairs were selected according to their different oxygen transport capacity and the heat distribution between the reduction and oxidation reactors: 1) endothermic reduction reaction, redox pair NiO/Ni, 2) exothermic reduction reaction, redox pairs CuO/Cu and Fe_2O_3/Fe_3O_4; however, the largest heat release occurs in the oxidation reaction for the pair CuO/Cu, while most of the heat for the pair Fe_2O_3/Fe_3O_4 is released in the reduction reactor (see Figure 5).

Some of the assumptions considered for the mass and heat balances were as follows:

- The reference of the power plant using anhydrous bioethanol is 1 MW_{th}.
- The temperature in the fuel reactor is constant at 900 °C.
- The reagents enter at ambient temperature, and the products exit at the temperatures of the respective reactors.
- The active phase for every oxygen carrier considered is 20 wt%.
- The system is considered adiabatic.

Two different trends can be observed depending on the heat distribution in relation to the redox pair considered. If the reduction reaction is endothermic (redox pair NiO/Ni), the circulating solids coming from the air reactor at higher temperature are responsible for providing the heat necessary for ethanol combustion in the fuel reactor. Therefore, air reactor temperature has to be higher than 900 °C, which will depend on the solids circulation rate and, therefore, on the variation of conversion, ΔX_S. In addition, it is necessary to extract heat from the system to prevent the temperature from exceeding the operational limits, which could lead to sintering/agglomeration taking place in the oxygen carriers in some cases.

In contrast, if both oxidation and reduction reactions are exothermic, the temperature in the air reactor decreases with respect to the fuel reactor temperature: the more exothermic the reduction reaction, the higher that decrease (Fe_2O_3/Fe_3O_4 versus CuO/Cu). In both cases, the solids circulation rate is not limited by the heat balance, and the minimum solids circulation rate should be selected according to the mass balance.

It is also observed that the higher the air reactor temperature, the lower the heat extracted from the system. However, it should be considered that the latent heat of the gaseous outlet streams from the fuel and air reactors can also be used for heat production.

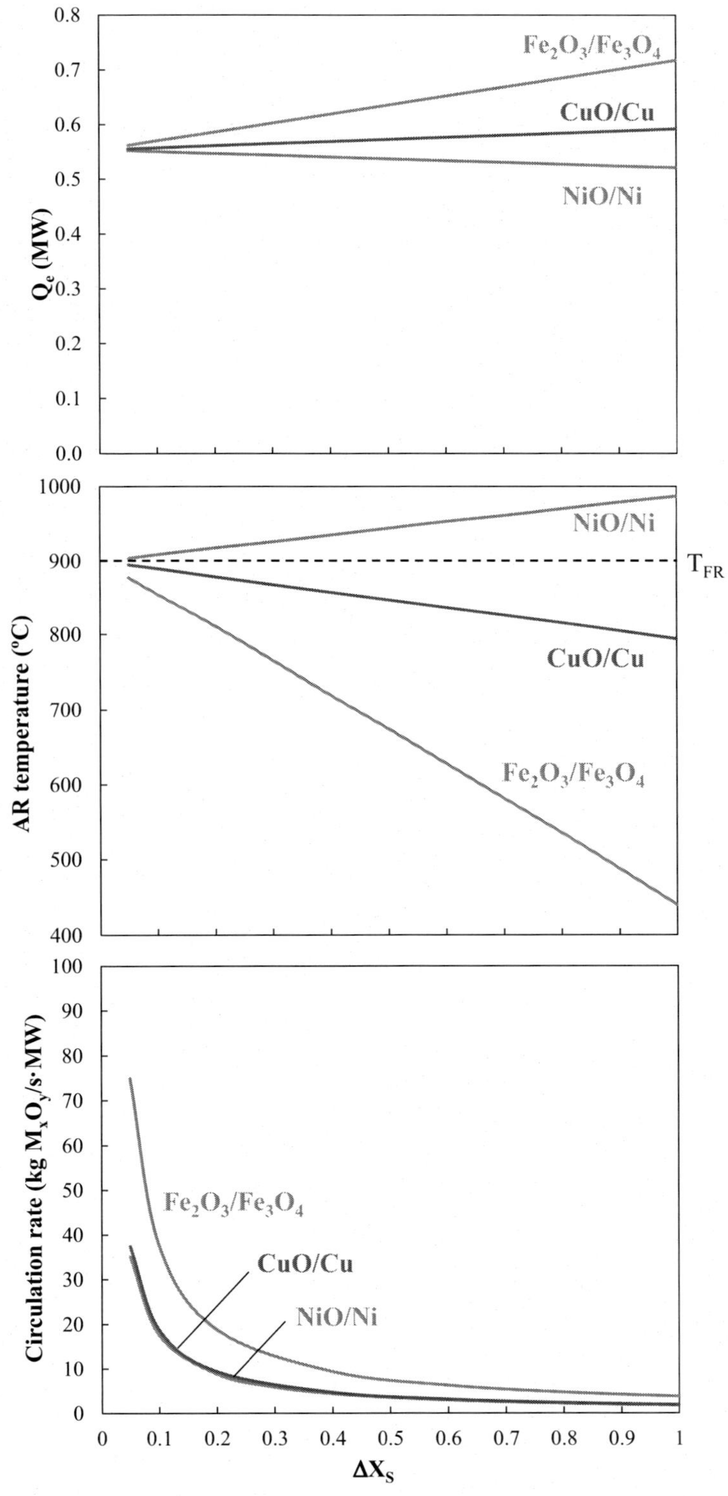

Figure 6. Heat extracted from the system and temperature in the air reactor as a function of the solids circulation rate for selected oxygen carriers.

3.3. Experience with Bioethanol CLC

There is relatively little experience in the use of bioethanol in the CLC process in continuous units. Most of this experience corresponds to the work by de Diego et al. (2016), who presented the experimental results obtained in a 1 kW_{th} CLC unit using ethanol as fuel for more than 100 h of continuous operation. Figure 7 is a schematic diagram of the unit, which comprised two interconnected fluidized bed reactors separated by a fluidized loop-seal; a riser to transport the solids from the air reactor, AR, to the fuel reactor, FR; a cyclone; and two solids valves to measure and fit the solids circulation flow rate in the whole system. Ethanol and water feeds were controlled by peristaltic pumps. The fuel flow was fully evaporated in a furnace before it reached the FR. N_2 and air flows were controlled by specific mass flow controllers.

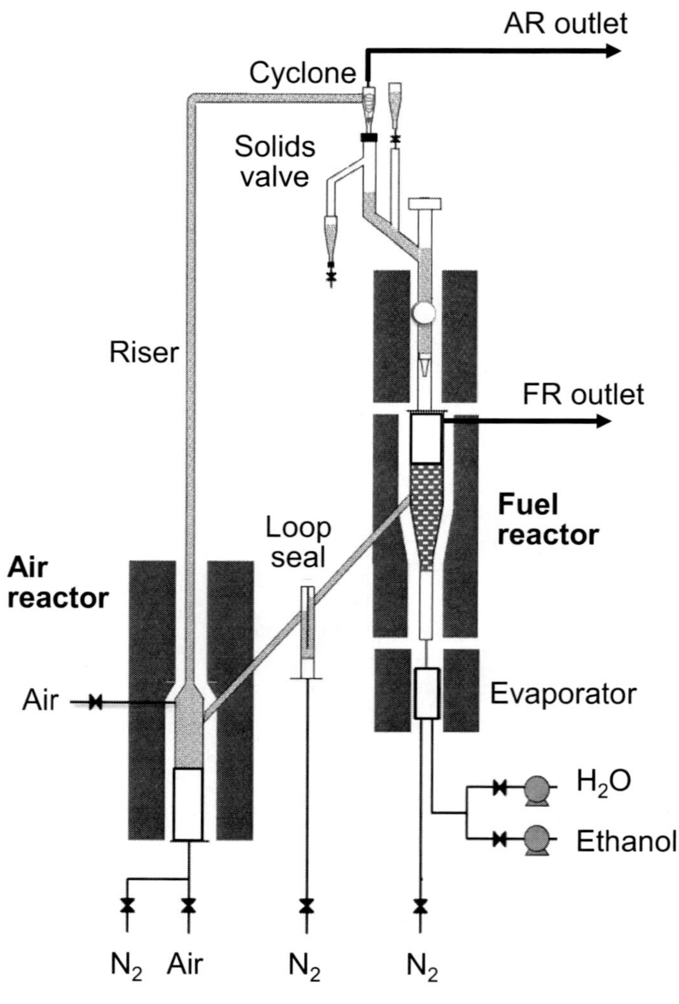

Figure 7. Schematic diagram of the 1 kW_{th} CLC unit.

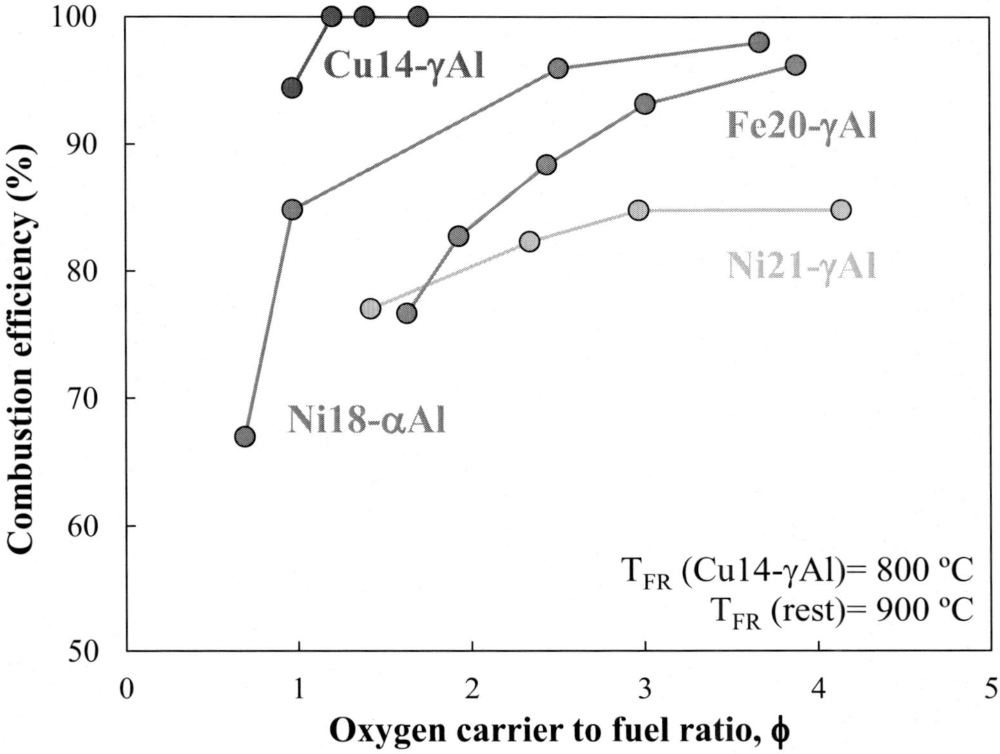

Figure 8. Combustion efficiency obtained for different oxygen carriers using ethanol in a 1 kW_{th} CLC unit.

These authors demonstrated the viability of the use of ethanol in the CLC process with different oxygen carriers prepared by impregnation on γ- and α-Al_2O_3: Cu14–γAl, Fe20–γAl, Ni21–γAl and Ni18–αAl. These last two materials have been patented by CSIC (CSIC Patent 2009a) (CSIC Patent 2009b). Figure 8 shows the combustion efficiency found for the different oxygen carriers as a function of the oxygen carrier-to-fuel ratio, ϕ.

The suitability of the use of the oxygen carriers was finally decided by taking into account oxygen transport capacity and hydrodynamic criteria. Cu14–γAl and Ni18–αAl met all the requirements to successfully conduct a CLC process and almost 100% CO_2 capture efficiency was obtained. On the contrary, the Ni21–γAl oxygen carrier was discarded due to the low reactivity of the spinel $NiAl_2O_4$ under normal operating conditions. Iron-based materials are suitable for the CLC process, although higher Fe-content or higher operating temperatures would be necessary to make use of the Fe20–γAl material.

Additional tests carried out in the CLC unit feeding ethanol and water, simulating bioethanol with different purities, did not show any water effect either on combustion efficiency or on the CO_2 capture process. However, water presence will affect negatively the energy balance in the entire system.

4. Chemical Looping Reforming

The most common processes for obtaining syngas/H_2 from bioethanol are steam reforming (SR), partial oxidation (POX) and autothermal reforming (ATR) (Ahmed and Krumpelt 2001).

SR is probably the most common method for producing hydrogen in the chemical process industry. However, this process needs external power supply for the endothermic reactions and CO_2 emissions can be generated in the process.

POX has the advantage of reaction exothermicity (R.2) and consequently, external energy input is not required. However, the efficiency of this process for hydrogen production is low (Krumpelt et al. 2002). Consideration should also be given to the fact that the process needs pure oxygen if high-quality syngas is required, which increases its cost.

ATR is a combination of both SR and POX processes. This process involves reacting bioethanol with steam/CO_2 and oxygen, leading to both the fuel reforming (R.3 and R.4) and partial oxidation (R.2). The energy required for the first reaction is provided by the second one, and adiabatic conditions are achieved without external energy input. This process presents a high efficiency in the H_2 production process without the need for large reactors and without the use of high pressure gradients (Farrauto et al. 2003), although the use of pure oxygen increases the economic cost of the process. Autothermal conditions are usually reached at oxygen to ethanol ratios of 0.3–0.45 and water to ethanol ratios of 2.1–2.4.

In all these cases, the incorporation of CCS technologies to the use of renewable bioethanol fuel in the hydrogen production process offers the possibility of achieving negative CO_2 emissions.

In contrast to the above process, autothermal chemical looping reforming (a-CLR) allows hydrogen production with CO_2 capture at a lower cost. a-CLR is a similar process to ATR, but instead the use of expensive gaseous oxygen, the high-quality syngas is produced by the lattice oxygen provided by a solid oxygen carrier, which decreases the economic costs derived from the use of an ASU plant. In fact, theoretical calculations made by Cormos (2014) comparing the use of different technologies for CO_2 capture during the hydrogen production process at industrial scale (300 MW_{th}) and using bioethanol as fuel concluded that the a-CLR process exhibited highly energy efficiencies including 100% CO_2 capture.

The scheme of a CLR process is similar to that previously shown for CLC, since both are based on the transfer of oxygen from air to the fuel by means of a solid oxygen carrier. However, the oxygen to fuel ratio in a CLR process is kept low to prevent the complete fuel oxidation to CO_2 and H_2O.

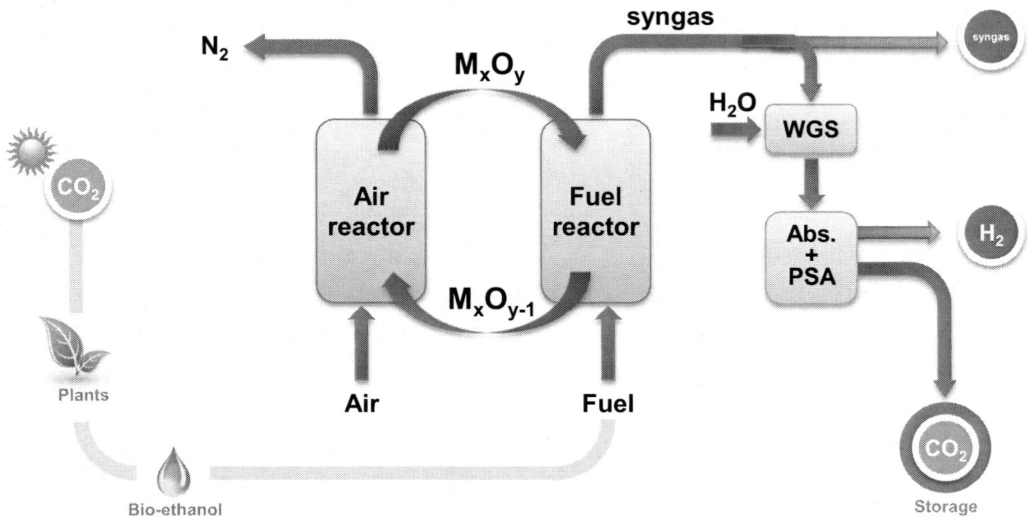

Figure 9. CLR process using bioethanol as fuel.

Figure 9 is a schematic diagram of a typical CLR process using bioethanol as fuel. A stream of N_2-free gas, i.e., a syngas of highly concentrated H_2 and CO, is obtained at the fuel reactor outlet. Moreover, the ASU required by conventional ATR is omitted here. In addition, a higher H_2 yield is obtained when a water-gas shift (WGS) reactor is used downstream of the CLR system, but with the advantage that the CO_2 capture is achieved in the overall process with no need for additional energy from an external source. A purification system comprising an absorption step for CO_2 separation is subsequently used, followed by a PSA (Pressure Swing Adsorption) process, to obtain a high purity hydrogen stream. PSA-off gas can be sent to the fuel reactor to improve the process yield.

One of the key points to be considered in any CL process is the selection of suitable oxygen carriers. The steam reforming reaction is quite relevant to the whole CLR process. Therefore, oxygen carriers that work to catalyse this reaction would be preferred. In this regard, Ni-based oxygen carriers can be considered a good choice for any CLR process owing to the presence of metallic Ni in the fuel reactor (Adánez et al. 2012).

Ni-based oxygen-carriers are one of the most extensively analysed materials in CL processes. They are highly reactive and well behaved at high temperatures up to 1100 °C, and they exhibit the longest lifetime among the materials tested. Working with CH_4 as fuel, Ni-based oxygen-carriers achieve almost complete conversion and only small concentrations of CO and H_2 (<1%) are detected in the FR gas outlet stream as the result of thermodynamic restrictions. The main drawbacks are those concerning cost, environmental and operational health and safety, and resistance to the presence of sulphur in the fuel (Adánez et al. 2012) (Dueso et al. 2012). However, the latter is not a problem when using bioethanol as fuel.

4.1. Control of the Oxygen Used in the System

The control of the oxygen available for reaction in the FR is very important in a CL process. The solids circulation rate between the reactors should be high enough to transfer the oxygen necessary for the desired reactions. In CLC, control of the oxygen fed into the fuel reactor normally involves controlling the solids circulation rate in the system for a given oxygen carrier. An excess of oxygen over the stoichiometric requirement is necessary in order to reach high combustion efficiencies. However, two different methods can be used to control the oxygen reacted in CLR. As previously mentioned, the oxygen-to-ethanol ratio should be comprised between 1 and 6. Figure 10 gives schematic diagrams of both methods, including the gas composition obtained at the air reactor outlet.

1- By controlling the solids circulation rate between reactors, and keeping a constant oxygen flow fed into the AR, (F_S=variable, $F_{O2,AR}$=cte); see Figure 10a. This method is similar to that of a typical CLC process, with the use of low values of the oxygen to fuel molar ratio, ϕ. The lack of oxygen in the fuel reactor produces unburnt products such as H_2 and CO (de Diego et al. 2016) (García-Labiano et al. 2014). However, in order to increase the H_2 and CO content, it is necessary to reduce the solids circulation rate, which can produce hydrodynamic instabilities in the interconnected fluidized beds.

2 - By controlling the oxygen flow fed into the AR and keeping the solids circulation rate constant (F_S=cte, $F_{O2,AR}$=variable); see Figure 10b. This method has important consequences from an industrial point of view because it allows easy operation, accurate control of the oxygen reacted in the FR, and particularly, a pure N_2 stream can be obtained at the AR outlet (García-Labiano et al. 2015) (García-Díez et al. 2016).

This last method allows operation with optimum solids circulation rates, averting hydrodynamic instabilities in the system. The method is based on an oxygen mass balance:

$$O_{captured,AR} = O_{transferred,FR-AR} = O_{reacted,FR} \tag{4}$$

The two terms on the left represent the oxygen mass balance in the air reactor, in such a way that all the oxygen fed into the air reactor reacts with the reduced oxygen carried coming from the fuel reactor.

$$F_{O_2,in} \Delta X_{O_2} = \frac{F_S x_{NiO,AR}}{M_{NiO}} \Delta X_S \tag{5}$$

where F_S is the solids circulation rate, $x_{NiO,AR}$ the NiO content at the air reactor outlet, M_{NiO} the molecular weight of NiO, and ΔX_{O2} the gaseous oxygen conversion in the air reactor. This parameter, defined in equation (6), takes the value of 1 when using this control method for oxygen and when no carbon enters from the fuel to the air reactor.

$$\Delta X_{O_2} = \frac{F_{O_2,in} - F_{O_2,out} - F_{CO_2,out}}{F_{O_2,in}} \tag{6}$$

where $F_{O2,in}$ and $F_{O2,out}$ are the molar flows of oxygen of the air reactor inlet and outlet, respectively, and $F_{CO2,out}$ the molar flow of CO_2 at the air reactor outlet.

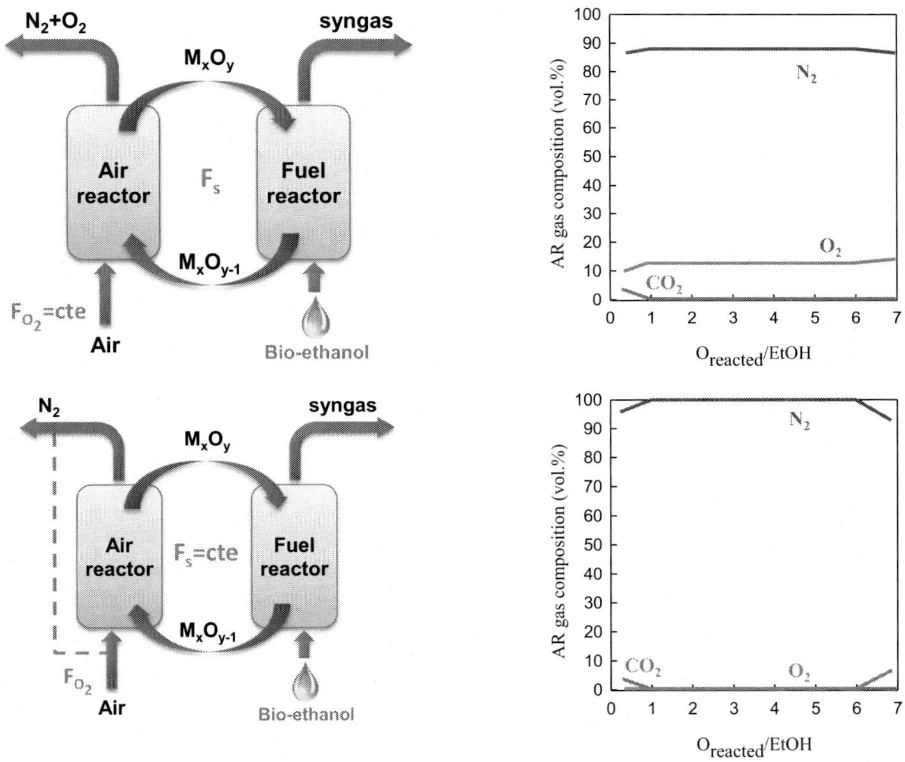

Figure 10. Different operational methods to control the oxygen reacted with the fuel.

The second part of the equality in equation (4) indicates that all the oxygen transferred by the oxygen carrier reacts with the bioethanol in the fuel reactor.

$$\frac{F_S x_{NiO,AR}}{M_{NiO}} \Delta X_S = F_{EtOH} \cdot (O_{NiO} / EtOH) \tag{7}$$

where F_{EtOH} is the molar flow of ethanol fed into the fuel reactor. The molar ratio between the oxygen transferred by the oxygen carrier and the bioethanol, O_{NiO}/EtOH, represents the most important parameter to be considered in a CLR process.

4.2. Reactions

Table 1 compiles the main reactions taking place in the fuel and air reactors. In the FR, the direct oxidation of the ethanol with the NiO, or some other intermediate gaseous compounds, takes place to give CO_2 and H_2O as final products (R.12-R.15). These reactions are all exothermic and help to provide the heat necessary to generate the endothermic reforming reactions in an autothermal process.

Table 1. Main reactions in a CLR process using Ni-based oxygen carriers and ethanol as fuel

Reaction		
Fuel Reactor		
Oxidation		
	$C_2H_5OH + 6\ NiO \rightarrow 6\ Ni + 2\ CO_2 + 3\ H_2O$	(R.12)
	$CH_4 + 4\ NiO \rightarrow 4\ Ni + CO_2 + 2\ H_2O$	(R.13)
	$H_2 + NiO \rightarrow Ni + H_2O$	(R.14)
	$CO + NiO \rightarrow Ni + CO_2$	(R.15)
Partial oxidation		
	$C_2H_5OH + NiO \rightarrow Ni + 2\ CO + 3\ H_2$	(R.16)
Steam reforming catalysed by Ni		
	$C_2H_5OH + H_2O \rightarrow 2\ CO + 4\ H_2$	(R.17) (R.18)
	$CH_4 + H_2O \rightarrow CO + 3\ H_2$	
Dry reforming catalysed by Ni		
	$C_2H_5OH + CO_2 \rightarrow 3\ CO + 3\ H_2$	(R.19)
	$CH_4 + CO_2 \rightarrow 2\ CO + 2\ H_2$	(R.20)
Carbon chain breakage		
	$C_2H_5OH \rightarrow C_2H_4O + H_2$	(R.21)
	$C_2H_4O \rightarrow CH_4 + CO$	(R.22)
	$CH_4 \rightarrow C + H_2$	(R.23)
Carbon gasification		
	$C + H_2O \rightarrow CO + H_2$	(R.24)
	$C + CO_2 \rightarrow 2\ CO$	(R.25)
Water gas shift		
	$CO + H_2O \leftrightarrow CO_2 + H_2$	(R.26)
Air Reactor		
Oxygen carrier oxidation		
	$Ni + \frac{1}{2} O_2 \rightarrow NiO$	(R.27)
Carbon combustion		
	$C + O_2 \rightarrow CO_2$	(R.28)

Syngas can be produced via partial oxidation (R.16), steam fuel reforming (R.17-R.18) or dry fuel reforming (R.19-R.20). Ethanol can undergo decomposition to produce lighter hydrocarbons and carbon (R.21-R.23) (Fatsikostas and Verykios 2004). This carbon can be gasified inside the reactor in the presence of H_2O or CO_2 (R.24-R.25). Finally, the water gas shift reaction is considered in the gaseous phase (R.26).

The main reaction taking place in the air reactor is Ni oxidation to give NiO (R.27). The carbon formed in the fuel reactor can be transported to the air reactor where it is burnt to CO_2 (R.28), decreasing the efficiency of the CO_2 capture process.

4.3. Experience with CLR Processes Using Bioethanol

Various studies can be found in the literature on the use of bioethanol in CLR processes. Most of them correspond to discontinuous tests carried out at laboratory scale for the development or testing of new materials and to analyse certain operating conditions, including pressure (Wang and Cao 2013) (Wang 2014) (Trevisanut et al. 2015a) (Trevisanut et al. 2015b) (Jiang et al. 2016) (Wang X. et al. 2016) (Wang K. et al. 2016).

The major experience in CLR processes using ethanol as fuel corresponds to the works of García-Labiano et al. (2015) and García-Díez et al. (2017a). These authors reported more than 100 hours of continuous operation in a 1 kW_{th} continuous CLR unit using two Ni-based oxygen carriers, NiO21-γAl_2O_3 and NiO18-αAl_2O_3 (CSIC Patent 2009a) (CSIC Patent 2009b). These materials, prepared by using the impregnation method on γ- and α-Al_2O_3 (Gayán et al. 2009) (Dueso et al. 2010), were previously tested for syngas production using CH_4 as fuel with very good results (de Diego et al. 2008) (de Diego et al. 2009).

The authors García-Labiano et al. (2015) and García-Díez et al. (2017a) experimentally demonstrated the viability of the CLR process using ethanol as fuel, with perfect control of the oxygen reacted in the fuel reactor for syngas production. For this purpose, use was made of the control of the flow of oxygen fed into the fuel reactor at a fixed solids circulation rate, which allowed a pure N_2 stream to be obtained at the air reactor outlet.

They also analysed the effect of the main parameters affecting syngas production, such as the O_{NiO}/EtOH and H_2O/EtOH molar ratios, and temperature. Figure 11 offers an example of the syngas obtained as a function of the ONiO/EtOH molar ratio for two H_2O/EtOH values using the oxygen carrier NiO18-αAl_2O_3. Each set of syngas data corresponds to at least 45 min of steady state operation in the CLR unit. Lines correspond to thermodynamic equilibrium data obtained using the HSC Chemistry 6.1 software (HSC Chemistry 2008).

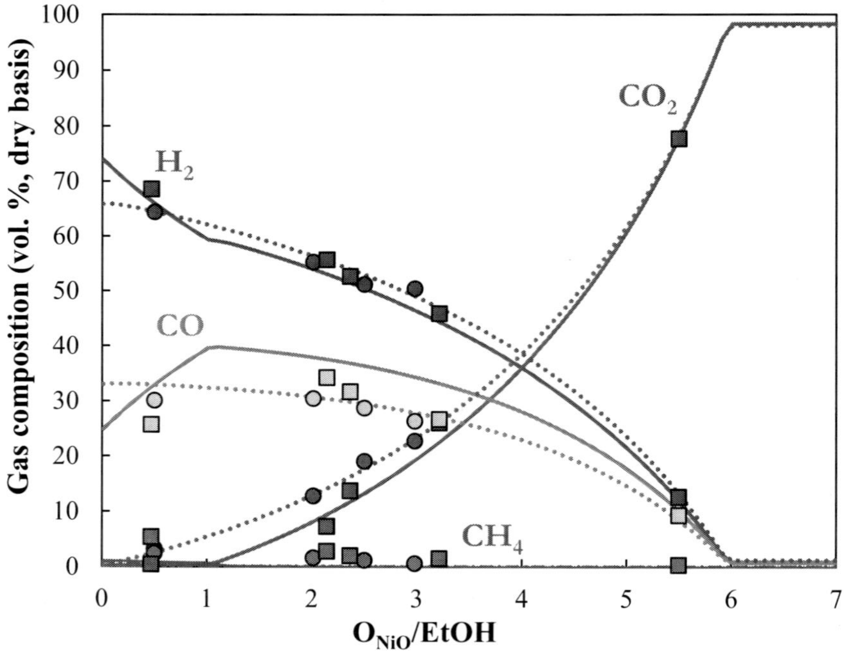

Figure 11. Gas composition obtained during tests in the 1 kW_{th} CLR unit. Experimental results (dots) and thermodynamic equilibrium data (lines). NiO18-αAl_2O_3. T = 900ºC. F_s=9 kg/h. H_2O/EtOH=0, —□ ; H_2O/EtOH=1, ······○ .

The complete conversion of the bioethanol was achieved in all experimental results. As theoretically predicted, values of O_{NiO}/EtOH equal to 6 gave CO_2 as the main product on a dry basis as a consequence of the complete oxidation of the ethanol. A small presence of H_2 and CO was detected as the result of thermodynamic restrictions that exist when Ni-based oxygen carriers are used (Adánez et al. 2012). Decreasing values of this ratio produced an increase in syngas (H_2 and CO) production, according to the more relevant partial oxidation and reforming reactions. Values of O_{NiO}/EtOH below 1 produced two different behaviours depending on water addition. When no water was fed into the fuel reactor, there was not enough oxygen for the partial oxidation reaction and ethanol decomposition took place. This fact was experimentally corroborated by the increase in H_2 concentration and the decrease in CO, as well as by the presence of CO_2 at the air reactor outlet proceeding from the carbon formed in the fuel reactor and burnt in the air reactor. In contrast, this behaviour was prevented when water was fed into the fuel reactor. In this case, the reforming reaction of the ethanol prevented fuel decomposition, and it was possible to operate at very low O_{NiO}/EtOH ratios without carbon formation. In all cases, small amounts of CH_4 were detected at the fuel reactor outlet, which indicates that this acts as an intermediate product during the reaction (García-Labiano et al. 2015).

Moreover, similar results were found during tests carried out at temperatures between 850 and 950 ºC, which revealed a minor effect of this variable on syngas production as a result of the high reactivity of the oxygen carrier, NiO18-αAl_2O_3.

As can be observed in Figure 11, experimental data were close to thermodynamic values. This means that equilibrium was almost achieved under the different operating conditions. Similar conclusions were previously reached by de Diego et al. (2009) during a CLR process using CH_4 as fuel.

Tests in the 1 kW_{th} CLR unit demonstrated the possibility of obtaining high H_2 concentrations in the syngas, which it is advantageous if hydrogen is the desired final product. Moreover, the pure N_2 stream obtained at the outlet of the air reactor makes the process perfect for the ammonia production.

In addition, the use of other gasifying agents such as CO_2, or the combination of H_2O and CO_2 during steam-dry reforming, opens up the possibility of obtaining a wide range of H_2/CO ratios valid for the Fischer-Tropsch process. This possibility was demonstrated by García-Díez et al. (2017b) in CLR tests in the same CLR unit using diesel as liquid fuel.

Although different syngas qualities and different H_2/CO ratios can be obtained from bioethanol, it is necessary to determine the O_{NiO}/EtOH molar ratio that produces autothermal conditions under each operating condition.

5. Autothermal CLR Processes

The use of a WGS reactor located downstream of the reforming reactor is preferred in any process of hydrogen production from syngas. The water added to the WGS reactor modifies the gaseous equilibrium to maximize hydrogen production. However, the CLR process offers the possibility of feeding water inside the fuel reactor during syngas production at autothermal conditions because the heat for the endothermic reactions in the fuel reactor mainly comes from the hot solids returned from the air reactor. Therefore, it is necessary to find those operational conditions that make the CLR process autothermal (a-CLR), doing away with the need for external energy.

Kale et al. (2013) conducted an initial study with Fe-based material to determine the optimal temperature for syngas production using bioethanol. Later, López Ortiz et al. (2015) carried out a study to determine autothermal conditions with NiO or CuO as oxygen carriers. García-Díez et al. (2016) performed a deeper analysis of the process in which they considered the possibility of using different bioethanol purities with a Ni-based oxygen carrier. It should be taken into account that bioethanol with different water contents is obtained during destillation. High-grade (high-purity) bioethanol is required for combustion processes, including motor vehicles and other mobile sources, which makes the use of costly distillation processes necessary (Pimentel et al. 2007). In contrast, the possibility of using low-grade bioethanol in a-CLR processes would entail significant energy savings during production, and therefore important economic benefits.

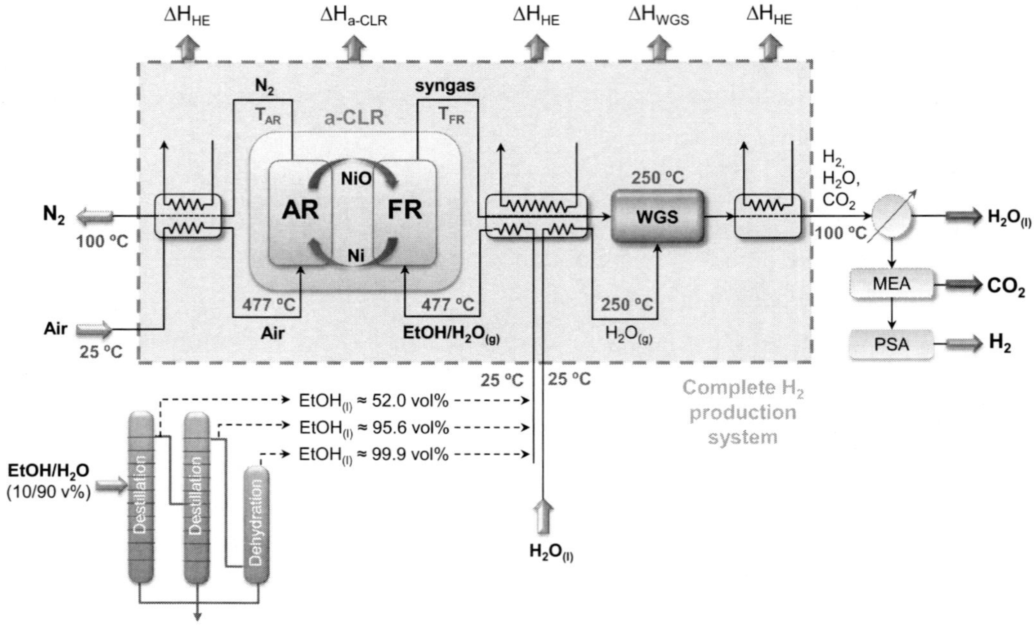

Figure 12. Complete hydrogen production process including the a-CLR system.

Figure 12 is a schematic diagram of a complete hydrogen production process using bioethanol as fuel. This incorporates an a-CLR system, a WGS reactor and several heat exchangers. The sensible heat from the gaseous outlet streams is used to evaporate the liquids and heat up the gases fed to the reactors. Depending on the operational conditions, extra heat can be recovered from the system. Purification processes after the WGS reactor (absorption and PSA) are not considered here.

García Díez et al. (2016) calculated the mass and heat balances required to optimize the process at two different scales. First, those accounting for the CLR process only were considered.

$$\Delta H_{CLR} = \Delta H_{FR} + \Delta H_{AR} \tag{8}$$

where ΔH_{CLR} is the enthalpy variation in the CLR system, and ΔH_{FR} and ΔH_{AR} the variation of enthalpy associated with all gases and solids inside the fuel and air reactors, respectively. The CLR system was autothermic when ΔH_{CLR} was equal to 0, which is denominated $\Delta H_{a\text{-}CLR}$. This state was reached at only one specific O_{NiO}/EtOH ratio and gave a specific syngas composition as a function of the other operating conditions in relation to the H_2O/EtOH molar ratio and reactor temperature. $\Delta H_{CLR} > 0$ meant that the energy produced in the air reactor was not sufficient to complete the energy balance, and extra heat was needed for the endothermic reactions in the fuel reactor. Otherwise, where $\Delta H_{CLR} < 0$, excess heat was obtained in the system and H_2 production was reduced.

The complete hydrogen production process, including the a-CLR system, was subsequently considered through the energy balance:

$$\Delta H_G = \Delta H_{a-CLR} + \Delta H_{WGS} + \Sigma \Delta H_{HE} \tag{9}$$

where ΔH_{WGS} and ΔH_{HE} are the variation of enthalpy associated with the WGS reactor and the different heat exchangers, respectively.

5.1. Use of Different Bioethanol Purities in an a-CLR Process

Bioethanol production involves several steps. Bioethanol obtained after biomass fermentation has a concentration of 10 vol%. This liquid is sent to a distillation column where a hydrous bioethanol is obtained, with a concentration of 95.6 vol%. The production of 99.9 vol% anhydrous bioethanol requires the use of a final dehydration process. However, these two processes require a large amount of energy. More specifically, 37% of the total amount of energy needed to obtain anhydrous bioethanol is consumed during the distillation and dehydration processes, 23% and 14%, respectively (Pimentel et al. 2007). However, there is also the possibility of modifying the distillation column to obtain bioethanol with different purities or water content. Bioethanol of lower purity could be fed into an a-CLR system with important energy and economic savings.

García-Díez et al. (2016) analysed the possibility of using 3 different bioethanol purities in an a-CLR process. These had different corresponding H_2O/EtOH molar ratios: anhydrous (99.9 vol%, H_2O/EtOH=0), hydrous (95.6 vol%, H_2O/EtOH=0.2) and diluted (52.0 vol%, H_2O/EtOH=3). The latter was chosen because it corresponded to the stoichiometry of the ethanol reforming reaction. The same amount of water (H_2O/EtOH=3) was always fed into the complete hydrogen production system and it was distributed either to the fuel reactor or to the WGS reactor. Figure 13 shows the sets of temperature and O_{NiO}/EtOH molar ratios that fit autothermal conditions for different H_2O/EtOH molar ratios, including those corresponding to the above bioethanol purities. The minimum operating temperature in each case (dots) was delimited by those conditions where carbon formation took place (represented by discontinuous lines).

The increase of the water content in bioethanol allows operation at decreasingly lower temperatures, with no restrictions by carbon formation at H_2O/EtOH ratios above 3. In addition, no restrictions were encountered during tests carried out with ethanol in the 1 kW_{th} CLR unit at temperatures as low as 650 °C derived from excesively low reaction rates in the fuel reactor (García-Díez et al. 2017a).

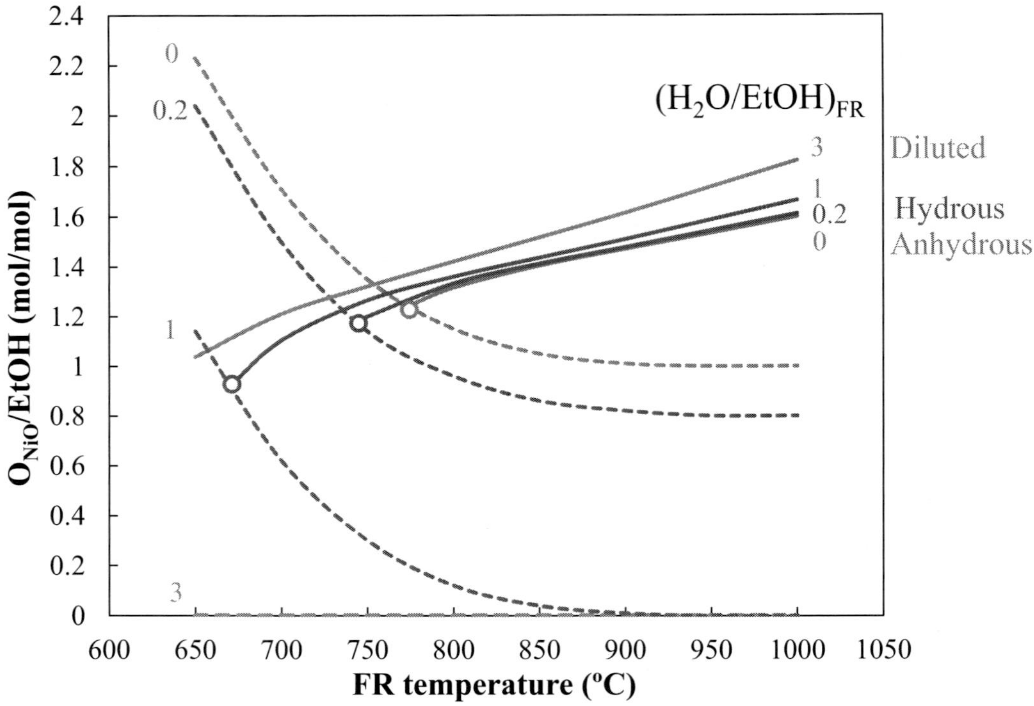

Figure 13. Autothermal conditions as a function of temperature for different H_2O/EtOH molar ratios. Discontinuous lines delimit carbon formation conditions for different bioethanol purities.

However, a more realistic process should consider a complete process that integrates the a-CLR system, WGS and the different heat exchangers used in the process (see Figure 12). Figure 14 shows hydrogen production and the variation of enthalpy in the complete process as a function of the fuel reactor temperature for the different bioethanol purities (diluted, hydrous and anhydrous).

A maximum value of H_2 production was found as a function of the FR temperature for each bioethanol purity. The lower the bioethanol quality and temperature, the higher the maximum value. Hydrogen production values of 4.65, 4.51 and 4.50 mol H_2/mol EtOH were found for diluted, hydrous and anhydrous bioethanol, respectively. However, these values corresponded to allothermal conditions, i.e., when external heat was added to the system.

In contrast, all the bioethanol purities resulted in the same hydrogen production values, about 4.4 mol of H_2 per mol of ethanol, when autothermal conditions in the complete system were considered. The only difference lay in the fuel reactor temperature, which was lower when the purity of the bioethanol fed into the FR was lower. This was a remarkable result considering the significant energy savings that could be achieved in the bioethanol production process.

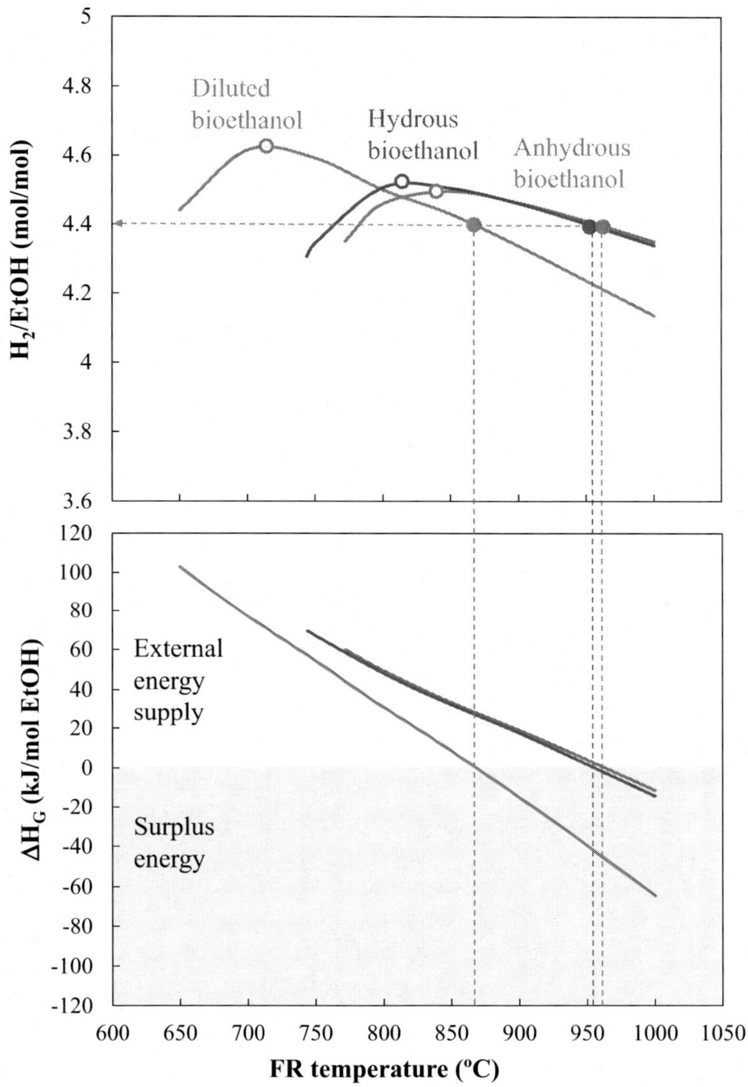

Figure 14. Hydrogen production as a function of fuel reactor temperature for different bioethanol purities.

REFERENCES

Abad, A., Adánez, J., García-Labiano, F., de Diego, L. F., Gayán, P. and Celaya, J. 2007. "Mapping of the range of operational conditions for Cu-, Fe-, and Ni-based oxygen carriers in chemical-looping combustion." *Chemical Engineering Science* 62:533-49.

Adánez, J., Abad, A., García-Labiano, F., Gayán, P. and de Diego, L. F. 2012. "Progress in chemical-looping combustion and reforming technologies." *Progress in Energy and Combustion Science* 38:215-82.

Adánez, J., Abad, A., Mendiara, T., Gayán, P., de Diego, L. F. and García-Labiano, F. 2018. "Chemical looping combustion of solid fuels." *Progress in Energy and Combustion Science*. 65:6-66.

Ahmed, S. and Krumpelt, M. 2001. "Hydrogen from hydrocarbon fuels for fuel cells." *International Journal of Hydrogen Energy* 26:2941-301.

Bhave, A., Taylor, R. H. S., Fennell, P., Livingston, W. R., Shah, N., Dowell, N. M., Dennis, J., Kraft, M., Pourkashanian, M., Insa, M., Jones, J., Burdett, N., Bauen, A., Beal, C., Smallbone, A. and Akroyd, J. 2017. "Screening and techno-economic assessment of biomass-based power generation with CCS technologies to meet 2050 CO_2 targets." *Applied Energy* 190:481–9.

Cormos, C. C. 2014. "Renewable hydrogen production concepts from bioethanol reforming with carbon capture." *International Journal of Hydrogen Energy* 39:5597-606.

CSIC patent. 2009a. Adánez, J., de Diego, L. F., García-Labiano, F., Gayán, P. and Abad, A. *Solid NiO/Al_2O_3 oxygen carrier that is useful for methane reforming, Method for producing same and applications thereof.* Reference: WO2009/101233.

CSIC patent. 2009b. Adánez, J., de Diego, L. F., García-Labiano, F., Gayán, P. and Abad, A. *NiO/Al_2O_3 oxygen carrier, method for obtaining same and use thereof.* Reference: WO2009/022046.

de Diego, L. F., Ortiz, M., Adánez, J., García-Labiano, F., Abad, A. and Gayán, P. 2008. "Synthesis gas generation by chemical-looping reforming in a batch fluidized bed reactor using Ni-based oxygen carriers." *Chemical Engineering Journal* 144:2289-98.

de Diego, L. F., Ortiz, M., García-Labiano, F., Adánez, J., Abad, A. and Gayán, P. 2009. "Hydrogen production by chemical-looping reforming in a circulating fluidized bed reactor using Ni-based oxygen carriers." *Journal Power Sources* 192:27-34.

de Diego, L. F., Serrano, A., García-Labiano, F., García-Díez, E., Abad, A., Gayán, P. and Adánez, J. 2016. "Bioethanol combustion with CO_2 capture in a 1 kW_{th} chemical Looping Combustion prototype: Suitability of the oxygen carrier." *Chemical Engineering Journal* 283:1405-1413.

Dueso, C., Abad, A., García-Labiano, F., de Diego, L. F., Gayán, P., Adánez, J. and Lyngfelt, A. 2010. "Reactivity of NiO/Al_2O_3 oxygen carrier prepared by impregnation for chemical-looping combustion." *Fuel* 89:3399-3409.

Dueso, C., Izquierdo, M. T., García-Labiano, F., de Diego, L. F., Abad, A., Gayán, P. and Adánez, J. 2012. "Effect of H_2S on the behaviour of an impregnated NiO-based oxygen-carrier for chemical-looping combustion (CLC)." *Applied Catalysis B: Environmental* 126:186-99.

Fan L. S. *Chemical Looping Systems for Fossil Energy Conversion*: John Wiley & Sons; 2010. AIChE. ISBN: 978-04708752-9.

Fan, L. S., Zeng, L., Wang, W. and Luo, S. 2012. "Chemical looping processes for CO_2 capture and carbonaceous fuel conversion-prospect and opportunity." *Energy Environmental Combustion Science* 5:7254-80.

Farrauto, R., Hwang, S., Shore, L., Ruettinger, W., Lampert, J., Giroux, T., Liu, Y. and Ilinich, O. 2003. "New material needs for hydrogen fuel processing: Generating Hydrogen for the PEM Fuel Cell." *Annual Review of Materials Research* 33:1-27.

Fatsikostas, A. N. and Verykios, X. E. 2004. "Reaction network of steam reforming of ethanol over Ni-based catalysts." *Journal of Catalysis* 225:439-52.

García-Díez, E., García-Labiano, F., de Diego, L. F., Serrano, A., Abad, A., Gayan, P., Adánez, J., Ruiz, J. A. C. 2016. "Optimization of hydrogen production with CO_2 capture by autothermal chemical-looping reforming using different bioethanol purities." *Applied Energy* 169: 491-498.

García-Díez, E., García-Labiano, F., de Diego, L. F., Serrano, A., Abad, A., Gayán, P. and Adánez, J. 2017a. "Autothermal chemical looping reforming process of different fossil liquid fuels." *International Journal of Hydrogen Energy* 42:13633-40.

García-Díez, E., García-Labiano, F., de Diego, L. F., Abad, A., Gayán, P., Adánez, J. and Ruíz, J. A. 2017b. "Steam, dry, steam-dry chemical looping reforming of diesel fuel in a 1 kW_{th} unit." *Chemical Engineering Journal* 325:369-77.

García-Labiano, F., de Diego, L. F., García-Díez, E., Serrano, A., Abad, A., Gayán, P. and Adánez, J. 2014. "Combustion and reforming of ethanol in a chemical looping continuous unit." *Energy Procedia* 63:53-62.

García-Labiano, F., García-Díez, E., de Diego, L. F., Serrano, A., Abad, A., Gayán, P., Adánez, J. and Ruíz, J. C. 2015. "Syngas/H_2 production from bioethanol in a continuous chemical-looping reforming prototype." *Fuel Processing Technology* 137:24-30.

García-Labiano, F., de Diego, L. F., García-Díez, E., Serrano, A., Abad, A., Gayán, P. and Adánez, J. 2017. "Combustion and reforming of liquid fossil fuels through chemical looping Process-Integration of chemical looping processes in a refinery." *Energy Procedia* 114:325-33.

Gasser, T., Guivarch, C., Tachiiri, K., Jones, C. D. and Ciais, P. 2015. "Negative emissions physically needed to keep global warming below 2ºC." *Nature communications*. Doi: 10.1038/ncommms8958.

Gayán, P., Dueso, C., Abad, A., Adánez, J., de Diego, L. F. and García-Labiano, F. 2009. "NiO/Al_2O_3 oxygen carriers for chemical-looping combustion prepared by impregnation and deposition-precipitation methods." *Fuel* 88:1016-23.

Gough, C. and Upham, P. 2011. "Biomass energy with carbon capture and storage (BECCS or Bio-CCS)." *Greenhouse Gas* 1:324–34.

HSC Chemistry 6.1. 2008. *Chemical reaction and equilibrium software with thermochemical database and simulation module.*

IEA. International Energy Agency. 2013. *CO_2 emissions from fuel combustion - highlights*. 2013 Edition.

IPCC Climate Change 2014: *Synthesis Report. Contribution of Working Groups I, II and III to the Fifth Assessment Report of the Intergovernmental Panel on Climate Change*; Geneva (Switzerland). 2014; p 151.

Jiang, B., Dou, B., Wang, K., Zhang, C., Song, Y., Chen, H. and Xu, Y. 2016. "Hydrogen production by chemical looping steam reforming of ethanol using NiO/ montmorillonite oxygen carriers in a fixed-bed reactor." *Chemical Engineering Journal* 298:96-106.

Kale, G. R., Kulkarni, B. D. and Bharadwaj, K. V. 2013. "Chemical looping reforming of ethanol for syngas generation: A theoretical investigation." *International Journal of Energy Research* 37:645-56.

Krumpelt, M., Krause, T. R., Carter, J. D., Kopasz, J. P. and Ahmed, S. 2002. "Fuel processing for fuel cell systems in transportation and portable power applications." *Catalysis Today* 77:3-16.

López Ortiz, A., Meléndez Zaragoza, M. and Collins-Martínez, V. 2015. "Thermodynamic analysis of the ethanol chemical looping autothermal reforming with CO_2 capture." *International Journal of Hydrogen Energy* 40:17180-91.

Lyngfelt, A. and Mattisson, T. Chemical looping materials for CO_2 separation. In: Scherer DSaV, editor. *Efficient carbon capture for coal power plants*, First ed. Germany: Wiley-VCH Verlag GmbH & Co KGaA; 2011. p. 475-504.

Moldenhauer, P., Rydén, M., Mattisson, T., Hoteit, A., Jamal, A. and Lyngfelt, A. 2014. "Chemical-looping combustion with fuel oil in a 10 kW pilot plant." *Energy Fuels* 28:5978-87.

Moldenhauer, P., Rydén, M., Mattisson, T., Jamal, A. and Lyngfelt, A. 2017. "Chemical-looping combustion with heavy liquid fuels in a 10 kW pilot plant." *Fuel Processing Technology* 156:124-37.

Pimentel, D., Patzek, T. and Cecil, G. 2007. "Ethanol production: energy, economic, and environmental losses." *Rev Environ Contam Toxicol* 189:25–41.

Rubin, E. S., Mantripragada, H., Marks, A., Versteeg, P. and Kitchin, J. 2012 "The outlook for improved carbon capture technology." *Progress in Energy and Combustion Science* 38:630-71.

Sánchez, D. L., Nelson, J. H., Johnston, J., Mileva, A. and Kammen, D. M. 2015. "Biomass enables the transition to a carbon-negative power system across western North America." *Nature Climate Change* 5:230–34.

Serrano, A., García-Labiano, F., de Diego, L. F., Gayán, P., Abad, A. and Adánez, J. 2017. "Chemical looping combustion of liquid fossil fuels in a 1 kW_{th} unit using a Fe-based oxygen carrier." *Fuel Processing Technology 160:47-54.*

Trevisanut, C., Mari, M., Millet, J. M. M. and Cavani, F. 2015a. "Chemical-loop reforming of ethanol over metal ferrites: An analysis of structural features affecting reactivity." *International Journal of Hydrogen Energy* 40:5264-71.

Trevisanut, C., Bosselet, F., Cavani, F. and Millet, J. M. M. 2015b. "A study of surface and structural changes of magnetite cycling material during chemical looping for hydrogen production from bio-ethanol." *Catalysis Science and Technology* 5:1280-9.

UNFCC. United Nations Framework Convention on Climate Change. "United Nations Treaty Collection. Paris Agreement 2015."

Wang, W. 2014. "Thermodynamic and experimental aspects on chemical looping reforming ethanol for hydrogen production using a Cu-based oxygen carrier." *International Journal of Energy Research* 38:1192-200.

Wang, W. and Cao, W. 2013. "A combined thermodynamic and experimental study on chemical-looping ethanol reforming with carbon dioxide capture for hydrogen generaton." *International Journal of Energy Research* 37:25-34.

Wang, K., Dou, B., Jiang, B., Zhang, Q., Li, M., Chen, H. and Xu, Y. 2016. "Effect of support on hydrogen production from chemical looping steam reforming of ethanol over Ni-based oxygen carrier." *International Journal of Hydrogen Energy* 41:17334-47.

Wang, X., Gao, J. and Jiang, L. 2016. "Feasibility analysis of H_2 production by calcium looping process based on coal gasification in a transport reactor." *International Journal of Hydrogen Energy* 41:12000-18.

In: Bioethanol and Beyond
Editor: Michel Brienzo

ISBN: 978-1-53613-478-0

Chapter 13

BIOETHANOL USAGE IN DIESEL ENGINES

Breda Kegl* and Marko Kegl
Faculty of Mechanical Engineering,
University of Maribor, Maribor, Slovenia

ABSTRACT

Diesel engines for personal and commercial vehicles hold a significant market share worldwide and their share and popularity are both increasing. Diesel engines can be run on mineral diesel, on various biofuels, and on their blends. In this context an attractive option is to blend bioethanol with various diesel engine fuels. This chapter deals with blends of bioethanol and mineral diesel as well as bioethanol and biodiesel. At first, chemical and physical properties of bioethanol and its blends with mineral diesel as well as with biodiesel are compared. Investigated fuel properties are stability, density, viscosity, cetane index, cetane number, cloud point, pour point, cold filter plugging point, flash point, corrosiveness, lubricity, FT–IR spectrum, carbon–hydrogen–nitrogen composition, water content, and sound of velocity. Then the bioethanol effects on fuels injection characteristics, especially on injection pressure, duration and timing, and on fuel injection system components, like surface roughness of injection elements and nozzle hole discharge coefficient are analyzed. Finally, the effects of bioethanol on fuel spray and engine characteristics are addressed briefly. The presented bioethanol effects are derived either numerically or experimentally on adequate equipment and on a bus diesel engine. Based on the injection and fuels spray characteristics of tested fuels, blends of 15% bioethanol in mineral diesel and 15% bioethanol in biodiesel seem to be an attractive option for diesel engines, provided that the engine is tuned and adapted properly.

Keywords: bioethanol, biodiesel, mineral diesel, fuel properties, injection characteristics, fuel spray characteristics, diesel engine performances

* Corresponding Author Email: breda.kegl@um.si.

INTRODUCTION

Diesel engines hold a significant market share worldwide. Because their share and popularity are both increasing, significant efforts are put into the improvement of diesel engine characteristics. In order to reduce harmful emissions and specific fuel consumption, and increase engine performance, special attention is focused on engine management, exhaust gas aftertreatment, and fuel. Exhaust gas after treatment reduces harmful emissions already produced by the engine, meanwhile engine management and alternative fuels offer a possibility to reduce the formation of harmful emissions and keep engine performance as well as specific fuel consumption within desired levels, Figure 1 (Kegl et al. 2013).

Diesel engine management is mostly related to the fuel injection process. In general, modern fuel injection systems must fulfil the following requirements:

- high injection pressure, which contributes to decreased fuel droplet size and improved combustion process, resulting in a reduction of smoke emissions,
- injection pressure control, which enables optimal injection pressures at various engine operating regimes,
- injection timing control, which influences to a great extent the combustion process and consequently the formation of harmful emissions, especially nitrogen oxides (NO_x) and particulate (PM) emissions,
- injection rate control, which influences engine performance, fuel consumption, and formation of harmful emissions.

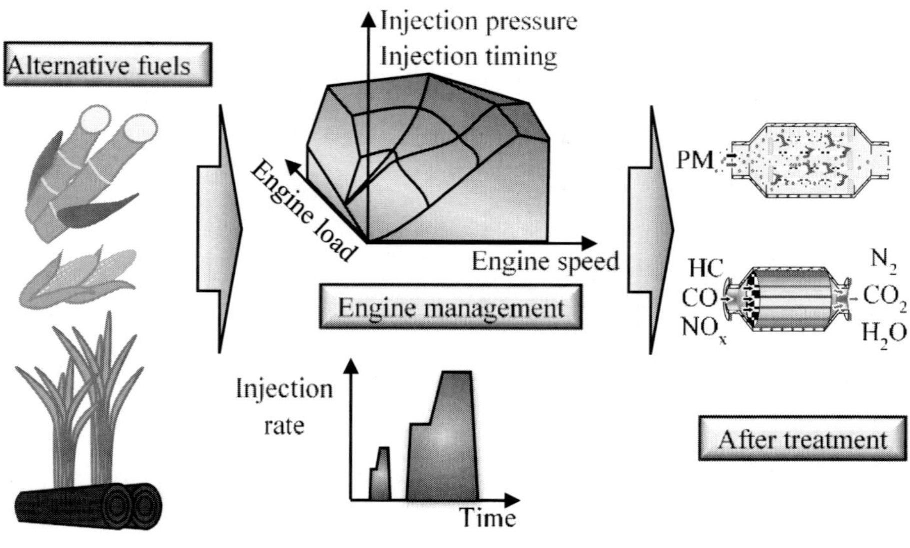

Figure 1. Options for diesel engine characteristics improvement.

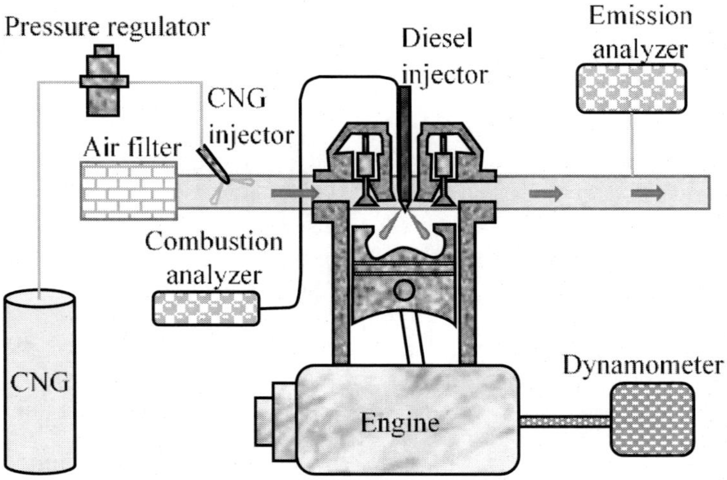

Figure 2. Dual fuel injection system.

Exhaust gas after treatment of diesel engines employs catalytic convertors, fuel filters and exhaust gas recirculation. Diesel catalytic convertors are oxidation catalysts, which remove carbon monoxide and unburned hydrocarbons and SCR catalysts, which remove NO_x. Particulate filter removes diesel particulate matter or aerosolized diesel exhaust pollution particles. Exhaust gas recirculation is used to reduce NO_x emission.

Traditionally, diesel engines run on mineral diesel, which is produced from crude oil. Due to the limited crude oil reserves and harmful effects of mineral diesel combustion on environmental pollution, alternative fuels for diesel engine are an interesting option to be investigated. Today's most frequently investigated alternative fuels are water in diesel emulsion, natural and liquified petroleum gas, methane and propane, dimethyl ether and dimethyl carbonate, Fischer–Tropsch, hydrogen, vegetable oils, biodiesel, and bioethanol. Diesel engines, which use mineral diesel and some gaseous fuel, like as methanol, compressed natural gas, methane, propane, liquefied petroleum gas and hydrogen, are referred to as dual fuel engines (Carlucci et al. 2011, Karagöz et al. 2016, Khan et al. 2016, Lilik et al. 2010, Lounici et al. 2015, Poompipatpong and Cheenkachorn 2011, Sahoo et al. 2009, Saravanan and Nagarajan 2010, Surawski et al. 2010, Wei and Geng 2016). Such engines need two fuel injection systems and a mixture of air and gaseous fuel has to be prepared in an external mixing device. In this context, especially the natural gas and bio–derived gas appear to be quite attractive alternative fuels, Figure 2, due to their environment–friendly nature (Wei and Geng 2016). Furthermore, dimethyl carbonate and bioethanol can be used as an additive to mineral diesel fuel.

The usage of mineral diesel is already quite well investigated in various diesel engines with various types of fuel injection systems. However, when biofuels are planned to be used, it has to be kept in mind that engine characteristics are influenced by the combination of engine type, fuel injection type, and used biofuel. Furthermore, biofuel production technology as well as the raw material used influence the chemical and physical properties of biofuels and consequently the injection and combustion process as well as the ecology, economy, and engine performance.

CHEMICAL AND PHYSICAL FUEL PROPERTIES

Chemical and physical fuel properties define engine performance to a great extent. Of all properties, those which influence most significantly the injection and combustion processes need to be most carefully examined. Because we target a diesel engine, it makes sense to compare these properties of various fuels to those of mineral diesel (D100). In this view, the properties of bioethanol and mineral diesel blends (E5D95, E10D90, E15D85), biodiesel (B100), and blends of bioethanol and biodiesel fuels (E5B95, E10B90, E15B85) are discussed. Moreover, the properties of tested fuels are presented together with appurtenant standards. The tested bioethanol was produced from the fermentation of sugars and the tested biodiesel was produced from rapeseed oil.

The most important chemical and physical properties of a fuel to be engaged in a diesel engine are (Torres et al. 2010, Kegl 2011, Kegl et al. 2013):

- stability,
- density,
- viscosity,
- cetane number, cetane index,
- cloud point, pour point, and cold filter plugging point,
- flash point,
- corrosiveness,
- lubricity,
- CHN content,
- water content,
- sound velocity.

Stability

Fuel stability is related to the ability of fuel to preserve its properties and quality over longer time periods. The first problem related to fuel stability is possible phase separation. For example, in case of bioethanol–mineral diesel blends this can happen because the nonpolar hydrocarbons present in mineral diesel has no affinity with the polar bioethanol (Reyes et al. 2009). Apart from phase separation, an increase of contaminants is also possible; for example, peroxides may emerge from the oxidation of the fuel. These contaminants can increase deposits in fuel pumps and injectors, as well as increase filter plugging. Therefore, the stability of fuels for diesel engines has to be studied carefully (Tan et al. 2017).

In order to estimate possible problems related to phase separation, the fuel has to be observed for a period of several weeks at various temperatures. The results may be used in order to determine, if additional mixing, additives, or heating (in the case that low temperatures produce solid aggregation of fuel) are necessary.

The data presented in the following is based on the phase separation tests conducted as follows: the fuel samples were tested at 25°C (ambient temperature), +30°C (representative temperature of summer), +8°C (representative temperature of not critical winter) and –18°C (representative temperature of critical winter). Each fuel sample was checked once a week during a 5 weeks period. A test period of 5 weeks is estimated to be long enough to know the fuel behavior during longer periods of storage (Torres et al. 2010, Torres et al. 2011).

Figure 3 shows some results for the blend of 15% bioethanol and 85% mineral diesel (E15D85) after five weeks. At temperature of +30°C, no changes in stability, color, and aggregation could be observed. At temperature of +25°C, bioethanol separates slowly from mineral diesel, which can be observed by bioethanol bubbles travelling up to the surface. At temperature of +8°C, bioethanol separated from mineral diesel in such a form that two fuel layers could be observed. At temperature of –18°C, the fuel blend separated into three layers; the upper layer contained mainly bioethanol and the lower two layers resulted mainly from mineral diesel.

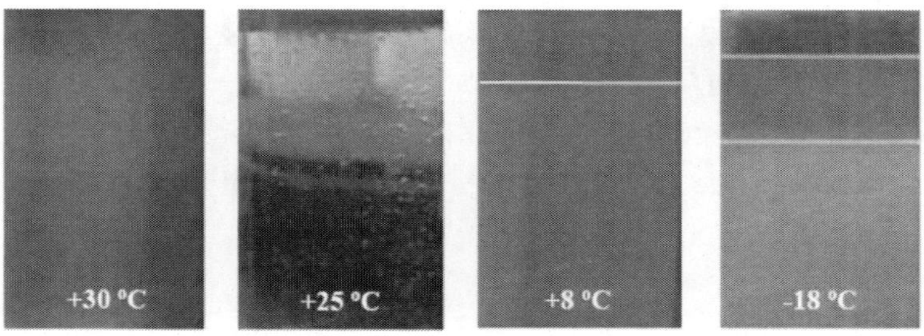

Figure 3. Fuel stability observation of E15D85.

Figure 4. Fuel stability observation of E15B85.

Figure 4 shows some results for the blend of 15% bioethanol and 85% biodiesel (E15B85) after five weeks. At temperatures of +30°C, +25°C and +8°C no changes in stability, color, and aggregation could be observed. At temperature of –18°C, however, all biodiesel blends became solid. At this point it might be worth mentioning that when pure solid biodiesel warms up and becomes liquid, usually white particles (paraffins) appear. In the case of E15B85, bioethanol changes this melting process in such a way that the paraffin particles do not appear. In this view bioethanol improves the fuel in regard to its winter properties; namely, cloud point and cold filter plugging point are related to the moment in which visible paraffins appear in the fuel. Therefore, bioethanol acts as a winter additive and bioethanol–biodiesel blends need less time to become liquid during warm-up than pure biodiesel.

Density

The density of fuel affects injection characteristics, fuel spray penetration, and atomization quality. In this way it has a significant impact on fuel economy, engine power, deposits, wear, and exhaust smoke (Lapuerta et al. 2010, Zaharin et al. 2017).

The data presented in the following is based on density measurements performed according to EN ISO 12185 standard at temperature of 15°C. The densities corresponding to other temperatures were obtained by using the petroleum measurement correction tables of the American Petroleum Institute (API). The testing device engaged was an APAAR density meter DMA 48. This device measures the oscillation frequency of a fluid sample contained in a temperature controlled sample cell. After the frequency is known, the density of the fluid can be calculated as $\rho = A\,T^2 - B$, where ρ is the density (kg/m^3), T is the oscillation period (s), and $A,\ B$ are cell constants; these constants are determined by measuring the oscillation frequency of a calibration fluid with known density (Torres et al. 2010, Torres et al. 2011).

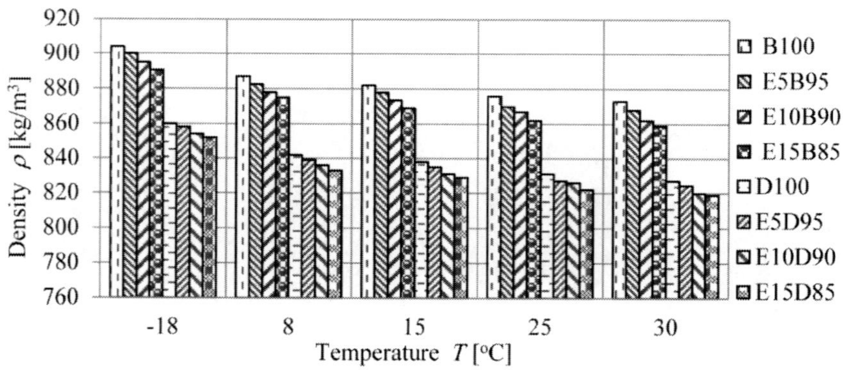

Figure 5. Fuel density of tested fuels with respect to temperature.

Figure 5 shows the measured and derived densities of tested fuels. It can be seen that the density decreases by increasing bioethanol content and temperature. In a mechanically controlled fuel injection system, fuel density decrease generally leads to retarded injection timing, which can cause a decrease of injection pressure, in-cylinder pressure, and in-cylinder temperature. This is especially true in the first part of injection and combustion process and it typically results in lower pollutant emissions, especially of NO_x (Kegl 2011).

Viscosity

Fuel viscosity is an important property which impacts the injection and fuel spray characteristics. Furthermore, if fuel viscosity is too low, it may lead to excessive wear of injection pumps and to power loss due to injector or pump leakage. On the other hand, if fuel viscosity is too high, it may cause too much pump resistance, filter damage, and may adversely affect fuel spray angle and atomization. In general, fuels with low viscosity tend to have poorer lubrication properties (Lapuera et al. 2017).

The data given in the following is based on kinematic viscosity measurements done according to ISO 3104, which defines this property as the resistance to flow of a fluid under gravity at 40°C. The testing device engaged was a Walter Herzog GmbH MP-480. This device measures the time needed by some fixed volume of liquid to flow under gravity through a capillary at some specified temperature. After this time is known, the kinematic viscosity can be calculated as $\upsilon = C\,t$, where υ is the kinematic viscosity (mm^2/s), C is a calibration constant (mm^2/s^2) and t is the mean flow time (s) (Torres et al. 2011).

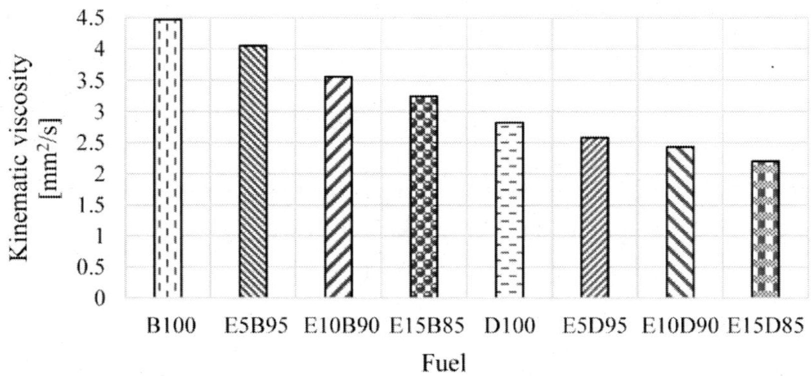

Figure 6. Fuel viscosity of tested fuels.

Figure 6 shows the kinematic viscosities of tested fuels. It was found that when bioethanol concentration increases, kinematic viscosity decreases proportionally. In other words, by increasing the content of bioethanol, the viscosity of biodiesel blends approaches the viscosity of neat mineral diesel. Viscosity of mineral diesel also decreases as bioethanol is added. However, all considered fuels blends exhibit viscosities still within the range defined by the standard for mineral diesel.

Cetane Number and Cetane Index

Fuel cetane number is a measure of the ignition quality of the fuel. It is defined as the inverse of fuel's ignition delay, which means that a fuel with higher cetane number will ignite more easily. Cetane number affects significantly the ignition and consequently the combustion process. The cetane index is a substitute or estimate for the cetane number, used when a test engine is not available to determine the cetane number directly. It can be calculated based on the fuel's density and boiling range.

The data presented here was obtained by a distillation process performed according to the standard test method EN 3405. The testing device engaged was the HERZOG Atmospheric distillation HAD 627. This device was usable only for mineral diesel and it blends E5D95 and E10D90. The data obtained was used to calculate the cetane index according to standard (EN ISO 4264) (Torres et al. 2010, Torres et al. 2011).

The obtained cetane index for pure mineral diesel is 51.8, for E5D95 it is 51.7, and for E10D90 it is 49.8. According to (Kuszewski et al. 2017) the cetane index of a blend of mineral diesel and bioethanol (E12D88) is around 42.7. Obviously the cetane index decreases with bioethanol content, which is not very desirable since the increase in the ignition delay tends to produce more NO_x and particulate emissions (Torres et al. 2011, Kwanhareon et al. 2007). It is true, however, that the cetane index increase is not very dramatic and that its values are approximately still within the standards for mineral diesel.

Cloud Point, Pour Point, and Cold Filter Plugging Point

Cloud point, pour point, and cold filter plugging point of a fuel are used to estimate its low temperature operability. The cloud point (CP) is the temperature at which the amount of precipitated wax crystals becomes large enough to make the fuel appear cloudy or hazy. Namely, as the temperature of the fuel is lowered, the paraffins that occur naturally in mineral diesel fuel become less soluble and precipitate out as wax crystals. Pour point (PP) is the lowest temperature at which the fuel will flow and is therefore closely related to the lowest temperature at which the fuel can be pumped. Cold filter plugging point (CFPP) is the lowest temperature, at which a given volume of fuel still passes through a standardized filter within some specified time when cooled under some defined conditions (Lapuerta et al. 2010).

The CP data given here is obtained from measurements performed in a Herzog CPP 97-2 device. The fuel sample was cooled at a specified rate and examined at regular intervals. The CP was recorded as the temperature at which a cloud was first observed at the bottom of the test jar. The PP test was performed in a Herzog HCP852 Combi device. Starting at 45°C, a fuel sample was cooled down at some specified rate. At temperature intervals of 3°C, the sample was checked, if the fluid is still moving. The PP is recorded as the lowest temperature at which the fuel was still moving. The CFPP test was performed on a Herzog CFPP analyzer, model MC 840. This test consists of passing a given volume of fuel through a standardized filter under specified conditions (Torres et al. 2010, Torres et al. 2011).

Figure 7 shows the CP, PP and CFPP for all tested fuels. Regarding the CP, one can see that for bioethanol–biodiesel blends its value is significantly lowered as bioethanol concentration increases. For mineral diesel blends this could not be observed. The CP of the E5D95 is practically the same as for D100. For higher concentrations of bioethanol, reliable measurements could not be obtained. This was probably caused by initiated phase separation in bioethanol–mineral diesel fuel mixtures as the temperature decreased. This disturbed the automatic testing device enough to deliver unstable and unreliable results.

Regarding PP, for bioethanol–biodiesel blends one can also observe that the PP value decreases as bioethanol concentration increases. For mineral diesel blends the situation seems to be similar, although the measurement for E15D85 was again problematic. The cause for the testing device having difficulties was again the phase separation.

Regarding CFPP, the data in Figure 7 indicates that the influence of bioethanol addition is rather minor. This seems to be true for biodiesel as well as for mineral diesel. Only for the E15B85 blend it seems that bioethanol addition might improve the CFPP value slightly.

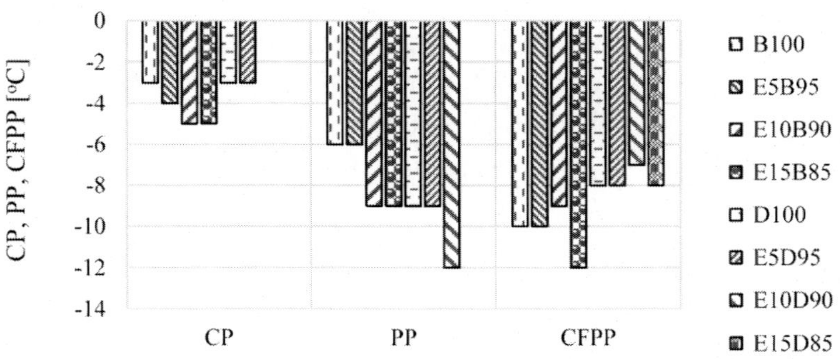

Figure 7. Cloud point, pour point, and cold filter plugging point of tested fuels.

Flash Point

By definition, a flash point (FP) is the lowest fuel temperature at which the vapor above a fuel sample will momentarily ignite under the prescribed test conditions. So, FP is not directly related to engine performance. Instead, it has to be kept within some limits in order to meet safety requirements for fuel handling and storage. A higher FP value makes the fuel more reliable regarding usage, transport, and storage (Aydın & Öğüt 2017).

The data given here were obtained by engaging a Walter Herzog MP-329 test device. To determine the FP, a fuel sample is placed into a test cup of a Pensky-Martens apparatus and heated at specified rate during continuous stirring. An ignition source is activated at an opening in the test cup lid at regular temperature intervals with simultaneous interruption of stirring (Torres et al. 2010).

The measurements have shown that FP values of bioethanol–biodiesel blends are close to 25°C, while the FP of pure biodiesel is around 138,5°C. During the heating and stirring process, bioethanol evaporates first at about 25°C. Therefore, ignition is recorded at this FP temperature for all blends containing bioethanol, irrespective of bioethanol concentration. Obviously, any bioethanol concentration results in a very low FP of the blend and this problem has to be circumvented somehow to avoid problems in shipping and storage. Similarly, bioethanol addition also lowers the FP of bioethanol–mineral diesel blends, but the resulting FP is not so low. Its values are around 66°C which is much better than the FP of bioethanol–mineral diesel blends.

Corrosiveness

Corrosiveness is obviously also not related to engine performance, but may influence significantly engine durability and required maintenance. A copper strip corrosion test may be used to estimate possible problems with copper, brass or bronze parts of a fuel system.

In order to estimate the influence of bioethanol on corrosiveness of tested fuel blends a polished copper strip was immersed into a specified volume of a fuel sample. The sample was then heated up to 50°C for a period of 3 h. At the end of the heating period, the strip was removed and washed. Its color was then assessed against the corrosion standards (Torres et al. 2010).

According to the results, all tested fuel blends could be classified as 1a, which is an appropriate classification from the point of view of corrosion. So, bioethanol addition does not worsen corrosiveness of neat mineral diesel fuel (Torres et al. 2011) or of neat biodiesel (Torres et al. 2010).

Lubricity

Fuel lubricity is also a very important property in view of engine durability and required maintenance. Namely, a diesel fuel injection system relies on the fuel to lubricate the moving parts. If the lubricating properties are insufficient, injection system operation will result in increased wear on the injectors and pumps. In such cases, lubricity additives are often required to avoid catastrophic fuel pump or injector failures. As little as one tank full of poor lubricity fuel can cause such catastrophic failures (Dharma et al. 2016).

The lubricity data presented in the following were derived in a measurement, where a test ball was oscillated and pushed against a stationary steel plate. The interface with the plate was fully immersed into a fuel sample. The test was performed in a HFR2 D481 lubricity and fretting test system. The ball wear scar diameter is a measure of the fluid lubricity. Its mean value can be calculated as $d_{MWSD} = \frac{x+y}{2}$, where x is the scar dimension, perpendicular to the oscillation direction (in μm) and y is the scar dimension parallel to the oscillation direction (in μm). Lower d_{MWSD} means better lubricity (Torres et al. 2010).

It is known that bio-based oils appear to be effective in improving fuel lubricity (Torres et al. 2010). So, the lubricity of biodiesel is quite good and it seems that the addition of bioethanol does not changes this significantly, Figure 8. On the other hand, the addition of bioethanol to mineral diesel fuel seems to improve slightly the lubricity of the fuel, as the d_{MWSD} of the blends tends to be somewhat lower compared to neat mineral diesel (Torres et al. 2011). Note that this trend may be enhanced when the engine operates at high temperatures. In this case, the presence of bioethanol and increased evaporation of bioethanol from the lubricating layer, may influence beneficially the lubricity (Lapuerta et al. 2010).

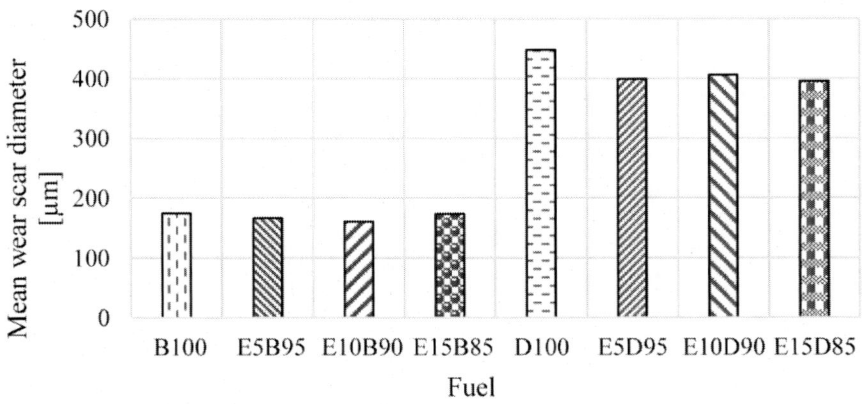

Figure 8. Mean wear scar diameter in dependence of tested fuels.

Contaminants and CHN Content

Contaminants content in the fuel are important in order to estimate possible problems, especially the engine deposit-forming tendency of fuels (Dharma et al. 2016). Meanwhile carbon and hydrogen content is also interesting to evaluate some fuel parameters like its specific energy content. The data presented in the following is based on spectroscopy (Fourier transformation infrared or FT-IR) analysis and on carbon, hydrogen, and nitrogen (CHN) analysis.

For the spectroscopy, the IR spectrum was obtained by engaging a Perkin-Elmer FT-IR spectroscopy device (model MSP 9250 PE). This device lightens a fuel sample with IR light. Then the absorption of the light in the IR spectrum is detected and the absorption zones are observed in order to identify patterns, characteristic for a specific molecule in the sample. The CHN analysis was performed by using a Leco CHNS 932 device, which can be used for elemental analysis determination of various materials. In this device a sample is combusted in a furnace under suitable conditions. Then IR detectors are used to analyze the content (Torres et al. 2010).

The IR spectrum for B100 and its blends is shown in Figure 9, wherefrom bioethanol presence is clearly exposed by the pits near the wavelength of 3400 cm^{-1} (which corresponds to O-H bonds of alcohols and phenols) and the pits at wavelength near 2900 cm^{-1} (which corresponds to the C-H bond absorption range). The rest of the spectrum shows typical biodiesel components like the pit at 1740 cm^{-1} (ester double bonds C = O) and the pit near around 1100 cm^{-1} (C-O bonds present in ester and ethanol). According to this spectrum, the content of undesirable components was negligible (Torres et al. 2010).

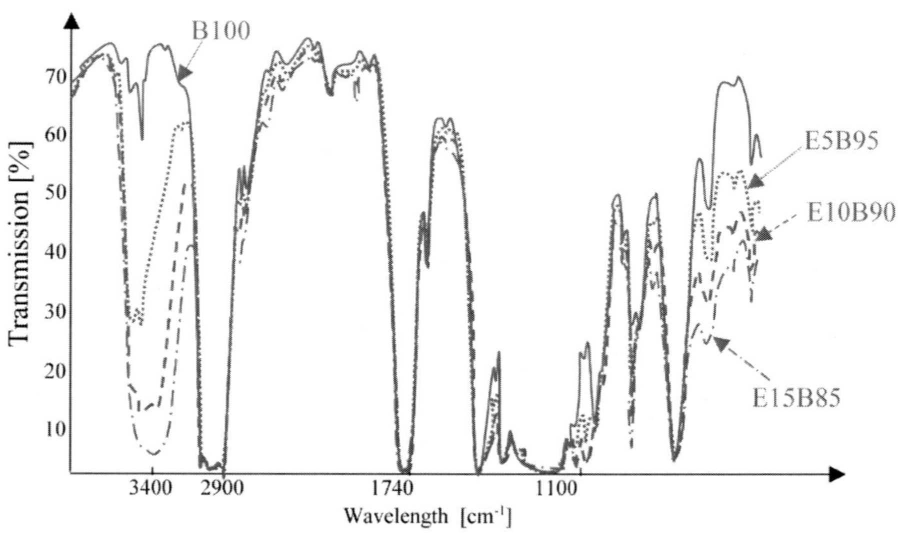

Figure 9. IR specter of bioethanol–biodiesel blends.

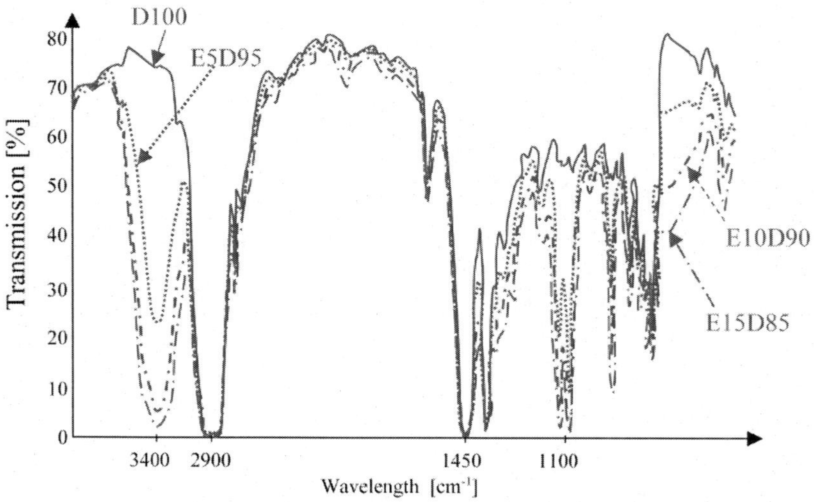

Figure 10. IR specter of bioethanol–mineral diesel blends.

Measurements of CHN content revealed that the bioethanol presence does not significantly changes the C (carbon) content biodiesel blends. Slightly lower C and H (hydrogen) concentrations are a consequence of higher oxygen presence in bioethanol. Lower C and H concentration decrease the specific energy content of the blends. However, the increased of oxygen content can be expected to have a beneficial effect on the emissions profile with lower soot emission, because it improves the combustion process. The presence of bioethanol in the blends also slightly reduces the nitrogen content of the blends (Torres et al. 2010).

The IR spectrum for D100 and its blends is shown in Figure 10, wherefrom bioethanol presence is clearly exposed. Otherwise, the spectrum shows typical mineral diesel components; so, undesirable components were not found (Torres et al. 2011).

Measurements of CHN content revealed that addition of bioethanol to mineral diesel somewhat reduces carbon and hydrogen content of the blend. Oxygen, which is not present in mineral diesel, increases as a consequence of oxygen presence in bioethanol. Nitrogen was not present in the tested blends. Lower C (carbon) and H (hydrogen) result in lower specific energy content of the blend. Increased oxygen concentration is expected to have a beneficial effect on the combustion process and emissions profile (Torres et al. 2011).

Water Content

Water and sediment contamination can contribute to filter plugging and fuel injection system wear. Additionally, water may also lead to problems related to phase separation, corrosion and aqueous microorganisms growth (Torres et al. 2011).

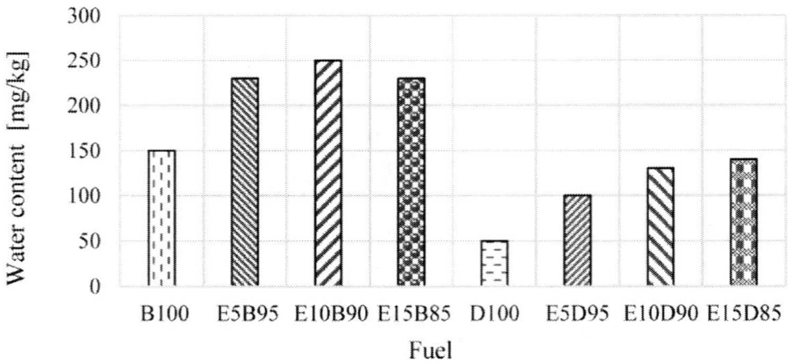

Figure 11. Water content in tested fuels.

The data presented hereafter were obtained by performing measurements with a Metrohm 684KF coulometer. To determine the water content, a fuel sample is taken to react with iodine. Afterwards, the excess iodine is detected, and on the basis of the stoichiometry of the reaction (1 mol of iodine reacts with 1 mol of water), the mass of water can be determined. Water content, given as the mass fraction in percent, is then calculated as $w_{H_2O} = \frac{m_2}{10^4 m_1}$, where m_1 is the mass of the sample injected into the KF device (in grams), and m_2 is the mass of water obtained by the KF device (in micrograms) (Torres et al. 2010).

As expected, water content increases with the bioethanol concentration in biodiesel, Figure 11. Namely, bioethanol and biodiesel are very hygroscopic. In the case of biodiesel, this fact comes from the strong polar interactions between esters and water. Free glycerol (coproduct during biodiesel production) also has a high affinity to water. Consequently, bioethanol–biodiesel blends are much more hydrophilic than mineral diesel. Therefore, some problems may occur during storage if the tank is not adequately sealed. In this case, the blend will absorb water from ambient humidity (Torres et al. 2010). For mineral diesel blends, the picture is quite similar, Figure 11. Water content also increases with bioethanol concentration because bioethanol is very hygroscopic; consequently, bioethanol–mineral diesel fuel blends are more hydrophilic than mineral diesel fuel. So, in case of inadequately sealed fuel tanks these blends will also absorb water from ambient humidity.

Sound Velocity

Sound velocity is a fuel property directly related to injection timing, which is a very important injection characteristics. Higher sound velocity results in advanced fuel injection timing and vice versa. Changed injection timing may cause significant variation in engine emissions (Kegl 2007).

The measurement of sound velocity in fuel is usually based on the principle of pressure wave propagation. The time difference between two pressure waves is typically measured on some specified length of the diesel engine high pressure tube. The sound velocity at various pressures up to 800 bars for neat fuels, bioethanol–biodiesel and bioethanol–mineral diesel fuel blends are shown in Figure 12.

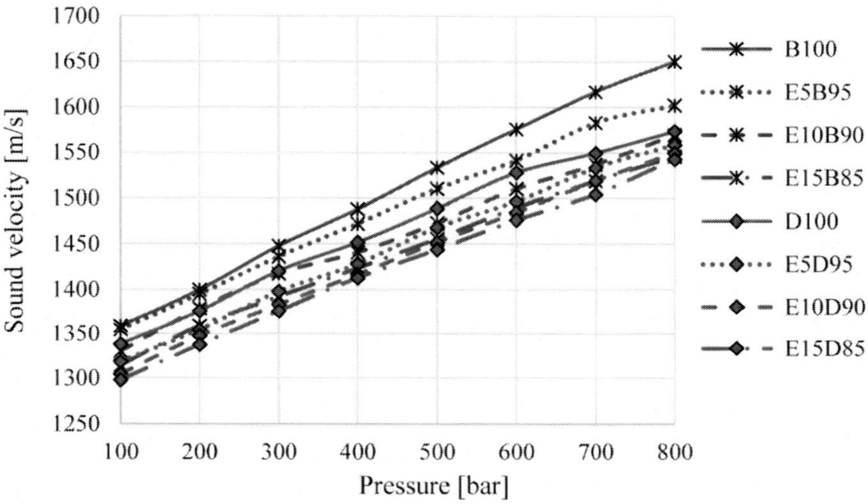

Figure 12. Sound velocity in tested fuels.

As can be seen, higher concentrations of bioethanol will cause the sound velocity to decrease. This leads to retarded injection timing in a mechanically controlled in-line injection system. Retarded injection timing typically results in decreased combustion temperature, which, among other consequences, may be one of the reasons for decreased NO_x emission.

Summary of Tested Fuels Properties

The physical and chemical properties of tested mineral diesel and its blends with bioethanol are summarized in Table 1. The values, corresponding to the standard EN590 for mineral diesel, are also listed for convenience.

The physical and chemical properties of tested biodiesel and its blends with bioethanol are summarized in Table 2. For convenience, the values of corresponding standard EN14214 for biodiesel, are also given.

All fuel properties influence fuel injection and combustion characteristics, and consequently engine emissions, fuel consumption, and engine performance. Their values are therefore indispensable for numerical simulation of injection processes.

Table 1. Physical and chemical properties of tested fuels with mineral diesel

Property	D100	Tested fuel			
	EN590	D100	E5D95	E10D90	E15D85
Density at 15 °C (kg/m^3)	820-845	837.3	834.3	831.7	829.4
CP (°C)	-	-3	-3	19	21
PP (°C)	-	-9	-9	-12	-36
CFPP (°C)	-	-8	-8	-7	-8
FP (°C)	>55	66	25	25	25
Lubricity (μm)	<460	448	399	406	395
Cetane index	>46	51.8	-	-	-
Sulphur content (mg/kg)	<50	31	-	-	-
Neat calorific value (MJ/kg)	-	42.9	-	-	-
Total contamination (mg/kg)	<24	2	-	-	-
Carbon residue (% m/m)	>0.3	<0.01	-	-	-
Element analysis (% w/w)		86.1 C 13.9 H 0 O	84.6 C 13.8 H 1.6 O	82.9 C 13.8 H 3.3 O	81.3 C 13.8 H 4.9 O
Kinematic viscosity at 40 °C (mm^2/s)	2-4.5	2.78	2.53	2.31	2.19
Water content (mg/kg)	<200	50	100	130	140
Corrosion Cu, 3 h at 50 °C	Class 1	1a	1a	1a	1a

Table 2. Physical and chemical properties of tested fuels with biodiesel

Property	B100	Tested fuel			
	EN14214	B100	E5B95	E10B90	E15B85
Density at 15 °C (kg/m^3)	860-900	882.6	878	873.5	869
CP (°C)	-	-10	-10		
PP (°C)	-	-3	-4		
CFPP (°C)	-	-6	-6		
FP (°C)	> 120	138.5	25	25	25
Lubricity (μm)	-	175	167	161	174
Sulphur content (mg/kg)	< 10	5.8	-	-	-
Neat calorific value (MJ/kg)	-	42.4	-	-	-
Total contamination (mg/kg)	< 2	14	-	-	-
Element analysis (% w/w)	-	76.7 C 11.1 H	76.9 C 11.4 H	77.0 C 11.3 H	76.4 C 10.9 H
Kinematic viscosity at 40 °C (mm^2/s)	3.5-5	4.477	4.041	3.575	3.244
Water content (mg/kg)	< 500	150	230	250	230
Corrosion Cu, 3 h at 50 °C	Class 1	1a	1a	1a	1a
Ester content (% m/m)	> 96.5	97.3	-	-	-
Iodine number (g/100g)	< 120	112	-	-	-
Property	B100	Tested fuel			
	EN14214	B100	E5B95	E10B90	E15B85
Acidity number (mg KOH/g)	<0.5	0.27	-	-	-
Linolenic acid methyl ester (% m/m)	<12	6.8	-	-	-
Phosphorous content (% m/m)	<10	<5	-	-	-
Monoglicerides content (% m/m)	<0.8	0.59	-	-	-
Diglycerides content (% m/m)	<0.2	0.14	-	-	-
Triglycerides content (% m/m)	<0.2	<0.05	-	-	-
Free glycerol (% m/m)	<0.02	0.006	-	-	-
Total glycerol (% m/m)	<0.25	0.176	-	-	-

Fuel Injection Characteristics

To achieve optimal engine performance and low engine emissions, the fuel injection system plays a very important role since it defines the fuel injection characteristics like the injection pressure, injection timing, fuelling, injection rate, and injection duration. Fuel injection characteristics, however, depend significantly on fuel properties (Tan et al. 2017, Kegl et al. 2008, Kegl et al. 2013, Kegl 2008a, Kegl 2008b). This is especially true for mechanically controlled in-line injection systems, although electronically controlled common rail injection systems are also affected (McCormick et al. 2001).

Introduction of new fuel types therefore requires careful investigation of the consequences caused by changed fuel properties. In this context, numerical simulation is very important and useful while experimental work is actually indispensable. For example, as already discussed, one of the main problems of blending two different fuels is that their chemical composition can lead to phase separation. If this happens, randomly variable fuel will be injected into the combustion chamber. Because the engine management is optimized for specific fuel properties, the engine characteristics will vary with respect to time. Such phenomena are too tricky to rely on simulation only; experimental work is here unavoidable.

The experimental data presented in the following was obtained by running tests with all considered fuels on a mechanically controlled fuel injection M system. The tests were run at various operating regimes, at ambient temperature, and at a constant pump injection timing. The considered injection system consists of a plunger-in-barrel assembly, high pressure (HP) tubes, and injectors. The scheme of the investigated system, which includes transducers for injection pressure, needle lift, and injection timing determination, is shown in Figure 13. The main specifications of this system are given in Table 3.

Table 3. Test injection system main specification

Injection model	**Direct injection system**
Fuel injection type	Bosch PES 6A 95D 410 LS 2542
Pump plunger (diameter x lift)	9.5 mm x 8 mm
Fuel tube (length x diameter)	1024 mm x 1.8 mm
Injector (number x nozzle hole diameter)	1 x 0.68 mm
Needle (maximal lift)	0.3 mm
Needle (opening pressure)	175 bar
Pump injection timing	30 CA BTC

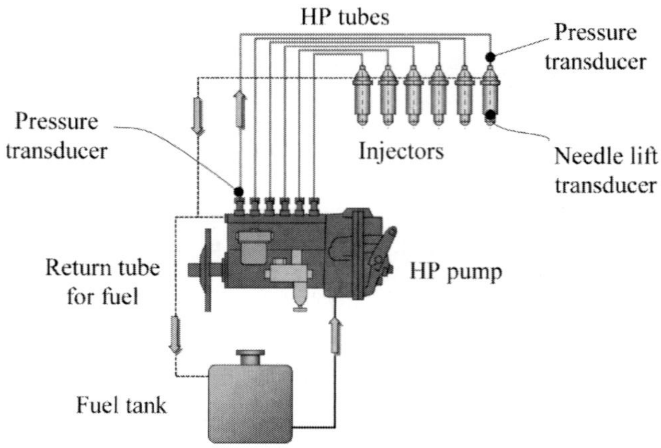

Figure 13. Scheme of the tested fuel injection system.

Injection Pressure

Although tests were run on all operating conditions, only two of them, being perhaps the most illustrative and interesting, are presented in the following: peak torque and rated condition.

Injection pressure histories for all tested fuels at peak torque condition (full load, 850 rpm) are presented in Figure 14. It can be seen that the addition of bioethanol decreases the maximal injection pressure. It is also evident that maximal injection pressure decreases more, when bioethanol is added to biodiesel than to mineral diesel. Furthermore, the history of injection pressure of E15B85 blend is very close to that of neat D100.

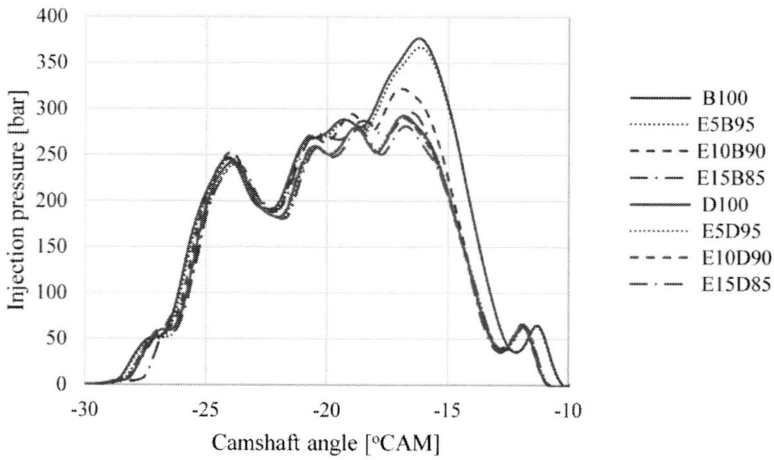

Figure 14. Injection pressure at peak torque condition.

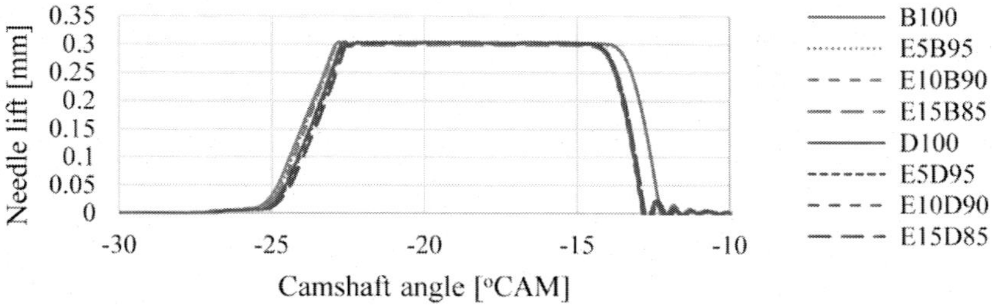

Figure 15. Needle lift at peak torque condition.

The influence of bioethanol in biodiesel and mineral diesel fuels at rated condition (full load, 1100 rpm) is similar to that obtained at peak torque condition. From Figures 16 and 17, the injection characteristics comparison confirms that E15B85 blend gives very close injection system behavior to D100.

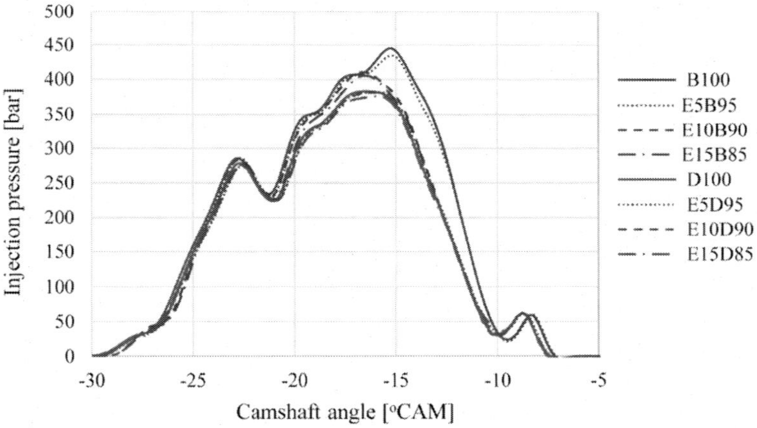

Figure 16. Injection pressure at rated condition.

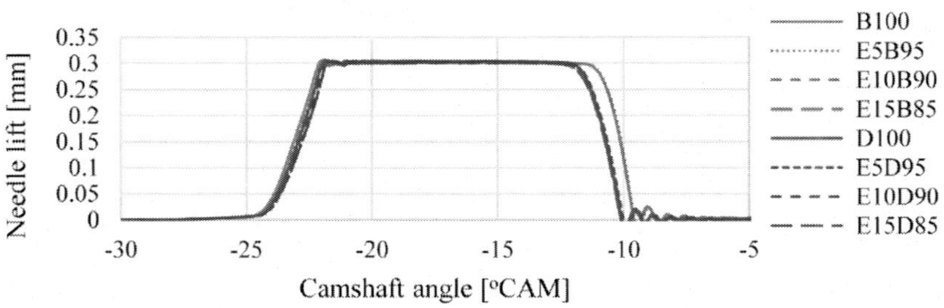

Figure 17. Needle lift at rated condition.

Experimental results of injection pressures at other operating conditions reflect quite well the findings at peak torque and rated conditions. In general, it can be observed that the maximum injection pressure decreases with increasing bioethanol content in biodiesel, which is expected to have a beneficial effect on NO_x emissions. Meanwhile, the variation of this injection characteristic is not significant or the pressure is only slightly lower when bioethanol is added to mineral diesel fuel.

The difference between the injection pressures of different fuels arises due to different fuel density, viscosity, bulk modulus, and sound velocity. The maximum injection pressure for bioethanol–biodiesel blends is always somewhat higher than that for mineral diesel. In general, mean injection pressure has to be as high as possible while keeping the maximal injection pressure at a low level in order to reduce harmful NO_x emissions (Kegl 2006, Kegl 2007). Therefore, a change from a mineral diesel to biodiesel necessitates some tuning of the injection system to lower the maximum injection pressure. Adding enough bioethanol, however, might make even this tuning unnecessary.

Injection Timing

Injection timing or start of injection is a very important parameter that influences significantly all engines characteristics due to the fact that it influences the mixing quality of the air–fuel mixture and, consequently, the combustion process and harmful emissions.

Injection timing is obtained from the needle lift history. Two most illustrative needle lift histories are shown in Figure 16 (peak torque condition) and in Figure 17 (rated condition). It can be observed that biodiesel injection timing is advanced with respect to that of mineral diesel and biodiesel injection duration is longer. The addition of bioethanol retards the injection timing in mineral diesel and biodiesel. It has to be pointed out, however, that this influence is more significant in biodiesel. Furthermore, the needle lift history of E15B85 is practically the same as that of D100. This means that the injection timing, injection duration, and injection delay of E15B85 is very close to those of D100.

The variations in injection timing are mainly due to the differences in density and consequently in bulk modulus, which affects the sound velocity. According to this, it is expected that higher concentrations of bioethanol will cause retarded injection timing. Nevertheless, viscosity also affects injection timing. Lower viscosity means lower friction while the fuel travels through the high pressure tube and through the nozzle, leading to an easier propagation and consequently to an advanced injection timing. Lower density of bioethanol, compared to that of biodiesel, is however the predominant property that affects the injection timing because bioethanol addition to biodiesel leads to retarded injection timing. In case of bioethanol–mineral diesel blends, low bioethanol concentrations lead to negligible variations or a slight advance in injection timing. This is probably caused by the lower viscosity of bioethanol compared to that of mineral diesel. Nevertheless, this tendency is not very clear. Due to the fact that injection timing is advanced for biodiesel with respect to mineral diesel, adding bioethanol to biodiesel has the advantage of bringing it nearer to the mineral diesel. At rated condition, a low concentration of bioethanol causes biodiesel to reach mineral diesel fuel injection timing, meanwhile at other tested operating regimes, higher concentrations of bioethanol in biodiesel are needed. With the advanced injection timing of biodiesel, the pressure in the cylinder increases and the NO_x emissions increase also. So, from the results obtained, it can be expected that the addition of bioethanol to biodiesel offers a possibility to reduce NO_x emissions, because the injection timing is retarded. Nevertheless and for the same reason, it can be expected that NO_x emissions will be higher in case of bioethanol–mineral diesel fuel blends. Moreover, this increment is also enhanced by the improved combustion efficiency, provided by the bioethanol oxygen content.

Fuelling

Fuelling is the amount of fuel injected during one cycle. Therefore, it is obviously directly related to injection timing and duration.

Figure 18 shows the fuelling at 100%, 75%, 50% and 25% loads and at various pump rotational speeds. At constant pump injection timing and a constant rack position, the fuelling of B100, bioethanol–biodiesel and bioethanol–mineral diesel blends is higher or equal than the fuelling of D100. B100 always shows the highest fuelling at all tested operating regimes. One can observe no significant changes or a slightly decrease in fuelling with increased bioethanol content in biodiesel. Meanwhile, the fuelling increases for higher concentration of bioethanol in mineral diesel. At all tested loads, except the 25% load, the bioethanol addition influence on fuelling is more evident in biodiesel than in mineral diesel fuel. At all tested operating regimes, the fuelling obtained with E15B85 closes to that of D100.

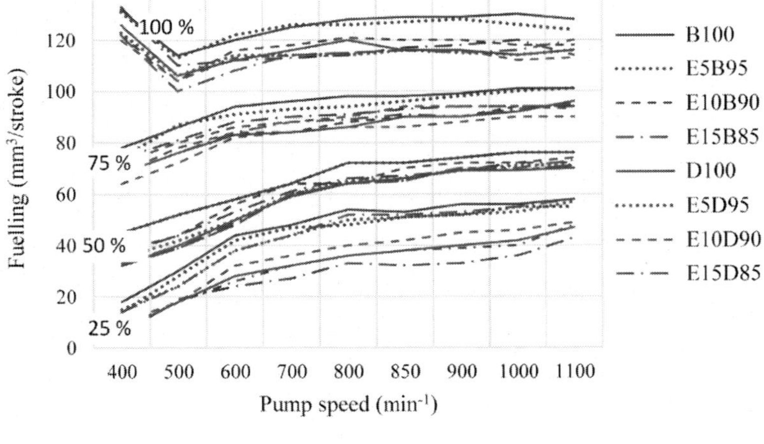

Figure 18. Fuelling of tested fuels at various operating regimes.

The observed fuelling variation is caused by the variation in densities and viscosities of the tested fuels. The higher fuelling of B100 is a consequence of its higher bulk modulus and its higher kinematic viscosity.

Mean Injection Rate

Mean injection rates at various operating conditions are shown in Figure 19. In case of mineral diesel fuel blends, the higher fuelling caused by bioethanol addition is also reflected in a slight increase of the mean injection rate. In case of bioethanol addition to biodiesel, the mean injection rate decreases. In all cases tested the influence of bioethanol addition on mean injection rate is more evident in biodiesel than in mineral diesel fuel. The differences in mean injection rate among tested fuels are higher at lower loads.

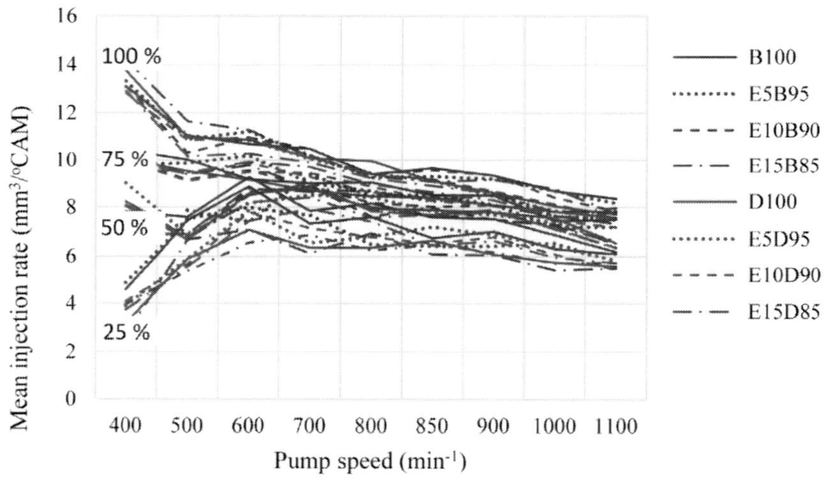

Figure 19. Mean injection rate of tested fuel at various operating regimes.

Injection Duration

Figure 20 shows the injection duration of tested fuels at various engine operating regimes. At all regimes the tendency is that an increment of bioethanol concentration in biodiesel is resulting in decreased injection duration; meanwhile the duration increases or it does not change substantially when bioethanol is added to mineral diesel.

The results show that the injection duration of pure biodiesel is always higher than that of mineral diesel. The decreased injection duration of bioethanol–biodiesel blends is mainly due to the injection timing variations caused by bioethanol addition. On the other hand, slightly advanced injection timing, caused by the addition of bioethanol to mineral diesel, leads to longer injection duration.

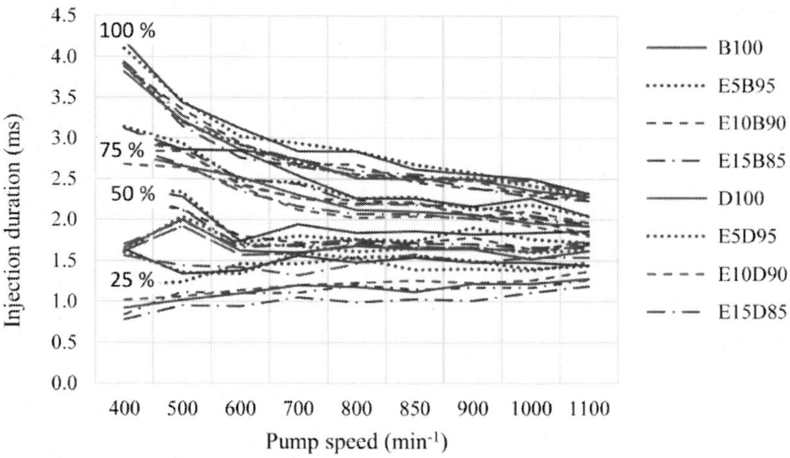

Figure 20. Injection duration of tested fuels at various operating regimes.

The obtained experimental results of injection characteristics show that the influence of bioethanol in biodiesel is much more significant that in mineral diesel. By addition of 15% of bioethanol to biodiesel, the injection characteristics at most operating regimes are relatively close to those of mineral diesel, which offers an opportunity to minimize the required engine modifications when switching from one fuel to another one.

WEAR AND DEPOSITS IN FUEL INJECTION SYSTEM COMPONENTS

A change from mineral diesel to bioethanol-biodiesel blends obviously influences injection characteristics. Besides of this, it may also have mechanical consequences for the injection system. Namely, fuel change may cause variations in wear and deposits observed on the system after some period of operation. Therefore, new fuels to be introduced have to be investigated carefully also in this context.

The data presented here is based on experimental work done with the B100, E15B85, D100, and E15D85 fuels at one operating condition. To get some insight into wear and deposits consequences, caused by the tested fuels, the nozzle discharge coefficient and the surface roughness of various moving parts were determined. Changes in the nozzle discharge coefficient may indicate deposits in the system or wear due to erosion. On the other side, surface roughness of moving parts can indicate wear due to relative movement of various parts.

In order to ensure comparable results, the testing of various fuels (B100, E15B85, D100, E15D85) was performed at the same operating conditions. For this reason, a device for simultaneous testing of two various fuels was developed. After running the fuel injection system on this device for some period of time, the nozzle discharge coefficient and surface roughness of inner injection system parts were investigated.

The device for simultaneous testing of two fuels is shown schematically in Figure 21. The low pressure gallery of the high pressure (HP) pump is divided into two partitions so that two fuels can enter and be pumped separately. Both tested fuels are delivered from two separate tanks by two low pressure pumps at 1 bar. The HP pump is driven by an electromotor at a constant pump speed of 750 rpm. Each tested fuel flows through 3 sections of its partition of the HP pump and through 3 HP tubes to the corresponding 3 injectors. After injection, each fuel is returned to its corresponding tank.

The nozzle discharge coefficient is defined as the ratio between the measured actual volume flow and theoretical volume flow, injected through the nozzle. The flow coefficient-measuring device was used to measure the fuel flow coefficients at steady state conditions. The procedure is commonly known as the Bosch procedure for measuring flow coefficients.

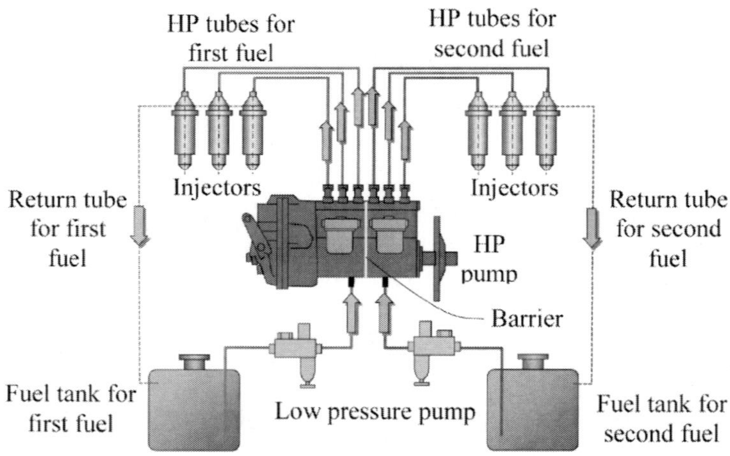

Figure 21. Scheme of the device for simultaneous testing of two fuels.

It was expected that the nozzle discharge coefficient value would change, if there would some deposits start to build up in the vicinity of the nozzle, or, if there would be some erosion in the nozzle, e.g., as a consequence of cavitation phenomena. After running the system for 500 hours, however, no significant changes in the nozzle discharge coefficient could be observed. So, the obtained results indicate, that the addition of bioethanol to biodiesel and to the mineral diesel seems not to cause any serious wear-related problems in the nozzle.

In order to examine wear of the moving parts of the injection system, a device consisting from a pick-up RHT 3/6 and a rotary drive unit PURV 3-100 was used. The pick-up RHT 3/6 enables surface roughness measurements inside bores with a diameter greater than 3 mm and depth of up to 10 mm as well as bores with a diameter greater than 6 mm and depth of up to 20 mm. The rotary drive unit PURV 3-100 can be used for measurements on the internal and external surfaces of axis-symmetrical components with diameters between 1 and 100 mm (Kegl et al. 2013).

To determine the influence of tested fuels usage, the roughness of various moving parts was analyzed before and after running the system for a specified period of time. The surface roughness was estimated by: (a) the arithmetic roughness average R_a, (b) the average peak to valley height R_z, and (c) the maximum peak to valley height R_{max}. These parameters were determined in the circular and longitudinal directions of the inspected parts (Kegl et al. 2013). The inspected parts were: the pump plunger, pump plunger sleeve, injector nozzle, and injector nozzle sleeve. The following measurements were done: (i) before usage (of new tested parts), (ii) after 250 hours of running the system with B100 and with D100 and at the same operating conditions, (iii) after 250 hours of running the system with E15B85 and E15D85 at the same operating conditions.

Relative average surface roughness parameters of the pump plunger, measured in the circular direction, are shown in Figure 22. It is evident that the usage of B100 as well as D100 after 250 hours reduces all surface roughness parameters by about 40% to 50%. The added bioethanol to biodiesel as well as to mineral diesel reduces furthermore the parameters by about 10%.

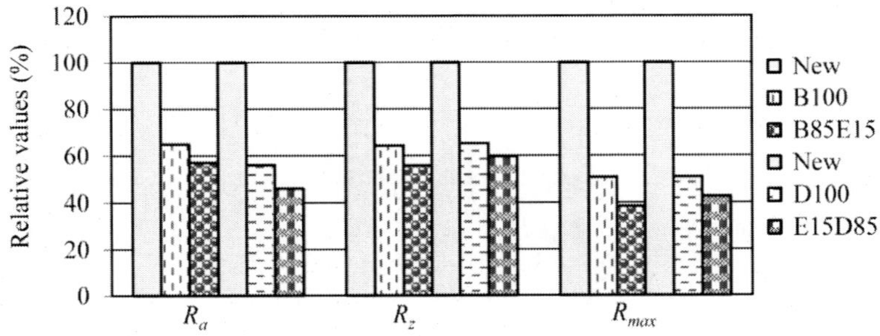

Figure 22. Relative average surface roughness parameters - circular measurement of the pump plunger.

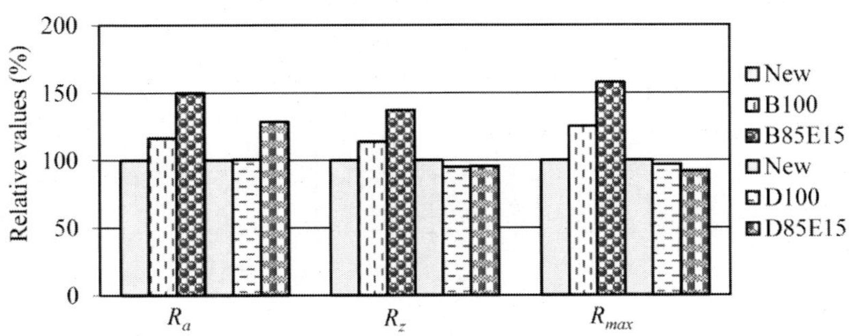

Figure 23. Relative average surface roughness parameters - linear measurement of the pump plunger.

Relative average surface roughness parameters of the pump plunger, measured in the linear direction, are shown in Figure 23. It is evident that the usage of B100 increases the average values of roughness parameters by about 20%. By adding bioethanol to biodiesel increases these parameters by further 30%. The influence of D100 after 250 hours is negligible. But, by adding bioethanol, the parameter R_a is increased by about 30%; meanwhile, the parameters R_z and R_{max} did not changed significantly.

Relative average surface roughness parameters of the needle, measured in the circular direction, are shown in Figure 24. It is evident that the usage of B100 increases the parameters R_a and R_{max} by about 40% and the parameter R_z by about 10%. The added bioethanol in biodiesel doesn't change R_a, but increases R_z and R_{max} by further 20% with respect to B100. The usage of D100 reduces all parameters by about 15%. The added bioethanol in mineral diesel has no effect on the parameter R_a, meanwhile it reduces the other two parameters by 15% with respect to D100.

Relative average surface roughness parameters of the needle, measured in the linear direction, are shown in Figure 25. It is evident that the usage of B100 and D100 decreases R_a, R_z and R_{max}. The added bioethanol in biodiesel decreases these parameters further. Meanwhile, bioethanol in mineral diesel increases the roughness parameters.

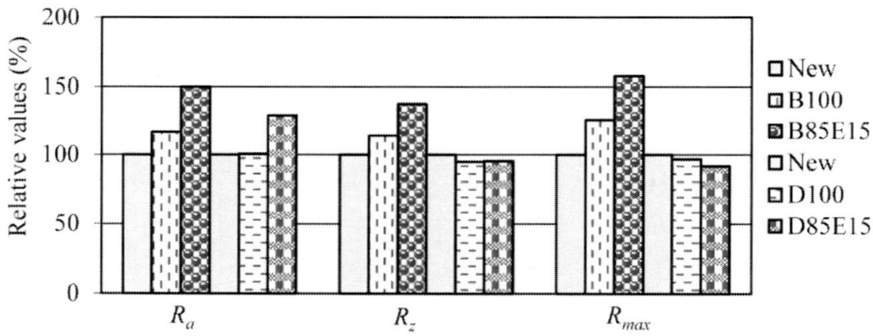

Figure 24. Relative average surface roughness parameters - circular measurement of the needle.

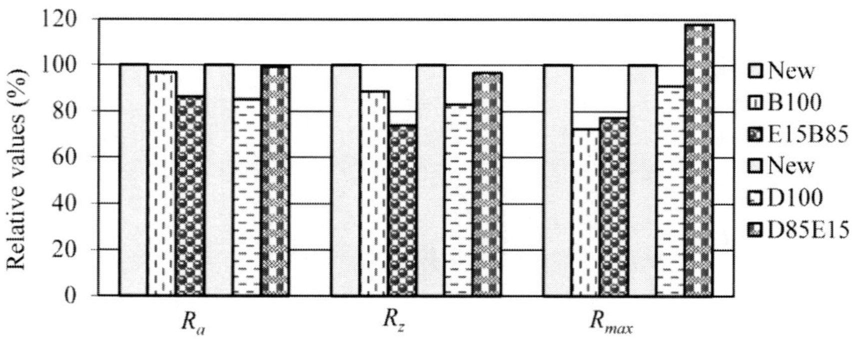

Figure 25. Relative average surface roughness parameters - linear measurement of the needle.

In general, it can be concluded that the influence of bioethanol addition on surface roughness is not uniform. In some cases the indicators are improved and in some cases they are worse. In any case, it looks that bioethanol addition should not have any serious negative consequences related to wear of injection system components.

A GLANCE AT FUEL SPRAY CHARACTERISTICS

Diesel engine fuel spray development and its characteristics are determined by sophisticated physical and chemical processes. Experimental and numerical investigation of fuel spray characteristics is therefore a wide and difficult scientific field.

In this work no attempt is made to cover this field rigorously. Instead, a modest amount of experimental data is presented, in order to offer just a few of the most basic observations that can be made when mineral diesel is blended with bioethanol.

As already thoroughly discussed, addition of bioethanol to mineral diesel changes fuel properties and consequently injection characteristics. Both of these variations influence fuel spray development and to get some basic insight, some experimental results obtained with bioethanol-mineral diesel blends are presented in the following.

Figures 26-33 show photographs of fuel spray development along with corresponding needle lift histories. The fuels considered are D100 and E15D85; the operating conditions are peak torque and rated conditions; the fuel is injected into a glass chamber with inner pressure of either 40 or 60 bar.

A general observation is that the D100 fuel spray angle is narrower and the penetration length is longer than in the case of E15D85. It also seems that E15D85 disintegrates more easily and its Sauter mean diameter is actually lower. Some of the most important reasons for this are higher fuel vaporization rate of bioethanol blend, better atomization properties, and lower injection pressure of E15D85. Better atomization of E15D85 is a consequence lower kinematic viscosity and density compared to D100. This also explains the narrower spray angle and shorter spray tip penetration lengths.

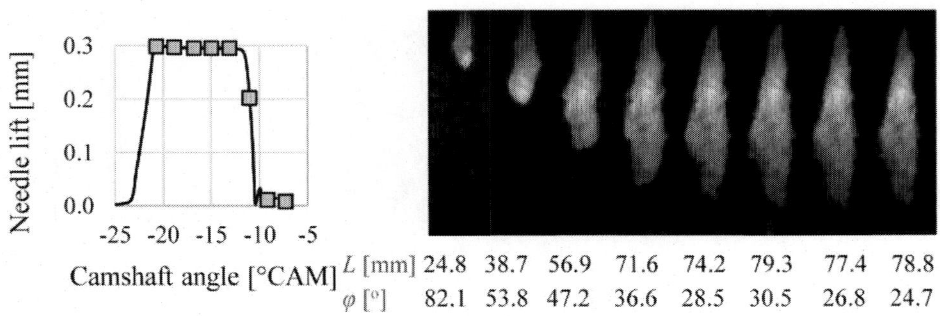

Figure 26. Fuel spray of D100; chamber at 40 bar; peak torque condition.

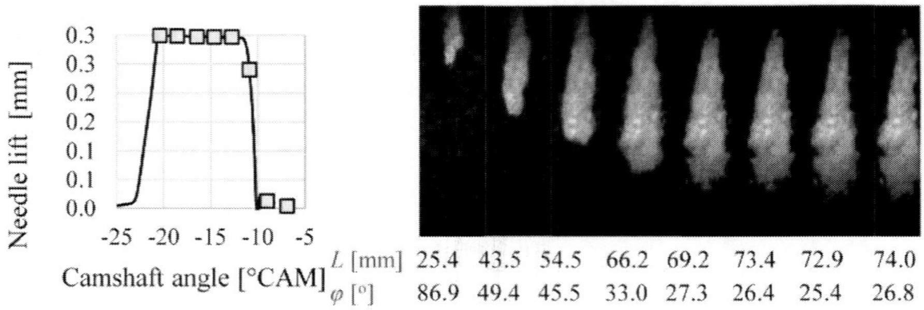

Figure 27. Fuel spray of E15D85; chamber at 40 bar; peak torque condition.

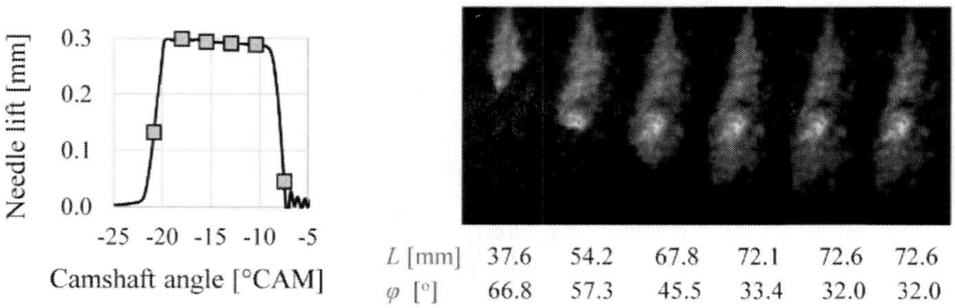

Figure 28. Fuel spray of D100; chamber at 40 bar; rated condition.

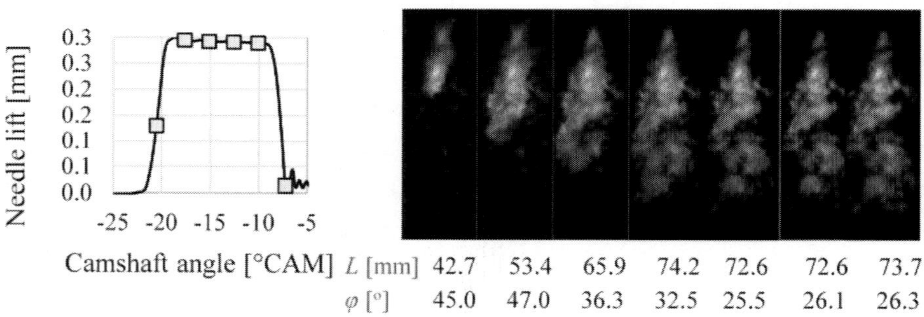

Figure 29. Fuel spray of E15D85; chamber at 40 bar; rated condition.

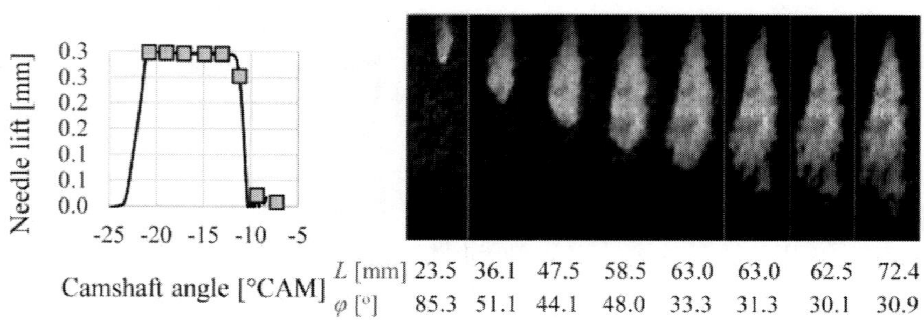

Figure 30. Fuel spray of D100; chamber at 60 bar; peak torque condition.

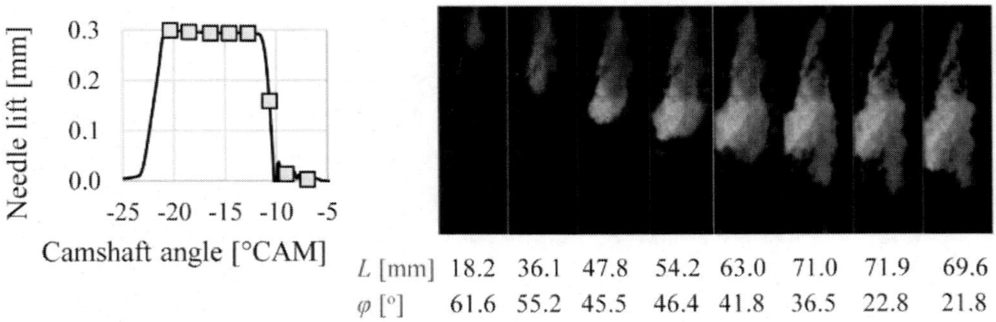

Figure 31. Fuel spray of E15D85; chamber at 60 bar; peak torque condition.

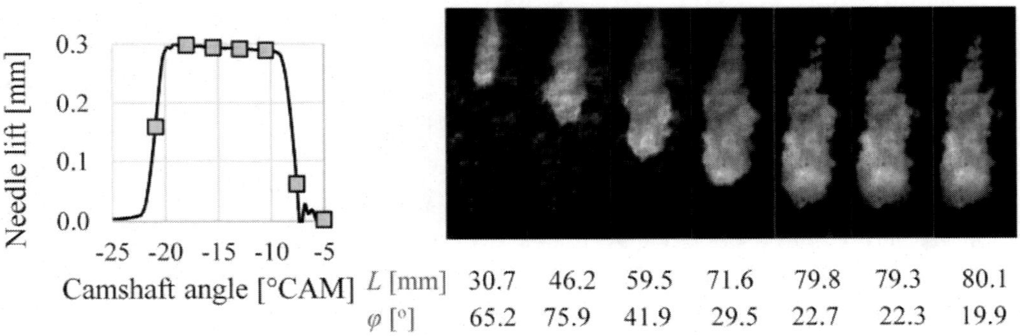

Figure 32. Fuel spray of D100; chamber at 60 bar; rated condition.

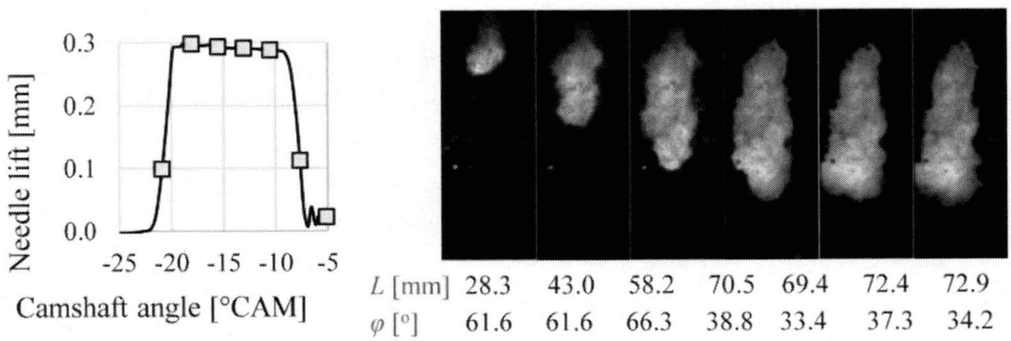

Figure 33. Fuel spray of E15D85; chamber at 60 bar; rated condition.

Experiments with bioethanol-biodiesel blends are not shown here. One may expect however, that the trends observed with these blends would be roughly the same as the addition of bioethanol to biodiesel has a similar effect on the kinematic viscosity, density and injection pressure.

A Glance at Engine Characteristics

Diesel engine characteristics are determined by a series of sophisticated processes, including the fuel spray development (Kegl 2008a). In this work no attempt is made to go into any details. Instead, bioethanol addition is illustrated very briefly and only in the context of engine emission characteristics.

In general, one has to say clearly that there are no simple and universal conclusions, except maybe in some quite rare cases. For example, bioethanol–mineral diesel fuel blends and bioethanol–biodiesel blends may undoubtedly bring benefits in the quest to lower sulfur emissions in order to meet the ever more demanding regulations. This is a consequence of the simple fact that bioethanol sulfur content is virtually zero.

As said there are no universal conclusions about bioethanol addition because there are simply too many influencing factors and too many variable circumstances to be taken into account. Therefore, just to illustrate the possibilities of bioethanol addition, one may first take a look at interesting experimental investigations performed on a single cylinder, four-stroke, and air-cooled diesel engine by using mineral diesel–biodiesel–bioethanol blends (Tan et al. 2017).

The experimentally obtained results show that, compared to mineral diesel, a reduction of NO_x up to about 45% is possible by using proper mineral diesel–biodiesel–bioethanol fuel blends in the lower range of engine speeds, Figure 34. On the other hand, it was observed that nitrogen oxides (NO_x) emissions gradually increase at high engine speed, and eventually meet the emission delivered by mineral diesel. Obviously, the addition of bioethanol reduces the cetane number and the low heating value consequently results in better air and fuel mixing during ignition and lower brake power output. Furthermore, the fuel vaporization property is enhanced, which reduces the in-cylinder combustion temperature. This is known to lower the NO_x emission. At higher engine speeds, however, the blended fuel is burning faster with higher in-cylinder combustion temperature; this increases the formation of NO_x emission.

Figure 34 illustrates also the carbon monoxide (CO) emissions for mineral diesel and diesel–biodiesel–bioethanol blend at various engine speeds. At lower speeds, the CO emission of mineral diesel was significantly higher than that of the fuel blend. However, at higher engine speeds the CO emission of tested fuels becomes comparable. The inverse dependence of mineral diesel CO emission on engine speed is probably due to the variation of the air to fuel ratio. When this ratio is low, incomplete combustion occurs and the amount of generated CO is relatively high. With increasing engine speed, however, the air to fuel ratio increases and more CO is converted to CO_2. The blended fuel, on the other hand, contains more oxygen. This improves the oxygen to fuel ratio in the fuel rich regions, which results in more complete combustion. In contrast to mineral diesel, the CO emission of the blended fuel increases with the increasing engine. Namely, at higher speeds the evaporative cooling effect of bioethanol in the emulsion becomes significant. This leads to reduced flame temperature and burning velocity, which indirectly increases the CO emission.

Regarding the total hydrocarbons (THC) one may notice that many researchers found significant reduction in THC emission resulting from mineral diesel–biodiesel–bioethanol blends, compared to mineral diesel (Park et al. 2012, Randazzo & Sodré 2011, Lu et al. 2008, Lapuerta et al. 2008, Nabi et al. 2006) while some of them found moderate decrease (Pang et al. 2006). Fuel blends of bioethanol and mineral diesel are found to increase THC, but adding biodiesel to this blends was found to reduce THC compared to mineral diesel (Shahir et al. 2015). Probably, this is because biodiesel's high cetane number causes shorter ignition delay and improves efficiency of the combustion, which in turn reduces the unburned hydrocarbon. In addition to this, the presence of oxygen in the molecules of biodiesel intensifies the post flame oxidation process of still unburned hydrocarbons.

Regarding the particulate matter (PM) emission, some authors have suggested that this varies from vehicle to vehicle (Park et al. 2010). But most investigations have shown reduced PM emission, even up to about 50% of mineral diesel–biodiesel–bioethanol blends, compared to mineral diesel. This was observed irrespective of the operating conditions (Randazzo & Sodré 2011, Shahir et al. 2015). This reduction might be due to the fact that by the formation of hydroxyl radical, the oxygen component consumes the soot precursors, which results in a lower soot formation.

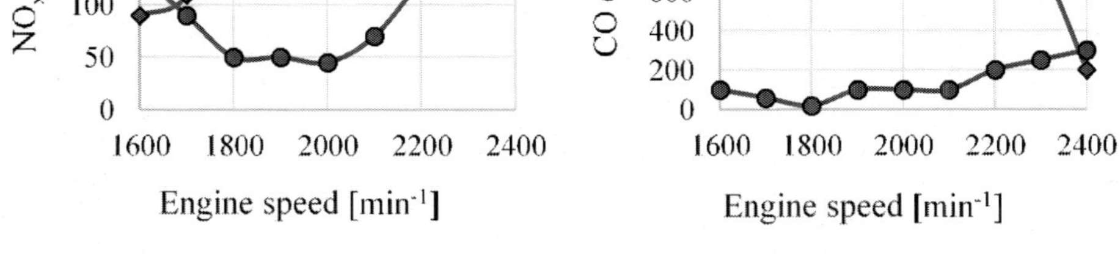

Figure 34. NO_x and CO emissions for diesel–biodiesel–bioethanol blends.

Conclusion

Based on the comparison of tested fuels properties and their usage in diesel engines in the considered experiments, one can conclude that:

- at moderately low temperatures, bioethanol separates from mineral diesel fuel, meanwhile bioethanol-biodiesel blends remain in one phase with no changes in stability, color and aggregation; bioethanol added to biodiesel acts as a winter additive,
- addition of bioethanol to mineral diesel and to biodiesel decreases fuel density, kinematic viscosity, cetane index, pour point, flash point, and sound velocity, meanwhile the water content increases with addition of bioethanol,
- bioethanol in mineral diesel and in biodiesel decreases the injection pressure, leading to lower NO_x emissions,
- injection timing is advanced for biodiesel with respect to mineral diesel fuel; by adding bioethanol to biodiesel has the advantage of bringing the injection timing of the blend closer to the one of mineral diesel,
- addition of bioethanol to biodiesel slightly decreases fuelling. Meanwhile the fuelling increases for higher concentration of bioethanol in mineral diesel, leading to a slight increase of the mean injection rate; in the case of addition of bioethanol to biodiesel, the mean injection rate decreases,
- the presented experimental results of injection characteristics show that the influence of bioethanol in biodiesel is much more significant that in mineral diesel; by addition of 15% of bioethanol to biodiesel, the injection characteristics at most operating regimes become closer to those of mineral diesel,
- addition of bioethanol to mineral diesel and to biodiesel has moderate or minor effects on average surface roughness parameters; the results vary in dependence on the measuring type and measured component of the injection system,
- addition of bioethanol to mineral diesel leads to shorter spray tip penetration length at rated and peak torque condition (at injection chamber pressures of 40 bar and 60 bar). Bioethanol addition causes a narrower spray at 40 bar at peak torque and rated condition; meanwhile, at 60 bar the spray is wider,
- finally, bioethanol to biodiesel addition has a potential to reduce the main harmful emissions. For more precise conclusions, however, one would need to investigate thoroughly the specific problem at hand.

REFERENCES

Aydın, F., Öğüt, H., (2017). Effects of using ethanol-biodiesel-diesel fuel in single cylinder diesel engine to engine performance and emissions. *Renewable Energy* 103, 688–694.

Carlucci, A. P., Laforgia, D., Saracino, R., Toto, G., (2011). Combustion and emissions control in diesel–methane dual fuel engines: The effects of methane supply method combined with variable in-cylinder charge bulk motion. *Energy Conversion and Management* 52/8-9, 3004–3017.

Dharma, S., Ong, H. C., Masjuki, H. H., Sebayng, H. H., Silitonga, A. S., (2016). An overview of engine durability and compatibility using biodiesel–bioethanol–diesel blends in compression-ignition engines. *Energy Conversion and Management* 128, 66–81.

Karagöz, Y., Sandalcı, T., Yüksek, L., Dalkılıc, A. S., Wongwises, S., (2016). Effect of hydrogen–diesel dual-fuel usage on performance, emissions and diesel combustion in diesel engines. *Advances Mechanical Engineering* 8/7, 1–13.

Kegl, B., Kegl, M. Pehan, S., (2013). *Green diesel engines: Biodiesel usage in diesel engines*. London: Springer.

Kegl, B., (2011). Influence of biodiesel on engine combustion and emission characterisitcs. *Applied Energy* 88, 1803–1812.

Kegl, B., Kegl, M., Pehan, S., (2008). Optimization of a fuel injection system for diesel and biodiesel usage. *Energy & Fuels* 22, 1046–1054.

Kegl, B., (2008a). Effects of biodiesel on emissions of a bus diesel engine. *Bioresource Technology* 99, 863–873.

Kegl, B., (2008b). Biodiesel usage at low temperature. *Fuel* 87, 1306–1317.

Kegl, B., (2007). NO_x and particulate matter (PM) emissions reduction potential by biodiesel usage. *Energy & Fuels* 21, 3310–3316.

Kegl, B., (2006). Numerical analysis of injection characterisitcs using biodiesel fuel. *Fuel* 85, 2377–2387.

Khan, M. I., Yasmeen, T., Shakoor, A., Khan, N. B., Wakeel, M., Chen, B., (2016). Exploring the potential of compressed natural gas as a viable fuel option to sustainable transport: A bibliography (2001-2015). *Journal of Natural Gas Science and Engineering* 31, 351–381.

Kuszewski, H., Jaworski, A., Ustrzycki, A., Lejda, K., Balawender, K., Wos, P., (2017). Use of the constant volume combustion chamber to examine the properties of autoignition and derived cetane number of mixtures of diesel fuel and ethanol. *Fuel* 200, 564–575.

Kwanchareon, P., Luengnaruemitchai, A., Jai-In, S., (2007). Solubility of a diesel-biodiesel-ethanol blend, its fuel properties, and its emission characteristics from diesel engine. *Fuel* 86/7–8, 1053–1061.

Lapuerta, M., Rodríguez-Fernández, J., Fernández-Rodríguez, D., Patiño-Camino, R., (2017) Modelind viscosity of butanol and ethanol blends with diesel and biodiesel fuels. *Fuel* 199, 332–338.

Lapuerta, M., García-Contreras, R., Campos-Fernández, J., Dorado, M. P., Stability, lubricity, viscosity, and cold-flow properties of alcohol-diesel blends. *Energy & Fuels* 24, 4497–4502.

Lapuerta, M., Armas, O., Herreros, J. M., (2008). Emissions from a diesel–bioethanol blend in an automotive diesel engine. *Fuel* 87/1, 25–31.

Lilik, G. K., Zhang, H., Herreros, J. M., Haworth, D. C., Boehman, A. L., (2010). Hydrogen assisted diesel combustion. *International Journal of Hydrogen Energy* 35, 4382–4398.

Lounici, M. S., Boussadi, A., Loubar, K., Tazerout, M., (2015). Experimental investigation on NG dual fuel engine improvement by hydrogen enrichment. *International Journal of Hydrogen Energy* 39, 21297–21306.

Lu, X., Ma, J., Ji, L., Huang, Z., (2008). Simultaneous reduction of NOx emission and smoke opacity of biodiesel-fueled engines by port injection of ethanol. *Fuel* 87/7, 1289–96.

McCormick, R. L., Graboski, M. S., Alleman, T. L., Herring, A. M., Tyson, K. S., (2001). Impact of biodiesel source material and chemical structure on emissions of criteria pollutants from a heavy-duty engine. *Environmental Science & Technology* 35/9, 1742–1747.

Nabi, M. N., Akhter, M. S., Zaglul Shahadat, M. M., (2006). Improvement of engine emissions with conventional diesel fuel and diesel–biodiesel blends. *Bioresource Technology* 97/3, 372–8.

Pang, X., Shi, X., Mu, Y., He, H., Shuai, S., Chen, H., Li, R., (2006). Characteristics of carbonyl compounds emission from a diesel-engine using biodiesel–ethanol–diesel as fuel. *Atmospheric Environment* 40, 7057–7065.

Park, S. H., Cha, J., Lee, C. S., (2012). Impact of biodiesel in bioethanol blended diesel on the engine performance and emissions characteristics in compression ignition engine. *Applied Energy* 99, 334–43.

Park, S. H., Cha, J., Lee, C. S., (2010). Effects of bioethanol-blended diesel fuel on combustion and emission reduction characteristics in a direct-injection diesel engine with exhaust gas recirculation (EGR). *Energy & Fuels* 24, 3872–83.

Poompipatpong, C., Cheenkachorn, K., (2011). A modified diesel engine for natural gas operation: Performance and emission tests. *Energy* 36/12, 6862–6866.

Randazzo, M. L., Sodré, J. R., (2011). Exhaust emissions from a diesel powered vehicle fuelled by soybean biodiesel blends (B3–B20) with ethanol as an additive (B20E2–B20E5). *Fuel* 90, 98–103.

Reyes, Y., Aranda, D. A. G., Santander. L. A. M., Cavado, A., Belchior, C. R. P., (2009). Action principles of cosolvent additives in ethanol-diesel stability studies. *Eenrgy & Fuels* 23/5, 2731–2735.

Sahoo, B. B., Sahoo, N., Saha, U. K., (2009). Effect of engine parameters and type of gaseous fuel on the performance of dual-fuel gas diesel engines – A critical review. *Renewable and Sustainable Energy Reviews* 13/6-7, 1151–1184.

Saravanan, N., Nagarajan, G., (2010). Performance and emission studies on port injection of hydrogen with varied flow rates with Diesel as an ignition source. *Applied Energy* 87, 2218–2229.

Shahir, S. A., Masjuki, H. H., Kalam, M. A., Imran, A., Ashraful, A. M., (2015). Performance and emission assessment of diesel–biodiesel–ethanol/bioethanol blend as a fuel in diesel engines: A review. *Renewable and Sustainable Energy Reviews* 48, 62–78.

Surawski, N., Miljevic, B., Roberts, B. A., Modini, R., Situ, R., Brown, R. J., Bottle, S. E., Ristovski, Z. D., (2010). Particle emissions, volatility, and toxicity from an ethanol fumigated compression ignition engine. *Environmental Science and Technology* 44, 229–235.

Tan, Y. H., Abdullah, M. O., Nolasco-Hipolito, C., Zauzi, N. S. A., Abdullah, G. W., (2017). Engine performance and emissions characteristics of a diesel engine fueled with diesel-biodiesel-bioethanol emulsions. *Energy Conversion and Management* 132, 54–64.

Torres-Jimenez, E., Svoljšak-Jerman, M., Gregorc, A., Lisec, I., Dorado, M. P., Kegl, B., (2010). Physical and chemical properties of ethanol−biodiesel blends for diesel engines. *Energy & Fuels* 24, 2002–2009.

Torres-Jimenez, E., Svoljšak-Jerman, M., Gregorc, A., Lisec, I., Dorado, M. P., Kegl, B., (2011). Physical and chemical properties of ethanol−diesel fuel blends. *Fuel* 90, 795–802.

Wei, L., Geng, P., (2016). A review on natural gas/diesel dual fuel combustion, emissions and performance. *Fuel Processing Technology* 142, 264–278.

Zaharin, M. S. M., Abdullah, N. R., Najafi, G., Sharudin, H., Yusaf, T., (2017). Effects of physicochemical properties of biodiesel fuel blends with alcohol on diesel engine performance and exhaust emissions: A review. *Renewable and Sustainable Energy Reviews* 79, 475–493.

In: Bioethanol and Beyond
Editor: Michel Brienzo
ISBN: 978-1-53613-478-0

Chapter 14

HYDROGEN PRODUCTION FROM BIOETHANOL ON MIXED OXIDE CATALYSTS

Lidiya Yu. Dolgikh*, Yuri I. Pyatnitsky and Peter E. Strizhak
L. V. Pisarzhevskii Institute of Physical Chemistry,
National Academy of Sciences of Ukraine, Kyiv, Ukraine

ABSTRACT

Ethanol steam reforming (ESR) is a promising way for the hydrogen production from renewable resources. This paper describes the thermodynamic analysis of the ESR and presents the experimental study of the process over the following catalysts: (1) Cu and Cu-M (M = Ni, Co, Fe) on several types of zirconia supports; (2) MFe_2O_4 (M = Ni, Co, Fe, Mn, Cu, Mg, Zn) ferrites. Some of these catalysts are characterized by the highest hydrogen productivity which is fairly close to the theoretically achievable value.

Keywords: hydrogen, bioethanol, oxide catalyst, reforming

INTRODUCTION

Ethanol steam reforming (ESR) is being intensively explored during a long time to produce hydrogen (Haryanto 2005; Mattos et al. 2012; Ni et al. 2007; Piscina and Homs 2008; Pyatnitsky et al. 2013; Vaidya and Rodrigues 2006). The use of ethanol has several advantages derived from its renewable, nontoxic and liquid nature. Moreover, the ESR

* Corresponding Author Email: ldolgykh@inphyschem-nas.kiev.ua; Dolgih.L.Ju@nas.gov.ua.

reaction gives a possibility to extract hydrogen not only from ethanol but also from water according to the following reaction stoichiometry:

$$C_2H_5OH + 3H_2O = 2CO_2 + 6H_2,\ \Delta H^o_{298} = 173.4\ kJ\ mol^{-1} \quad (I)$$

providing completely 6 moles of hydrogen per one mole of ethanol reacted.

Many solid catalysts have been studied in the ESR reaction. Mainly, they contain supported metals, including Ni, Co, Cu, Rh, Ru, Pd, Ir, and Pt. Only a limited number of another type of catalysts have been investigated in this reaction, particularly, the simple and complex oxides (Piscina and Homs 2008; Pyatnitsky et al. 2013). In this paper, we review the results of our studies of the catalytic properties in the ESR reaction of copper containing oxide catalysts and mixed oxides – MFe_2O_4 (M = Ni, Co, Fe, Mn, Cu, Mg, Zn) ferrites of the spinel structure. We focus our attention on the influence of the chemical composition of the catalysts on their performance in the hydrogen production by the steam reforming of bioethanol.

EXPERIMENTAL

Catalyst Preparation

The copper-zirconia catalysts were prepared by impregnation of the support with an aqueous solution of $Cu(NO_3)_2 \cdot 3H_2O$ (chemically pure grade) followed by drying at 80°C and calcination at 300°C for 6 h. The supports used for this study were commercial zirconium dioxide ZrO_2 with specific surface area S_{sp} = 16 m^2/g (State Scientific Enterprise "Zirconium," Ukraine, pure grade, TU-05.20.144-92) and laboratory samples of yttrium-stabilized zirconium dioxide (3 mol % Y_2O_3-ZrO_2) obtained by coprecipitation of the hydroxides from solutions of the corresponding chlorides followed by pulse wave treatment and calcination at 400°C (YSZ_{400}, S_{sp} = 128 m^2/g) and 1000°C (YSZ_{1000}, S_{sp} = 17 m^2/g) in order to vary the textural characteristics (Donetsk Institute for Physics and Engineering named after O.O. Galkin, NAS of Ukraine). Catalysts with different copper content (n = 2, 5, 10, 20 and 40 wt %) were prepared.

The catalysts M-Cu/ZrO_2 were prepared by the incipient wetness impregnation method, using aqueous solutions of the salts $Cu(NO_3)_2 \cdot 3H_2O$ (chemically pure grade), $Co(NO_3)_2 \cdot 6H_2O$ (chemically pure grade), $Ni(NO_3)_2 \cdot 6H_2O$ (chemically pure grade), and $Fe(NO_3)_3 \cdot 9H_2O$ (chemically pure grade) followed by drying (80°C) and calcination in air at 300°C for 6 h. As the support, the commercial zirconium dioxide with specific surface area 4 m^2/g (Donetsk Chemical Reagent Plant) was used. The content of the active components in the Cu/ZrO_2 and M-Cu/ZrO_2 samples was 10 wt %, calculated on the basis of the metal (M : Cu = 1 : 1).

The preparation of MFe_2O_4 catalysts has been performed by two different procedures. The ferrites MFe_2O_4 (M = Mn, Co, Ni) were synthesized by decomposition of the heteronuclear complex $[MFe_2O(CH_3COO)_6(H_2O)_3]\cdot 2H_2O$ by the method (Vasilenko et al. 2007). The use of these complexes permits the formation of ferrites with the precise stoichiometric ratio of Fe/M = 2. The ferrites obtained were calcined at 250°C for 5 h, and the manganese ferrite was denoted as $MnFe_2O_4$-HC.

The manganese ferrite, labeled as $MnFe_2O_4$-CP, was prepared by a chemical co-precipitation method. Accordingly, the ammonia water solution was dropped to a solution of Fe(III) and Mn(II) nitrates with molar ratio 2:1 under stirring. The produced brown reaction mixture was kept at 90°C during 5 h under stirring. The obtained precipitate was separated by magnetic decantation, washed with water, ethanol and diethyl ether, dried at room temperature and calcinated at 400°C for 2 h. Commercial chemically pure grade and analytical grade reagents were used without additional purification.

Ferrites MFe_2O_4 (M = Mg, Fe, Cu, Zn) were synthesized by similar procedure described for each ferrite in (Kim et al. 2003; Liu et al. 2007; Panda et al. 2011; Philip et al. 2007). The aqueous solutions of Fe(III) and M(II) nitrates were used (chlorides for M = Fe). Finally, the prepared samples were calcinated at 400°C for 2 h (in nitrogen flow for M = Fe to avoid oxidation of Fe(II)).

Catalyst Characterization

X-ray diffraction studies (XRD) of the Cu/ZrO_2 and $M\text{-}Cu/ZrO_2$ catalysts were conducted on an HZG-4C apparatus in copper radiation, monochromatized by a flat graphite monochromator. Phase analysis was performed by comparing the positions and intensities of the diffraction peaks with the ICSD and JCPDS databases. In order to estimate the average crystallite size in the identified phases, we used the size of the coherent scattering region (CSR), which was determined from the width of the corresponding diffraction peaks using the Scherrer formula. The crystallized phase content (by weight) was determined by full-profile analysis of the diffraction patterns (the Rietveld method) using the X'Pert HighScore Plus software package.

X-ray diffraction measurements of the ferrites MFe_2O_4 (M = Mg, Mn, Fe, Co, Ni, Cu, Zn) were carried out with a Bruker D8 Advance diffractometer, with a Co-anode λ = 0.179 nm (for catalyst prepared from the heteronuclear complex) and a Cu-anode, λ = 0.154 nm (for catalyst prepared by co-precipitation), step 2θ = 0.050, exposition time 5 s/step. Identification of crystalline phases was performed by matching with the ICDD files in the PDF-2 Version 2.0602 (2006) database. BET surface areas were measured by using a Sorptomatic 1990 instrument by adsorption of nitrogen at -196°C.

Temperature-programmed desorption of ammonia (NH_3-TPD) was used to characterize the number and strength of acid sites. Typically, 0.4 g catalyst sample was placed in the reactor and was heated up to 550°C at a rate 15°C/min under the He flow (40 ml/min). At this temperature, the sample was further heated for 0.5 h. Adsorption of dry ammonia took place at 100°C in the 1 ml/min flow. Prior to desorption, samples were flushed with He at 100°C (until no ammonia was detected in the flow) to remove the physically adsorbed ammonia. Desorption of ammonia was done at a rate of 15°C/min from 100 to 550°C under a He flow (40 ml/min). The desorbed ammonia was detected on the thermal conductivity detector and then titrated by 10^{-5} M HCl aqueous standard solution.

In order to evaluate the basicity of the oxide surface, CO_2-TPD experiments were carried out with a conventional setup equipped with a TCD detector. The catalyst samples were first purged with He at 350°C for 2 h, then cooled down to room temperature. The adsorption of CO_2 was carried out using a feed mixture of 50 vol % CO_2/He for 30 min. The reactor was then purged with He for 2 h. After that, the temperature was increased to 550°C with a heating rate of 10°C/min under a He flow (40 ml/min). The change in thermal conductivity is due to the concentration change of CO_2 in the effluent was recorded. The CO_2-TPD signals were referred to 1 m^2 of the surface area of the samples.

Hydrogen temperature-programmed reduction experiments (H_2-TPR) was performed in a conventional atmospheric quarts flow reactor. A catalyst sample was pretreated at 200-250°C for 2 h under flowing Ar (50 ml/min). After cooling to room temperature, the sample was exposed to a flow of 10 vol % H_2/He (50 ml/min), and the temperature was increased linearly at a rate 10°C/min.

Adsorptive decomposition of N_2O with subsequent hydrogen temperature-programmed reduction of oxidized surfaces (*s*-TPR) was used to evaluate textural characteristics of the reduced copper particles on the support surface. Prior to N_2O decomposition, the supported CuO particles were pre-reduced by heating in H_2/Ar flow (10 vol % H_2, 50 ml/min) from 25 to 500°C at 10°C/min. The surface of the reduced Cu was oxidized by N_2O in an isothermal mode at 60°C for 45 min. Subsequent temperature-programmed reduction of the surface Cu_2O formed was carried out under the same conditions as in the first reduction cycle (Gervasisni and Bennici 2005). Quantitative analysis of the absorbed hydrogen was carried out by integrating the experimental *s*-TPR profiles basing on previous calibration measurements with a known amount of CuO powder. After determination of the amount of hydrogen consumed in the *s*-TPR analysis, we calculated the specific surface of copper S_{Cu} (m^2/g_{Cu}) in the catalyst samples assuming that the total number of copper atoms in the surface layer is $1.46 \cdot 10^{19}$ atom/m^2 (Gervasini and Bennici, 2005). The mean size of the copper particles d_{Cu} (nm) was evaluated using a formula obtained assuming that the particles are cubic in shape:

$$d_{Cu} = \frac{6 \cdot 10^3}{S_{Cu} \cdot \rho_{Cu}} \tag{1}$$

where ρ_{Cu} is the density of metallic copper equal to 8.92 g/cm^3.

The dispersion is characterized by the ratio of the number of copper atoms on the surface to the total number of copper atoms in the sample.

To elucidate the influence of catalyst calcination on a catalyst state, the thermal gravimetric method was used. The curves of mass loss (TG) as well as of differential thermal analysis (DTA) and differential gravimetric analysis (DTG) were registered with a thermal gravimetric analysis system Q-1500 (supplied by MOM instruments, Budapest) with an air flow from 25°C to 1000°C at a heating rate of 10°C/min; a sample mass was from 0.4 to 0.6 g.

Catalytic Experiments

The ESR reaction was carried out in a fixed-bed tubular quartz reactor at atmospheric pressure with the temperature increased stepwise at molar ratio H_2O/C_2H_5OH=19 (2.7 vol % C_2H_5OH, 50 vol % H_2O, N_2 balance) that is close to water/ethanol ratio in bioethanol obtained by fermentation. The procedure of catalytic tests and reaction mixture analysis were as described elsewhere (Dolgykh et al. 2006). A catalyst, approx. 1 g, of particle size between 1 and 2 mm, was placed in a reactor between two layers of quartz grains of the same size; it was held at each temperature for 1 h, followed by gas chromatography analysis. At the end of a catalytic test, the flow of reaction mixture was stopped and a catalyst was cooled under N_2 stream and stored for characterization.

Ethanol conversion, X, and selectivity to carbon-based reaction products, S_{Cn}, were evaluated according to the following expressions:

$$X = \frac{\sum n \cdot F_{Cn}}{2 \cdot F_{Et,in}} \cdot 100 \qquad (2)$$

$$S_{Cn} = \frac{n \cdot F_{Cn}}{\sum n \cdot F_{Cn}} \cdot 100 \qquad (3)$$

where n is a number of C-atoms in a product C_n; $F_{Et,in}$ is an inlet feed of ethanol, mol h^{-1}; F_{Cn} is a feed of corresponding product, mol h^{-1}.

The selectivity to hydrogen was defined as 100% when 6 mol of H_2 were formed per 1 mol of reacted C_2H_5OH. Then:

$$S_{H2} = \frac{F_{H2}}{3 \cdot \sum n \cdot F_{Cn}} \cdot 100 \qquad (4)$$

where F_{H2} is an outlet feed of hydrogen, mol h^{-1}.

Hydrogen yield, Y_{H2}, was calculated as:

$$Y_{H2} = X \cdot S_{H2}/100 \quad (5)$$

The hydrogen selectivity was also calculated by an alternative formula which expresses S_{H2} as a function of selectivity of carbon-based reaction products (Pyatnitsky et al. 2013):

$$S_{H2} = S_{CO2} + \frac{2}{3}S_{CO} - \frac{1}{3}S_{CH4} + \frac{1}{6}S_{CH3CHO} + \frac{1}{9}S_{CH3COCH3} \quad (6)$$

where all selectivities expressed in fractions or percent.

It is important to note that selectivity to C_2H_4 and C_3H_6 are not included in Eq. (6) explicitly. However, the formation of these compounds influences the yield of hydrogen due to the lower selectivity to other carbon-containing products.

The calculations of hydrogen selectivity by the alternative equations (4) and (6) must give the same results if there are no errors in experimental data (Pyatnitsky et al. 2013). Taking this into account, we compared some literature and calculated values of hydrogen selectivity (Table 1).

Table 1 shows that the selectivity to hydrogen calculated via formulas (4) and (6) using the data from (Llorca et al. 2004) completely coincide with the reported in (Llorca et al. 2004) hydrogen selectivity. A different result was found in the analogous comparison using the data from (Biswas and Kunzru 2007). In this case, a complete correspondence was observed between the literature values for the hydrogen selectivity and values calculated via formula (4), whereas there was considerable disagreement between the literature values and those calculated via formula (6). This indicates that the values of the selectivity to hydrogen cited in (Biswas and Kunzru 2007) is required being corrected. Particularly, the selectivity of 96.7% is clearly too high.

Table 1. Comparison of the hydrogen selectivity reported in the literature and the hydrogen selectivity calculated according to Eqs. (4) and (6)

Product distribution, mol %				**S_{H2}, %**		
H_2	CO	CO_2	CH_4	Ref.	(4)	(6)
73.9	0	25.0	1.1	94.3[a)]	94.3	94.3
72.3	6.8	17.8	3.2	86.7[b)]	86.7	76.7
74.3	6.9	17.3	1.5	96.7[b)]	96.7	83.3

[a)] reported in (Llorca et al. 2004) for the 0.23NaCoZn catalyst.

[b)] reported in (Biswas and Kunzru 2007) for the 30% Ni/ZrO_2 catalyst.

Another expressive example of a similar discrepancy can be found in (Ni et al. 2007) where the yield of hydrogen of 96% is cited for the catalyst Ni/γ-Al_2O_3 for 800°C according to the data from (Fatsikostas and Verykios 2004). In this case, the selectivities for CO, CO_2, and CH_4 are equal to 65, 30 and 5%, respectively, then the hydrogen selectivity calculated via equation (6) is equal to 72% which is considerably less than that cited in (Ni et al. 2007).

These examples show clearly that Eq. (6) offers an additional possibility to estimate the reliability of the value of the hydrogen selectivity (as well as hydrogen yield) achieved in the process of steam reforming of ethanol. The considerable differences between the values of the hydrogen selectivity calculated via formula (6) and the corresponding values presented in several publications can be apparently explained by the insufficiently good accounting of the material balance. We believe that a more precise control of the material balance in the steam reforming of ethanol can be carried out by using the alternative methods for calculating the yield of hydrogen considered in this paper.

THERMODYNAMICS

Equilibrium constants, K_p, of separate reactions of the ESR reaction at a given temperature were evaluated according to the following equation:

$$\ln K_p = AT^{-1} + B\ln T + CT + DT^2 + E \tag{7}$$

The numerical values of the coefficients *A*, *B*, *C*, *D*, and *E* were determined from the thermodynamic parameters listed elsewhere (Stull et al. 1969) for temperatures of 500, 700 and 900 K. The composition of the reaction mixture for the given conditions was obtained by means of the program described in detail in (Pyatnitsky and Strizhak 2011).

Let us consider the case if only reaction (I) takes place. Figure 1 shows the temperature dependence of the equilibrium yield of hydrogen ($Y_{H2,eq}$) for various compositions of the water-alcohol mixture. For bioethanol (H_2O/C_2H_5OH = 19), the 100% yield of hydrogen is achieved at the temperature of 150°C, and the temperature for the achievement of a 100% yield increases with a decrease in the water/ethanol ratio.

In general case, the yield of hydrogen is affected by the occurrence of side reactions during ESR. The main side reactions are:

$$C_2H_5OH = CH_3CHO + H_2,\ \Delta H^o_{298} = 68.5\ \text{kJ mol}^{-1} \tag{II}$$

$$CH_3CHO + 3\ H_2O = 2\ CO_2 + 5\ H_2,\ \Delta H^o_{298} = 104.9\ \text{kJ mol}^{-1} \tag{III}$$

$CH_3CHO + H_2O = 2\ CO + 3\ H_2$, $\Delta H^o_{298} = 187.2$ kJ mol^{-1} (IV)

$CO + H_2O = CO_2 + H_2$, $\Delta H^o_{298} = -41.2$ kJ mol^{-1} (V)

$CH_3CHO = CO + CH_4$, $\Delta H^o_{298} = -19.1$ kJ mol^{-1} (VI)

$CO + 3\ H_2 = CH_4 + H_2O$, $\Delta H^o_{298} = 206.3$ kJ mol^{-1} (VII)

$2\ CH_3CHO + H_2O = CH_3COCH_3 + CO_2 + 2\ H_2$, $\Delta H^o_{298} = -36.6$ kJ mol^{-1} (VIII)

$C_2H_5OH = C_2H_4 + H_2O$, $\Delta H^o_{298} = 45.3$ kJ mol^{-1} (IX)

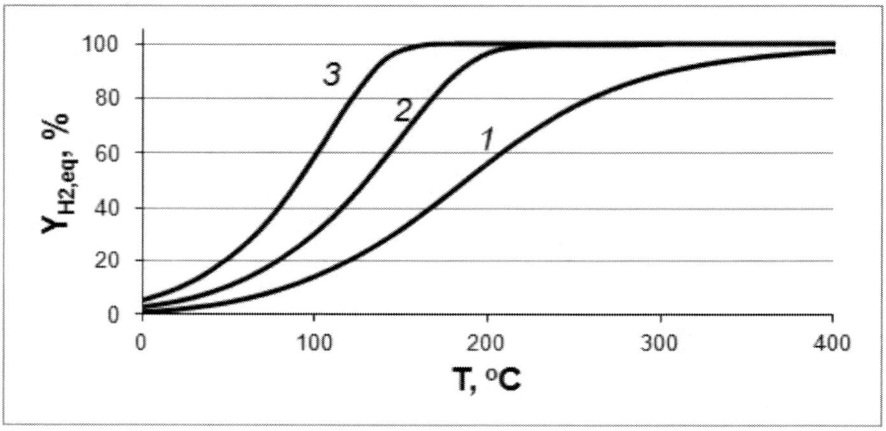

Figure 1. The temperature dependence of the equilibrium yield of hydrogen in reaction (I): 1) $H_2O/C_2H_5OH = 3$, stoichiometric mixture; 2) $H_2O/C_2H_5OH = 8$; 3) $H_2O/C_2H_5OH = 19$, bioethanol.

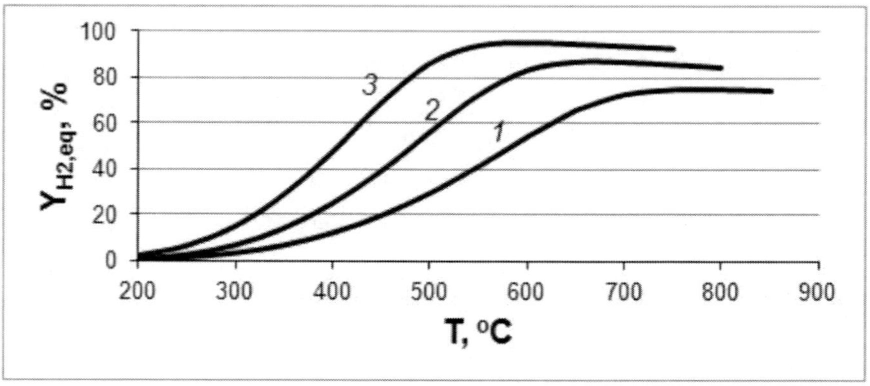

Figure 2. The temperature dependence of the equilibrium yield of hydrogen calculated for reaction scheme (II)-(IX): 1) $H_2O/C_2H_5OH = 3$, stoichiometric mixture; 2) $H_2O/C_2H_5OH = 8$; 3) $H_2O/C_2H_5OH = 19$, bioethanol.

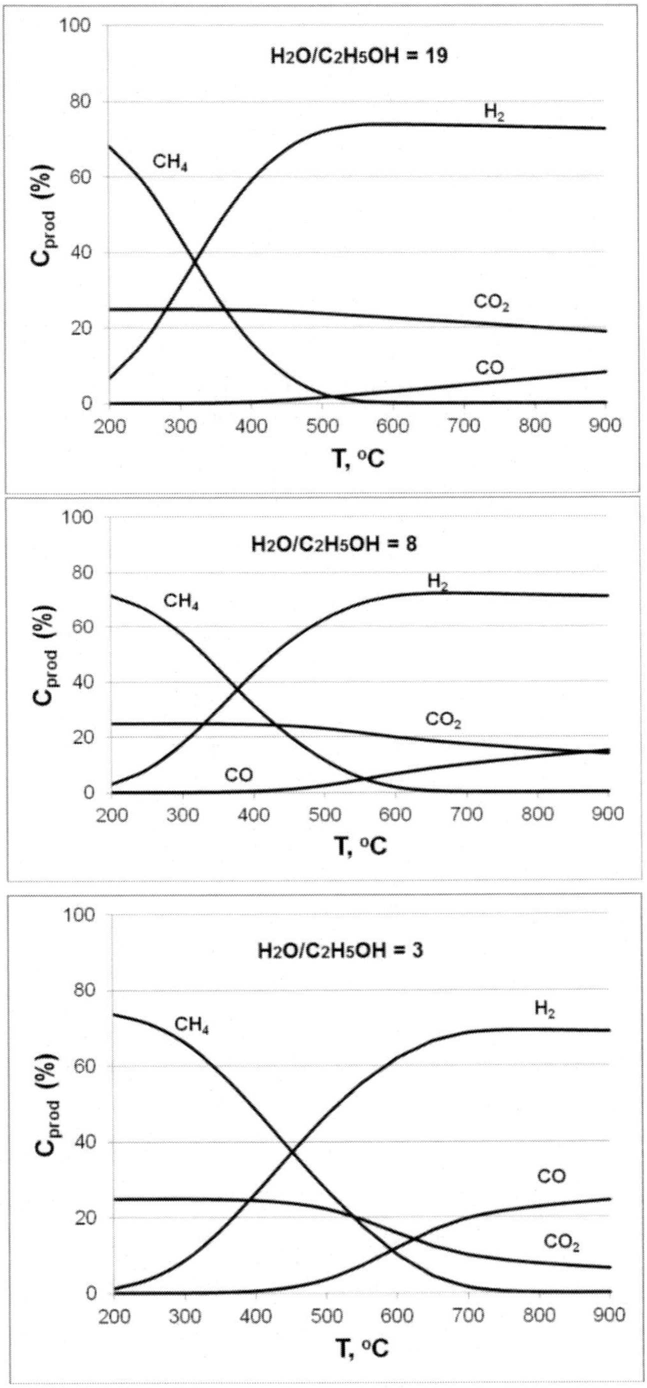

Figure 3. The temperature dependence of the equilibrium concentration of the ESR reaction products at different relations H_2O/C_2H_5OH.

The ESR realization by reactions (II) and (III) only, i.e., through the formation of acetaldehyde, leads to practically the same yields of hydrogen as for the direct reaction (I) because the equilibrium content of acetaldehyde in the reaction products is extremely small. Its maximum value is about 0.8% at 300°C for the stoichiometric mixture of ethanol and water and 0.01% at 100°C for bioethanol (H_2O/C_2H_5OH = 19). The occurrence of the reactions in which all side products are formed (reactions (II)-(IX)) leads to a decrease in the equilibrium yield of hydrogen and to a greater degree the smaller the water/ethanol ratio (Figure 2). It is worth noting that the scheme of reactions (II)-(IX) includes six linearly independent equations, and such equations are sufficient for a complete description of the thermodynamic process. One of the reactions (III), (IV), and (V), for example, corresponds to linear depend equations since any of these reactions can be obtained by linear combination of the other two. Therefore, the exclusion of one of these equations from the general scheme is not reflected in any way in the results from the thermodynamic calculation of the equilibrium concentration of the components of the reaction mixture.

Figure 3 shows the equilibrium values for the concentrations of the products for reaction scheme (II)-(IX) and the yield of hydrogen in dependence on temperature. The equilibrium conversion of ethanol is 100% over the whole temperature range; the equilibrium concentrations of acetaldehyde and ethylene are negligibly small. If bioethanol is used, the highest yield of hydrogen (95.3%) is achieved at 600°C, and the equilibrium concentrations of H_2, CO, CO_2, and CH_4 are 74.10%, 22.65%, 3.14%, and 0.12% respectively; if the stoichiometric mixture is used, the maximum yield of hydrogen is obtained at 800°C, and the concentrations of H_2, CO, CO_2, and CH_4 are 69.16%, 7.86%, 22.85%, and 0.13% respectively.

However, these parameters can be improved if the catalyst does not accelerate reaction (IV), which leads to the CO formation and appearance of the water shift reaction (V). This results in the absence of CO in the reaction products. The most favorable reaction pathway involves a transformation of ethanol into hydrogen and CO_2 without the intermediate formation of CO and CH_4, i.e., when all the transformations into the final products (CO_2 and H_2) take place on the surface of a catalyst and the water-gas shift reaction does not occur.

Another typical by-side product is a carbon which may be formed on a catalyst surface by the Boudouard reaction, methane cracking and/or the reverse reaction of steam gasification of carbon:

$$2\,CO = CO_2 + C,\ \Delta H^o_{298} = -172.5\ kJ\ mol^{-1} \tag{X}$$

$$CH_4 = 2\,H_2 + C,\ \Delta H^o_{298} = 74.9\ kJ\ mol^{-1} \tag{XI}$$

$$CO + H_2 = H_2O + C,\ \Delta H^o_{298} = -131.4\ kJ\ mol^{-1} \tag{XII}$$

Figure 4 shows the region of carbon deposition in dependence on temperature and reaction mixture composition at thermodynamic equilibrium of the ESR process. The carbon appears at lower temperatures (below 236°C for the stoichiometric mixture). A position of dividing line between regions of the presence and absence of carbon is quite the same if any of (X), (XI) and (XII) reactions, as well as any their combination, is used for calculations, because in terms of the set of stoichiometric equations (I)-(XII) only one of the equations (X), (XI), and (XII) is linearly independent.

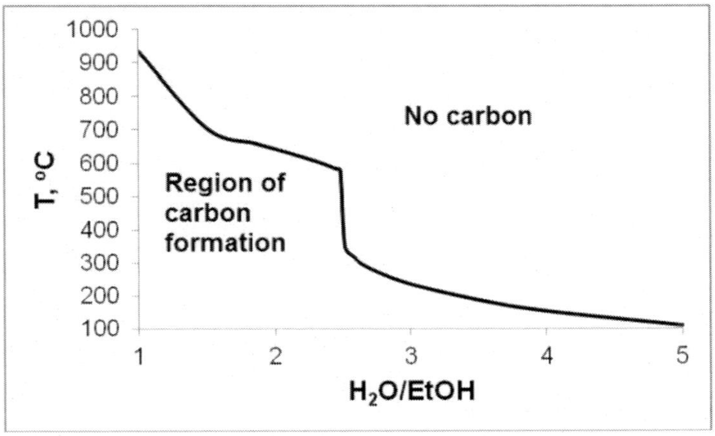

Figure 4. Influence of temperature and reaction mixture composition on carbon formation at the ESR process (atmospheric pressure, no diluents).

RESULTS AND DISCUSSION

Catalytic Properties of Nanosized Copper Catalysts Supported on Zirconia and Yttrium-Stabilized Zirconia in Bioethanol Steam Reforming

Copper catalysts, both individually and in combination with more active nickel ones are among the systems often used in steam reforming of ethanol. Nature and structural-functional properties of a support play an important role in catalyst performance because its active sites can favor breaking a bond in a water molecule with OH-groups formation and promote their migration to metal particles, they can also promote dehydrogenation pathway for decreasing deactivation caused by coke formation. Moreover, the essential role of the support is in a stabilization of the active metal particles at elevated temperatures and high steam concentrations in the reaction conditions (Aupretre et al. 2005). Several researchers showed that such oxides as ZnO (Llorca et al. 2002), La_2O_3-Al_2O_3 (Fatsikostas et al. 2004), CeO_2 (Fatsikostas et al. 2004; Deluga et al. 2004), MgO (Freni et al. 2003; Cavallaro et al. 2001) were effective as the supports for cobalt and nickel-containing catalysts in the ethanol steam reforming. Pure and modified zirconium oxides were also reported as the perspective catalyst supports (Bergamaschi et al. 2005; Benito et al. 2007; Resini et al. 2008; Bellido et al. 2008). There are several reports in the literature concerning structure and behavior of ZrO_2-supported Cu catalysts for ethanol dehydrogenation (Wang et al. 2010; Sato et al. 2013; Freitas et al. 2014). Recently, it was revealed the effect of zirconia phase on the dispersion and structure of supported copper oxide species (Pakharukova et al. 2009; 2015).

The catalytic activity and selectivity of copper catalysts supported on the pure and yttrium-stabilized zirconia with different structures were studied in the steam reforming of a dilute aqueous solution of the ethanol. The obtained results were analyzed with respect to the state of supported copper species.

Figure 5a shows XRD patterns of the support ZrO_2 and Cu/ZrO_2 catalysts with various copper loadings. The XRD phase analysis revealed that the support contained a monoclinic zirconia (m-ZrO_2) with unit cell parameters: a = 0.5146 nm, b = 0.52052 nm, c = 0.5313 nm, β = 99.18° (ICSD №89426). The mean size of crystallites of this phase (CSR dimension) was 46 nm. In addition, a negligible impurity of tetragonal zirconia (t-ZrO_2) with CSR dimension about 30 nm is observed. The diffraction peaks of the m-ZrO_2 phase are predominant in the XRD patterns of the catalysts. A monoclinic CuO phase (a = 0.4684 nm, b = 0.3425 nm, c = 0.5129 nm, β = 99.47°, ICDD PDF #05-0661) was detected in catalysts with the copper content ≥ 5 wt %. Difference curves between the normalized XRD patterns of the catalysts and support illustrate reflections from the CuO phase at 2θ = 35.5 and 38.6° (Figure 5b).

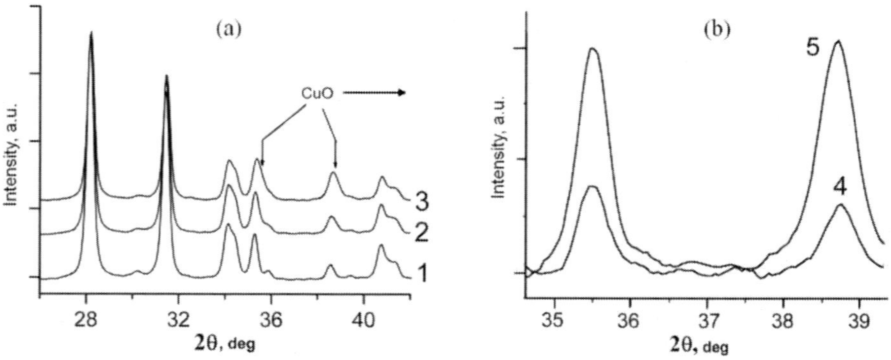

Figure 5. Diffraction patterns of the ZrO_2 (1) and 5Cu/ZrO_2 (2), 10Cu/ZrO_2 (3) catalysts (a) and difference curves between the normalized XRD patterns of the catalysts 5Cu/ZrO_2 (4), 10Cu/ZrO_2 (5) and support (b).

Increase in the copper loading up to 40 wt % results in a progressive enlargement of the CuO crystallites. Diminishing microstrain within CuO crystallites with an increase in the copper loading from 10 to 40 wt % is observed. Thus, the XRD data revealed a predominant formation of the bulk crystalline CuO particles on the m-ZrO_2 surface, in agreement with the literature reports (Sato et al. 2013; Rhodes et al. 2005, Ma et al. 2005; Wang et al. 2007).

The diffraction patterns of YSZ_{400} and nCu/YSZ_{400}, where n – copper loading (Figure 6a), indicate that the YSZ_{400} support has unit cell parameters a = 0.3629 nm, c = 0.5106 nm, c/2a = 0.995 and forms crystals with fluorite structure. The diffraction patterns of YSZ_{400} and nCu/YSZ_{400} studied in this work are analogous to the patterns obtained in the previous investigation (Pakharukova et al. 2009). Weak diffraction peaks for bulk CuO were observed only for $10Cu/YSZ_{400}$ (the positions of the lines at 2θ = 35.5° and 38.6° which are characteristic for CuO are shown by arrows). The XRD data show that the tetragonal zirconia phase predominates for YSZ_{1000} and nCu/YSZ_{1000} with unit cell parameters a = 3.608 Å, c = 5.167 Å, c/2a = 1.013 (Figure 6b). The samples also contain a trace of monoclinic zirconia (< 5 wt %). In catalysts with copper content up to 5 wt % inclusive, no peaks corresponding to individual copper compounds were found. Increasing the copper content to 10 wt % leads to the appearance of reflections in the diffraction pattern related to the bulk copper oxide phase.

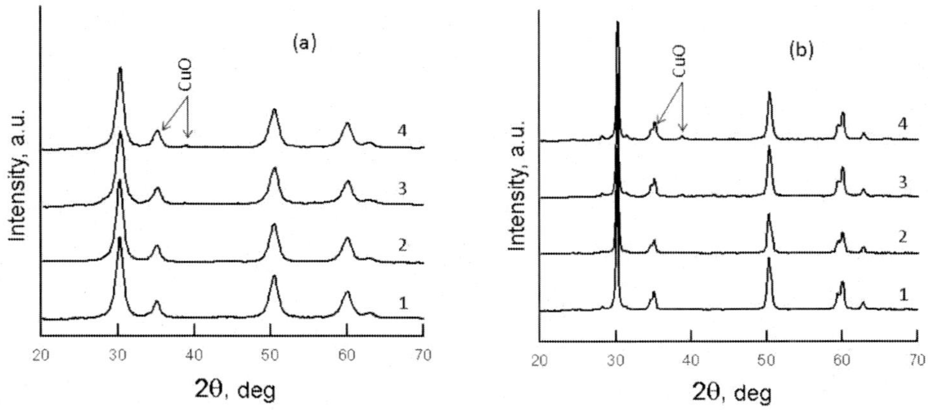

Figure 6. Diffraction patterns for nCu/YSZ_{400} (a) and nCu/YSZ_{1000} (b) with different copper loadings n: *1* – 0, *2* – 2, *3* – 5, *4* – 10 wt %.

The temperature-programmed reduction profiles for Cu/ZrO_2 and Cu/YSZ catalysts obtained after prior oxidation of the metallic copper surface by N_2O (*s*-TPR) permitted us to determine not only the textural characteristics of the supported copper particles but also evaluate the reactivity of these particles as shown below.

Table 2 summarizes the data on the temperatures of the reduction maxima during *s*-TPR as well as the textural characteristics of the reduced copper particles on the ZrO_2 and YSZ calculated from the *s*-TPR curves.

For the same copper content, the values of Cu/S_{sp}, giving the ratio of the amount of copper atoms per unit of surface area of the support, for nCu/YSZ_{400} catalysts on the support with high specific surface area and for nCu/YSZ_{1000} as well as for nCu/ZrO_2 on the supports with low specific surface areas, differ significantly.

In the range of copper loadings from 2 to 10 wt % the values of Cu/S_{sp} for YSZ_{400} do not exceed 7.3 atom/nm^2, which is lower than the dispersion capacity of the support, which is 8.6 Cu^{2+} ion/nm^2 for tetragonal zirconia and characterizes the state when all the possible vacant sites on the zirconia surface are occupied by cupric ions (Liu et al. 2005). For catalysts derived from YSZ_{1000}, the values of Cu/S_{sp} range from 11.1 to 55.3 atom/nm^2 and for monoclinic ZrO_2 from 17.6 to 58.8 atom/nm^2 which is considerably greater than the dispersion capacity of the support.

Table 2 shows that the mean size of the copper particles for samples based on monoclinic ZrO_2 drops from 33 to 3 nm and the dispersion increases from 3 to 31% with a decrease in the copper content from 40 to 3 wt %. For equal copper content, the copper particles have somewhat higher S_{Cu} and lower d_{Cu} on the surface of yttrium-stabilized tetragonal zirconium dioxide in comparison with monoclinic ZrO_2.

Table 2. Copper loading (Cu/S_{sp}), temperature of reduction maximum during *s*-TPR (T_s), dispersion (D_{Cu}), specific surface area (S_{Cu}), mean particle size (d_{Cu}) for Cu/ZrO_2 and Cu/YSZ catalysts

Catalyst	Cu/S_{sp}, atom/nm^2	T_s, °C	D_{Cu},%	S_{Cu}, m^2/g_{Cu}	d_{Cu}, nm
2Cu/YSZ_{400}	1.5	200	96	615	1
5Cu/YSZ_{400}	3.7	197	85	543	1
10Cu/YSZ_{400}	7.3	211	53	336	2
2Cu/YSZ_{1000}	11.1	179	38	246	3
5Cu/YSZ_{1000}	27.4	195	20	131	5
10Cu/YSZ_{1000}	55.3	227	10	70	10
3Cu/ZrO_2	17.6	189	31	200	3
5Cu/ZrO_2	29.4	199	19	120	6
10Cu/ZrO_2	58.8	209	9	59	11
20Cu/ZrO_2	117.6	202	6	38	17
40Cu/ZrO_2	235.2	212	3	20	33

The *s*-TPR profiles of the surface-oxidized copper particles for samples based on monoclinic and tetragonal ZrO_2 are given by a single well-pronounced reduction peak in the temperature range 140-285°C. Golodets (Golodets 1983) has proposed that the reduction of the oxide catalysts by hydrogen serves as a test reaction for determining their reactivity in oxidation-reduction reactions. Upon summarizing the data for the reduction of nCu/YSZ samples taking account the results (Pakharukova et al. 2009), we may conclude that copper oxide nanoclusters with chain structure are mostly formed on yttrium-stabilized zirconia with low Cu/S_{sp} ratios (1.5-3.7 atom/nm^2). The reduction of these nanoclusters leads to the formation of active phase particles with mean diameter ~1 nm. The reductive treatment of nCu/YSZ samples with Cu/S_{sp} $\geq$ 7 atom/nm^2 leads to the formation of copper nanoparticles with diameter 2-10 nm depending on the value of Cu/S_{sp}. The temperatures for the maximum reduction rate for Cu/ZrO_2 samples with different copper contents 5-40 wt % (Table 2) have similar values (T_s = 199-212°C). The proximity of the reduction temperatures of 5Cu/YSZ$_{400}$, 5Cu/YSZ$_{1000}$ and 5Cu/ZrO_2 in the *s*-TPR process may indicate similar reactivity of the reduced copper particles in these catalysts. These results suggest that the reduced copper particles on the zirconium dioxide surface have similar reactivity without regard to the crystalline modification.

Figure 7 and Tables 3-5 illustrate the catalytic properties of Cu/ZrO_2 and Cu/YSZ samples in the steam reforming of bioethanol. Predominant dehydrogenation of ethanol to acetaldehyde (reaction II) with the release of hydrogen is characteristic for the conversion of ethanol-water mixtures on Cu/ZrO_2 at T < 350°C. Hydrogen formation at higher temperatures results from an increased rate of subsequent transformation of acetaldehyde and the involvement of water vapor to the reaction.

The specific rate of acetaldehyde formation (the key product of the steam reforming of bioethanol) R(CH_3CHO) at 300°C hardly depends on the particle size in the range 6-33 nm (Figure 7), indicating that the dehydrogenation of ethanol is structurally insensitive. The activity of pure supports at 300°C was insignificant. Reducing the particle size to less than 3 nm leads to a decrease in R(CH_3CHO), indicating the existence of size effect.

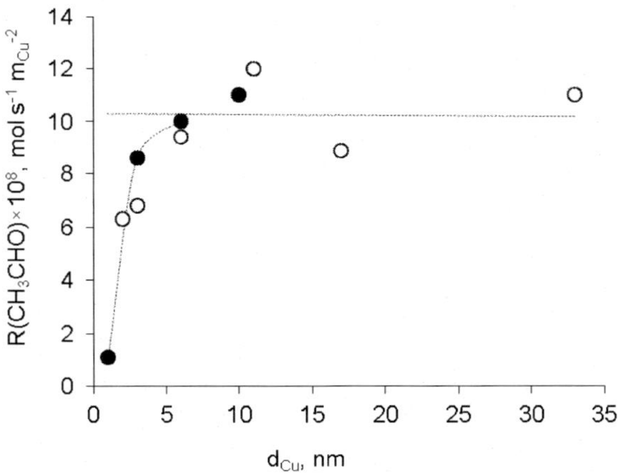

Figure 7. The specific rate of the conversion of ethanol to acetaldehyde R(CH_3CHO) depending on the mean diameter of the copper particles d_{Cu} for Cu/ZrO_2 (○) and Cu/YSZ (●) catalysts.

We should note that this effect is seen for Cu/S_{sp} < 7 atom/nm^2, which is less than the dispersion capacity of zirconia. The XRD and TPR data indicate that formation of bulk cupric oxide is not observed in this region and the formation of clusters of reduced active phase atoms occurs most likely from chain oxide species. An increase in Cu/S_{sp} enhances the probability of formation of larger disperse cupric oxide particles and cupric oxide microcrystallites and, as a consequence, an increase in the size of the reduced metallic phase particles (Dolgikh et al. 2011, Deinega et al. 2014, Pakharukova et al. 2009; 2015). The nCu/YSZ$_{1000}$ catalysts (n = 5, 10 wt %) and nCu/ZrO_2 (n = 5, 10, 20 and 40 wt %), containing such particles, have higher specific catalytic activity. The observed dependence may be explained in the framework of the differentiation of small particles according to whether these particles include clusters without a fully-formed surface and dispersed particles with a well-formed surface (Patrylak et al. 2013). Thus, the decrease in the catalytic activity of the samples with cluster-size active phase copper particles (~1 nm) is most likely related to features of formation of the surface of such particles.

Tables 3-5 give a comparison of the ethanol conversion (X) at different temperatures for catalysts with the same copper content $5Cu/ZrO_2$, $5Cu/YSZ_{1000}$ and $5Cu/YSZ_{400}$. It should be noted that the ethanol conversion on $5Cu/ZrO_2$ drops at temperatures above 300°C and does not exceed 44%. This behavior is probably related to sintering of the copper particles. No drop in conversion is observed on the $5Cu/YSZ_{1000}$ and X reaches 53%. Upon an increase in the temperature from 300 to 450°C, the value of X for $5Cu/YSZ_{400}$ increases from 5.8 to 30.1%, which is much less than the values of X for the YSZ_{1000} samples.

It may be proposed that the lack of a sufficient quantity of dispersed cupric oxide particles with a well-formed surface on the $5Cu/YSZ_{400}$, which are then involved in the formation of the reduction active phase nanoparticles, may account for the lower catalytic activity of this catalyst in comparison with $5Cu/YSZ_{1000}$. Comparison of the data given in Tables 3-5 shows 99% selectivity to acetaldehyde at T = 250°C on the $5Cu/YSZ_{1000}$ catalyst, while this value is 92.8% on $5Cu/YSZ_{400}$ and decreases with increasing temperature. It was observed an increase in the acetone selectivity to 22.9% on $5Cu/YSZ_{1000}$ (T = 450°C) and 32.6% (T = 350°C) with a subsequent decrease to 13.7% on $5Cu/YSZ_{400}$. For both catalysts at T > 350°C, we observed an increase in selectivity to ethylene from 22.5 to 25.3% for $5Cu/YSZ_{1000}$ and $5Cu/YSZ_{400}$, respectively.

The observed products suggest that the initial conversion of ethanol proceeds predominantly by dehydrogenation of this alcohol to acetaldehyde. At elevated temperatures, acetaldehyde undergoes the further transformation with steam, leading to the formation of hydrogen, carbon dioxide, acetone, and methane in the products according to reaction (VIII) and (XIII):

$$CH_3CHO + H_2O = CH_4 + CO_2 + H_2 \qquad \text{(XIII)}$$

Table 3. Temperature dependencies of ethanol conversion (X) and product selectivity for $5Cu/ZrO_2$

T, °C	X, %	Selectivity, %							
		CH_3CHO	$(CH_3)_2CO$	CO	CO_2	CH_4	C_2H_4	C_3+	H_2
250	26.7	100	0.0	0.0	0.0	0.0	0.0	0.0	16.7
300	44.0	100	0.0	0.0	0.0	0.0	0.0	0.0	16.0
350	27.1	100	0.0	0.0	0.0	0.0	0.0	0.0	16.9
400	15.7	66.3	22.9	0.0	6.7	0.1	3.6	0.4	20.7
450	23.1	17.7	37.3	0.0	19.5	0.2	18.3	7.0	25.9

Table 4. Temperature dependencies of ethanol conversion (X) and product selectivity for 5Cu/YSZ_{1000}

T, °C	X, %	Selectivity, %							
		CH_3CHO	$(CH_3)_2CO$	CO	CO_2	CH_4	C_2H_4	C_3+	H_2
250	21.9	99.0	0.0	0.0	1.0	0.0	0.0	0.0	16.9
300	37.8	85.9	10.4	0.0	3.7	0.0	0.0	0.0	18.4
350	40.3	48.3	17.8	0.0	33.2	0.0	0.6	0.1	38.8
400	47.0	9.3	22.9	0.0	55.3	0.1	9.0	3.4	58.9
450	52.7	1.3	21.9	0.0	46.8	2.7	22.5	4.8	45.6

Table 5. Temperature dependencies of ethanol conversion (X) and product selectivity for 5Cu/YSZ_{400}

T, °C	X, %	Selectivity, %							
		CH_3CHO	$(CH_3)_2CO$	CO	CO_2	CH_4	C_2H_4	C_3+	H_2
250	5.8	92.8	0.0	0.0	7.2	0.0	0.0	0.0	23.0
300	19.3	81.4	7.4	0.0	11.1	0.0	0.1	0.0	27.1
350	22.0	36.5	32.6	0.0	30.2	0.1	0.4	0.2	37.2
400	30.1	1.6	28.6	0.0	44.8	0.6	8.2	16.2	48.4
450	26.2	0.5	13.7	0.0	39.3	3.2	25.2	18.1	40.5

The formation of ethylene is the result of the dehydration of ethanol (reaction (IX)) with the participation of the acid sites of the support (Zhao et al. 2002).

Comparison of the selectivity of the steam reforming of ethanol on monoclinic and yttrium-stabilized tetragonal zirconium dioxide shows that higher CO_2 selectivity, which serves as an important indicator for the rapid reforming of ethanol with the participation of water vapor, is found on 5Cu/YSZ. Under studied conditions, the catalyst 5Cu/YSZ_{1000} demonstrates a higher selectivity to CO_2 (55.3% at 400°C). The selectivity to hydrogen reaches 58.9% on 5Cu/YSZ_{1000} and 48.4% on 5Cu/YSZ_{400} at 400°C. This indicates a greater intensity of reforming process over 5Cu/YSZ_{1000}.

The enhanced selectivity of the catalyst based on YSZ_{1000} may be associated with higher oxygen mobility of the support that facilitates the oxidation of the carbon-containing intermediates to CO_2 by oxygen on the zirconia surface (Bellido and Assaf 2008). This is favored by the higher concentration of oxygen vacancies per unit of the support surface, which are formed upon the introduction of yttrium oxide as a stabilizer and the possible insertion of cupric ions into the zirconia crystal lattice as well as the formation of the support at high temperatures.

Therefore, our studies have shown a size effect in the dehydrogenation of ethanol to acetaldehyde on copper catalysts supported on zirconia. The differences in the activity of the catalysts are probably due to features of the surfaces of the clusters and dispersed species of copper in the samples. The enhanced selectivity of the catalysts obtained with yttrium-stabilized zirconia (YSZ) roasted at 1000°C (S_{sp} = 17 m^2/g) in the steam reforming of bioethanol relative to YSZ roasted at 400°C (S_{sp} = 128 m^2/g) may be related to the higher mobility of oxygen in the support.

The obtained structural data (Pakharukova et al. 2015) allow explaining the existing structure-performance relationships. At equal copper loading, the copper particles have somewhat lower surface area and large average size on the monoclinic ZrO_2 surface in comparison with yttrium-stabilized tetragonal zirconia (with the nearly identical surface area of the supports). However, the Cu/ZrO_2 catalysts showed catalytic performance comparable to that of Cu/YSZ systems. The high activity seems to be related to stabilization of highly dispersed copper species on the Cu/ZrO_2 surface.

Catalytic Properties of Binary Copper Catalysts Supported on Zirconia in Bioethanol Steam Reforming

Analysis of the literature shows (Haryanto et al. 2005; Ni et al. 2007; Pyatnitsky et al. 2013; Vaidya and Rodrigues 2006) that the ethanol steam reforming process has been studied mainly at high temperature (T > 600°C). There is significant interest in low-temperature ethanol steam reforming, but under these conditions, we observe decreased hydrogen selectivity, increased yield of secondary products, and also carbonization of the catalyst. As catalysts for ethanol steam reforming at low temperatures, it seems promising to use systems containing two or more active components, allowing to apply the advantages of each of them to improve the activity and selectivity of the catalyst. Such systems have not yet been sufficiently studied in the literature; the best studied include Ni-Cu (Mariño et al. 1998, 2004), and Co with additives of Ni, Cu, Fe, Mn, Cr, as well as precious metals (Homs et al. 2006, Casanovas et al. 2010).

Therefore, the activity and selectivity of the ESR catalysts may be improved by using the bicomponent supported systems M-Cu/ZrO_2, where M = Fe, Co, Ni are known components for their capability for low-temperature activation of ethanol, breaking C—C and O—H bonds, and acceleration of the water-gas shift reaction (Ni et al. 2007; Vaidya and Rodrigues 2006; Auprêtre et al. 2002).

Figure 8 shows the results of X-ray phase analysis of the support ZrO_2 and the supported catalysts based on it: Cu/ZrO_2 and M-Cu/ZrO_2 (M = Fe, Co, Ni). Zirconium dioxide is characterized by a monoclinic structure with unit cell parameters corresponding to the standard values (JCPDS No. 37-1484, a = 5.147 Å, b = 5.212 Å, c = 5.312 Å, b = 99.22°). The average crystallite size (the CSR size) is 65 nm. The structural parameters of the support do not change after depositing the active components. An overlap of the diffraction peaks for the phases ZrO_2 and CuO complicates quantitative determination of the copper oxide content in the samples of the supported catalysts. The separate diffraction peaks are observed for the CuO phase only on the difference curve between the X-ray diffraction patterns of the catalyst and the support, which are shown in Figure 8.

According to XRD data for the Cu/ZrO_2 sample, a characteristic feature is the presence of the CuO phase with parameters matching the standard values (JCPDS No. 05-0661, a = 4.684 Å, b = 3.425 Å, c = 5.129 Å, b = 99.47°) and average crystallite size 43 nm (Figure 8a). The composition of the catalyst $Ni\text{-}Cu/ZrO_2$ (Figure 8b) includes a copper oxide phase in an amount no greater than 2 wt % and a nickel oxide phase which has a disordered character. The CSR size for the nickel oxide phase as determined from different lines differs by a factor of two: according to the broadening of the 111 line, the CSR is 39 nm; according to the broadening of the 200 line, the CSR is 15 nm. This may be due to reaction with copper to form a mixed oxide such as $Ni_{0.8}Cu_{0.2}O$ (JCPDS No. 25-1049) with unit cell parameter 4.186 Å. According to XRD data, the sample of the catalyst $Co\text{-}Cu/ZrO_2$ (Figure 8c) contains a mixed cobalt/copper oxide with a spinel type structure.

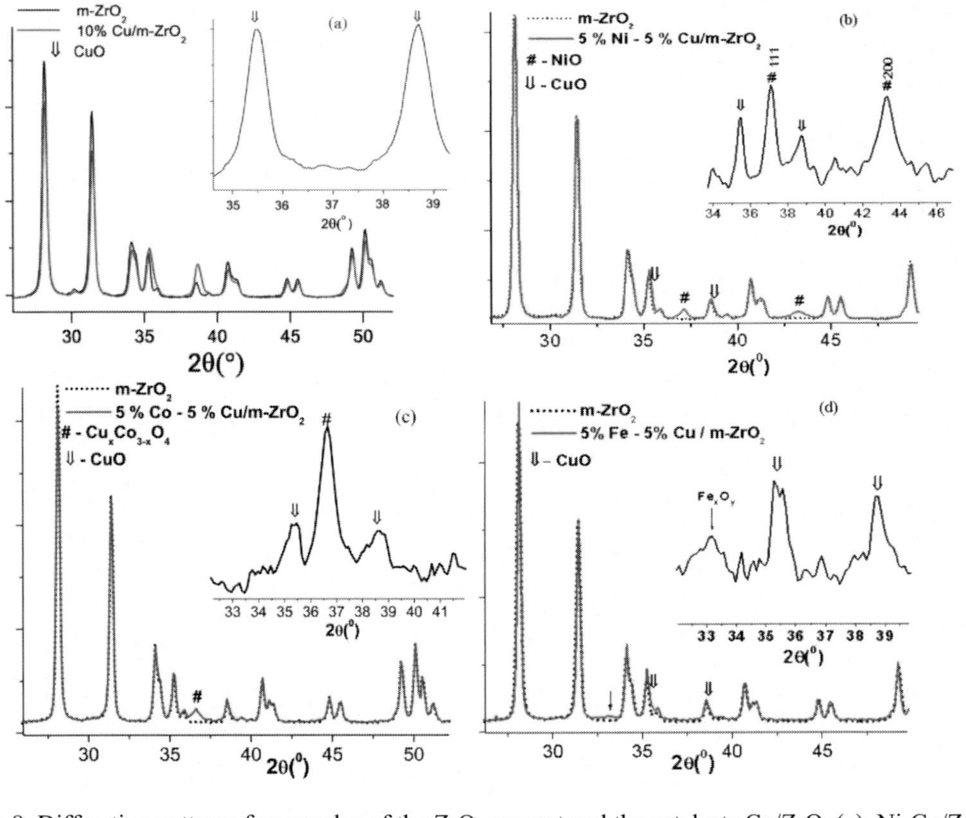

Figure 8. Diffraction patterns for samples of the ZrO_2 support and the catalysts Cu/ZrO_2 (a), $Ni\text{-}Cu/ZrO_2$ (b), $Co\text{-}Cu/ZrO_2$ (c) $Fe\text{-}Cu/ZrO_2$ (d). The inset shows the difference between the diffraction patterns of the catalyst and the support, which illustrates the presence of peaks for the phase CuO (a-d), NiO (b), $Cu_xCo_{3-x}O_4$(c), Fe_xO_y (d).

Based on the unit cell parameter for a cubic unit cell, a = 8.125 Å, we can assume that in the oxide formed, $Cu_xCo_{3-x}O_4$, the copper content x is close to 0.9 (JCPDS No. 37-0878: $Cu_{0.92}Co_{2.08}O_4$, a = 8.105 Å; ICSD No. 63387: $Cu_{0.95}Co_{2.05}O_4$, a = 8.133 Å). The content of this oxide in the sample is about 8.5 wt %; the CSR size is 20 nm. In the composition of the catalyst, we also detect a small amount of the phase CuO (≤ 2 wt %). Analysis of the XRD results for the catalyst Fe-Cu/ZrO_2 (Figure 8d) shows that the composition of the sample includes the phase CuO, the content of which can be estimated as about 4 wt %. The CSR size for the CuO particles is 29 nm. In the range 2θ = 33°, we observe a broadened low-intensity peak, which may indicate the presence of a hematite phase Fe_2O_3, for which a diffraction maximum in the 2θ = 33.18° region is characteristic. The lack of reflections from crystallized oxide phases allows us to assume that in Fe-Cu/ZrO_2 iron compounds are found in an X-ray amorphous or highly dispersed state.

Therefore, according to XRD data for samples of the supported catalysts M-Cu/ZrO_2 (M = Fe, Co, Ni), we observe the presence on their surface of crystallites of CuO and oxide phases of the second metal, with different extents of interaction with the copper oxide: formation of a cobalt/copper mixed oxide with a spinel type structure for Co-Cu/ZrO_2; disordering of the NiO crystal lattice, due to possible formation of a nickel/copper mixed oxide for Ni-Cu/ZrO_2. In contrast to the indicated catalysts, for Fe-Cu/ZrO_2 we do not observe the interaction between the metal oxides on the surface of the support.

The specific surface areas of the studied catalysts are given in Table 6. We see that the highest value of S_{sp} occurs in the Fe-Cu/ZrO_2 sample, for which (according to XRD data), an amorphous or highly dispersed state is possible for the supported particles of the active components.

Table 6. Specific surface area of samples (S_{sp}), reduction onset temperatures (T_i), and temperatures of the reduction maxima (T_{Mi}), quantity of H_2 absorbed (m_i) during temperature-programmed reduction of the catalysts Cu/ZrO_2 and M-Cu/ZrO_2 (M = Fe, Co, Ni)

Catalyst	S_{sp}, m²/g	Oxygen content in the supported phase, mmol O/g	T_i, °C	Temperatures of the maxima, T_{Mi}, °C			Quantity of absorbed H_2, mmol/g		
				T_{M1}	T_{M2}	T_{M3}	m_1	m_2	m_3
Cu/ZrO_2	4	1.56	250**	220	326	-	0.04	1.56	-
Fe-Cu/ZrO_2	11	2.13*	190	226	490	600	0.75	0.31	1.16
Co-Cu/ZrO_2	5	1.92*	150	260	-	-	1.95	-	-
Ni-Cu/ZrO_2	3	1.64*	200	240	-	-	1.34	-	-

*The values include 0.78 mmol O/g due to CuO; ** the onset temperature is indicated for reduction of crystallites of the CuO phase.

The TPR profiles for the catalysts Co-Cu/ZrO_2 and Ni-Cu/ZrO_2 are represented by a single reduction peak with a temperature of the maximum at 260°C and 240°C respectively (Figure 9). Contrary to the indicated samples, reduction of Fe-Cu/ZrO_2 is characterized by the presence of three peaks with temperatures of the maximum at 226°C, 490°C, and 600°C. From the data presented in Table 6, it follows that the onset temperatures for reduction of the catalysts M-Cu/ZrO_2 are found within the range 150-200°C, which is considerably lower than the value T_i = 250°C for reduction of the supported copper oxide phase.

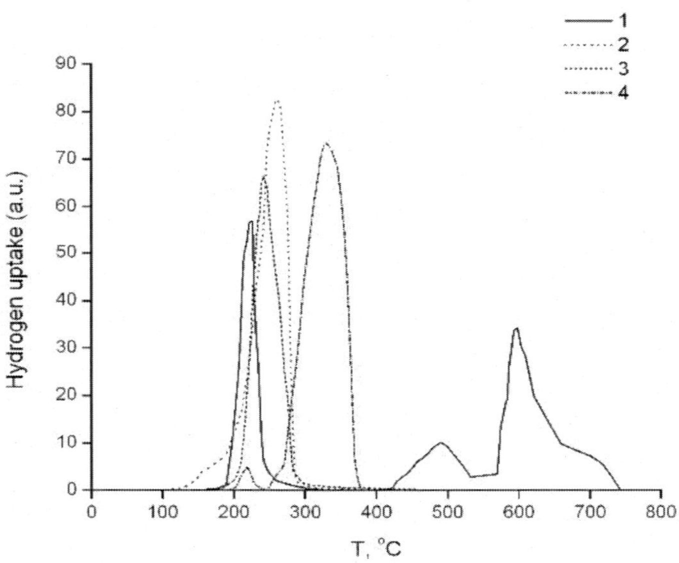

Figure 9. Profiles for temperature-programmed reduction of Fe-Cu/ZrO_2 (1), Co-Cu/ZrO_2 (2), Ni-Cu/ZrO_2 (3), and Cu/ZrO_2 (4).

We can arrange the studied catalysts in a series according to the temperature of the reduction maximum as follows, in order of increase in T_{M1} (for the copper catalyst, we used T_{M2}): Fe-Cu/ZrO_2 < Ni-Cu/ZrO_2 < Co-Cu/ZrO_2 < Cu/ZrO_2. This series shows that addition of a second metal to a monometallic copper catalyst leads to a decrease in the reduction temperature for the copper, nickel, and cobalt oxides, while we observe the greatest effect when we add iron. The established fact that the reduction temperature decreases can be caused by the following possible factors: an increase in the dispersity of the copper oxide in the presence of the second metal, formation of mixed oxides (leading to weakening of the metal–oxygen bonds), and possible hydrogen spillover from the reduced copper particles to the oxide of the second metal.

The data obtained are in qualitative agreement with the XRD results, according to which for Co-Cu/ZrO_2 and Ni-Cu/ZrO_2 we observe the formation of mixed oxides of the metals, and we observe a decrease in the CSR sizes of the CuO particles for Fe-Cu/ZrO_2.

In (Homs et al. 2006), the formation of mixed oxides $Co_{3-x}M_xO_4$ (M = Ni, Cu) was also observed after calcination of cobalt catalysts supported on ZnO and containing 1 wt % Cu or Ni. It is indicated that after reduction, such particles may evolve to metallic aggregates containing Co-M solid solutions. The authors of (Vizcaino et al. 2007) noted a decrease in the reduction temperature for metal oxides, in particular, NiO, in the presence of CuO, which is also observed in this work. According to XRD data for Fe-Cu/ZrO_2, no interaction is observed for metal oxides on the surface of the support, which allows us to assume that the low-temperature peak recorded for Fe-Cu/ZrO_2 characterizes reduction of copper oxide particles while the high-temperature maxima most likely reflect a reduction of highly dispersed particles of iron oxides Fe_xO_y. This assumption is also supported by the closeness of the values given in Table 6 for the oxygen content of the supported phase of the catalyst and the amount of absorbed hydrogen (calculated from the TPR data) for the condition that the reduction process proceeds according to the equation $M_xO_y + yH_2 = xM + yH_2O$.

The data obtained also allow us to conclude that the preliminary treatment of the catalysts with a hydrogen/nitrogen mixture at 350°C leads to complete reduction of the supported metal oxide particles of the active phase Co-Cu/ZrO_2 and Ni-Cu/ZrO_2. In this case, Fe-Cu/ZrO_2 is reduced only partially, and the reduced catalyst represents highly dispersed copper particles and iron oxide particles co-existing on the surface of the support. The results are in agreement with the literature data (Fu et al. 2010, Chen et al. 2010), according which it is established a formation of Fe_3O_4 and, possibly, coordination-unsaturated Fe atoms on the surface of the reduced at 200°C for 0.5 h iron-containing catalyst. Regarding the copper-containing catalyst, according to the TPR data, the maximum reduction rate for crystallites of the phase CuO supported on ZrO_2 is achieved at temperatures above 300°C. The selected conditions for preliminary reduction of Cu/ZrO_2 at 250°C ensure low reduction rates, promoting the formation of highly dispersed copper catalysts.

Tables 7-10 show the ethanol conversion (X), the selectivities with respect to the products and hydrogen (S_i and S_{H2} respectively) in ethanol steam reforming on the catalysts Cu/ZrO_2 and M-Cu/ZrO_2 (M = Fe, Co, Ni) at different temperatures. At 300°C, ethanol conversion on Cu/ZrO_2 reaches the maximum value, equal to 61.2%, and decreases in the following series of catalysts: Cu/ZrO_2 > Co-Cu/ZrO_2 > Ni-Cu/ZrO_2 > Fe-Cu/ZrO_2. The lowest activity for Fe-Cu/ZrO_2 is probably due to the minimal amount of reduced active phase, which is shown by the TPR method.

For T > 300°C, ethanol conversion on Cu/ZrO_2 decreases considerably, probably due to sintering of Cu particles. This is supported by XRD data for the samples after treatment with the reaction mixture (not shown in this paper), for which the presence of large crystallized copper particles was observed. In bicomponent catalysts, X reaches maximum values in the range 40–57% at 450-550°C.

Table 7. Temperature dependencies of ethanol conversion (X) and product selectivity for Cu/ZrO₂

T, °C	X, %	Selectivity, %							
		CH_3CHO	$(CH_3)_2CO$	CO	CO_2	CH_4	C_2H_4	C_{3+}	H_2
300	61.2	100	0.0	0.0	0.0	0,0	0.0	0.0	16.7
350	22.7	99.6	0.0	0.0	0,0	0,0	0.4	0.0	16.6
400	15.7	60.8	16.3	0.0	12.6	0.1	9.1	1.1	23.7
450	20.9	3.5	28.4	0.0	31.7	0.4	31.3	4.7	33.1

Table 8. Temperature dependencies of ethanol conversion (X) and product selectivity for Ni-Cu/ZrO₂

T, °C	X, %	Selectivity, %							
		CH_3CHO	$(CH_3)_2CO$	CO	CO_2	CH_4	C_2H_4	C_{3+}	H_2
300	16.4	13.7	0.0	32.5	10.3	42.8	0.3	0.4	20.0
350	25.5	11.3	0.0	31.8	19.7	36.3	0.3	0.6	30.7
400	26.1	61.0	0.0	6.8	9.2	17.0	0.5	5.5	18.2
450	30.7	36.2	34.3	0.0	19.4	1.6	6.8	1.7	28.7
500	40.3	6.1	40.2	0.0	41.8	0.9	9.1	1.9	47.0
550	33.6	2.8	41.0	0.0	42.7	1.8	10.8	0.9	47.2

Table 9. Temperature dependencies of ethanol conversion (X) and product selectivity for Co-Cu/ZrO₂

T, °C	X, %	Selectivity, %							
		CH_3CHO	$(CH_3)_2CO$	CO	CO_2	CH_4	C_2H_4	C_{3+}	H_2
300	31.6	93.9	0.0	0.0	0.0	4.4	0.2	1.5	14.2
350	39.5	89.9	0.0	0.0	3.0	5.2	0.4	1.5	16.3
400	28.0	80.6	0.0	0.0	13.0	4.0	0.6	1.8	25.0
450	39.9	26.9	36.5	0.0	33.3	0.7	1.9	0.7	41.6
500	52.1	0.8	46.7	3.5	45.0	1.4	2.2	0.4	52.2
550	54.5	0.1	40.2	3.4	50.7	2.1	2.8	0.7	56.8

At the indicated temperatures, the activity series for the catalysts can be represented as follows in order of decrease in X: Fe-Cu/$ZrO_2 \geq$ Co-Cu/$ZrO_2 >$ Ni-Cu/ZrO_2. We should point out some decrease in ethanol conversion which can be observed for Ni- and Fe-containing samples for $T \geq 500$°C, which may be due to changes in the catalyst when exposed to the reaction medium.

At the initial temperatures (300-350°C), on all catalysts, except for Ni-Cu/ZrO_2, ethanol is converted to a mixture of acetaldehyde and hydrogen; carbon dioxide, methane, and C_2-C_3 hydrocarbons are present in considerably smaller amounts. The high selectivity to acetaldehyde (S_{CH3CHO}) suggests that primary conversion of ethanol on Co-Cu/ZrO_2 and Fe-Cu/ZrO_2 catalysts occurs preferentially via the dehydrogenation route (reaction II). Formation of carbon dioxide and methane is possible as a result of the steam reforming of acetaldehyde (according to reaction XIII).

Table 10. Temperature dependencies of ethanol conversion (X) and product selectivity for Fe-Cu/ZrO_2

T, °C	X, %	Selectivity, %							
		CH_3CHO	$(CH_3)_2CO$	CO	CO_2	CH_4	C_2H_4	C_{3+}	H_2
300	8.2	99.4	0.0	0.0	0.0	0.6	0.0	0.0	16.3
350	13.2	51.1	30.8	0.0	17.6	0.2	0.3	0.0	29.5
400	39.1	7.3	56.0	0.0	36.0	0.2	0.5	0.0	43.4
450	57.2	1.2	45.4	0.0	52.7	0.2	0.5	0.0	57.9
500	53.9	0.3	45.1	0.0	51.8	0.5	2.0	0.3	56.7
550	45.9	0.1	35.4	0.0	59.6	2.1	2.4	0.4	62.9

On Ni-Cu/ZrO_2 at 300-400°C, in the reaction products mainly CO and CH_4 are observed along with the hydrogen; CO_2 is present in small amounts. High initial selectivity with respect to acetaldehyde is not characteristic for this catalyst, which allows to assume that for the indicated temperatures, the reaction of decomposition of the acetaldehyde to form methane and carbon monoxide (reaction VI) predominates, which is catalyzed by nickel (according to literature data, Ni is especially effective in breaking the C—C bond (Vaidya and Rodrigues 2006)). We should note that at 300-400°C, on Ni-Cu/ZrO_2 the value of S_{CH4} exceeds the values of S_{CO}, which may be due to occurrence of the methane formation reaction on the active sites of the nickel (reaction VII) and also the water-gas shift reaction, evidence for which is the presence of CO_2 in the conversion products (V).

For Co-Cu/ZrO_2 and Fe-Cu/ZrO_2, a decrease in S_{CH3CHO} with an increase in temperature was observed; for Ni-Cu/ZrO_2, the value of S_{CH3CHO} passes through a maximum at 400°C and decreases with a further rise in temperature, which suggests an acceleration of further transformations of acetaldehyde. In the region T > 400°C, a characteristic feature for catalysis of ethanol steam reforming on bicomponent catalysts is preferential formation of carbon dioxide and acetone along with the hydrogen, indicating that hydrogen evolution occurs as a result of further conversion of acetaldehyde according to the reaction (VIII) (Nishiguchi et al. 2005).

It should be noted that the formation of acetone on the iron-containing catalyst is observed at temperatures that are 100°C lower than on the nickel- and cobalt-containing samples; among the studied systems, the maximum selectivities are achieved on Fe-Cu/ZrO_2: $S_{CH3COCH}$ = 56.0% (400°C) and S_{CO2} = 59.6% (550°C).

The maximum hydrogen selectivity in the temperature range 400-550°C reaches 62.9%, 56.8%, and 47.2% for Fe-Cu/ZrO_2, Co-Cu/ZrO_2, and Ni-Cu/ZrO_2 respectively. In this case, the nickel/copper catalysts are distinguished by elevated selectivity for C_1-C_3 hydrocarbon formation, which reaches 13%; in the presence of the cobalt/copper catalysts, the hydrocarbon selectivity does not exceed 5.5%, and on the iron/copper samples the overall contribution of hydrocarbon formation is no greater than 5%. The appearance of CO in the reaction products on Co-Cu/ZrO_2 at 500-550°C indicates that under these conditions, ethanol steam reforming is accompanied by the reaction of ethanol decomposition (Casanovas et al. 2010).

The elevated activity of the binary catalyst Fe-Cu/ZrO_2 in the steam reforming of bioethanol may be associated with the presence of iron oxides in the composition of the catalyst. The iron oxide containing solids are known to be active catalysts for the water-gas shift reaction. This ability of the catalysts leads to an increase of the CO conversion to CO_2 and an increase in the hydrogen yield at low temperatures. The promoting effect of iron additives on the water-gas shift reaction under ethanol steam reforming conditions has also been noted for the supported catalysts 10%Co-1%Fe/ZnO (Casanovas et al. 2010), 1%Rh-10%Fe/Ca-Al_2O_3 (Chen et al. 2010), 10%Ni-5%Fe/$La_2O_2CO_3$ (Shi et al. 2011).

Therefore, bicomponent catalysts M-Cu/ZrO_2 (M = Ni, Co, Fe) are considerably more efficient in ethanol steam reforming at 350-550°C than monometallic Cu/ZrO_2, which is unstable in the reforming process. A characteristic feature of the binary catalysts studied in this work is a formation of acetone as the only secondary product, which is condensed in the liquid phase after the reforming reactor, and evolution of predominantly H_2 and CO_2 into the gas phase. This makes it possible to realize the concept of ethanol steam reforming with selective formation of hydrogen and the target secondary product at low temperatures (400-550°C). We have found that the activity and selectivity of the catalyst Fe-Cu/ZrO_2 increases in the bioethanol steam reforming due to the presence of highly dispersed copper particles and iron oxide particles, accelerating the water-gas shift reaction, which plays an important role in the final distribution of ethanol steam reforming products. In contrast to Co- and Ni-containing systems, on Fe-Cu/ZrO_2 we do not observe a formation of such a secondary product as CO, which is important for obtaining pure hydrogen and for using it subsequently in low-temperature fuel cells.

Ethanol Steam Reforming over MFe_2O_4 (M = Ni, Co, Fe, Mn, Cu, Mg, Zn) Ferrites

Catalyst Stability

The XRD patterns of ferrites MFe_2O_4 (M = Ni, Co, Mn), prepared from the heteronuclear complex, before and after catalytic experiments are shown in Figure 10.

The data indicate that the initial catalyst samples are pure phases of ferrites with crystal structures of cubic spinel with the elemental cell parameters a = 8.4 Å (Figure 10, curves 1) (ICDD No. 01-074-2403, 00-022-1086, 01-070-8730). The diffraction reflections at 2θ = 21.6°, 34.8°, 41.1°, 49.8°, 66.3°, and 73.3° were assigned to planes (111), (220), (311), (400), (333), and (440), respectively.

Calculated using the Scherrer equation for reflection (311), the size of the crystallites of the fresh samples of catalysts was 8 nm, which was in excellent agreement with electron microscopic data (Figure 11), which revealed the same values for the average size of the ferrite particles and their sharp distribution with respect to a size. This indicates a high degree of crystallinity and the absence of an amorphous layer at the surface of the nanoparticles of the catalyst.

After catalysis, the XRD patterns were significantly changed in the case of $NiFe_2O_4$ and $CoFe_2O_4$. $NiFe_2O_4$ has the least stable phase which is completely reduced to metallic iron and a Fe-Ni alloy; the peaks corresponding to the spinel phase were not observed after catalysis (Figure 10a, curve 2).

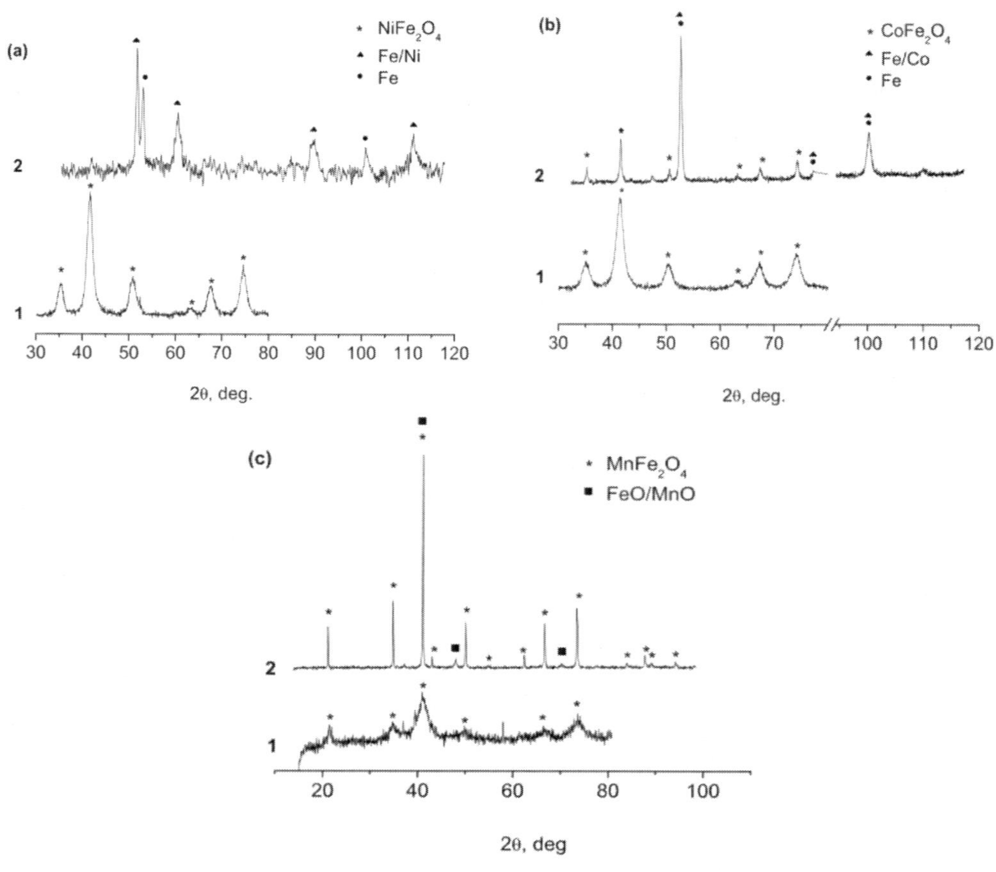

Figure 10. Diffraction patterns (Co-anode) of $NiFe_2O_4$ (a), $CoFe_2O_4$ (b), and $MnFe_2O_4$-HC (c): curves 1 give the XRD patterns of fresh catalyst samples; curves 2 represent the XRD patterns for the same solids after catalysis.

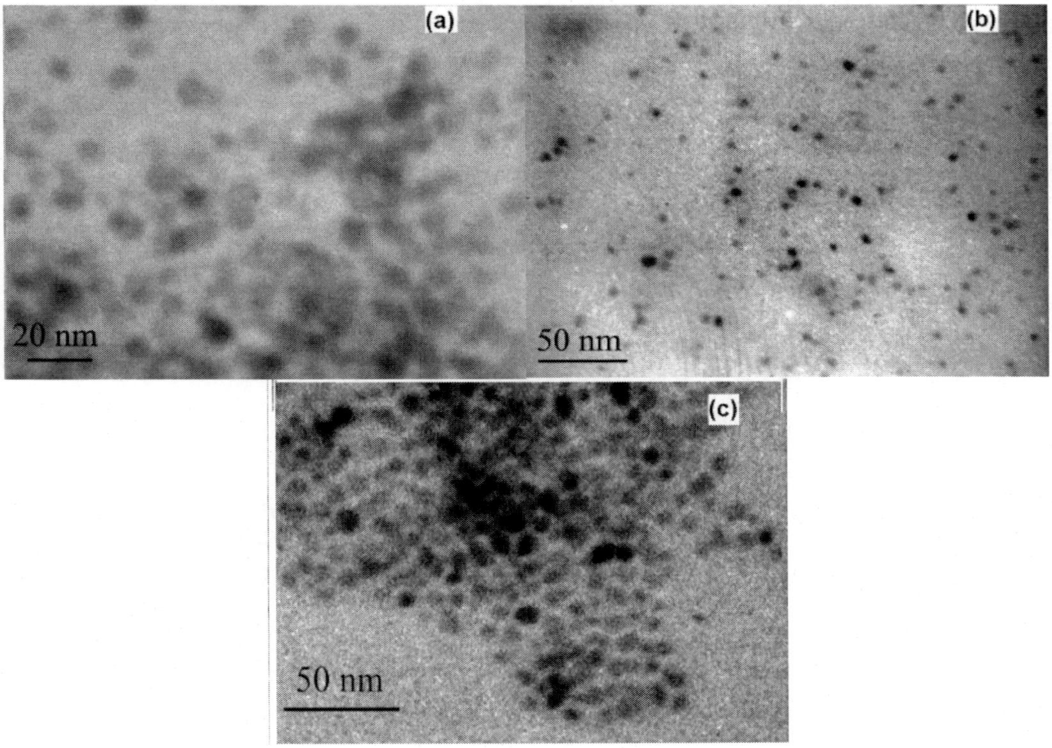

Figure 11. TEM microphotographs of fresh nanoparticles of $NiFe_2O_4$ (a), $CoFe_2O_4$ (b), and $MnFe_2O_4$-HC (c).

For $CoFe_2O_4$ sample after carrying out the ESR reaction, it was established that most of the complex oxide was reduced to metallic iron and a Fe-Co alloy, and only a small amount of the initial ferrite remained (Figure 10b, curve 2). In contrast to the nickel and cobalt ferrites, the phase composition of $MnFe_2O_4$-HC remains practically unchanged, only a small amount of the complex oxide FeO/MnO (ICDD No. 01-077-2360) was observed (Figure 10c, curve 2); the XRD peaks became more narrow that indicates the increase in size of the crystallites of the catalysts. The size increase of the crystallites was directly confirmed by TEM (Figure 12).

Similar to the $NiFe_2O_4$ and $CoFe_2O_4$ ferrites, the $CuFe_2O_4$ ferrite is also reduced during the catalysis (Figure 13). The fresh sample of $CuFe_2O_4$ consisted of a crystalline ferrite phase with spinel structure (ICDD N 01-077-0427). The sample is reduced under the catalysis conditions with the formation of metallic copper (ICDD N 01-085-1326). The reduction of copper ferrite, as well as of nickel and cobalt ferrites, proceeds through the action of hydrogen, which is a product of the ESR reaction, most likely, according to reaction $CuFe_2O_4 + H_2 \rightarrow Cu + Fe_2O_3 + H_2O$ followed by $Fe_2O_3 + H_2 \rightarrow FeFe_2O_4 + H_2$. It is difficult to distinguish between the reflections of $CuFe_2O_4$ and $FeFe_2O_4$, thus the diffraction patterns given in Figure 13 may indicate either partial or complete reduction of the copper ferrite.

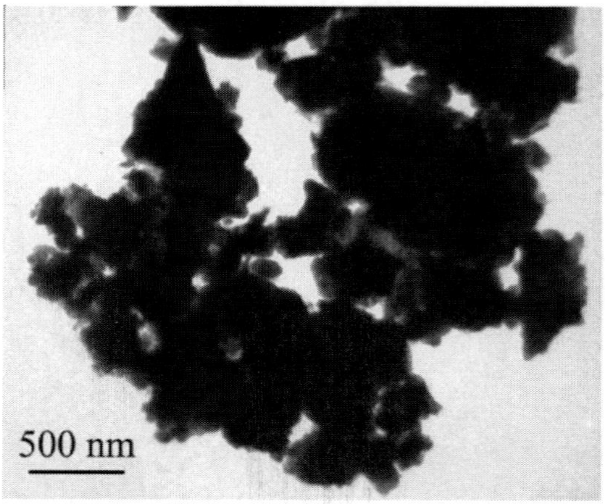

Figure 12. TEM microphotographs of $MnFe_2O_4$ after catalysis.

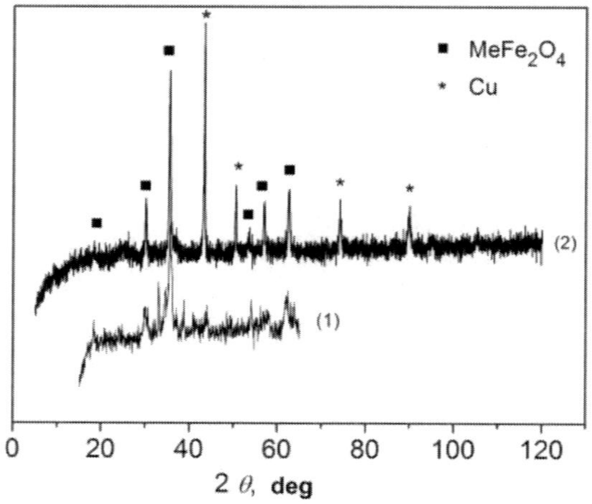

Figure 13. Diffraction patterns (Cu-anode) of $CuFe_2O_4$ before (curve 1) and after (curve 2) catalysis.

According to X-ray diffraction data of other investigated ferrites, their fresh samples calcined at 400°C were amorphous ($FeFe_2O_4$, $MnFe_2O_4$, $MgFe_2O_4$) or poorly crystallized ($ZnFe_2O_4$). After increasing of calcination temperature to 600°C in air manganese ferrite remained in an amorphous state during 4 h, and it was transformed to Mn_2O_3 (ICDD N 01-089-2809) and Fe_2O_3 (ICDD N 01-085-0599) after 10 h calcination at the same temperature due to ferrite oxidation. In contrary, after catalysis, all ferrite samples possessed a clear crystalline image with a structure of cubic spinel (Figure 14). For identification of ferrite crystalline phases, we used ICDD N 00-019-0629 for $FeFe_2O_4$, ICDD N 01-074-2403 for $MnFe_2O_4$, ICDD N 01-088-1942 for $MgFe_2O_4$ and ICDD N 00-022-1012 for $ZnFe_2O_4$. Crystallization of ferrites in the catalytic conditions was accompanied with a significant decreasing of its specific surface area (Table 11). This effect may be associated with alternating acts of reduction and oxidation of a catalyst during the ethanol steam reforming reaction; oxygen defects formed in a reduction step make more easy migration of ions in the catalyst lattice and thus promote catalyst crystallization in the reaction conditions. Note, that similar behavior is inherent also to magnesium ferrite. Indeed, the specific surface area of $MgFe_2O_4$ was decreased from 121 to only 115 m^2/g after calcination at 550°C in an air, while it was observed dramatically strong falling down of the specific surface area to 14 m^2/g when $MgFe_2O_4$ was explored as a catalyst for the ESR reaction (Table 11).

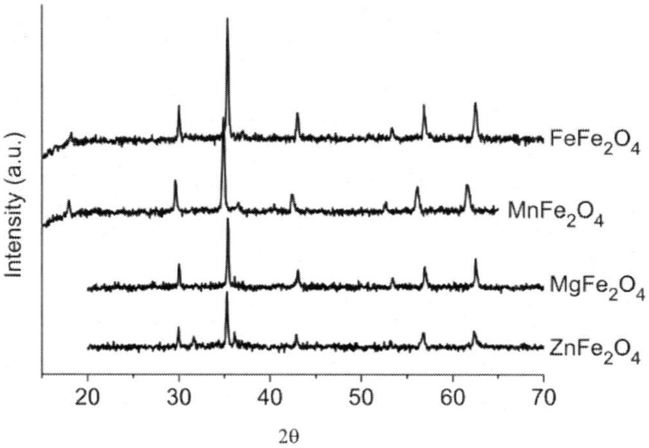

Figure 14. XRD profiles (Cu-anode) of $FeFe_2O_4$, $MnFe_2O_4$, $MgFe_2O_4$, and $ZnFe_2O_4$ after catalysis.

Table 11. Influence of the ESR reaction (550°C, $H_2O/C_2H_5OH = 19$) on the catalyst specific surface area

Catalyst	Specific surface area, m^2/g	
	Before catalysis	After catalysis
$FeFe_2O_4$	220	18
$MnFe_2O_4$	140	10
$MgFe_2O_4$	121	14
$ZnFe_2O_4$	18	6

On the whole, the stability of MFe_2O_4 ferrites depends on a reducing ability of M^{2+} cations in the conditions of the ESR reaction where hydrogen is reducing agent.

A thermal stability of the MFe_2O_4 (M = Fe, Mn, Zn, Mg) ferrites was also estimated by a thermal gravimetric analysis method; Figure 15 presents the obtained thermograms. There are two temperature ranges of a mass loss. The first associated with water desorption was observed at 50-200°C (323-473 K) with a maximum at 100-120°C (373-393 K) for all studied ferrites; it was accompanied with an endothermic peak in the DTA curves. The mass loss at this temperature range was 11.6%, 12.7 %, 11.9% and 0.8% for $FeFe_2O_4$, $MnFe_2O_4$, $MgFe_2O_4$ and $ZnFe_2O_4$ respectively. The second temperature range of a mass loss was observed at 250-350°C (523-623 K) for $FeFe_2O_4$ and at 300-400°C (573-673 K) for $MgFe_2O_4$, it may be associated both with surface dehydroxylation as well as with decomposition of nitrate groups and organic compounds (Giri et al. 2005). The mass loss was 5.7% $FeFe_2O_4$ and 5.0% for $MgFe_2O_4$. The exothermic peaks at 280-320°C (553–593 K) in DTA curves may be related to ferrospinel crystallization (Rana et al. 2010). At T > 400°C (673 K), the process of a mass loss was close to breakdown that may be an evidence of a crystallization completion. In these conditions, the cation migration from octahedral sites to tetrahedral ones may take place (Rana et al. 2010). The total mass loss of the ferrite samples after their heating from 20°C (293 K) to 1000°C (1273 K) was 21.2% for $FeFe_2O_4$; 15.8% for $MnFe_2O_4$; 20.3% for $MgFe_2O_4$ and 2.2% for $ZnFe_2O_4$.

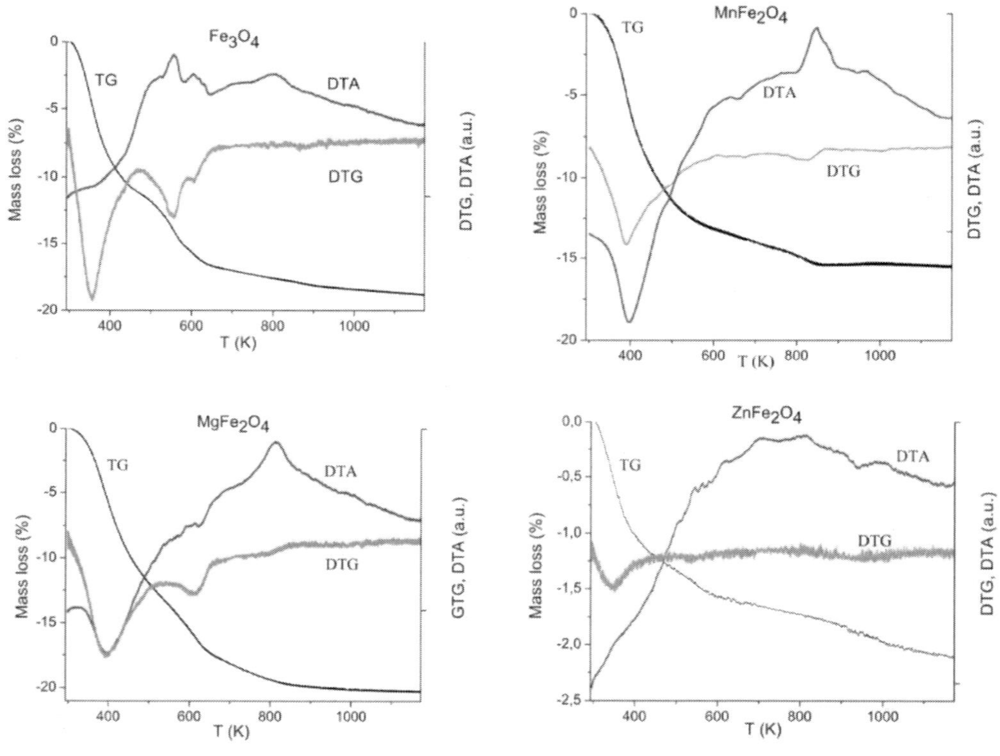

Figure 15. Thermograms of $FeFe_2O_4$, $MnFe_2O_4$-CP, $MgFe_2O_4$, and $ZnFe_2O_4$.

Our studies allow one to introduce two different groups of ferrites according to their structural changes during the catalysis. The first group refers to ferrites with remaining unchanged their crystal structure during the catalysis. The following ferrites belong to the first group: $FeFe_2O_4$, $MnFe_2O_4$, $MgFe_2O_4$, and $ZnFe_2O_4$. The second group represents the ferrites, which are reduced during the catalysis. Our studies show that $NiFe_2O_4$, $CoFe_2O_4$, and $CuFe_2O_4$ belong to the second group of ferrites. The ferrites from the second group are reduced by hydrogen, which is the main ESR product. These results in a reduction of

Ni^{2+}, Co^{2+}, and Cu^{2+} ions into corresponding metals. A difference to be reduced during the catalysis between the ferrites of the first and second groups is in a qualitative agreement with the well-known electrochemical series according to which Ni^{2+}, Co^{2+}, and Cu^{2+} are easier reducible comparing to Fe^{2+}, Mn^{2+}, Mn^{2+}, and Zn^{2+}.

Catalyst Reducing Ability

Catalyst reducing ability was estimated by the hydrogen temperature-programmed reduction method (H_2-TPR) for ferrites which do not change their spinel phase during catalytic tests. The obtained results are shown in Figure 16 and Table 12. For comparison, the H_2-TPR of iron oxide Fe_2O_3 was carried out (not shown in this chapter). It is known that reduction of Fe_2O_3 proceeds in two steps via Fe_3O_4 intermediate (Sastri et al. 1982). The two peaks with maximum reduction temperatures 375 and 633°C were observed that characterized reductive transformation $Fe^{3+} \rightarrow Fe^{2+,3+}$ ($Fe_2O_3 \rightarrow Fe_3O_4$) and $Fe^{2+,3+} \rightarrow Fe^{2+}$ ($Fe_3O_4 \rightarrow FeO$) respectively. Under studied experimental conditions, complete reduction of Fe_2O_3 to metallic iron was not observed up to 700°C, only partial reduction took place.

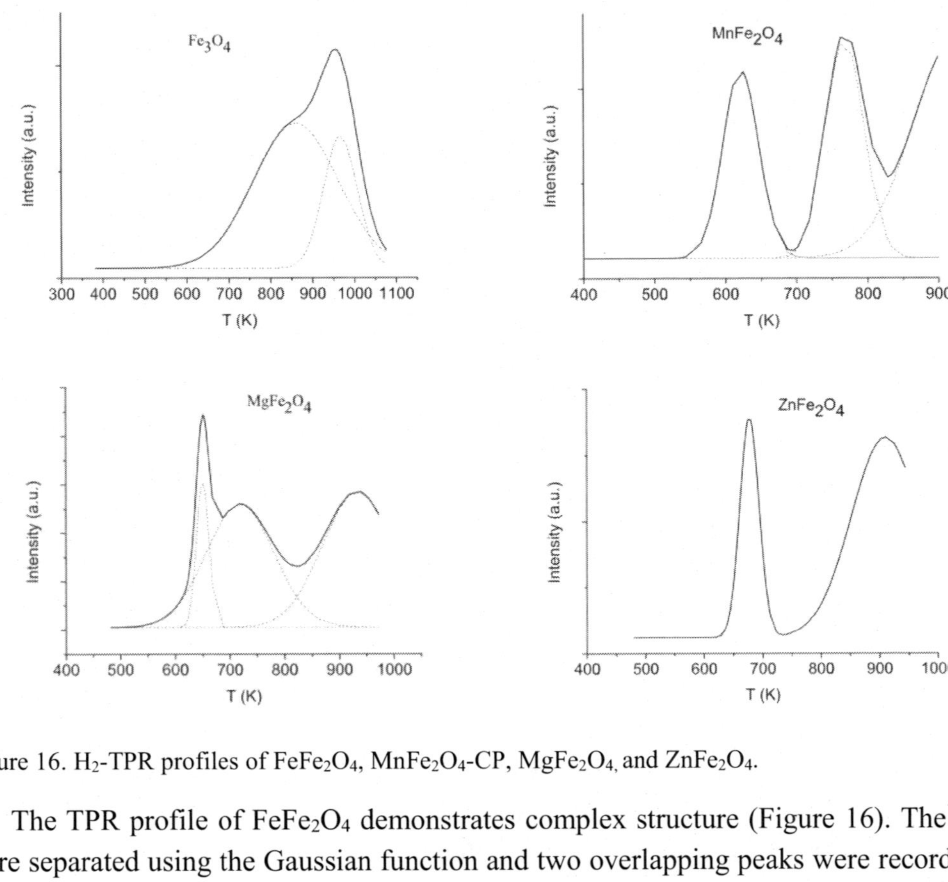

Figure 16. H_2-TPR profiles of $FeFe_2O_4$, $MnFe_2O_4$-CP, $MgFe_2O_4$, and $ZnFe_2O_4$.

The TPR profile of $FeFe_2O_4$ demonstrates complex structure (Figure 16). The peaks were separated using the Gaussian function and two overlapping peaks were recorded for the $FeFe_2O_4$ with maximum reduction temperatures 590°C (873 K) and 693°C (966 K) that may be attributed to reductive transformation $Fe^{3+} \rightarrow Fe^{2+}$ and $Fe^{2+} \rightarrow Fe^0$ respectively. The profile of $MnFe_2O_4$ consists of two peaks at 359 and 495°C (632 and 768 K). Peaks at 377°C (650 K) and 446°C (719 K) were found for TPR profile of $MgFe_2O_4$. It may be assumed that these peaks characterize the reduction of Fe^{3+} ions in octahedral (first peak) and tetrahedral (second peak) coordination. In addition to that, it was observed a shift of peaks to the low-temperature range in a comparison with Fe_3O_4. Thus, a presence of Mn in ferrite spinel facilitates a Fe^{3+} ion reduction.

Table 12. The H_2-TPR data for $FeFe_2O_4$, $MnFe_2O_4$, $MgFe_2O_4$, and $ZnFe_2O_4$

Catalyst	Peak maximum, °C	
	$Fe_2O_3 \rightarrow Fe_3O_4$	$Fe_3O_4 \rightarrow FeO$
$FeFe_2O_4$		590
$MnFe_2O_4$	349, 495	> 650
$MgFe_2O_4$	377, 446	677
$ZnFe_2O_4$	405	637

Hydrogen consumption at T > 550°C (823 K) is related to subsequent sample reduction to FeO. The reduction peak at 403°C (676 K) was observed for $ZnFe_2O_4$ that may be associated with reduction $Fe^{3+} \rightarrow Fe^{2+,3+}$, while hydrogen consumption at T > 500°C (773 K) may be a result of the reduction of Zn^{2+} to Zn^o, as well as $Fe^{2+,3+}$ to Fe^{2+}.

The TPR data highlight that the nature of metal M in the MFe_2O_4 ferrospinels influence significantly on the reductive ability of spinel Fe^{3+} ions. The order of reductive facility of Fe^{3+} to Fe^{2+} in spinels during $Fe_2O_3 \rightarrow Fe_3O_4$ transformation is as follows: $MnFe_2O_4 > MgFe_2O_4 > ZnFe_2O_4$.

Catalytic Properties

Catalytic properties of the investigated ferrites are presented in the Tables 13-20. All data in these tables correspond to the following experimental conditions: initial reaction mixture composed of 2.7 vol % C_2H_5OH, 50 vol % H_2O and N_2 balance, total feed 0.17 mole h^{-1}. More details can be found in our works (Stolyarchuk et al. 2012, 2016; Dolgikh et al. 2013, 2014, 2015).

We found that no CO was detected in the ESR reaction products up to 500°C for all explored ferrites. At higher temperatures, CO was detected for $MnFe_2O_4$ only at 700°C, for $NiFe_2O_4$ and $CoFe_2O_4$ at 550-700°C and 600-700°C, respectively. In the case of $NiFe_2O_4$ and $CoFe_2O_4$, CO appearance may be mainly attributed to Ni and Co metals which are result of corresponding ferrite reduction during catalysis because these metals promote CO formation in the ESR reaction (see reviews Haryanto 2005; Ni et al. 2007; Pyatnitsky et al. 2013; Vaidya and Rodrigues 2006).

Table 13. Temperature dependencies of ethanol conversion (X) and product selectivity for $NiFe_2O_4$

		Selectivity, %							
T, °C	X, %	CH_3CHO	$(CH_3)_2CO$	CO	CO_2	CH_4	C_2H_4	C_{3+}	H_2
250	2.6	100	0.0	0.0	0.0	0.0	0.0	0.0	18.2
300	5.2	99.0	0.0	0.0	0.0	1.0	0.0	0.0	17.9
350	10.8	93.1	0.0	0.0	4.6	2.3	0.0	0.0	20.6
400	19.4	92.8	0.0	0.0	3.5	3.7	0.0	0.0	19.0
450	48.5	95.2	0.0	0.0	1.8	2.7	0.3	0.0	17.8
500	62.8	90.5	0.0	0.0	5.2	3.0	1.3	0.0	20.4
550	97.8	72.4	0.0	14.3	3.3	7.3	2.7	0.0	23.9
600	100	62.2	0.0	24.3	5.0	6.2	2.3	0.0	30.8
650	100	15.2	0.0	57.4	2.4	13.9	10.8	0.3	39.9
700	100	2.3	0.0	59.8	3.1	24.5	10.2	0.1	38.2

Table 14. Temperature dependencies of ethanol conversion (X) and product selectivity for $CoFe_2O_4$

		Selectivity, %							
T, °C	X, %	CH_3CHO	$(CH_3)_2CO$	CO	CO_2	CH_4	C_2H_4	C_{3+}	H_2
250	8.8	100	0.0	0.0	0.0	0.0	0.0	0.0	17.3
300	10.8	65.9	0.0	0.0	34.0	0.1	0.0	0.0	45.4
350	15.1	62.5	0.0	0.0	37.2	0.1	0.2	0.0	49.1
400	11.5	95.1	0.0	0.0	4.4	0.2	0.3	0.0	22.0
450	15.9	86.3	0.0	0.0	12.9	0.4	0.4	0.0	28.7
500	12.6	82.5	0.0	0.0	16.0	0.6	0.9	0.0	31.7
550	13.5	92.1	0.0	0.0	6.3	0.7	0.9	0.0	22.8
600	27.3	71.0	0.0	20.2	3.3	3.1	2.4	0.0	29.0
650	47.7	34.3	0.0	50.0	8.2	4.7	2.8	0.0	48.2
700	84.0	1.8	0.0	84.8	7.7	3.8	1.9	0.0	64.1

Table 15. Temperature dependencies of ethanol conversion (X) and product selectivity for $MnFe_2O_4$ – HC

		Selectivity, %							
T, °C	X, %	CH_3CHO	$(CH_3)_2CO$	CO	CO_2	CH_4	C_2H_4	C_{3+}	H_2
300	0.6	43.2	0.0	0.0	51.3	0.2	5.3	0.0	59.6
350	1.5	24.9	0.0	0.0	66.8	0.1	8.2	0.0	72.2
400	14.3	1.7	54.5	0.0	37.8	0.1	5.8	0.1	44.9
450	26.5	1.4	44.3	0.0	48.6	0.2	5.4	0.1	54.6
500	45.6	0.8	34.2	0.0	59.9	0.3	4.7	0.1	64.4
550	47.6	0.8	16.4	0.0	81.4	0.4	1.0	0.0	84.1
600	80.4	2.6	3.9	0.0	89.8	1.6	2.1	0.0	91.0
650	100	0.3	0.0	0.0	92.5	4.3	2.9	0.0	94.6
700	100	0.0	0.0	11.1	83.8	3.3	1.8	0.0	93.0

Another common feature of the catalytic action of all investigated ferrites (with exception of unstable $NiFe_2O_4$ and $CoFe_2O_4$ ferrites) is a production of significant amounts of acetone at temperatures near 400°C. Therefore, acetone is an important intermediate product in the transformation of ethanol to H_2 and CO_2 in the ESR reaction over ferrite oxide surface.

The primary reaction in the ESR process pathway is the reaction of ethanol dehydrogenation to acetaldehyde (II). Figure 17 shows the scheme of a probable heterolytic (non-redox) mechanism of reaction (II) over ferrite catalysts. The main features of this mechanism were adopted from the work (Schweitzer et al. 2014; Hu et al. 2015a and 2015b) that gives the similar mechanism for the propane to propene dehydrogenation over single site M^{II} (M = Zn, Co, Fe) on silica catalysts.

The surface metal cations in ferrites are attributed to Lewis acid centers and oxygen anions to Brensted base centers. Interaction of ethanol molecule with an acid-base pair leads to heterolytic O-H cleavage in ethanol with a formation of ethoxy intermediate located on metal ion (Lewis acid center) and proton located on surface oxygen ion (Bronsted base center).

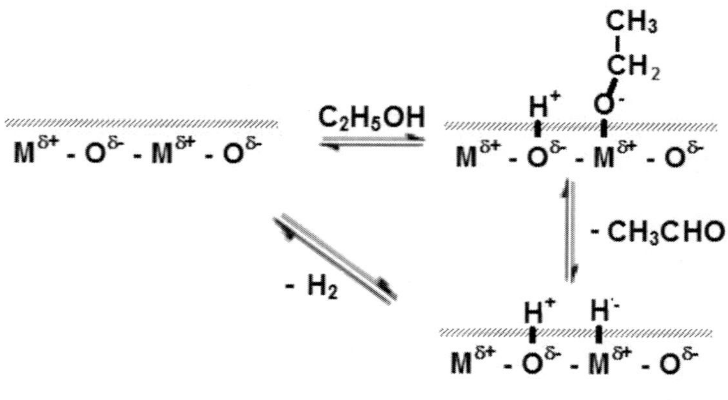

Figure 17. Probable catalytic reaction pathway for ethanol dehydrogenation to acetaldehyde on ferrites.

Next step is a β-elimination of hydride ion H^- from C-H bond in an ethoxy particle with the transfer of H^- to the surface metal ion and desorption of acetaldehyde. Finally, a recombination of hydride ion and proton take place leading to the formation of gaseous molecular hydrogen.

This heterolytic mechanism appears less likely for $CuFe_2O_4$ since it does not readily account for the observed sharp difference in the catalytic activities of copper ferrite and other ferrites studied in the present work with respect to ethanol conversion to acetaldehyde. The high activity of $CuFe_2O_4$ is probably related to the reversible $Cu^{2+} \leftrightarrow Cu^{+}$ transitions at the copper ferrite surface, and homolytic (redox) mechanism of ethanol dehydrogenation takes place.

Desorption of acetaldehyde occurs most likely through intermediate formation of adsorbed $CH_3CHO_{(a)}$ species which give rise to further reaction pathways, in particular, to acetone formation via the aldol-type mechanism that is similar to the mechanism proposed by Elliott and Pennella (1989) and confirmed by Nishiguchi et al. (2005):

$$2\ CH_3\text{-}CHO_{(a)} + O_{(s)} \leftrightarrow CH_3\text{-}CH(OH)\text{-}CH_2\text{-}COO_{(a)} + H_{(a)} \quad \text{(XIV)}$$

$$CH_3\text{-}CH(OH)\text{-}CH_2\text{-}COO_{(a)} + H_{(a)} \leftrightarrow CH_3\text{-}CO\text{-}CH_3 + CO_2 + H_2 \quad \text{(XV)}$$

where $O_{(s)}$ is a lattice oxygen, and (a) indicates an adsorbed species.

Table 16. Temperature dependencies of ethanol conversion (X) and product selectivity for $CuFe_2O_4$

		Selectivity, %							
T, °C	X, %	CH_3CHO	$(CH_3)_2CO$	CO	CO_2	CH_4	C_2H_4	C_{3+}	H_2
300	92.7	97.2	0.0	0.0	2.5	0.3	0.0	0.0	18.6
350	95.9	87.3	7.5	0.0	4.7	0.3	0.0	0.2	20.0
400	95.4	34.6	31.6	0.0	32.2	0.9	0.0	0.7	41.2
450	91.1	20.7	40.4	0.0	32.8	4.6	0.9	0.6	39.2
500	97.1	9.7	16.2	0.0	63.8	7.5	1.8	1.0	64.7
550	98.9	1.7	7.2	0.0	60.2	28.1	2.2	0.6	52.0

The aldol condensation is catalyzed by acids or bases. Therefore, for the ESR over ferrites, it occurs on acid or basic surface centers of ferrite catalyst. The overall reaction of acetaldehyde transformation includes also the redox-type reactions with the participation of surface lattice oxygen. The removal of the surface oxygen according to the step (XIV) leads to the formation of the surface oxygen vacancies (with simultaneous decrease of the oxidation state of the nearest metal ions). Oxygen loss is compensated by the interaction of water with the reduced surface centers according to the following reaction

$$H_2O + \square_{(s)} \leftrightarrow O_{(s)} + H_2 \quad \text{(XVI)}$$

where $\square_{(s)}$ denotes a surface oxygen vacancy.

In more details, an oxygen vacancy may be presented as $[Fe^{2+}\square Fe^{2+}]_s$ which is converted in reaction (XVI) to $[Fe^{3+}O^{2-}Fe^{3+}]_s$. The reaction (XVI) plays a very important role in the steam reforming of ethanol, first of all, because it gives up to 50% of hydrogen produced in the ESR process.

The oxygen vacancies appear as a result of the reduction of the surface metal ions by gaseous and adsorbed organic compounds or their fragments like $CH_{3(a)}$, $CH_{2(a)}$, $CO_{(a)}$, and by hydrogen which is the main reducing agent at an elevated temperature.

The overall reductive step in the ESR reaction can be expressed as

$$CH_3CH_2OH + 3\ O_{(s)} \leftrightarrow 2\ CO_2 + 3\ H_2 + 3\ \square_{(s)} \qquad \text{(XVII)}$$

The surface concentration of oxygen vacancies may change in a definite range without change of the ferrite crystalline phase, but if it becomes too large the phase is destroyed as it was observed for $NiFe_2O_4$, $CoFe_2O_4$, and $CuFe_2O_4$.

The ESR process over stable ferrites ($FeFe_2O_4$, $MnFe_2O_4$, $MgFe_2O_4$, $ZnFe_2O_4$) and $CuFe_2O_4$ predominantly proceed via consecutive reaction pathway $CH_3CH_2OH \leftrightarrow CH_3CHO \leftrightarrow CH_3COCH_3 \leftrightarrow CO_2, H_2$. This conclusion is derived from the temperatures dependencies of the reaction product selectivity (Tables 15-20). A selectivity to acetaldehyde diminishes with increasing the ethanol conversion whereas the selectivity to acetone goes through a maximum; such type tendency is typical for consecutive reactions.

Transformation of ethanol to acetone can be expressed by the reaction:

$$2\ CH_3CH_2OH + H_2O = CH_3COCH_3 + CO_2 + 4\ H_2 \qquad \text{(XVIII)}$$

According to the reaction stoichiometry, a maximum theoretical selectivity of 75% can be expected for acetone and 25% for CO_2 since the ¾ part of ethanol molecules converts to acetone and ¼ to CO_2. It is worth noting that the experimental values of the acetone selectivity (71.3%) and CO_2 selectivity (24.3%) for $FeFe_2O_4$ at 400°C (Table 17) are fairly close to their theoretical values. Besides, the selectivity to hydrogen is also very close to its limit theoretical value 33.3% (in accordance with the definition of the hydrogen selectivity adopted in the present work). Obtained results point out also that further transformation of acetone to CO_2 and H_2 proceeds yet slow at 400°C, steam reforming of acetone becomes significant at more high temperatures. Therefore, the ferrite $FeFe_2O_4$ may serve as a promising catalyst for acetone producing from ethanol.

The results discussed above were obtained for the reaction mixture with the great excess of water towards ethanol (H_2O/C_2H_5OH=19). Our preliminary study of the ESR reaction on $MnFe_2O_4$ at 400°C for input reaction mixture 12 mol % C_2H_5OH, 36 mol % H_2O, N_2 balance (i.e., with stoichiometric ratio H_2O/C_2H_5OH = 3 for the ESR reaction) gave the following results: ethanol conversion 79.1%, selectivity to acetaldehyde, acetone, CO_2, methane and C_2-C_3 hydrocarbons 0.8, 71.4, 24.1, 0.1 and 3.6%, respectively. Besides, no remarkable changes in catalytic activity and selectivity were observed during 10 h. Therefore, selectivity to acetone remains to be high when the composition of the reaction mixture is significantly changed.

Table 17. Temperature dependencies of ethanol conversion (X) and product selectivity for FeFe₂O4

		Selectivity, %							
T, °C	X, %	CH_3CHO	$(CH_3)_2CO$	CO	CO_2	CH_4	C_2H_4	C_{3+}	H_2
300	9.6	18.8	60.1	0.0	20.2	0.1	0.8	0.0	30.0
350	41.6	4.4	61.9	0.0	32.9	0.1	0.7	0.0	40.5
400	88.7	2.9	71.3	0.0	24.3	0.2	1.2	0.1	32.7
450	94.2	4.4	36.7	0.0	50.8	6.8	1.2	0.1	53.3
500	97.8	1.4	18.1	0.0	60.7	18.6	0.8	0.4	56.7
550	97.4	1.4	5.5	0.0	76.5	15.8	0.6	0.2	72.0

Table 18. Temperature dependencies of ethanol conversion (X) and product selectivity for $MnFe_2O_4$ – CP

		Selectivity, %							
T, °C	X, %	CH_3CHO	$(CH_3)_2CO$	CO	CO_2	CH_4	C_2H_4	C_{3+}	H_2
350	36.8	0.8	62.7	0.0	33.1	0.0	3.3	0.1	40.2
400	86.9	2.6	58.7	0.0	35.7	0.1	2.8	0.1	42.7
450	98.3	3.8	40.0	0.0	47.6	2.8	5.6	0.2	51.8
500	98.9	3.9	16.5	0.0	66.1	6.3	6.8	0.4	66.4
550	98.9	1.1	4.7	0.0	85.0	4.2	4.6	0.4	84.3

Table 19. Temperature dependencies of ethanol conversion (X) and product selectivity for $MgFe_2O_4$

		Selectivity, %							
T, °C	X, %	CH_3CHO	$(CH_3)_2CO$	CO	CO_2	CH_4	C_2H_4	C_{3+}	H_2
300	7.9	83.3	0.0	0.0	16.5	0.0	0.2	0.0	30.4
350	23.7	21.6	55.9	0.0	21.3	0.0	1.2	0.0	31.1
400	42.1	3.9	45.2	0.0	48.8	0.2	1.9	0.0	54.4
450	85.9	0.9	43.2	0.0	52.8	0.5	2.5	0.1	57.6
500	99.8	0.5	27.1	0.0	68.6	0.9	2.7	0.2	71.4
550	99.9	0.3	14.1	0.0	81.2	1.3	2.9	0.2	82.4

At relatively high temperatures, methane was detected in reaction products. According to (Torres et al. 2007), methane is formed by reaction of acetaldehyde decomposition with C-C-bond breakage on a catalyst surface (reaction (VI)). The obtained results (Tables 17-20) show that ability to break C-C-bond is higher for the studied ferrites MFe_2O_4 where M is a transition metal in a comparison with nontransition ones.

Analyzing the data for the stable ferrites (Tables 15, 17-20), one can deduce that an increase of temperature leads to a shift of the reaction product composition toward CO_2

and H_2, which is the target product of the ESR reaction. The maximum hydrogen yield is achieved at 650°C for $MnFe_2O_4$-HC (94.6%). Therefore, this catalyst exhibits the highest performance with respect to the hydrogen yield as follows from the data presented in Table 21.

Table 20. Temperature dependencies of ethanol conversion (X) and product selectivity for $ZnFe_2O_4$

		Selectivity, %							
T, °C	X, %	CH_3CHO	$(CH_3)_2CO$	CO	CO_2	CH_4	C_2H_4	C_{3+}	H_2
300	5.3	99.7	0.0	0.0	0.0	0.0	0.3	0.0	16.6
350	21.1	23.6	54.9	0.0	20.9	0.1	0.5	0.0	30.9
400	28.9	21.7	56.2	0.0	21.4	0.4	0.3	0.0	31.2
450	42.1	17.6	38.7	0.0	42.6	0.6	0.1	0.4	49.7
500	46.0	17.3	22.4	0.0	58.2	0.6	0.4	1.1	63.4
550	68.4	17.7	15.1	0.0	61.4	2.7	0.9	2.2	65.1

Table 21. Examples of catalytic systems in which a hydrogen yield of not less than 85% was obtained (the designations of the catalysts are the same as in the original papers)

Catalyst	T, °C	Volume flow rate, h^{-1}	EtOH:H_2O:inert the molar ratio	Y_{H2}, %	Reference
18%Ni/MgO	650	10 000	1:8.4:3	87	Freni et al. 2003
30%Ni/CeO_2	650		1:8.4:3	87	Frusteri et al. 2006
10%Ni/La_2O_3	650		1:8	89.3	Yang et al. 2006
10%Co/ZnO	350	5 000	1:4:25	91.8	Llorca et al. 2003
1%Rh/γ-Al_2O_3	700		1:3:7.8	92	Auprêtre et al. 2002
10%Co/ZrO_2	550	5 000	1:10:75	92	Song et al. 2009
$MnFe_2O_4$	650		1:18.5:17.5	94.6	Present work
0.78%Na10%CoZn	400	5 000	1:13:70	95.8	Llorca et al. 2004
1%Fe/Co_3O_4	400	5 200	1:6:21	96	O'Shea et al. 2008
10%Co/CeO_2	550	10 000	1:10:2.3	97.4	Song et al. 2010

The data reported here show that the studied ferrites with spinel structure are active for the ESR reaction. Catalytic properties of spinels containing ions of transition metals depend on their redox properties as well as ion distribution between octahedral and tetrahedral sites in a spinel crystalline structure. Octahedral sites are exposed almost exclusively at the surface of the spinel oxides and the cations of these sites play the main role in the catalytic reaction (Jacobs et al. 1994; Ramankutty et al. 2001, 2002). As shown by (Boreskov et al. 1970), Fe^{2+} and Fe^{3+} ions localized in the octahedral sites function as the redox pair and the catalyst on the base of magnetite $FeFe_2O_4$ are highly efficient for water dissociation to hydrogen and adsorbed oxygen. Thus, high activity of $FeFe_2O_4$ spinel in the ESR reaction may be a result of the presence of octahedral Fe^{2+} and Fe^{3+} ions in the magnetite structure that facilitates redox transformations in the steps of organic intermediate oxidation and water decomposition with the formation of final products (CO_2 and H_2).

High activity of $MnFe_2O_4$ in the ESR reaction could be caused by the octahedral ions Fe^{3+} and Mn^{2+} that facilitate the redox transfer in M^{3+}/M^{2+} pairs. According to the TPR results, $MnFe_2O_4$ should more easy supply oxygen for oxidation of organic surface intermediates in comparison with $MgFe_2O_4$ and $ZnFe_2O_4$ thus selectivity to the mostly oxidized reaction product CO_2 and hydrogen is in whole higher for $MnFe_2O_4$ (as well as for $FeFe_2O_4$).

It is well-known that a distribution of the products of the ethanol steam reforming depends also on the acid-base properties of the catalysts (Song et al. 2012; Garbarino et al. 2015; Busca 2010). Acetaldehyde is the primary product of the ethanol conversion on the ferrite catalysts. Many literature data point out that selectivity to aldehydes close relates to the oxide basic strength at enough amounts of the weak surface acid sites (Kibby and Hall 1973; Tanabe et al. 1989; Canesson and Blanchard 1979). The presence of weak acid sites is considered as a key factor in the alcohol dehydrogenation reactions. Among the stable studied ferrites, the high selectivity to acetaldehyde at low ethanol conversions was observed for Zn and Mg containing catalysts and the order of selectivity change is as follows: $ZnFe_2O_4 > MgFe_2O_4 > FeFe_2O_4 > MnFe_2O_4$. The temperature rise enhances conversion of acetaldehyde to acetone. Selectivity to acetone reaches 71% for $FeFe_2O_4$ and 63% for $MnFe_2O_4$ while it does not exceed 56% for $ZnFe_2O_4$ and $MgFe_2O_4$. Aldol condensation of acetaldehyde includes the formation of carbanion intermediate located on the acid-base surface centers. Therefore, the relatively high selectivity to acetone obtained for $FeFe_2O_4$ and $MnFe_2O_4$ could be explained by the presence on their surface both the basic and more strength acid centers in a comparison with Zn- and Mg-containing ferrites.

A probable reaction network for the main reaction products may be presented by the following simplified scheme:

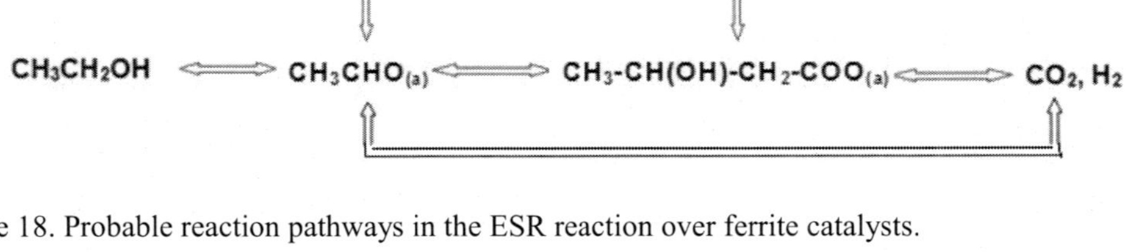

Figure 18. Probable reaction pathways in the ESR reaction over ferrite catalysts.

The presented type of the consecutive-parallel schemes is typical for the selective redox conversion of organic compounds on oxide catalysts (Golodets 1983).

Adsorption/desorption steps depend sufficiently on the acid-base properties of oxides, whereas transformation of surface intermediate organic species $CH_3CHO_{(a)}$ and CH_3-$CH(OH)$-CH_2-$COO_{(a)}$ (see Eq.(XIV)) includes reactions of their oxidation, i.e., depends to a great extent on the redox properties of oxide catalysts. Selectivity on reaction products should depend on a relation of the rates of adsorption and surface reactions; thus, for achievement desired selectivity the appropriate balance of acid-base and redox properties is needed.

Conclusion

A study of the ESR process at 250-500°C on copper catalysts supported on ZrO_2 of the monoclinic and yttrium-stabilized tetragonal crystalline modifications reveal a strong size effect. The specific rate of the ethanol dehydrogenation to acetaldehyde (the key product of ethanol reforming) on copper catalysts is almost independent of the average copper particle size in the range of 6-33 nm and significantly reduced for particles size less than 3 nm. The differences of the activities of the catalysts are probably associated with features of the surfaces of the clusters and dispersed species of copper in the catalysts. The enhanced selectivity to hydrogen of the catalyst based on yttrium-stabilized zirconia may be associated with higher oxygen mobility of the support that facilitates the oxidation of carbonaceous intermediates to CO_2 by oxygen on the zirconia surface.

The binary copper catalysts M-Cu/ZrO_2 (M = Ni, Co, Fe) are considerably more efficient in ethanol steam reforming at 350-550°C than monometallic Cu/ZrO_2, which is unstable in the ethanol steam reforming process. It was found that increased activity and selectivity of the catalyst Fe-Cu/ZrO_2 is related to the presence of highly dispersed copper particles and iron oxide particles on the surface of support, accelerating the water-gas shift reaction and as a result carbon monoxide (a secondary product of ethanol steam reforming) is converted to the target CO_2 and H_2.

The catalytic ethanol steam reforming was studied at 300-700°C over the MFe_2O_4 (M = Fe, Mn, Mg, Zn) ferrites with the spinel structure. The active phase of these catalysts is oxide phase because the catalysts remained in the unreduced state in the ESR conditions. Ethanol was firstly dehydrogenated producing acetaldehyde, which was then converted into acetone via condensation pathway. However, a direct way of the ethanol conversion to acetone is also possible without the intermediate formation of acetaldehyde in a gas phase. Perhaps, this possibility is realized for $FeFe_2O_4$ because only this catalyst has the highest selectivity towards acetone at 300°C. At higher temperatures, acetone steam reforming produces hydrogen and CO_2. No CO is detected in this case.

Finally, the $FeFe_2O_4$, $MnFe_2O_4$ and $MgFe_2O_4$ ferrites could be considered as promising catalysts for the hydrogen production via the ESR reaction. The target process over $MnFe_2O_4$ is characterized by the close to stoichiometry hydrogen yield of 94.6% (5.68 mole H_2 per one mole of inlet ethanol) at 650°C. Moreover, the $FeFe_2O_4$ ferrite is efficient for the ethanol conversion to acetone. Therefore, it can be used, for example, as the catalyst of the first step in a two-step ESR reaction process or in other two-step processes which involve an intermediate formation of acetone.

ACKNOWLEDGMENTS

Partial financial support received from the Target complex Programmes for Research of National Academy of Sciences of Ukraine "Hydrogen in alternative energy and new technologies" and "Fundamental problems of hydrogen and renewable energy and fuel cell technologies" is appreciated.

REFERENCES

Auprêtre, F., Descorme, C., Duprez, D. 2002. "Bio-ethanol catalytic steam reforming over supported metal catalysts." *Catal. Comm.* 3:263-67.

Auprêtre, F., Descorme, C., Duprez, D., Casavane, D., Uzio, D. 2005. "Ethanol steam reforming over $Mg_xNi_{1-x}Al_2O_3$ spinel oxide-supported Rh catalysts." *J. Catal.* 233:464-77.

Bellido, J. D. A., Assaf, E. M. 2008. "Nickel catalysts supported on ZrO_2, Y_2O_3-stabilized ZrO_2 and CaO-stabilized ZrO_2 for the steam reforming of ethanol: Effect of the support and nickel load." *J. Power Sources* 177:24-32.

Benito, M., Padilla, R., Rodriguez, L., Sanz, J. L., Daza, L. 2007. "Zirconia supported catalysts for bioethanol steam reforming: Effect of active phase and zirconia structure." *J. Power Sources* 169:167-76.

Bergamaschi, V. S., Carvalho, F. M. S., Rodrigues, C., Fernandes, D. B. 2005. "Preparation and evaluation of zirconia microspheres as inorganic exchanger in adsorption of copper and nickel ions and as catalyst in hydrogen production from bioethanol." *Chem. Eng. J.* 112:153-58.

Biswas, P., Kunzru, D. 2007. "Oxidative steam reforming of ethanol over Ni/CeO_2-ZrO_2 catalyst." *Int. J. Hydrogen Energy* 32:969-80.

Boreskov, G. K., Yurieva, T. M., Sergeeva, A. S. 1970. *Kinet. Catal.* 11:1476-78.

Busca, G. 2010. "Bases and Basic Materials in Chemical and Environmental Processes. Liquid versus Solid Basicity." *Chem. Rev.* 110:2217-49.

Canesson, P., Blanchard, M. 1979. "Catalytic dehydration of secondary alcohols over thoria: Infrared study and generalization to basic catalysts." *J. Catal*. 42:205-12.

Casanovas, A., Roig, M., de Leitenburg, C., Trovarelli, A., Llorca, J. 2010. "Ethanol steam reforming and water gas shift over Co/ZnO catalytic honeycombs doped with Fe, Ni, Cu, Cr, and Na." *Int. J. Hydrogen Energy* 35:7690–98.

Cavallaro, S., Mondello, N., Freni, S. 2001. "Hydrogen produced from ethanol for internal reforming molten carbonate fuel cell." *J. Power Sources* 102:198-204.

Chen, L., Choong, C., Zhong, Z., Huang, L., Ang, T., Hong, L., Lin, J. 2010. "Carbon monoxide-free hydrogen production via low-temperature steam reforming of ethanol over iron-promoted Rh catalyst." *J. Catal*. 276:197-200.

Deluga, G. A., Salge, J. R., Schmidt, L. D., Verykios, X. E. 2004. "Renewable hydrogen from ethanol by autothermal reforming." *Science* 303:993-97.

Dolgykh, L. Yu., Stolyarchuk, I. L., Staraya, L. A., Vasylenko, I. V., Pyatnitsky, Y. I. 2014. "Catalytic properties of MnO, Fe_2O_3, and $MnFe_2O_4$ in the steam reforming of ethanol." *Theor. Exp. Chem.* 50:245-49.

Dolgykh, L. Yu., Stolyarchuk, I. L., Staraya L. A., Vasylenko I. V., Pyatnitsky, Y. I., Strizhak, P. E. 2015. "Catalytic properties of $CuFe_2O_4$ in steam reforming of ethanol." *Theor. Exp. Chem.* 51:230-35.

Dolgykh, L. Yu., Stolyarchuk, I. L., Staraya L. A., Vasylenko I. V., Pyatnitsky, Y. I., Strizhak, P. E. 2015. "Steam Reforming of Ethanol over Manganese and Iron Oxides for Hydrogen Production." *Ads. Sci. Tech*. 33:715-21.

Dolgykh, L. Yu., Stolyarchuk, I. L., Vasylenko I. V., Pyatnitsky, Y. I., Strizhak, P. E. 2013. "Influence of the composition of nanosized MFe_2O_4 spinels (M = Ni, Co, Mn) on their catalytic properties in the steam reforming of ethanol." *Theor. Exp. Chem.* 49:185-92.

Dolgykh, L., Stolyarchuk, I., Deynega I., Strizhak, P. 2006. "The use of industrial dehydrogenation catalysts for hydrogen production from bioethanol." *Int. J. Hydrogen Energy* 31:1607-10.

Elliott, D. J., Pennella F. 1989. "The formation of ketones in the presence of carbon monoxide over $CuO/ZnO/Al_2O_3$." *J. Catal*. 119:359-67.

Fatsikostas, A. N., Verykios, X. E. 2004. "Reaction network of steam reforming of ethanol over Ni-based catalysts." *J. Catal.* 225:439-52.

Freitas, I. C., Damyanova, S., Oliveira, D. C., Marques, C. M. P, Bueno J. M. C. 2014. "Effect of Cu content on the surface and catalytic properties of Cu/ZrO_2 catalyst for ethanol dehydrogenation." *J. Mol. Catal. A. Chem.* 381:26-37.

Freni, S., Cavallaro, S., Mondello, N., Spadaro, L., Frusteri, F. 2003. "Production of hydrogen for MC fuel cell by steam reforming of ethanol over MgO supported Ni and Co catalysts." *Catal. Comm*. 4:259-68.

Frusteri, F., Freni, S., Chiodo, V., Donato, S., Bonura, G., Cavallaro, S. 2006. "Steam and auto-thermal reforming of bio-ethanol over MgO and CeO_2 Ni supported catalysts." *Int. J. Hydrogen Energy* 31:2193-99.

Fu, Q., Li, W. X., Yao, Y., Liu, H., Su, H. Y., Ma, D., Gu, X. K., Chen, I., Wang, Z., Zang, H., Wang, B., Bao, X. 2010. "Interface-confined ferrous centers for catalytic oxidation." *Science* 328:1141-44.

Garbarino, G., Wang, C., Valsamakis, I., Chitsazan, S., Riani, P., Finocchio, E., Flytzani-Stephanopoulos, M., Busca, G. 2015. "A study of Ni/Al_2O_3 and $Ni–La/Al_2O_3$ catalysts for the steam reforming of ethanol and phenol." *Appl. Cat. B*. 174-175:21-34.

Gervasini, A., Bennici, S. 2005. "Dispersion and surface states of copper catalysts by temperature-programmed-reduction of oxidized surfaces (s-TPR)." *Appl. Catal. A*. 281:199-205.

Giri, J., Sriharsha, T., Asthana, S., Tumkur, K., Rao, G., Nigam, A. K., Dhirendra, B. 2005. "Synthesis of capped nanosized $Mn_{1-}xZn_xFe_2O_4$ ($0\leqslant x\leqslant 0.8$) by microwave refluxing for bio-medical applications." *J. Magnet. Magnet. Mater*. 293:55-61.

Golodets, G. I. 1983. *Heterogeneous catalytic reactions involving molecular oxygen*. New York: Elsevier.

Haryanto, A., Fernando, S., Murali, N., Adhikari, S. 2005. "Current Status of Hydrogen Production Techniques by Steam Reforming of Ethanol: A Review." *Energy Fuels* 19:2098-106.

Homs, N., Llorca, J., Ramirez de la Piscina, P. 2006. "Low-temperature steam-reforming of ethanol over ZnO-supported Ni and Cu catalysts. The effect of nickel and copper addition to ZnO-supported cobalt-based catalysts." *Catal. Today* 116:361–66.

Hu, B., Getsoian, A., Schweitzer, N. M., Das, U., Kim, H., Niklas, J., Poluectov, O., Curtiss, L. A., P. C. Stair, P. C., Miller, J. T., Hock, A. S. 2015. "Selective propane dehydrogenation with single-site Co^{II} on SiO_2 by a non-redox mechanism. *J. Catal*. 322:24-37.

Hu, B., Schweitzer, N. M., Zhang, G., Kraft, S. J., Childers, D. J., Lanci, M. P., Miller, J. T., Hock, A. S. 2015. "Isolated Fe^{II} on Silica as a Selective Propane Dehydrogenation Catalyst." *ACS Catal*. 5:3494-503.

Jacobs, J. P., Maltha, A., Reitjes, J. G. H., Drimal, J., Ponec, V., Brongersma, H. H. 1994. "The surface of catalytically active spinels." *J. Catal*. 47:294-300.

Kibby, C. L., Hall, W. K. 1973. "Studies of acid catalyzed reactions: XII. Alcohol decomposition over hydroxyapatite catalysts." *J. Catal*. 29:144-59.

Kim, D. K., Mikhaylova, M., Zhang, Yu, Muhammed, M. 2003. "Protective Coating of Superparamagnetic Iron Oxide Nanoparticles." *Chem. Mater*. 15:1617-27.

Liu, C. P., Li, M. W., Cui, Z. Huang, J. R., Tian, Y. L., Lin, T., Mi, W. B. 2007. "Comparative study of magnesium ferrite nanocrystallites prepared by sol–gel and coprecipitation methods." *J. Mater. Sci*. 42:6133-38.

Liu, Z., Amiridis, M., Chen, Y. 2005. "Characterization of CuO Supported on Tetragonal ZrO_2 Catalysts for N_2O Decomposition to N_2." *J. Phys. Chem. B* 109:1251-55.

Llorca, J., Homs, N., Sales, J., G. Fierro, J., Piscina, P. R. 2004. "Effect of sodium addition on the performance of Co–ZnO-based catalysts for hydrogen production from bioethanol." *J. Catal*. 222:470-80.

Llorca, J., Piscina, P. R., Dalmon, J., Sales, J., Homs, N. 2003. "CO-free hydrogen from steam-reforming of bioethanol over ZnO-supported cobalt catalysts: Effect of metal precursor." *Appl. Catal. B*. 43:355-69.

Llorca, J., Homs, N., Sales, J., de la Piscina, P. R. 2002. "Efficient Production of Hydrogen over Supported Cobalt Catalysts from Ethanol Steam Reforming." *J. Catal*. 209:306-17.

Ma, Z. Y., Yang, C., Wei, W., Li, W. H., Sun, Y. H. 2005. "Catalytic performance of copper supported on zirconia polymorphs for CO hydrogenation." *J. Mol. Catal. A. Chem*. 231:75−81.

Mariño, F. J., Cerrella, E. G., Duhalde, S., Jobbagy, M., Laborde, M. A. 1998. "Hydrogen from steam reforming of Ethanol. Characterization and performance of copper-nickel supported catalysts." *Int. J. Hydrogen Energy* 23:1095–101.

Mariño, F., Boveri, M., Baronetti, G., Laborde, M. 2004. "Hydrogen production via catalytic gasification of ethanol. A mechanism proposal over copper-nickel catalysts." *Int. J. Hydrogen Energy* 29:67–71.

Mattos, L. V., Jacobs, G., Davis, B. H., Noronha, F. B. 2012. "Production of Hydrogen from Ethanol: Review of Reaction Mechanism and Catalyst Deactivation." *Chem. Rev*. 112:4094−123.

Ni, M., Leung, D. Y. C., Leung, M. K. H. 2007. "A review on reforming bio-ethanol for hydrogen production." *Int. J. Hydrogen Energy* 32:3238-47.

Nishiguchi, T., Matsumoto, T. Kanai, H., Utani, K., Marsumura, Y., Shen, W., Imamura, S. 2005. "Catalytic steam reforming of ethanol to produce hydrogen and acetone." *Appl. Catal. A Gen*. 279:273-77.

O'Shea, V. A., Nafria, R., Piscina, P. R., Homs, N. 2008. "Development of robust Co-based catalysts for the selective H_2-production by ethanol steam-reforming. The Fe-promoter effect." *Int. J. Hydrogen Energy* 33:3601–6.

Pakharukova, V. P., Moroz, E. M., Kriventsov, V. V., Larina, T. V., Boronin, A. I., Dolgikh, L. Yu., Strizhak, P. E. 2009. "Structure and State of Copper Oxide Species Supported on Yttria-Stabilized Zirconia." *J. Phys. Chem. C* 113:21368-75.

Pakharukova, V. P., Moroz, E. M., Zyuzin, D. A., Ishchenko, A. V., Dolgikh, L. Yu., Strizhak, P. E. 2015. "Structure of Copper Oxide Species Supported on Monoclinic Zirconia." *J. Phys. Chem. C,* 119:28828-35.

Panda, N., Jena, A. K., Mohapatra, S., Rout, S. R. 2011. "Copper ferrite nanoparticle-mediated N-arylation of heterocycles: a ligand-free reaction." *Tetrahedron Lett*. 52:1924–27.

Patrylak, K. I., Patrylak, L. K., Konovalov, S. V. 2013. "Thermodynamic approach to prediction of the effect of size on nanostructure properties." *Theor. Exp. Chem.* 49:35-39.

Philip, J., Gnanaprakash, G., Panneerselvam, G., Antony, M. P., Jayakumar, T., Raj, B. 2007. "Effect of thermal annealing under vacuum on the crystal structure, size, and magnetic properties of $ZnFe_2O_4$ nanoparticles." *J. Appl. Phys.* 102:054305.

Piscina, P. R., Homs, N. 2008. "Use of biofuels to produce hydrogen (reformation processes)." *Chem. Soc. Rev.* 37:2459-67.

Pyatnitsky, Y. I., Dolgykh, L. Yu., Stolyarchuk, I. L., Strizhak, P. E. 2013. "Production of hydrogen by steam reforming of ethanol." *Theor. Exp. Chem.* 49:277-97.

Pyatnitsky, Y. I., Dolgykh, L. Yu., Stolyarchuk, I. L., Strizhak, P. E. 2013. "Relationship between yield of hydrogen in steam reforming of ethanol and selectivity with respect to carbon-containing products." *Theor. Exp. Chem.* 49:109-14.

Pyatnitsky, Yu. I., Strizhak, P. E. 2011. *Calculations of Equilibrium and Kinetic Modeling of Heterogeneous Catalytic Reactions*. Kyiv: Vadym Karpenko.

Ramankutty, C. G., Sugunan, S., Thomas, B. 2002. "Study of cyclohexanol decomposition reaction over the ferrospinels, $A_{(1-x)}Cu_{(x)}Fe_{(2)}O_{(4)}$ (A = Ni or Co and x = 0.03, 0.5, 0.7 and 1) prepared by soft chemical methods." *J. Mol. Catal. A*. 187:105-17.

Ramankutty, C. G., Sugunan, S. 2001. "Surface properties and catalytic activity of ferrospinels of nickel, cobalt and copper, prepared by soft chemical methods." *Appl. Catal. A*. 218:39-51.

Rana, S., Philip, J., Raj, B. 2010. "Micelle based synthesis of cobalt ferrite nanoparticles and its characterization using Fourier Transform Infrared Transmission Spectrometry and Thermogravimetry." *Mater. Chem. Phys.* 124:264-69.

Resini, C., Delgado, M. C. H., Presto, S., Alermany, L. J., Riani, P., Marazza, R., Ramis, G., Busca, G. 2008. "Yttria-stabilized zirconia (YSZ) supported Ni-Co alloys (precursor of SOFC anodes) as catalysts for the steam reforming of ethanol." *Int. J. Hydrogen Energy* 33:3728-35.

Rhodes, M. D., Bell, A. T. 2005. "The effects of zirconia morphology on methanol synthesis from CO and H_2 over Cu/ZrO_2 catalysts: Part I. Steady-state studies." *J. Catal.* 233:198–209.

Sastri, M. V. C., Vishwanath, R. P., Vishwanath, B. 1982. "Studies on the reduction of iron oxide with hydrogen." *Int. J. Hydrogen Energy* 7:951-55.

Sato, A. G., Volanti, D. P., Meira, D. M., Damyanova, S., Longo, E., Bueno, J. M. C. 2013. "Effect of the ZrO_2 phase on the structure and behavior of supported Cu catalysts for ethanol conversions." *J. Catal.* 307:1-17.

Schweitzer, N. M., Hu, B., Das, U., Kim, H., Greeley, J., Curtiss, L. A., Stair, P. C., Miller, J. T., Hock, A. S. 2014. "Propylene Hydrogenation and Propane Dehydrogenation by a Single-Site Zn^{2+} on Silica Catalyst." *ACS Catal.* 4:1091-92.

Shi, Q., Peng, Z., Chen, W., Zhang, N. 2011. "$La_2O_2CO_3$ supported Ni-Fe catalysts for hydrogen production from steam reforming of ethanol." *J. Rare Earth* 29:861-65.

Song, H., Mirkelamoglu, B., Ozkan, U. S. 2010. "Effect of cobalt precursor on the performance of ceria-supported cobalt catalysts for ethanol steam reforming." *Appl. Catal. A*. 382:58-64.

Song, H., Tan, B., Ozkan, U. S. 2009. "Novel Synthesis Techniques for Preparation of Co/CeO_2 as Ethanol Steam Reforming Catalysts." *Catal. Lett.* 132:422-29.

Song, H., Zhang, L, Ozkan, U. S. 2012. "The Effect of Surface Acidic and Basic Properties on the Performance of Cobalt-Based Catalysts for Ethanol Steam Reforming." *Top. Catal*. 55:1324-31.

Stolyarchuk, I. L., Dolgykh, L. Yu., Vasilenko I. V., Pyatnitsky, Y. I., Strizhak, P. E. 2012. "Catalysis of steam reforming of ethanol by nanosized manganese ferrite for hydrogen production." *Theor. Exp. Chem.* 48:129-34.

Stolyarchuk, I. L., Dolgykh, L. Yu., Vasylenko I. V., Pyatnitsky, Y. I., Strizhak, P. E. 2016. "Ferrites MFe_2O_4 (M = Mg, Mn, Fe, Zn) as catalysts for steam reforming of ethanol." *Theor. Exp. Chem.* 52:246-51.

Stull, R. D., Westrum, E. F., Sinke, G. C. 1969. *The Chemical Thermodynamics of Organic Compounds*. New York: John Wiley.

Tanabe, K., Misono, M., Ono, Y., Hattori, H. 1989. *New Solid Acids and bases*. Tokyo: Elsevier.

Torres, J. A., Llorca, J. M. Cassanovas, A., Dominguez, M., Salvado, J., Montane, D. 2007. "Steam reforming of ethanol at moderate temperature: Multifactorial design analysis of Ni/La_2O_3-Al_2O_3, and Fe- and Mn-promoted Co/ZnO catalysts." *J. Power Sources* 169:158-66.

Vaidya, P. D., Rodrigues, A. E. 2006. "Insight into steam reforming of ethanol to produce hydrogen for fuel cells." *Chem. Eng. J.* 117:39-49.

Vasilenko, I. V., Gavrilenko, K. S., Kotenko, I. A. 2007. "Effect of formation conditions on the structure, morphology and magnetic properties of nanosized $M^{II}Fe_2^{III}O_4$ ferrites (M = Mn, Co, Ni) and Fe_2O_3." *Theor. Exp. Chem.* 43:353-58.

Vizcaino, A., Carrero, A., Calles, J. 2007. "Hydrogen production by ethanol steam reforming over Cu–Ni supported catalysts." *Int. J. Hydrogen Energy* 32:1450-61.

Wang, L., Zhu, W., Zheng, D., Yu, X., Cui, J., Jia, M., Zhang, W., Wang, Z. 2010. "Direct transformation of ethanol to ethyl acetate on Cu/ZrO_2 catalyst." *React. Kinet. Mech. Cat*. 101:365-75.

Wang, L. C., Liu, Q., Chen, M., Liu, Y. M., Cao, Y., He, H. Y., Fan, K. N. 2007. "Structural Evolution and Catalytic Properties of Nanostructured Cu/ZrO_2 Catalysts Prepared by Oxalate Gel-Coprecipitation Technique." *J. Phys. Chem.* C. 111:16549-57.

Yang, Y., Ma, J., Wu, F. 2006. "Production of hydrogen by steam reforming of ethanol over a Ni/ZnO catalyst." *Int. J. Hydrogen Energy* 31:877-82.

Zhao, Y., Li, W., Zhang, M, Tao, K. 2002. "A comparison of surface acidic features between tetragonal and monoclinic nanostructured zirconia." *Catal. Commun*. 3:239-45.

In: Bioethanol and Beyond
Editor: Michel Brienzo

ISBN: 978-1-53613-478-0

Chapter 15

INTEGRATION BETWEEN THE ETHANOL PRODUCTION AND THE ENERGY CO-GENERATION THROUGH THE ENERGETICAL DENSIFICATION OF LIGNOCELLULOSIC RESIDUES AND GASIFICATION PROCESS

***Andrea Cressoni De Conti*[1,*], *Cláudio De Conti*[1] *and Nestor Proeza Perez*[2]**

[1]Energy Engineering, São Paulo State University (UNESP),
Campus de Rosana, Rosana, São Paulo, Brazil
[2]Mechanical Engineering,
Federal Center of Technological Education Celso Suckow da Fonseca (CEFET/RJ),
Angra dos Reis Campus, Angra dos Reis, Rio de Janeiro, Brazil

ABSTRACT

This chapter presents the energetic densification as an alternative for the use of lignocellulosic residues, such as the ones from ethanol production, so that they can be used in gasifiers, aiming towards the energy cogeneration. The residues from ethanol's producing process (first or second generation) are valuable feedstocks for the energy cogeneration process, presenting, thus, a proposal to integrate the ethanol production and the energy cogeneration. The authors intend to address the concepts related to energetic densification, biomass characterization, as well as the kinds of energetic densification, like briquettes and pellets, and the respective machinery, not forgetting about the concepts of gasification and the kinds of gasifiers used for the energy cogeneration process.

---

* Corresponding Author Email: andrea@rosana.unesp.br, andrea.cdeconti@gmail.com.

Keywords: bioenergy, biomass, energy

1. Introduction

Biomass is a renewable energy source thanks to its big energetic potential, economic convenience, and to the several social and environmental benefits associated to its use. This feedstock may represent an important energy source for richer countries as well as for the countries in development.

Sugarcane is a lignocellulosic material that has been widely studied. By using sugarcane, it is possible to obtain first-generation ethanol (1G) by the sucrose, and second-generation ethanol (2G) by the lignocellulosic part, which is very present in the sugarcane bagasse. The process to obtain first-generation ethanol (1G) is well known, but there is still need for more investments in order to improve it and reduce its costs. The cellulosic ethanol (2G) represents a big challenge: an upside is that it explores the lignocellulosic part of the bagasse through enzymatic hydrolysis, chiefly trying to break the recalcitrant part found in it, mainly in order to better use the existing lignin. However, this process is expensive and becomes a great obstacle for the industry. Even if ethanol 2G's success is considered, the biomass's exploitation is not complete, generating a residue that is rich in lignin [1], and which can be used in the energy cogeneration process.

On the other hand, the sugarcane bagasse can be used in the gasification process to cogenerate energy, whether through the use of ethanol or through direct burnt. This technology may reduce the dependence that regions and countries have upon the fluctuations on the prices of imported fuels. The international political scenario has been very voluble lately, showing that the offer and the market process may change drastically in a short period, generating uncertainties in the international market. Therefore, the possession of a local renewable feedstock source like biomass would help diversify the energetic matrix, besides contributing to reduce the greenhouse gases and recover the environment, since biomass is capable of being transformed into a gaseous fuel – through gasification technologies – and possible to be used in internal combustion engines or gas turbines, producing mechanical work and electrical energy.

Just like the sugarcane bagasse, other lignocellulosic materials have a high lignin content and, consequently, a high calorific power content. Since the production of these residues is very high, an alternative would be to use them through the energetic densification process, so that this product is burned in a gasifier, increasing even more the energy production.

The gasification of densified biomass materials, like sugarcane bagasse and sugarcane straw, might be considered a highly interesting alternative conversion technology, besides being promising and viable. A possible alternative to diversify the use of these biomasses would be the electricity production in large scale by the introduction of technological improvements in the current conventional systems and the viabilization of new systems.

2. Energetic Densification

The energetic densification consists in compacting the feedstock in order to obtain a higher density, which is measured in kg/m^3, and a higher energetic density, which is measured in kcal/m^3, if compared to its initial values. This technological process is used mainly feedstocks with low density, such as residues from agriculture and agroindustry, like sawdust and wood chips, rice husk, and sugarcane bagasse, since, economically, it becomes inviable to transport this material long-distance to stock it without densification [2][3] - [6].

The two most common types of energetic densification products are pellets and briquettes (Figure 1a and Figure 1b, respectively). The pellets are highly dense and are formed through the extrusion process, with a cylindric size of 38mm (1,5in.) and a diameter of 7mm (0,3in.). The briquettes differ from the pellets due to their size, since they also have a cylindric format, but with a diameter of 25mm (2in.) or bigger, being produced when the biomass is pressed by a piston under high pressure [3].

(1a) (1b)

Source: (1a): author, (1b): [7]

Figure 1. Figure (1a) – Pellet; Figure (1b) – Briquette.

The process of characterizing the biomass consists in an important step in energetic densification, since it allows good briquettes and pellets to be manufactured, so they can supply enough energy for the process in which they are going to be used. In item 2.2, there is the description of how this characterization is made.

2.1. Biomass Characterization

For energetic finalities, the biomass must be characterized through the following analyses:

(i) Chemical composition;
 (a) elementary chemical composition;
 (b) immediate chemical composition;
(ii) Moisture Content;
(iii) Calorific power.

The chemical composition is characterized by two processes: elementary chemical composition and immediate chemical composition.

2.1.1. Elementary Chemical Composition

Through the elementary chemical composition, it is possible to analyze the existing amount of the main elements that compose the biomass. These elements are: carbon (C), hydrogen (H), sulfur (S), oxygen (O), nitrogen (N), and ashes (A). The ashes contain all the elements that are not relevant for the combustion reactions, such as potassium, phosphorus, and calcium [8]. In Table 1, there is the elementary chemical composition of some biomasses that are used for energy generation.

Table 1. Elementary chemical composition of several biomasses and residues

Kind of biomass and residues	Average Elementary Chemical Composition (%)						PCS
	C	H	O	N	S	A	kcal/kg
Conifers (Gymnosperm)	49,29	5,99	44,36	0,06	0,03	0,30	4700
Hardwood (Angiosperms)	49,0	5,87	43,97	0,30	0,01	0,72	4600
Rice Husk	38,3-40,96	4,0-4,3	35,86-38,60	0,40	0,02-0,50	18,34-18,60	3850
Sugarcane	44-47,9	6,0-6,7	44-48	-	-	1-2,5	4639
Sugarcane bagasse	44,8-47	5,35-6,5	39,55-35,8	0,38-4,1	0,01	9,79	4100
Coconut shell	48,23	5,23	33,19	2,98	0,12	10,25	4500
Elephant grass	41,16	5,55	45,91	1,78	-	5,60	3600
Wheat Straw	45,4	5,1	34,1	1,8	-	13,6	4132
Waste Paper	44,7	6,1	48,1	0,4	-	2,7	3917,4

Source: [8-10].

2.1.2. Immediate Chemical Composition

Through the immediate chemical composition ([8], [9], [11]), it is possible to analyze the fixed carbon content (F), the volatile content (V), and the ash content (A). The volatile content is related to the facility in burning the material, representing the fraction of the fuel mass which is volatized during the heating process that is applied to the sample during 7 minutes, at a temperature of 850°C. The fixed carbon content is an indirect measure that may be calculated through equation (a), given by the norm [12]:

$$FC = 100 - (AC + VM) \tag{1}$$

where:

- FC = fixed carbon content, in %
- AC = ashes content, in %
- VM = volatile material content, in %

The fixed carbon and the volatile contents are inversely proportional. Thus, fuels with a high fixed carbon content, i.e., low volatility, will be burned more slowly, remaining in the furnace for more time until they can be completely burned [8]. In Table 2, there are the immediate analysis data for some kinds of biomass.

2.1.3. Moisture Content

The moisture content has a very straight relation with the heat production, therefore, the smaller the moisture content, the bigger the heat production (by mass unit), since less energy is going to be used to vaporize the existing water in the biomass. This way, the calorific value of the material is reduced. Also, a high moisture content reflects in a bigger corrosion from the steam generator and in the accumulation of dirt in the heating surfaces. Usually, the moisture content for the burnt must be around 15% to 20% [7], [8], [13-16].

Table 2. Immediate chemical composition of several biomasses and residues

Kind of biomass and residues	Immediate Composition (%)			HHV
	Volatiles	Ashes	Fixed Carbon	Kcal/kg
Conifers (Gymnosperm)	82,54	0,29	17,10	4700
Hardwood (Angiosperms)	81,42	0,79	17,82	4600
Rice Husk	65,47	17,89	16,67	3850
Sugarcane Bagasse	73,78	11,27	14,95	4100
Coconut Shell	67,95	8,25	23,80	4500
Elephant Grass	76,69	5,6	17,70	3600

Source: [8-10].

2.1.4. Calorific Value

By studying the calorific value, it has been possible to verify the amount of energy (in this case, heat) that is released during the complete combustion, being measured in kcal/kg or kJ/Kg or kcal/m^3 or kJ/m^3 [8], [9], [11]. The calorific power value may vary a lot, depending on the moisture content found in the material. There are two kinds of calorific power: The High Heat Value (HHV) and the Low Heat value (LHV). HHV refers to how much energy the material has, i.e., how much energy can be released in the burnt, which is measured in a calorimetric bomb. LHV refers to the energy amount that the material has when the calculation computes the amount of energy that has been spent in order to cause the existing water in the material to evaporate [9].

In order to calculate the LHV, we may use equation 2 [8]:

$$LHV = HHV - 600 * 9\, H/100 \tag{2}$$

where:

- LHV = Low Heat Value
- HHV = High Heat Value
- H = % of hydrogen in the material

2.2. Equipment for Energy Densification

For the energetic densification it is often necessary to treat the feedstock through these processes: i) Crushing: a process where the biomass has its size reduced, so that it can be accommodated in the compaction equipment; ii) Drying: a process in which the moisture content in the biomass is reduced, avoiding the machinery's obstruction, besides making it possible to produce a thicker and better quality product; iii) Milling: a process that is necessary when there is no crushing or when the used machinery requires a smaller granulometry in order to form the pellet or the briquette; and iv) Particle classification: in this process, the biomass is separated, through sieving, according to the size of its particles, in order to obtain a more uniform material [15].

There are two kinds of equipment used to carry out the energetic densification process: the briquette machines and the pelleting machines. Below, there is a presentation of the most common kinds of each one of these.

2.2.1. Briquette Machines

The most common kinds of briquette machines ([7] [17]) are:

2.2.1.1. Screw Type Briquette Press

In Figure 2, there is the description of the pieces found in a briquette press, i.e., feeding hopper, screw conveyor, heating element, die, barrel, and cut-off knife. The briquetting process starts by the feedstock being thrown into the feeding hopper and transported through the cylinder, which is heated. The heated material is, then, pressed onto the mold and forced to go through it under high temperature and pressure, forming, then, the briquettes [17].

2.2.1.2. Piston Type Briquette Press

This other model, as seen in Figure 3, is composed by: piston, screw conveyor, die, and feeding hopper. The material is strangled through the mold with the piston's help, which forms the briquette [17].

2.2.1.3. Roller Type Briquette Press

Figure 4 presents the roll presses, which consist, basically, in two rolls with the same diameter, which move vertically in the same direction. The feedstock has to go through the opening between the two of them, which forms small bags: the briquettes [17].

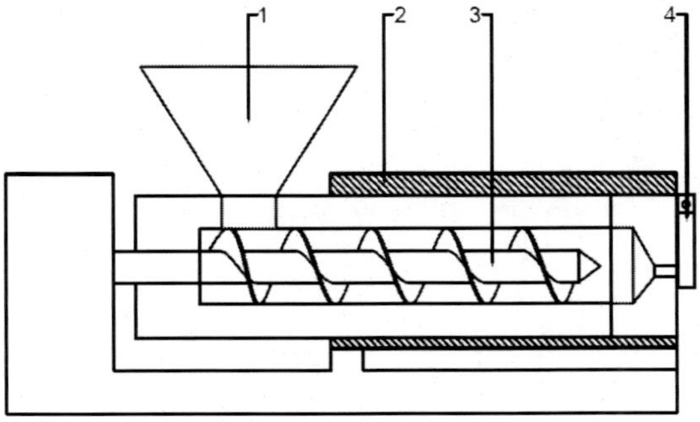

Figure 2. Screw-type Briquette Press.1. Feeding Hopper, 2. Heating Element, 3. Screw Conveyor, 4. Cut-off Knife. Adapted: [17].

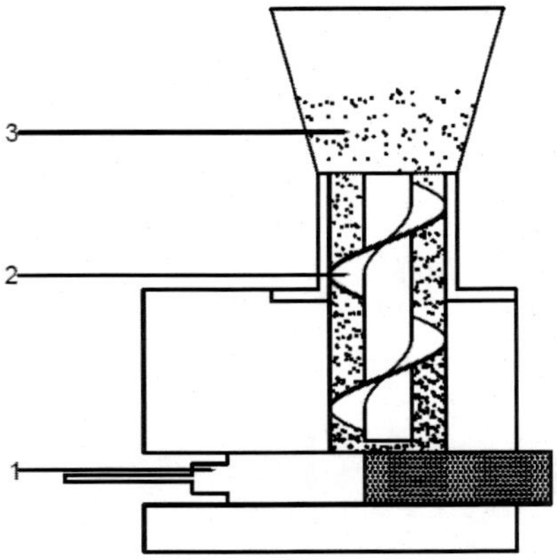

Figure 3. Piston-type Briquette Press. 1. Piston 2. Screw Conveyor 3. Feeding Hopper. Adapted: [17].

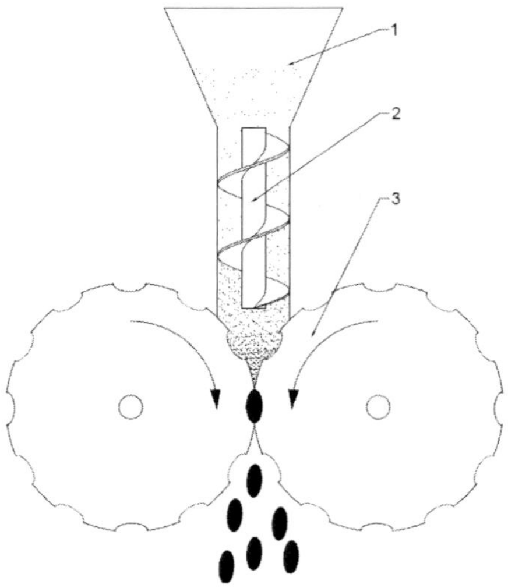

Figure 4. Roller-type Briquette Press. 1. Feeding Hopper 2. Screw Conveyor 3. Roller. Adapted: [17].

2.2.2. Pelleting Machines

For pellets, we have the following equipment:

2.2.2.1. Flat Die Pellet Mill

The feedstock is inserted into the feeding hopper, so it falls, by the gravitational force, into the plan molds, where two rolls, which move through the mold, strangle the material, forming the pellets (Figure 5 [17]).

2.2.2.2. Ring Die Pellet Mill

In this machine, as seen in Figure 6, the operation is similar to the one from the Flat Die Pellet Mill. However, in this case, the material is pressed by two rolls into a mold with small holes, and, therefore, is strangled. When the material comes out of this mold, it is cut by a knife (Figure 6a). Figure 6b presents the pelleting chamber, where the feedstock is compressed by the compression rolls, and strangled through the small holes [17].

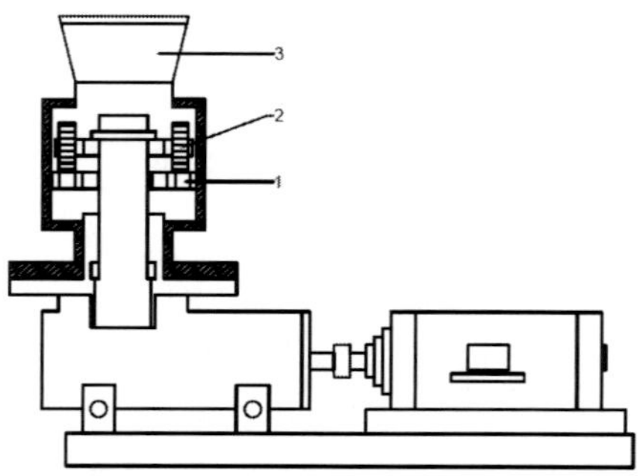

Figure 5. Flat Die Pellet Mill. 1. Flat Die 2. Pressing Rollers 3. Feeding Hopper. Adapted: [17].

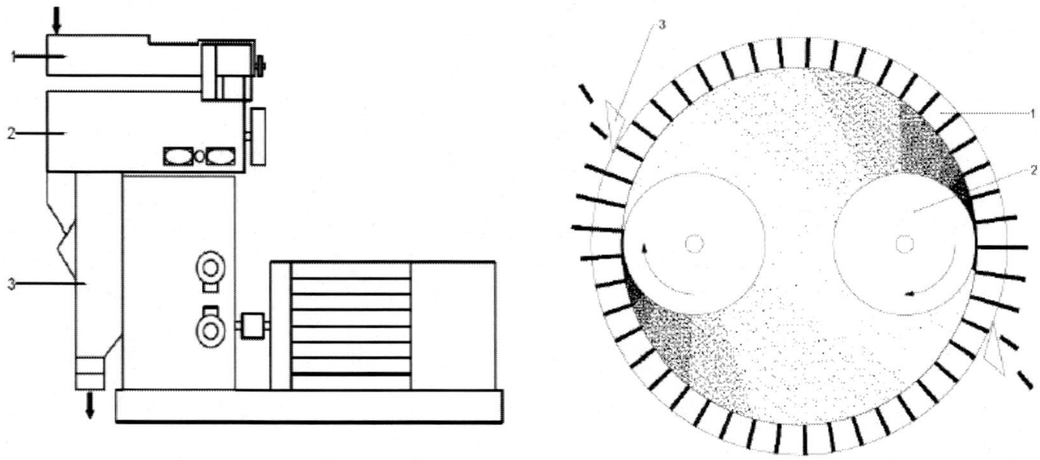

6a 6b

Figure 6. Figure (6a) Ring Die Pellet Mill. 1. Feeding hopper 2. Conditioner 3. Pellet Chamber. Figure (6b) Working process of Ring Die. 1. Ring Die 2. Pressing Roller 3. Cut-off Knife. Adapted: [17].

2.3. Characterization of the Briquettes and Pellets

After the processes of briquetting and pelleting, some tests are necessary in order to evaluate these products.

The main tests for this characterization are:

(i) Volumetric expansion;
(ii) Mechanic resistance;
(iii) Friability.

2.3.1. Volumetric Expansion

The volumetric expansion is conducted through mediation with a pachymeter according to the length, i.e., the area where the compaction has been carried out. This process is important because it shows how much briquette has been formed, i.e., if, with time, it expands but does not come apart. In Figure 7, it is possible to see the compaction process, in laboratorial level, conducted in a 30-ton hydraulic press (Figure 7a), and in Figure 7b there is the formed briquette.

Figure 8 presents the measurement of the longitudinal length, which is used to measure the volumetric expansion.

(7a) (7b)

Source: Author.

Figure 7. Figure 7a presents the briquette confection in a hydraulic press. In Figure 7b, there is the formed briquette.

Source: Author.

Figure 8. Briquette's length measurement.

In Figure 9, there are the volumetric expansion graphics. In Figure 9a, there is an analysis of the volumetric expansion values for the briquettes formed from two kinds of wood [14] (Leucaena and Pau-Jacaré[1]) with three different moisture contents: 10%, 12%, and 15%, disposed as it follows: T1 (leucaena 10%), T2 (leucaena 12%), T3 (leucaena 15%), T4 (pau-jacaré 10%), T5 (pau-jacaré 12%), and T6 (pau-jacaré 15%). In Figure 9b, there is the result of the volumetric expansion for the sugarcane bagasse, with two different particle granulometries: 20 mesh and 40 mesh. Both in Figure 9a and 9b, it is possible to verify that all the confectioned briquettes have suffered a volumetric expansion. In the case of Figure 9a, Yamaji et al. have verified that the briquettes with pau-jacaré have presented a more homogenous behavior. When analyzing the

[1] *Piptadenia gonoacantha*

granulometry influence in the briquettes confectioned with sugarcane bagasse, it was possible to see that the ones with a 40-mesh granulometry have presented a smaller volumetric expansion, since, during the first 48 hours, the 20-mesh briquette presented a volumetric expansion of 18,8%, while the 40-mesh briquette presented a volumetric expansion of 14%.

2.3.2. Mechanical Resistance Test

The mechanical resistance test, which is also called "traction resistance by diametric compression", consists in analyzing the briquette's mechanical resistance through mechanic compression. This process shall produce a parallel traction tension to the charging plan. The test is conducted in a universal test machine [5], [18], [19], and is mainly used to tell how much charge a briquette can stand before it breaks.

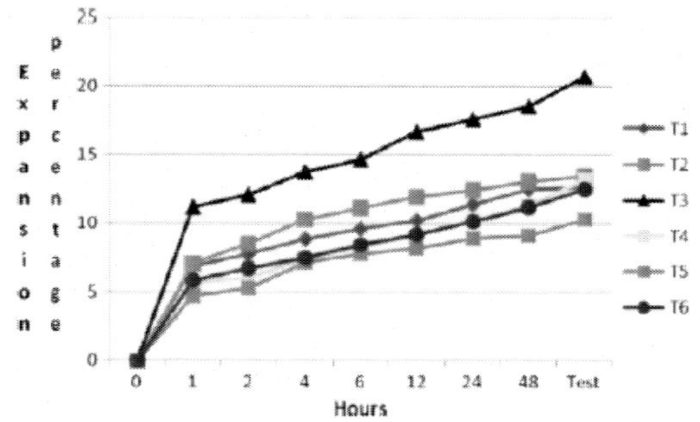

(9a)

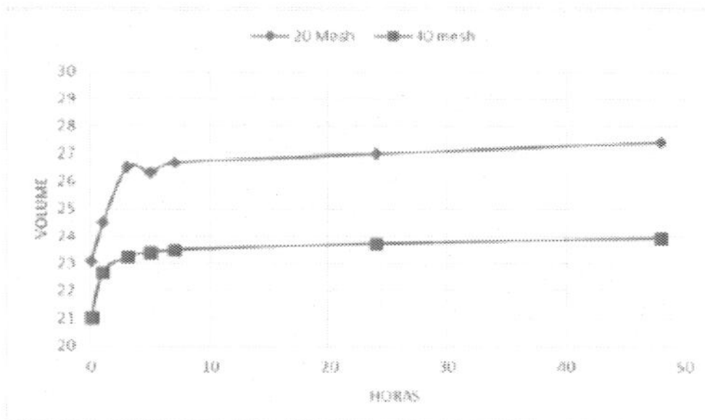

(9b)

(Source: [14]). Figure 9b: Sugarcane bagasse briquettes' volumetric expansion. Source: Author.

Figure 9. Figure 9a: Briquettes' volumetric expansion for two different kinds of wood.

(10a)

(10b)

Source: Author.

Figure 10. Figure 10a presents the universal test machine. Figure 10b makes it possible to visualize the test conducted with the briquette.

2.3.3. Friability Test (Tumbling Test)

The friability test [20], [21], also known as tumbling test, is conducted in order to verify the mechanical behavior that the compacted material presents when it is submitted to impacts and abrasive processes. The most important thing to be acknowledged in this test is that both the pellet and the briquette preserve their structure as much as possible with the minimum of fines (the amount of material that has been unattached from the briquette or the pellet). This test consists in putting the briquettes and pellets in the friabilometer, presented in Figure 11. The friability level is given by:

$$Fr = \frac{m_i - m_r}{m_i} * 100 \quad (3)$$

where:

- Fr = friability level, in %;
- m_i = initial mass in g;
- m_r = mass retained in the sieve, in g;
- 100 – conversion factor into %

2.4. Standards around the World (References, Embrapa, and Dorival's Thesis)

Some standards have been stablished so that all feedstock characterization is well done and the quality of the briquettes and pellets is verified. For that, each country has their norm. In Table 3, there are the main norms used around the world.

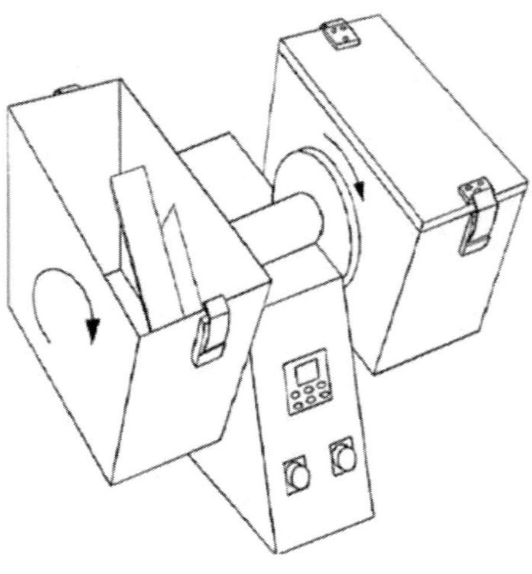

Figure 11. Friability or Tumbling Test. Adapted: [21].

Table 3. Technical Norms

Norm	Title
ASTM E 873	Standard Test Method for bulk Density of Densified Particulate Biomass Fuels
ASTM E 871	Standard Test Method for Moisture Analysis of Particulate Wood Fuels
ASTM D 1102	Standard Test Method for Ash in Wood
ASTM E 776	Standard Test Method for Forms of Chlorine in Refuse-Derived Fuel
ASTM D 4208	Standard Test Method for Total Chlorine in Coal by the Oxygen Bomb Combustions/Ions Selective Electrode Method
ASTM D 6721	Standard Test Method for Determining the Chlorine in the Coal by Oxidative Hydrolysis Microcoulometry
ASTM E 711	Standard Test Method for Gross Calorific Value of Refuse-Derived Fuel by the Bomb Calorimeter
BS EN 14961	Solid Biofuels. Fuel Specification and classes.
BS EN 14775	Solid Biofuels. Determination of Ash Content
BS EN ISO 17828	Solid Biofuels. Determination of Bulk Density.
BS EN ISO 18123	Solid Biofuels. Determination of the Content of Volatile Matter
BS EN ISO 17831-1	Solid Biofuels. Determination of mechanical durability of pellets and briquettes.
BS EN ISO 17831-1	Solid biofuels. Determination of mechanical durability of pellets and briquettes.
BS EN ISO 18135	Solid Biofuels. Sampling.
BS EN ISO 16559	Solid biofuels. Terminology, definitions and descriptions
BS EN ISO 16967	Solid biofuels. Determination of major elements Al, Ca, Fe, Mg, P, K, Si, Na, and Ti
BS EN ISO 16968	Solid biofuels. Determination of minor elements
BS EN ISO 18847	Solid biofuels. Determination of the particle density of pellets and briquettes
DD CEN/TS 15370-1	Solid biofuels. Method for the determination of ash melting behavior. Characteristic temperature method
DIN 51731	Testing of solid fuels - Compressed untreated wood - Requirements and testing
DIN EN 14918	Solid Biofuels. Determination of calorific Value
DIN EN 15270	Pellet burners for small heating boilers - Definitions, requirements, testing, marking
DIN 66165-1	Particle size analysis - Sieving analysis - Part 1: Fundamentals
DIN 66165-1	Particle size analysis - Sieving analysis - Part 2: Procedure
ISO 18134-1, -2, -3	Solid biofuels. Determination of Moisture content
BS EN ISO 17225-1	Solid biofuels. Fuel specifications and classes. General requirements
BS EN ISO 17225-1	Solid biofuels. Fuel specifications and classes. Graded wood pellets
BS EN ISO 17225-3	Solid biofuels - Fuel specifications and classes - Part 3: Graded wood briquettes
BS EN ISO 18846	Solid biofuels. Determination of fine content in quantities of pellets
BS EN ISO 17830	Solid biofuels. Particle size distribution of disintegrated pellets

Adapted from: [7], [22]

3. Gasification

Gasification is the thermochemical process of converting an organic solid into a gaseous fuel through a series of chemical reactions that are held under high temperatures in the presence of an oxidant agent, which can be air, oxygen, steam, or a mixture of the three [23]. This technology presents excellent results for the thermochemical conversion of biomass, answering to the possibility of using the gas produced in this process as fuel. It also ensures that this gas is energetically explored in different kinds of installations, like ovens (for direct heat generation), internal combustion engines, and gas turbines, in which mechanical work and electric energy (among others) may be generated.

The gasification process (Figure 12) is conducted in four steps. The first step is an endothermic process that accounts on drying the biomass under temperatures between 100-200ºC [23]. In the second step, there is the pyrolysis, process in which the biomass undergoes a thermic decomposition in the absence of oxygen through complex reactions that happen under temperatures around 350-600°C [25], giving room to a carbonaceous residue (*char*), formed by carbon (mainly) and condensable (light and heavy hydrocarbons) and non-condensable gases (CH_4, water steam, Co, H_2, CO). Later, the combustion occurs, which requires a minimum temperature of 700-900°C [26] when the process uses air as oxidant agent. With pure oxygen, the temperature may be around 1000-3000°C [27]. Finally, in the fourth and last step, there is the reduction, in which both heterogeneous (gas – solid) and homogeneous (gas – gas) gasification reactions happen. In these processes, the coal reacts with oxygen, water steam, carbon dioxide, and hydrogen, producing the final gas mixture. The process's final result is a gas, formed mainly by carbon monoxide, hydrogen, nitrogen, carbon dioxide, water steam, and hydrocarbons or tars.

The fuel gases (produced in the reaction) leave the gasifier at a temperature around 400 and 800°C [29], carrying powder, pyrolysis products (tar), and water steam with them. Depending on the final use, it is necessary to cool and clean the gas, so as to remove as much water steam, dust, and tar from it as possible, especially if it is to be used in an internal combustion engine or in a gas turbine.

3.1. Gasification Theory

Gasification is a technology that has been widely used for more than 50 years for producing fuel and chemical products. The most known gasification technology is the Fischer Tropsch process which produces, as a primary fuel, liquid hydrocarbons from the synthesis gas that are generated by the mineral coal gasification, but the number of gasifiers that use biomass as fuel is still limited.

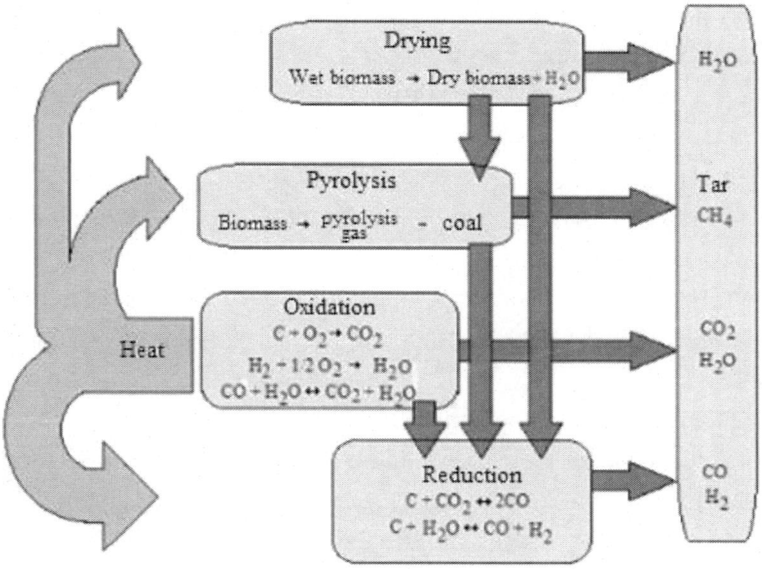

Figure 12. Steps of the gasification process. Adapted from [28].

The biomass gasification process is a consequence of two fundamental sub-processes.

- **Pyrolysis** (also called thermic decomposition, reduction or volatilization): Biomass + Heat = *char* (coal) + gases
- **Gasification:** *Char* + gasifying agent + heat = gases + ashes

Taking the sugarcane bagasse's chemical composition as an example, it may be considered approximately constant and equal to $CH_{1,488}O_{0,7}$, once it is dry. The general bagasse gasification reaction can be described in a schematic way by equation 1 [30]:

$$CH_{1,488} \cdot O_{0,7} + wH_2O + mO_2 + 3,76mN_2 \rightarrow X_1H_2 + X_2CO + X_3CO_2 + X_4H_2O + X_5CH_4 + 3,76mN_2 \qquad (1)$$

After the biomass is dehydrated and the pyrolysis process occurs, the products react with the gasification agent at a temperature of 1000 to 1200K. Initially, there is the combustion reaction. Afterwards, a series of reactions occurs, both heterogeneous (gas – solid) and homogeneous (gas – gas), which happen simultaneously inside the reactor. Therefore, the main reaction is the Boudouard carbon conversion reaction with CO_2 and with water steam (*water – gas reaction*), which is responsible for controlling the speed of the gasification process. In the gas-gas phase, there is the "shift reaction", also known as "water-gas shift reaction", which increments the hydrogen content in the resulting gases from the gasification into carbon monoxide expanses. This reaction is slightly exothermic and depends on the temperature, being able to happen in both ways, i.e., in the products or in the reagents, but without being influenced by the pressure. The main reactions that occur in the gasification process are presented in Table 4, as well as the secondary reactions that might happen simultaneously.

Since the C, H, and O reactions for different kinds of biomass are very similar, the main biomass parameter that influences the gas composition is the moisture content. So, the higher the moisture content in the biomass, the more gasifying agent it requires, because the water has to be heated and evaporate. A gas that comes from a humid biomass has relatively high steam, H_2, and N_2 amounts, if compared to a dry biomass. The water contribution in the gasifying agent is indeed necessary once there is the intention of heating the gas with H_2, producing a gas with an average calorific power [31].

It is difficult to determine the kinetics in these reactions under real gasification conditions, since the process is usually studied globally. This theory does not work always in the same way for all kinds of gasifiers. So, it is necessary to consider that the permanence time of all the present species is enough to allow that thermodynamic balance state in the system to be reached. In theory, there must be a complete biomass conversion, but, in reality, this does not happen.

The gas obtained in the gasification process, usually contains a variable amount of tar and ashes besides the typical compounds, which can cause incrustation problems when the gas is cooled; this gas may also contain other pollutants, such as H_2S, NH_3, HCl, HCN, COS (Carbonyl Sulfide), which have to be removed before the gas is used in the final application [32]. This gas's composition depends on the fuel characteristics, the gasification agent, and the process conditions. A very used approach consists in expressing the concentration of different species, depending on the temperature and the consumed oxygen; the latter is represented by what is called Equivalence Ratio (ER) (REED, DAS; 1998):

The ER is one of the most important parameters involved in operating biomass gasifiers when working with air and gasification agents [33]. According to Alauddin et al. [34], the ER may be defined as the relation between the real and the stoichiometric air/fuel ratios, so that there is the complete combustion, according to equation (2).

$$ER = \frac{\frac{oxidant's\ mass}{dry\ fuel's\ mass}}{\frac{oxidant's\ stoichiometric\ mass}{fuel's\ mass}} \tag{2}$$

Previous studies have shown that Equivalence Ratios with very low values are not convenient for gasifying biomasses because of the decrease in the reaction's temperature [35]. Therefore, there is a wide consensus in the scientific literature that the optimum ER value for gasifiers that work with biomass is around 0,2-0,4. So, they shall strongly depend on several operation parameters [33]. The selection of an adequate ER value will somehow also depend on the later use of the produced gas. When the produced gas is destined to be burned in ovens or furnaces, the tar content is not a serious problem and the gas shall have a high calorific power, considering that the gasifier may work with the minimum ER value possible, around 0,2. In case the gasification temperature is smaller than 850°C, the tar formation is high, what makes it possible to work with higher ER values, around 0,3-0,4, in order to compensate the negative effect of the low temperature [33].

Table 4. Main reactions by zones in the gasification systems

Drying Zone	**H_2O (Gaseous)**	
Pyrolysis Zone	H_2O (Gaseous) + Methanol + Acetic Acid + Tars	
Combustion Zone	$C + O_2 \rightarrow CO_2$ (Combustion reaction)	–394 kJ/mol
Reduction or Gasification Zone	$C + CO_2 \leftrightarrow 2CO$ (Bouduard Reaction) $C + H_2O \leftrightarrow CO + H_2$ (Water-gas Reaction) $CO + H_2O \leftrightarrow CO_2 + H_2$ (Shift reaction) $C + 2H_2 \leftrightarrow CH_4$ (Hydrogasification reaction)	+172,6 kJ/mol +131,4 kJ/mol –41,2 kJ/mol –75,0 kJ/mol
Secondary Reactions		
Oxidation Zone	$CO + ½O_2 \rightarrow CO_2$ $CH_4 + 2O_2 \leftrightarrow CO_2 + H_2O$ $H_2 + ½O_2 \rightarrow H_2O$	– 284 kJ/mol – 803 kJ/mol – 242 kJ/mol
Reduction Zone	$C + ½O_2 \rightarrow CO$	– 111 kJ/mol
Methanation reactions	$2CO + 2H_2 \rightarrow CH_4 + CO_2$ $CO + 3H_2 \leftrightarrow CH_4 + H_2O$ $CO_2 + 4H_2 \rightarrow CH_4 + 2H_2O$	– 247 kJ/mol – 206 kJ/mol – 165 kJ/mol
Steam-reforming reactions	$CH_4 + H_2O \leftrightarrow CO + 3H_2$ $CH_4 + ½O_2 \rightarrow 2H_2$	+206 kJ/mol –36 kJ/mol

The air gasification technology is one of the most employed techniques since it uses simpler and less expensive equipment. In spite of that, this process produces a fuel gas with low calorific power, usually around 4 – 7 MJ/Nm^3 [36]. On the other hand, the gasification with oxygen or water steam is less used, due to the high cost of obtaining the gasification agents, and since there are problems with installation security. However, this technology makes it possible to obtain a fuel gas with a higher quality when it comes to calorific power and tar content. In the case of oxygen utilization, the produced gas may have an average calorific power between 10 – 18 MJ/Nm^3. If the gasifying agent is water steam, the calorific power is around 14 -18 MJ/Nm^3 [36].

3.2. Types of Gasifiers

Currently, there are different gasification technologies involving the type of equipment known as gasifiers, that is, the area where the conversion process occurs. These gasifiers can be classified according to several parameters, as it follows:

According to the calorific power of the produced gas:

- Gas with low-calorific-power: up to 5 MJ/Nm^3 (gasification temperature ≈ 1123K);
- Gas with medium-calorific-power: 5 to 10 MJ/Nm^3 (between 1123K and 1400K);
- Gas with high-calorific-power: 10 to 40 MJ/Nm^3 (typically from 1500 to 1800K tops).

According to the kind of gasifying agent:

- Air;
- Water steam;
- Oxygen.

According to the working pressure:

- Low pressure (atmospheric);
- Pressurized (≈ up to 3 MPa).

According to the relative direction of the movement in the biomass:

- Updraft (counter-current), straight shot;
- Downdraft (co-current), inverted shot;
- Crossed-draft, transverse shot;
- Fluidized bed (Bubbling, Circling and Indirect, with double bed);
- Entrained flow.

According to the way of feeding the fuel and discharging the ashes

- Intermittent;
- Continuous or semi-continuous.

In general, these gasifiers can be classified into three big groups: fixed (or moveable bed) gasifiers, fluidized bed gasifiers, and incoming stream gasifiers. The ones with fixed bed are classified according to the direction of the fuel drainage and the gasification agent in the reactor. On the other hand, the fluidized bed gasifiers are different and depend upon the fluid-dynamic regime.

3.2.1. Fixed Bed Gasifiers

The fixed bed gasifiers can be classified into downdraft gasifiers (co-current) and updraft gasifiers (counter-current). Both are relatively easy to project and operate. They are efficient when working with high density and *biomasses* with unpolished granulometry, also being used in small or medium size energy plants [38].

3.2.1.1. Updraft Gasifiers, also Known as Straight Shot or Counter-Current Gasifiers

In this type of gasifier, the biomass is fed through the upper part of the gasifier. The gasification agent (air, oxygen and/or steam) is introduced through the inferior part. In consequence, the biomass and the generated gases move in opposite directions (counter-current). The combustion happens on the bottom of the bed, in the oxidation zone, where CO_2 and H_2O are formed at around 100ºC. The hot gases, afterwards, go up, into the reduction zone, where they are reduced into CO and H_2 and are also cooled down to around 750°C. Later, these gases pyrolyze the biomass that is going down, producing a high amount of tar and other products. The methane-rich gas and the tar exit, through the top of the gasifier, at 500°C approximately, while the ashes fall to a grate, in order to be gathered in the inferior part.

The temperature in the gasification zone is controlled by adding water steam to the air or by humidifying it. Due to the low temperature of the gas that leaves the gasifier, the total energetic profit in this process is high, but so is the tar content in the gas; for this kind of gasifier, the tar content varies between 10.000 and 100.000 g/Nm^3. As for the particles (particulate material), this content varies between 100 and 1000 mg/Nm^3 [Guigon and Large, 1990].

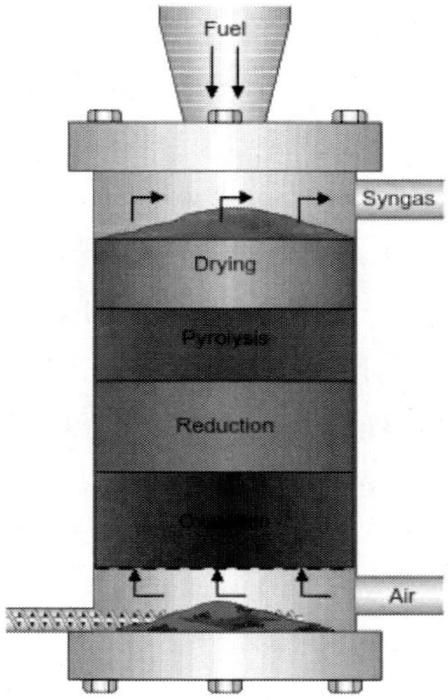

Figure 13. Updraft Gasifier [39].

Advantages:

- The particle size, shape, and humidity content in the biomass are not critical.
- The design is simple.
- The process results in a gas with elevated calorific power, due to the fact that the gasification agent passes through the oxidation zone and, afterwards, through the reduction and pyrolysis zones.

Disadvantages:

- The gas quality is usually very poor, since the produced gases go through the pyrolysis and the drying zones without cracking the tar.
- The high temperature near the reactor's grate can cause obstruction, due to the ash fusion.
- The optimization of the appropriated process is practically inviable, due to their complex and heterogeneous nature.

3.2.1.2. Downdraft Gasifiers, also Known as Inverted Shot or Co-Current Gasifiers

Basically, there are two configurations for this type of gasifier: The *Open-Core* one and those who have a kind of "*throat*". The latter is the most spread one. The fundamental difference is that, in the *Open-Core* ones, there are no significant diameter variations throughout their bodies, what makes it difficult to delimitate the different zones. Both the biomass and the gasifying agents go into the upper part of the gasifier.

On the other hand, in *downdraft throat-like* gasifiers, the fuel is fed through the top, the gasification medium is fed through the lateral part of the reactor (above the grate), and the fuel gas is removed below it. In downdraft reactors, it is possible to see the four regions as biomass goes down them: the drying region, the pyrolysis region, the oxidation region, and the reduction (gasification) region.

The pyrolysis process starts when the temperature reaches 250°C. This process generates gas, tar, and a carbonaceous residue (*char*). Since the pyrolysis occurs slowly at low temperatures, the biomass fraction converted into tar is very high. In sequence, comes the oxidation region, where there is the injection of the gasifying agent (air or oxygen). This agent reacts quickly with the fuel gases that precede the pyrolysis, elevating the temperature in this region up to levels around 1000 to 1200°C; these high temperatures favor the cracking of the present tars. Finally, the reduction region is reached. The *char* is gasified as the set of reactions that happen around 700°C takes place.

The typical potencies of these gasifiers are between 100 and 1000 kW of thermic potency. The fuel is biomass with a humidity content around 20%.

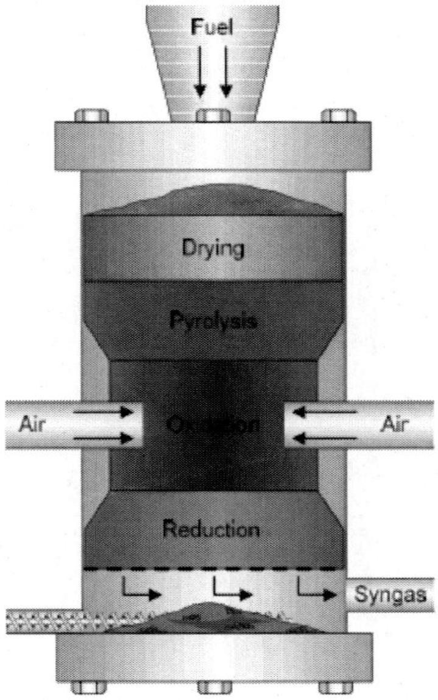

Figure 14. Downdraft Gasifier [39].

Advantages:

- Despite the simple and relatively cheap technology, is general, the produced gas has a relatively good quality and a very low tar content.
- Up to 99,9% of the formed tar can be removed with traditional cleaning methods.

Disadvantages:

- The gas has relatively high CO_2 levels, since a big part of the biomass is oxidized, thus, the calorific power is low.
- There is the possibility of ashes funding in the grate, which can cause the reactor to obstruct.
- The global efficiency is smaller here than in other updraft gasifiers.
- The humidity content and the particle sizes must be very strictly controlled.
- Another disadvantage in this system is that a big part of the energy is converted into heat, what results in a gas with a low calorific value.
- It presents around 4-7% of non-converted coal.

3.2.1.3. Cross-Draft Gasifiers

This kind of gasifier is an adaptation for the use of vegetal coal as fuel. The vegetal coal gasification occurs under high temperatures (over 1500°C) in the oxidation region, which may cause structural problems. In cross-draft gasifiers, the fuel itself (vegetal coal) is the thermic isolation (Figure 15).

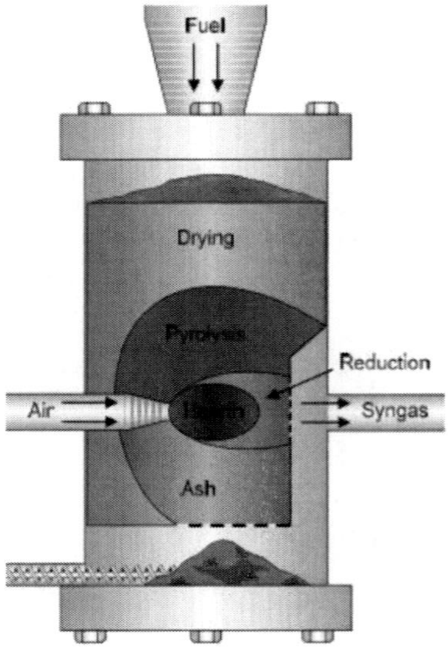

Figure 15. Crossdraft Gasifier [39].

The reactor is fed through its upper part, in a downward trajectory; the air is introduced in one side and the gases are removed on the opposite side, at the same level. The furnace region is close to the air entrance. The pyrolysis and the drying regions are found a little more in the upper part. The ashes are removed through the inferior part and the temperatures that the exiting gases reach are around 800 to 900°C, what results in a low total energetic profit for the process. The resulting gas has a tar content similar to the ones from *downdraft* gasifiers, what leads to a necessity of employing a high-quality kind of vegetal coal (low volatile product content).

The main advantage here is that the system is able to work in small scale, what viabilizes having installations under 10kW (shaft power).

Advantages:

- The design is simple.

Disadvantages:

- The gas quality is usually very poor. The calorific value of the gas is low and the tar content is high.
- The energy losses are frequently a problem, since the outlet temperature of the gas is high.

The fixed bed biomass gasifiers (updraft and downdraft) are operated dryly, which means that the ash does not reach the fusion point. This result is reached by keeping the operating temperature in the reactor below the fusion temperature of the ashes. Both gasifiers usually use air as gasifying agent. The synthesis gas that is produced has a high nitrogen content: about 50% of the mixture, if air is employed as gasifying agent.

In general, it is possible to say that the fixed bed gasifiers are limited to few MW of thermic capacity, due to the production of big amounts of ashes and tar. These residues are results from the low heat and from the mass transference between the solid biomass and the gasifying agent, but the excessive pressure drop in the gasifiers also play a part in their formation.

3.2.2. Fluidized Bed Gasifiers

The fluidized bed gasification was originally developed in order to solve the operational problems found in fixed bed reactors. The technology can be expanded to several tenths of MW. The first installations were developed before the Second World War in order to work with mineral coal gasifications systems that have high capacity (Winkler gasifiers). Later, this technology was adopted by petrochemical and chemical industries for the catalytic cracking of heavy hydrocarbons, for drying, and for an infinity of other applications. The fluidized bed gasifiers are more complicated to operate than the fixed bed ones and, thus, are majorly applied in big installations [41].

The fluidized bed gasifiers (atmospheric or pressurized) are divided in two categories: Bubbling Fluidized Bed Gasifier (BFBG), and Circulating Fluidized Bed Gasifier (CFBG) [36].

According to Basu [41], unlike the fixed bed gasifiers, the fluidized bed ones do not present differentiated reaction zones. The fluidized bed reactors have an isothermal bed that operates usually under temperatures around 700-900ºC [42]. In this kind of system, the gasifying agent enters through the particle bed with enough speed to be kept in suspension. The bed is heated and, when it reaches a high temperature, the fuel particles are introduced, quickly reaching the same temperature as the bed, while the exiting gas temperature is around 600-800°C [43]. This kind of installation also has an excellent mixing characteristic and high reaction speed, due to the increase in the contact between the gas and the solid, which is usually added to the gasifier's bed. The material of the bed is chosen with the objective of distributing and transporting the heat in the gasifier, allowing it to homogenize the temperature, mix the fuel with the gasifying agent and the produced gases, and, in the case of catalytic-active material (like dolomite or olivine) reduce the tar concentration [41]. This favors smaller installations to work with high volumes of gas, increasing their volumetric potency and conversion efficiency. Silica sand is usually used as inert material, but alumina and other refractory oxides are also used to stop the sintering, due to the reaction between the alkaline materials present in the biomass (for example, sodium and potassium carbonate) which makes mixtures with the silica sand which have a low fusion point, causing the bed to sinterize and the fluidization to be lost [44].

So that the biomass is gasified in fluidized bed, it is necessary, in general, that the fuel particles be small (< 6cm), with a humidity content preferably under 30% of the sample's weight [45].

It is also important to consider the operation pressure when working with fluidized-bed gasification systems. High pressure values are preferable in high capacity systems (>120 MW), since they contribute to increase the gas's calorific power. This is consequence of the fact that the reactions that happen with a reduction in the number of mols are favored, like the reaction that forms methane. It is also necessary to consider that there is no need of compressing the syntheses' gas in the entrance of the gas turbine, in case this gas is being used in this kind of installation. On the other hand, the equipment costs have increased for small scale systems. Also, a critical element that must be considered in pressurized systems is the way the biomass is fed into the reaction.

3.2.2.1. Bubbling Fluidized Bed Gasifier

A material bed with an inert fine is placed in the bottom of the gasifier, and the gasification agent (air, oxygen, or steam) is fed into the system through this area by being blown up through the bed at enough speed (1-3m/s) [46]. This happens through a distribution board, in order to fluidize the deposited materials, Figure 16. The biomass is usually fed through a space at the same height as the dense bed basis. This causes the biomass to get in contact with the hot material in the bed and this contact produces the syngas, which then leaves the gasifier through its upper part, carrying a low tar content, typically around 1 to 3 g/Nm^3. It operates at temperatures under 900°C [45] in order to avoid melting the ashes or forming agglomerations in the bed.

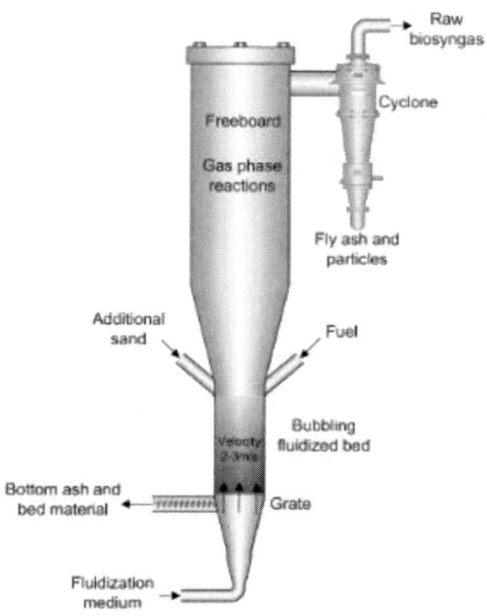

Figure 16. Bubbling Fluidized Bed Gasifier (BFBG) [39].

These gasifiers are normally employed in a scale inferior to 25 MW (thermic), when working with atmospheric pressure. The reason for this size limitation is the necessity of a good fuel distribution over the bed, which becomes more difficult when the reactor's diameter increases [47].

3.2.2.2. Circling Fluidized Bed Gasifier

In fluidized bed gasifiers, the fuel particles are kept in suspension on an inert-material bed (sand, alumina, dolomite, etc.) that is fluidized by the gasification agent (air, oxygen, oxygen-enriched air, etc.). However, the fluidization is supplied at speeds between 5-10m/s [46]. This recirculation creates better conditions to transfer heat and temperature homogeneity throughout the reactor, making it possible for most of the volatiles to be in contact with the particles from the heated bed. This contributes for a more complete gasification. The synthesis gas mixture is separated from the solid particles that were not completely converted by using a cyclone. After this, the gas goes back to the dense region of the bed (in the bottom of the gasifier) through the solid recirculation system, caused by a non-mechanical loop seal valve. This valve also has the function of preventing gas escapes (in the gasifier's inferior part) into the recirculation tube.

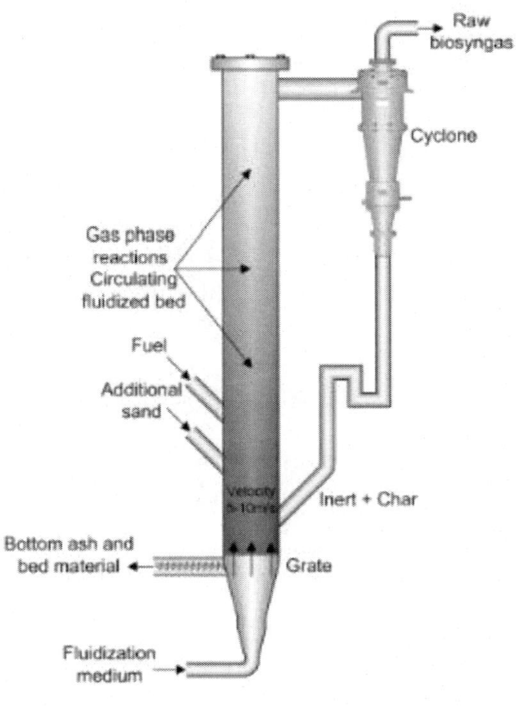

Figure 17. Circulating Fluidized Bed Gasifier (CFBG) [39].

3.2.2.3. Dual Fluidized Bed Gasifier (Indirect Gasification)

This kind of system has two chambers: a gasifier and a combustor. The biomass is fed into the gasification chamber, which can be CFB/BFB, like the combustor. After the feeding process, the biomass is converted into a syngas that is practically nitrogen and coal free, in a process that uses steam as gasifying agent, Figure 18. The coal is burned with air in the combustor. This burnt constitutes the heat source for the whole process, and is responsible for heating the bed's particles. The bed's hot material is, afterwards, fed through the inner part of the gasification chamber, becoming the heat exchange agent and supplying indirect-reaction heat for syngas production.

Advantages:

- The reactor operates isotermically. The average temperature in the bed is usually homogeneous, without zones with excessively high temperatures, which could propitiate the ashes' fusion and, consequently, the waste formation.
- High capacity equipment, with the possibility of conducting the "*scale-up*".
- Possibility to operate with different kinds of biomasses of different granulometries and different physicochemical characteristics.
- High rate of heat transference in the mixture of the inert material from the heated bed and the to-be-processed feedstock, in conditions of temperature uniformity in the reactor. This fact improves the conditions under which the main chemical reactions happen in the phases solid-gas, gas-gas, and others.
- High carbon conversion rates, with reduced residence times for the fuel particles. Since this gasifier works with small-sized fuel particles, it decreases the residence time, which leads to achieving a high gas production with reactors that have relatively small dimensions.
- High volumetric pressure, MW/m^3. It allows systems with higher energy conversion capacities with relatively small-dimension reactors.

Disadvantages:

- More complicated design and, thus, more elevated specific investment cost.
- Possibly higher tar and dust contents (except for the updraft-fixed-bed design, which can contain 10-20% of tar).
- More complex operation. It requires the control of the biomass, oxygen, and steam supplies, as well as the reactor's pressure.
- Higher energy consumption, due to the gas flux compression, if it is pressurized.
- Detrition in the reactor's inner walls and in the elements involved in it, caused by the fluid's relatively high speed and the attrition with the sand and the particles.

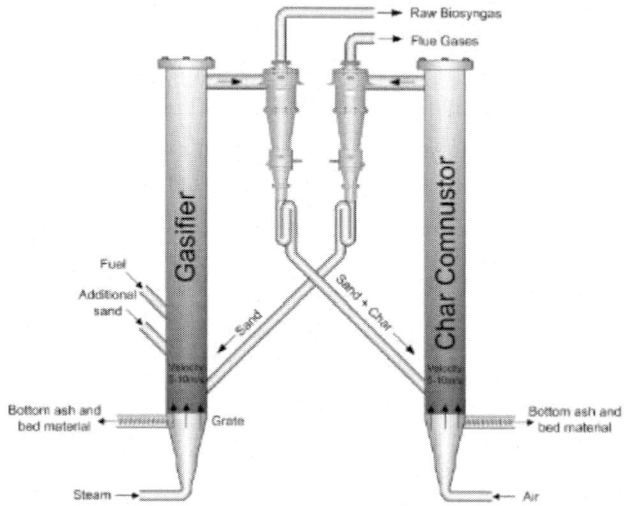

Figure 18. Dual Fluidized Bed Gasifier [39].

Nowadays, the gasification in fluidized bed reactors presents to be very promising, as is practically considered the most viable technology, technically speaking, for the thermic conversion of biomass residues with polydisperse character. Its integration in potency-combined cycles for large installations (BIG/GTCC technology) makes its use even more attractive [45] [47] [48].

3.2.2.4. Entrained Flow Gasifiers

The entrained flow gasifiers have been developed for applying the gasification of petroleum coal and mineral coal in high capacity plants that work with IGCC – Integrated Gasification Combined Cycle. The gasifiers that have already been tested were projected by Destec, Lurgi, Shell, and Texaco. They are constituted by an autothermic reactor in pneumatic transport, being a relatively compact equipment that operates at high temperatures (1.100-1.600°C) and pressures (20-50 bar) [48]. These gasifiers also present relatively small reaction times, producing a gas with little methane, but which, however, requires previous cooling before the cleaning, due to the fact that the fuel gas has a high temperature, which results in a loss of the system's thermic efficiency. The gasification fluid used in this kind of gasifier is pure oxygen or a mixture of water steam and oxygen or air. The fuel, which is finely pulverized, enters through the top of the reactor. The ash is removed in liquid state and the tar is completely cracked, what guarantees the production of a clean synthesis gas with high calorific power. Using this kind of technology with biomass has its biggest difficult in the high cost of triturating it into the required granulometry (because of biomass's low density and high energy consumption). The proposed solutions are the torrefaction before the milestone or the biomass's pyrolytic conversion into a mixture of oil and coal, which can be pulverized into the gasifier's entrance.

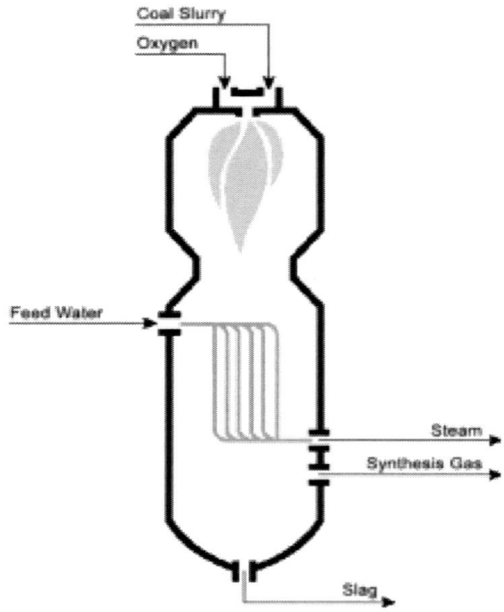

Figure 19. Entrained Flow Gasifier [39].

Advantages:

- High quality of the generated gas: this gas has low tar and methane contents, due to the high operation temperatures (over 1300°C) and the high levels of CO and H_2;
- Capacity of operating with liquid flows or particulate material with low granulometry (bio-oil and torrefacted and milled biomass);
- High pressures and potency;
- Capacity to remove the funded ashes.

Disadvantages:

- Necessity to pulverize the biomass particles between 1mm-200μm approximately, increasing the energy consumption.
- Necessity of conducting pretreatments in the biomass that is used.

3.2.3. Plasma Gasifier

The plasma gasification is a relatively new technology, developed fundamentally for gasifying non-conventional residues, like the ones from auto-shredder and, tyres, incinerator ashes, coal, and hazardous medical, industrial, and radioactive wastes. Other feedstocks like PCBs, asbestos, sewage sludge, oil, coal/water slurry, petroleum coke, paper, plastics, and metals have also been tested.

The plasma is usually generated by a straight DC current or an alternate AC current between two graphite electrodes (Figure 20), which partially ionize a gas that is capable of conducing electricity. The plasma torches operate flexibly, increasing or decreasing the entrance electrical potency or the plasma flux rate. This functioning allows that all the variations – whether they happen in the feedstock amount, humidity, and composition – are heeded, keeping a constant temperature in the gasifier. The temperature may reach values above 1500°C when the biomass or the residues are introduced in the plasma arc. In an atmosphere without oxygen, all the coal and the hydrogen are converted into CO and H_2. The possibility of converting the coal and the hydrogen from several kinds of biomasses into a high-quality gas makes this technology interesting, especially if they were unused or dangerous refuses.

In a general way, the process's and the reactor's technologies resemble the one in updraft-fixed-bed gasifiers, where the biomass is dried, pyrolyzed, and gasified in its downstream movement inside the reactor. Since there is no combustion, except for an extremely high-temperature flame, the whole problem related to tar and ashes is solved, so the organic content is gasified and the fuel's inorganic content is vitrified and recycled in the bottom of reactor.

The size reduction and the biomass drying usually are not necessary, and heterogeneous feedstocks are acceptable. However, in general, feedstocks with high humidity and inorganic content produce a smaller reaction rate and a smaller temperature in the process, generating a synthesis gas with low calorific power.

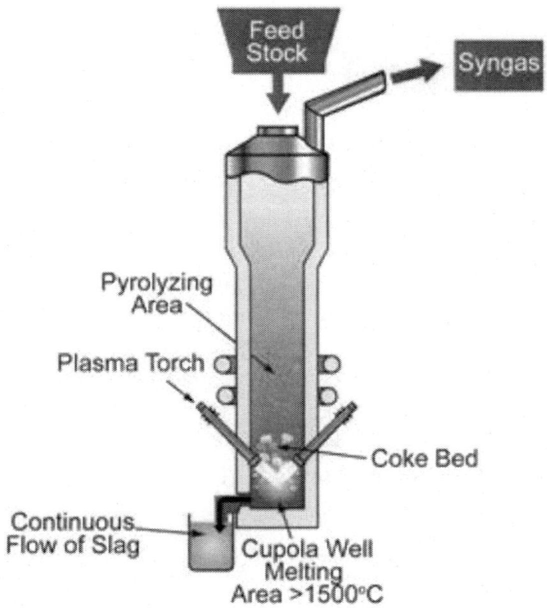

Figure 20. Plasma Gasifier. Adapted from [39].

Advantages:

- High quality gas when it comes to tar residues in it.
- Usually there is no residual carbon in the bed, reaching high coal conversion rates.
- Exploration of residues that are difficult to process.
- Accepting of a high range of charge.
- Very high working temperatures that may destroy the dioxins.

Disadvantages:

- High implementation cost.
- Gas with an average calorific power.

4. Cogeneration

In recent years, the Brazilian government has encouraged the use of such technologies, based on international experiences from countries like France, the UK, Denmark, Finland, the Netherlands, and the US, countries where incentive policies have been created for the use of fossil fuels and renewable sources. In Brazil, in 2000, the Cogeneration Incentive Program was created through the edict 212/2000, with the short-term objective of providing private investment incentives in CHP systems, leading to enhanced thermoelectric generation capacity installed in the country. Later, in 2002, the Alternative Energy Source Incentive Program (PROINFA) was created by Law 10.438, in order to stimulate the electricity generation from renewable energy sources, such as wind, biomass, and hydric. By this law, the use of cogeneration systems associated with the renewable ones is promoted through public financing incentives available in the Brazilian Development Bank (BNDES), as well as through direct subsidies, including taxes and import fees' reduction for cogeneration equipment, and the warranty of long-term contracts and attractive selling prices for cogeneration companies [49].

The term "cogeneration" has an American origin and is used to designate the production processes that combine thermic energy and potency (CHP), be it mechanical or electrical, with the use of energy that is released by an equal primary fuel source [49], whatever the thermodynamic cycle is. Usually, the companies choose to employ the Rankine cycles, the cycles that employ steam turbines, or the Brayton cycles, which use gas turbines.

Since two products with distinct values are obtained – thermic energy and potency – using an equal energy source, the cogeneration systems become attractive for presenting bigger first law efficiencies than the ones found when both energy kinds are produced in independent processes. These efficiencies can be around 75% to 90% [50]. Figure 21 illustrates the gain of aggregated efficiency through cogeneration systems. In the example, highly effective generation technologies (potency plant with a 44% efficiency) are assumed, what leads to a fuel economy of 58 fuel units when compared to the separated power and heat production, which is the mean through which the aggregated efficiency winds through cogeneration, around 37%.

4.1. Classification of the Cogeneration Systems

The "*topping*" cycle (Figure 22) is, in practice, the most frequently found one. In it, first, the electric potency us generated by the energy that comes from the fuel, so the resulting thermic energy is recovered and productively used. The existence of a motive machine is characteristic of these cycles, and the residual energy from this machine (usually thermic) satisfies the thermic charge of the process.

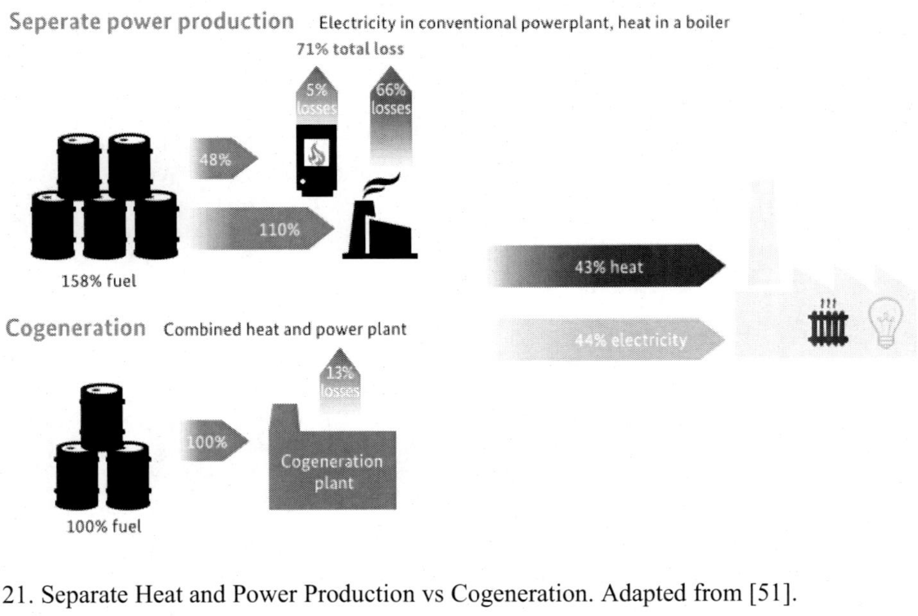

Figure 21. Separate Heat and Power Production vs Cogeneration. Adapted from [51].

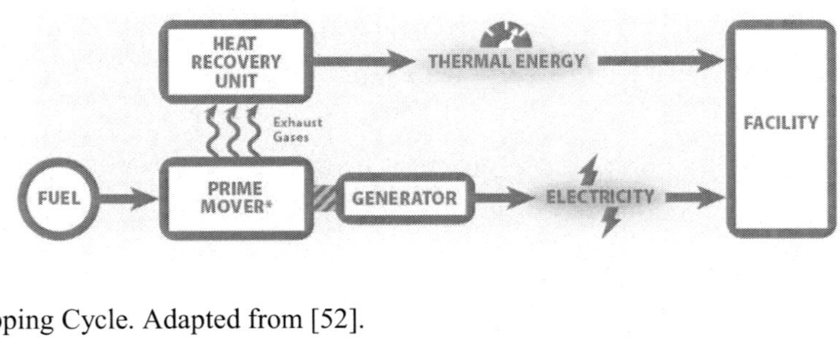

Figure 22. Topping Cycle. Adapted from [52].

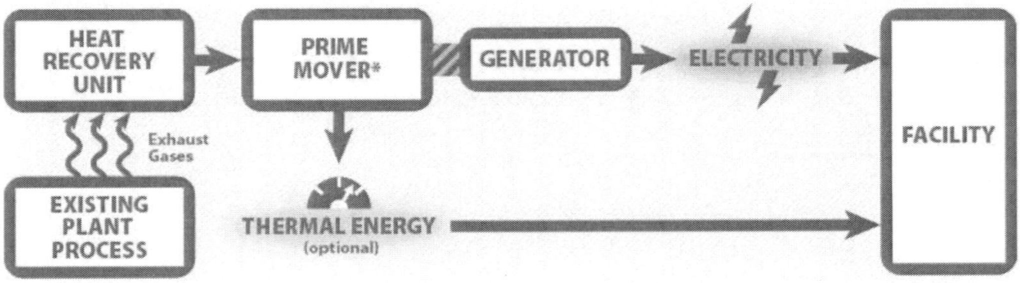

Figure 23. Bottoming Cycle. Adapted from [52].

According to Lizarraga (1994), in the cycles called *"bottoming"* (Figure 23), the residual thermic energy associated to industrial processes (metallurgical, ceramic ovens, cement plants, petroleum refineries) operates at high temperatures, around 1000~1200°C. Instead of being discarded in the atmosphere, this residual heat can be directed to a heat exchanger for the production of steam, which will feed a steam turbine in order to produce electric energy. In these processes, the energetic residue (in the form of heat) can reach very high temperatures, which favor the later conversion. However, it is much more frequent that this energetic effluent be highly infected by corrosive agents, which can significantly and negatively alter the cost of recovering the installation boilers as a whole. Nevertheless, this kind of cogeneration presents, in general, inferior electromechanical revenues than the ones found in the *topping* kind, which work with more elevated temperatures, obtained from the direct fuel burnt.

4.2. Steam Turbine Cogeneration (STCS)

The development of the thermoelectric cycles with steam turbines has reached a limit point of efficiency around 38 to 40 in centrals with hundreds of MWs of potency, besides reaching complex configurations with maximum steam temperatures around 540 to 560°C. Big efforts were made in the decade of 1960 with the objective of increasing the efficiency and improving the development of these cycles, however, there was no success [53].

The employ of steam turbines as thermic machines in cogeneration is entailed to three fundamental configurations, all of them operating in steam cycles. In them, the common aspect consists in exploring the energetic content from the steam that is generated in the generating station for putting the turbine in action and, consequently, generating potency. The condensation of a part of the exhaust steam or an extraction of steam from an extraction-condensation turbine guarantees the energy needs in the form of heat in the system (Figure 24). There are three configurations:

- Counter-pressure turbines;
- Combination of counter-pressure steam turbines with condensation turbines that use exceeding flux;
- Extraction-condensation turbines endowed with automatic regulation systems.

Given these conditions, it is possible to consider the sugarcane bagasse gasification, as well as other kinds of biomasses', as being a highly interesting, promising and viable alternative conversion technology, mainly in a country with ideal economic and geographic characteristics, like Brazil.

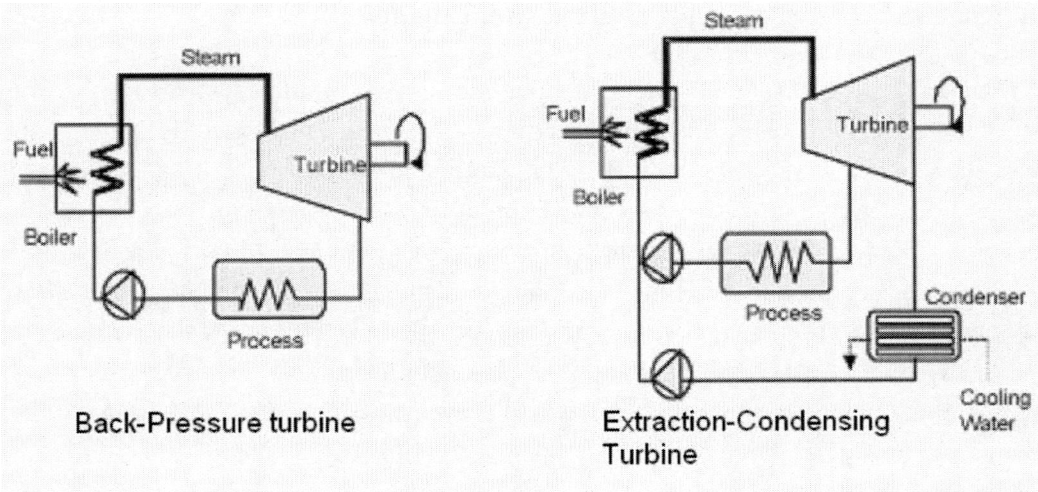

Figure 24. Cogeneration with steam turbines [54].

Key Facts: Steam Turbines	
Technology capacity range	45 kW_{el} – 500 MW_{el}
Typical cogen capacity range	300 kW_{el} – 50 MW_{el}
Electric efficiency	~ 20 – 30%
Temperature level required (steam)	> 400°C
Application focus	Industry, power plants (incl. solid biomass)

Steam turbine, capacity range 5 – 150 MW_{el}, also used for biomass applications

Figure 25. Main characteristics of steam turbines. Adapted from [51].

A possible diversification alternative for the use of sugarcane bagasse would be the large-scale electricity production through the introduction of technological improvements in the current conventional systems, along with the viabilization of newly proposed systems. The traditional boilers can be exchanged for ones that are more potent and more capable of generating a much bigger steam volume with a much higher temperature. This trade, associated to the condensation and extraction cycles, known as CEST (*Condensing Extraction Steam Turbine*), improves substantially the energetic efficiency when burning the bagasse and the amount of thermic energy that can be generated. This would allow a bigger electricity production, with about half the consumption of bagasse that there is nowadays [55], having as a consequence a bigger volume of surpluses from this byproduct. Some characteristics of steam turbines are presented in Figure 25.

4.3. Cogeneration with Inside Combustion Engines

The application of internal combustion engines in cogeneration systems results from their big practical usefulness in small industries, where the needs of electrical energy and steam water at low pressure prevail, highlighting the use of diesel-cycle engines or ignition by compression, where the combustion takes place through diffusion, started by self-ignition. Thus, this engine aspirates air with no restriction at all (absence of throttle body) and adjusts the amount of fuel according to the potency requirement, which allows the employment of high compression rates (around 20:1). These rates make the revenues in these systems surpass 45%, but they are also able to achieve 50% revenues. These engines have been employed in cases where the electric potency that is going to be consumed can be relatively large and also where the demand does not justify the installation of a steam or gas central. On the other hand, in *Otto* cycle engines, a pre-mixture of air and fuel is aspirated; the flame is pre-mixed and highly susceptible for detonation, and the burnt is started by a sparkling plug. In this case, the engine presents inferior compression rates, around 10:1, and the revenues reach values around 35%.

The fuel's chemical energy is liberated through chemical reactions from the combustion, and up to 40% is explored as mechanical work. Among the recoverable heat forms, two kinds are highlighted: the ones that come from the cooling system of the engine and the energy fraction contained in its exiting gases. For the first forms, they can reach up to 20% from the fuel's energy, meanwhile, in the case of the energy fraction, it is possible to carry up to 30%. In this situation, the temperature oscillates between 400 and 500°C (Figure 26). In general, the potency in the axle can be employed for the generation of electricity with a typical efficiency between 25-40%, or for putting a compressor or a bomb in action. The latter also works for the recovered heat, which is usually produced at a rate of 1-2 heat units for each electricity unit. This energy can be destined for several uses, including industrial or domestic heating, drying, absorption refrigeration, and low-pressure steam production, among other applications. The heat/power ratio generally decreases with size.

The internal combustion engines are available in a big potency rate, starting from only some kilowatts and going up to 10 MW, being typically used between 70-1500 kWe (Figure 27). They have a compact construction and may use a variety of liquid and gaseous fuels, with high efficiency in a simple cycle and a good availability factor (80~90%). Because of these characteristics, they are presented as the first option to be applied in small-size cogeneration systems for commercial buildings, hospitals, hotels, and supermarkets. They are usually used in packaged CHP units, along with heat exchangers, in order to recover the heat that comes from the numerous waste heat sources. These essentially include: engine cooling circuit, engine exhaust gases (that provide the highest temperatures), and oil circuit [56].

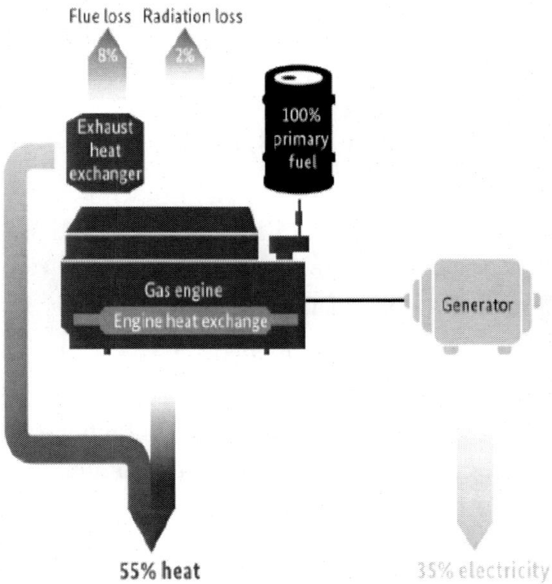

Key Facts: Gas Engines	
Typical capacity range	1 kW_{el} - 10 MW_{el}
Electric efficiency	~ 35 - 45%
Typical costs	From > 1,000 €/kW for small scale to < 500 €/kW for MW size
Application focus	Broad application fields

Figure 26. Cogeneration with internal combustion engine. Adapted from [51].

Figure 27. Engines from Small to Large Scale [51].

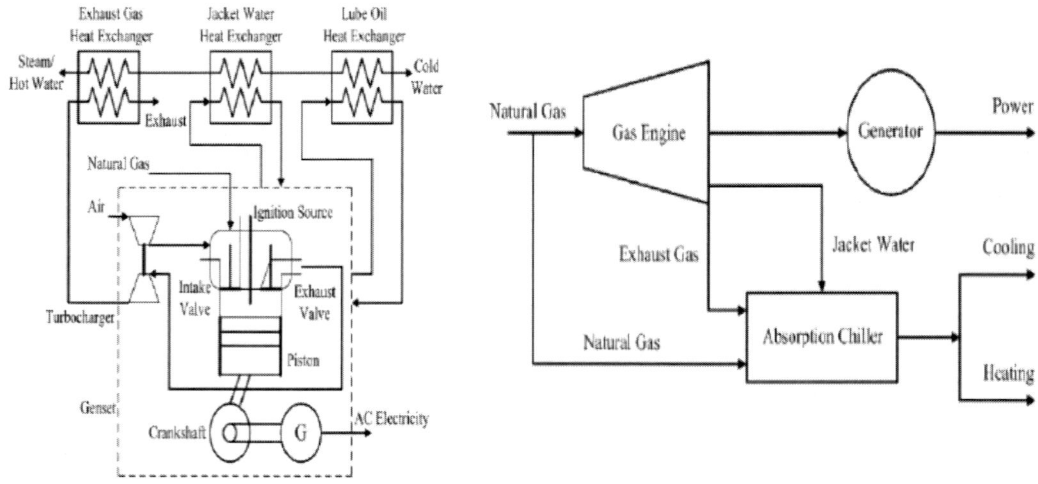

Figure 28. CHP and CCHP systems [57].

In the cogeneration concept, it is possible to consider that a headquarters produces two or more utilities, besides electromechanical energy and thermic energy (CHP systems) (Figure 28a). However, the low-temperature energy production (also called "cold") by refrigeration machines through heat absorption deserves to be highlighted as part of the cogeneration plant because of the importance of the role they have played in leveraging the viability of small size cogeneration. This kind of system is currently known as trigeneration. Like the name suggests, trigeneration systems – also called Combined Cooling, Heating as Power (CCHP) systems – are typically a combination of cogeneration plants and chillers in order to produce electricity, heat, and cooling energy in only one process (Figure 28b).

4.4. Gas Turbine Cogeneration

The gas turbines have been used in industrial and aeronautical areas as elements to generate electric energy and/or for mechanical activation, like the aircraft propellant. The technological evolution in its components implied a great contribution for incrementing its efficiency, which, along with the fact that there is a growing availability of natural gas at competitive prices, has been favoring the implantation of cogeneration systems with gas turbines. Probably, it has been the technology that has experienced the quickest development in the last few years, due to the significant reduction in the installation costs and the better environmental performance. Besides, the gestation period for the development of a project is shorter and the equipment can be delivered in modules.

The main elements of a gas turbine are:

- a compressor, which compresses the entrance air;
- a combustion chamber, where the air is mixed and burned along with the fuel;
- and the turbine itself, where the energy from the hot gases is converted into high electric potency pressure.

The Brayton cycle is the basic description how a gas turbine operates: air in environmental conditions is aspirated into the compressor, where the temperature and the pressure are increased. The following step is the combustion chamber, where the fuel is burned along with the air that comes in at high temperature and pressure. Then, the resulting exhaust gases enter the turbine, expanding themselves and moving across the blades, producing electric energy in the generator (Figure 29).

The exhaust gases carry between 60 and 80% of the primary energy that is consumed by the turbine. In fact, they constitute the fundamental heat source in these systems. When leaving the turbine, they are not recirculated, allowing the cycle to be classified as an open cycle. However, in some cases, this cycle can be modeled as a closed cycle. Frequently, the temperature of the exhaust gases at the exit of the turbine (400-600°C) is considerably more elevated than the temperature of the air that leaves the compressor. This also happens for the high oxygen content in the exhaust gases (13-18%), which happens due to the excess o fair used in the combustion. These two values constitute the starting point for some modifications done in the standard cycle, aiming at additions to the thermic efficiency values, and/or to the liquid work. In this sense, there are attempts of benefiting from the energetic potential that the turbine's exhaust gases have. For that, one of the first modifications to be studied and evaluated is the Regenerator Gas Turbine Cycle. This process uses the energy that comes from the gases (when leaving the expansion) to heat up the air before the combustion chamber. This happens through a heat exchanger, which is also known as a regenerator or recoverer (Figure 30).

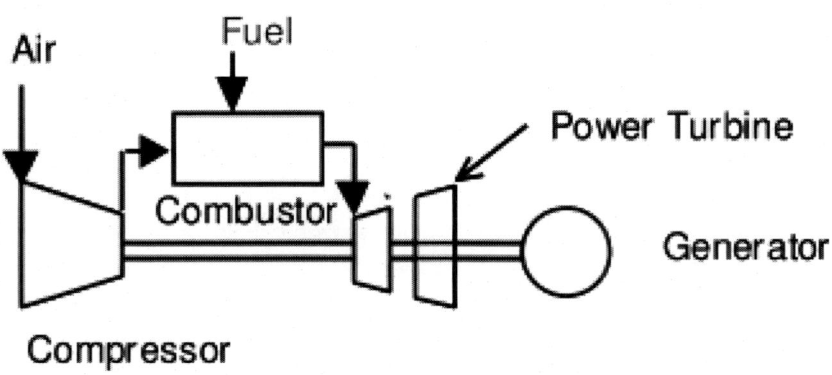

Figure 29. Brayton cycle [58].

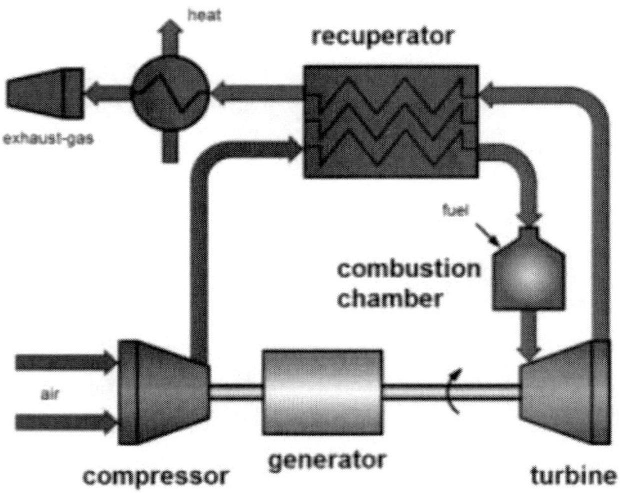

Figure 30. A CHP plant scheme with a gas turbine [56].

The steam injection in a gas turbine has gained plenty of advertisement in the past few years. This effect involves the combination of the standard cycle with the injection of steam generated in a recovery boiler. The energy from the gases that exit the turbine is recovered in a recovery boiler, producing steam that can be used in a process and is able to have a part of itself injected in the combustion chamber of the gas turbine. The capacity of producing steam could be enhanced by putting an additional post-combustion burner, which allows to alter the electricity and steam production. This gives the system a large operational flexibility (Figure 31).

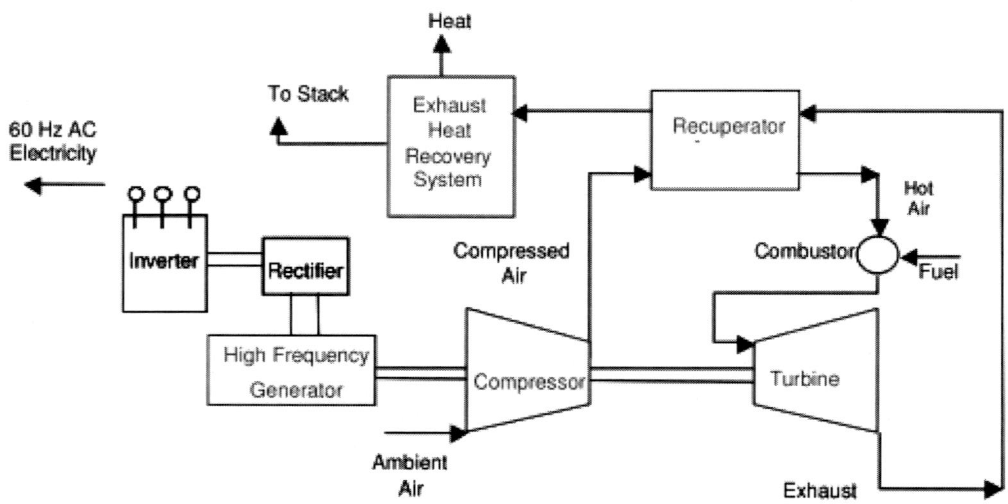

Figure 31. Microturbine CHP System (Single-Shaft Design) [58].

This technology is generally used in large scale installations (more than 1MW), although there are small scale turbines – between 80 and 100kWe – that are available as compact cogeneration systems. The electric efficiency in this technology varies from around 21% for turbines with smaller dimensions, to 25% for standard turbines (around 1MWe), and even up to 36% for very big turbines (over 100MW).

BIG/GT (*Biomass Integrated Gasification Gas Turbines*) is the name used to designate the technology that is used for generating electric energy from biomass and that integrates the fuel's gasification to gas turbines in simple cycles. This new technology is part of the sphere of technological innovations that have been proposed in order to viabilize the exploration of energetic resources from biomass, especially from sugarcane bagasse and straw [48].

The gasification technology integrated to the gas turbine for cogeneration can be divided in two main sets. The first set embodies the refinement of the BIG/GT cycle by using steam injection in gas-derivative aero turbines, increasing their potency, process known as BIG/STIG (*Biomass Integrated Gasification Steam Injected Gas Turbine*) or intermediary cooling in the air compression (BIG/ISTIG - *Biomass Integrated Gasification Intercooled Steam Injected Gas Turbine*) [59]. The second set is the thermoelectric generation cycle, which uses a combination of gas and steam turbines and is known as combined cycle, integrating a biomass gasifier for the production of fuel gas, which is called BIG/GT-CC system (*Biomass Integrated Gasification Gas Turbine Combined Cycle*) [30].

4.5. Cogeneration in Combined Cycle

The term "combined cycle" is usually reserved for the arrangement between two or more cycles, with the main finality of increasing the global revenue of the plant, which has as working essence the good use of the energy from the gases that leave the gas turbine (*Brayton* cycle). This energy is destined to generate steam under pressure in the heat recovery boiler and, afterwards, obtain more electric potency in a steam turbine (*Rankine* cycle). This constitutes, thus, a combination of the gas cycle and the steam cycle (Figure 32), in which the exhaust gases from the gas turbine, with temperatures over 550°C, are taken to the steam cycle's boiler, making the total electric revenue surpass 60%, against 35%, had they been operating separately. Considering the low temperature used in most industrial processes, in general, the cogeneration in combined cycles does not recover heat from the exhaust gases from the gas turbine. It does by means of intermediary extractions from the steam turbine, as well as in the heat that is rejected by it in the condenser. From that, the results are cogeneration systems with total efficiencies that can reach 85%.

The combined-cycle cogeneration is employed when there is need of maximizing the production of electric energy in relation to the heat of the process. In most cases, the energy that is made available by the fuel is explored in two consecutive thermodynamic cycles, and, only then, the heat fraction is obtained. It is important to stress that the combined-cycle cogeneration is only possible in base operative regimes (full-time), which have high capacity factors, given the contribution of the high initial value of investment that this enterprise has (Figure 33).

There are other cogeneration technologies in development, which are being refined and are coming out in the global market, like the Stirling engines, the fuel cells, and the Organic Rankine Cycles (ORC). Figure 34 and Table 5 show a summary of the main cogeneration technologies that are currently available.

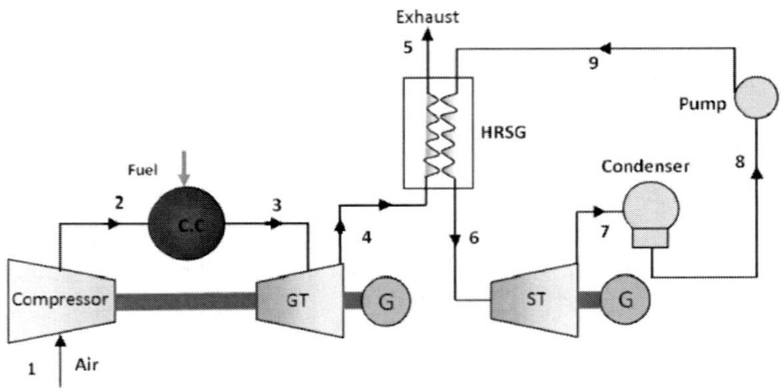

Figure 32. Combined-cycle gas turbine power plant (CCGT) [60].

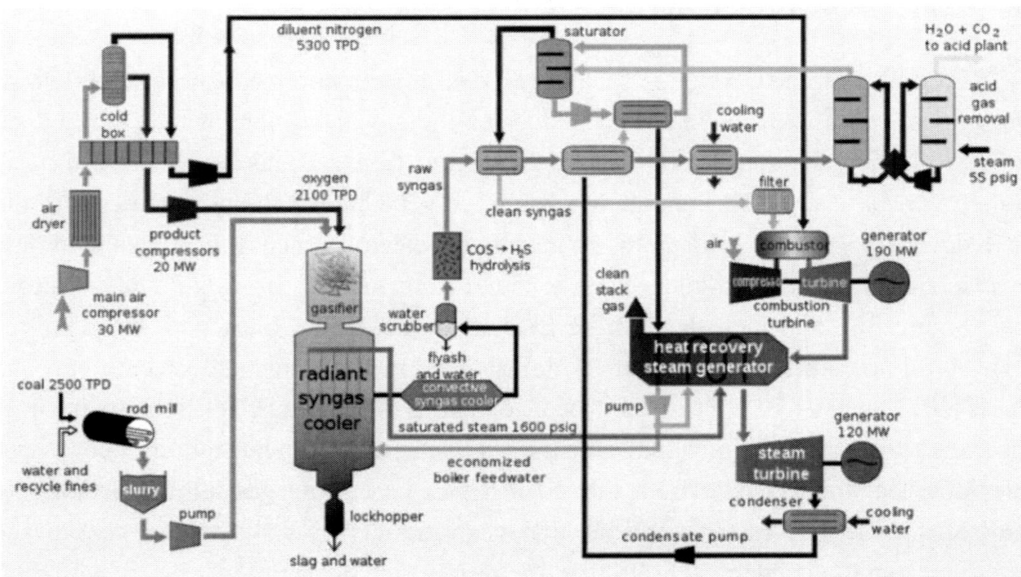

Figure 33. Integrated gasification combined cycle (BIG/GT-CC).

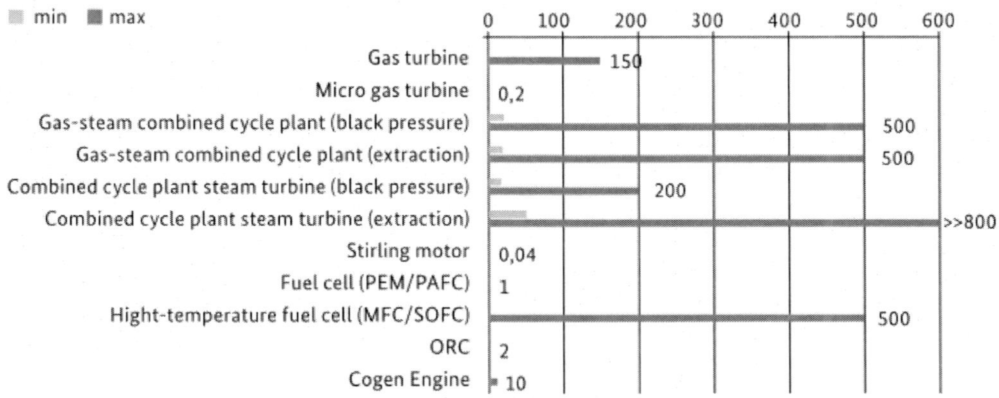

Figure 34. Cogeneration capacity range per technology (MWe) [51].

Table 5. Summary of the typical cost and performance characteristics by CHP technology [61]

	Steam Turbines	Int. Comb. Engines	Gas Turbines	Micro Turbines	Fuel Cells
Electrical Efficiency	15 – 38%	22 – 40%	22 – 36%	18 – 27%	30 – 63%
Overall Efficiency	80%	70 – 80%	70 – 75%	65 – 75%	55 – 80%
Typical Capacity (MWe)	0.5 – 250	0.01 – 5	0.5 - 250	0.03 – 0.25	0.005 – 0.2
Typical Power-to-Heat Ratio	0.1 – 0.3	0.5 – 1	0.5 – 2	0.4 – 0.7	1 – 2
Part Load Behaviour	ok	ok	poor	ok	good
CHP Installing Costs ($/kWe)	430 – 1,100	1,100 – 2,200	970 – 1,300	2,400 – 3,000	5,000 – 6,500
O&M Costs ($/kWhe)	< 0.005	0.009 – 0.022	0.004 – 0.011	0.012 – 0.025	0.032 – 0.038
Hours to Overhauls	> 50,000	25,000 – 50,000	25,000 – 50,000	20,000 – 40,000	32,000 – 64,000
Start-up Time	1 hr – 1 day	10 sec	10 min – 1 hr	60 sec	3 hr – 2 day
Fuels	All	Natural gas, biogas, propane, landfill gas	Natural gas, biogas, propane, oil	Natural gas, biogas, propane, oil	Hydrogen, natural gas, propane, methanol
Noise	High	High	Moderate	Moderate	Low
Power Density (kW/m^2)	> 100	35 – 50	20 – 500	5 – 70	5 – 20

REFERENCES

[1] Wallace, J., Brienzo, M., García-Aparicio, M. P. & J. F. (2016). Görgens, "Lignin enrichment and enzyme deactivation as the root cause of enzymatic hydrolysis

slowdown of steam pretreated sugarcane bagasse," *N. Biotechnol.*, vol. *33*, no. 3, pp. 361–371.

[2] Nogueira, E. E. S. & Luiz Augusto Hora, Lora. (2003). *Dendroenergia: Fundamentos e aplicações*, 2, ed. Rio de Janeiro: Interciência. [*Dendro-energy: fundamentals and aplications*, 2 ed. Rio de janeiro: Interciência.].

[3] Clarke, S., Eng, P. & Preto, F. (2011). "Biomass Densification for Energy Production," *Factsheet*, no. *11*, pp. 1–8.

[4] Balasubramani, P., Anbumalar, V., Nagarajan, M. S. & Prabu, P. M. (2016). "Biomass briquette manufacturing system model for environment," *J. Alloys Compd.*, vol. *686*.

[5] Richards, S. R. (1990). "Physical Testing of Fuel Briquettes," *Fuel Process. Technol.*, vol. *25*, pp. 89–100.

[6] Liu, Z., Quek, A. & Balasubramanian, R. (2014). "Preparation and characterization of fuel pellets from woody biomass, agro-residues and their corresponding hydrochars," *Appl. Energy*, vol. *113*, pp. 1315–1322, Jan.

[7] Agroenegia, E. (2017). "*Briquete, a lenha ecológica* [Briquette, ecological wood],"Accessed September 08, http://www.funverde.org.br/blog/briquete-a-lenha-ecologica/.

[8] Brand, M. A. (2010). *Energia de Biomassa Florestal*, 1st ed. Rio de Janeiro. [*Forest Biomass Energy*, 1st ed, Rio de janeiro.]

[9] Cortez, E. O., Luis, Agusto Barbosa., Lora, Electo Eduardo Silva Lora. & Gómez. (2008). *Biomassa para energia*. campinas. [*Biomass for energy*, campinas].

[10] Demirbas, A. (1998). "Evaluation of biomass residue 1. Briquetting waste paper and wheat straw mixtures." *Fuel Processing Technology*, *55*, pp. 175–183.

[11] Ramos, M. L., Luana, Elis., Trugilho, Paulo Fernando., Napoli, Alfredo. & Bianchi. (2011). "Characterization of Residues from Plant Biomass for use in Energy Generation," *Cern. Lavras*, vol. *17*, no. 2, pp. 237–246.

[12] "Associação brasileira de Normas técnicas (ABNT). NBR 8112 -1986: *Carvão Vegetal - Análise imediata*. Rio de janeiro. [Brazilian Association of Technical Standards (ABNT). NBR 8112-1986: *Charcoal - Immediate analysis*, Rio de Janeiro].

[13] Yamaji, Fábio Minoru., Vendrasco, Lais., Chrisostomo, Walbert. & Paula, Flores. (2013). "Análise do Comportamento Higroscópico de Briquetes," *Energ. na Agric.*, vol. *28*, no. 1, pp. 11–15. [Analysis of Higrosgopic Behavior of Briquettes, *Energ, na Agric*, v. *28*, no. 1, pp. 11-15.].

[14] Larissa, A., Hansted, S., Tami, G., Provedel, M., Yamamoto, H. & Minoru, F. (2016). "Comparative analyses of fast growing species in different moisture content for high quality solid fuel production," *Fuel*, vol. *184*, pp. 180–184.

[15] de S. Dias, J. M. C., et al. (2012). "Produção de briquetes e péletes a partir de resíduos agrícolas, agroindustrais e florestais," *agroindustriais e florestais.*

[Production of briquettes and pellets from agricultural, agroindustrial and forestry waste, *Agricultural and forestry* .].

[16] Serrano, C., Monedero, E., Lapuerta, M. & Portero, H. (2011). "Effect of moisture content, particle size and pine addition on quality parameters of barley straw pellets," *Fuel Process. Technol.*, vol. *92*, no. 3, pp. 699–706.

[17] O. U. R. Projects. (2017). ABC Machine. *"Biomass Fuel Densification Processing Methods And Equipment Pelletizing/Briquetting Methods And Equipment.,"* Accessed September 08. http://www.gcmec.com/latest-news/biomass-fuel-densification-processing-methods-and-equipment.html.

[18] Chrisostomo, W. (2011). *Estudo da compactação de Resíduos Lignocelulósicos para utilização como Combustível sólido*, Universidade Federal de São Carlos, Campus Sorocaba.[*Study of the compaction of Lignocellulosic Residues for use as Solid Fuel*, Universidade Federal de São Carlos, Campus Sorocaba].

[19] Guo, L., Tabil, L. G., Wang, D. & Wang, G. (2016). "Biomass and Bioenergy In fl uence of moisture content and hammer mill screen size on the physical quality of barley, oat, canola and wheat straw briquettes," *Biomass and Bioenergy*, vol. *94*, pp. 201–208.

[20] de Barros, J. L. (2014). *Caracterização de blendas e briquetes de carvões vegetal e mineral*, Universidade Federal de São Carlos, Campus Sorocaba. [*Characterization of mineral and vegetable charcoal blends and briquettes*, Universidade Federal de São Carlos, Campus de Sorocaba].

[21] Stelte, Wolfgang., Sanadi, Anand R., Shang, Lei., Holm, Jens k., Ahrenfeldt, J. & Henriksen, Ulrik B. (2012). *"Recent Developments in Biomass Pelletizations -A Review"*., v. *7*. 4451-4490.

[22] Garcia, D. P. (2017). *Ténicas Multivariadas (HCA e PCA) Aplicadas na avaliação de Biomassa Vegetais para a produção de pellets*., PhD diss, Universidade Estadual Júlio de Mesquita Filho. [*Multivariate Techniques (HCA and PCA) Applied in the evaluation of Vegetable Biomass for the production of pellets*. São Paulo State University (Unesp).

[23] Puig-Arnavat, M., Bruno, J. C. & Coronas, A. (2010). "Review and analysis of biomass gasification models," *Renew. Sustain. Energy Rev.*, vol. *14*, no. 9, pp. 2841–2851.

[24] Dong, L., Liu, H. & Riffat, S. (2009). "Development of small-scale and micro-scale biomass-fuelled CHP systems – A literature review," *Appl. Therm. Eng.*, vol. *29*, no. 11–12, pp. 2119–2126, Aug.

[25] Bridgwater, A. (2014). "Renewable fuels and chemicals by thermal processing of biomass," *Chem. Eng. J.*, vol. *91*, no. 2–3, pp. 87–102, Mar.

[26] Heidenreich, S. & Foscolo, P. U. (2014). "New concepts in biomass gasification," *Prog. Energy Combust. Sci.*, vol. *46*, pp. 72–95.

[27] Wang, Z. et al. (2015). "Gasification of biomass with oxygen-enriched air in a pilot scale two-stage gasifier," *Fuel*, vol. *150*, pp. 386–393.

[28] Andrade, R. V. (2007). *Gaseificação de Biomassa: uma análise teórica e experimentasl*. 205f. Tese (Doutorado em Engenharia Mecânica) - Inst. de Engenharia Mecânica. Universidade Federal de Itajubá. Itajubá. Minas Gerais. [*Biomass Gasification: A Theoretical Analysis and Experimental*. Mechanical Engineering Institute. Federal University of Itajubá.Itajubá. Minas Gerais].

[29] Cordiner, S., De Simone, G. & Mulone, V. (2012). "Experimental–numerical design of a biomass bubbling fluidized bed gasifier for paper sludge energy recovery," *Appl. Energy*, vol. *97*, pp. 532–542, Sep.

[30] Pérez, N. P., Machin, E. B., Pedroso, D. T., Antunes, J. S. & Silveira, J. L. (2014). "Fluid-dynamic assessment of sugarcane bagasse to use as feedstock in bubbling fluidized bed gasifiers," *Appl. Therm. Eng.*, vol. *73*, no. 1, pp. 238–244, Dec.

[31] CIRCE, *Course Expert in renewable energy systems research center for resources and energy consumption* (CIRCE. Centro Politécnico Superior). Spain: University of Zaragoza; April-August, 2005.

[32] Van der Meijden, C. M. & van der, MEIJDEN. (2010). "Development of the MILENA gasification technology for the production of Bio-SNG. PhD. Thesis.," *ECN Energy Res. Netherlands*, no. December, p. 206.

[33] Narvaez, I., Orio, A., Aznar, M. P. & Corella, J. (1996). "Biomass gasification with air in an atmospheric bubbling fluidized bed. Effect of six operational variables on the quality of the produced raw gas," *Industrial & Engineergin Chemistry Research*, vol. *5885*, no. 95, pp. 2110–2120.

[34] Alauddin, Z. A. B. Z., Lahijani, P., Mohammadi, M. & Mohamed, A. R. (2010). "Gasification of lignocellulosic biomass in fluidized beds for renewable energy development: A review," *Renew. Sustain. Energy Rev.*, vol. *14*, no. 9, pp. 2852–2862.

[35] Lv, P. M., Xiong, Z. H., Chang, J., Wu, C. Z., Chen, Y. & Zhu, J. X. (2004). "An experimental study on biomass air-steam gasification in a fluidized bed.," *Bioresour. Technol.*, vol. *95*, no. 1, pp. 95–101, Oct.

[36] Gómez-Barea, a. & Leckner, B. (2010). "Modeling of biomass gasification in fluidized bed," *Prog. Energy Combust. Sci.*, vol. *36*, no. 4, pp. 444–509, Aug.

[37] Higman, C. & Van Der Burgt, M. (2011). *Gasification*, Gulf Professional Publishing, Second Edi.

[38] Patra, T. K. & Sheth, P. N. (2015). "Biomass gasification models for downdraft gasifier: A state-of-the-art review," *Renew. Sustain. Energy Rev.*, vol. *50*, pp. 583–593.

[39] Olofsson, I., Nordin, A. & Söderlind, U. (2005). "Initial Review and Evaluation of Process Technologies and Systems Suitable for Cost-Efficient Medium-Scale Gasification for Biomass to Liquid Fuels," *Physics* (College. Park. Md)., p. 95.

[40] Abdoulmoumine, N., Kulkarni, A. & Adhikari, S. (2014). "Effects of temperature and equivalence ratio on pine syngas primary gases and contaminants in a bench-scale fluidized bed gasifier," *Ind. Eng. Chem. Res.*, vol. *53*, no. 14, pp. 5767–5777.

[41] Basu, P. (2006). *Combustion and Gasification in Fluidized Beds*, Taylor Fr. Group. LLC, vol. 1, p. 467.

[42] Gómez-barea, A., Leckner, B., Villanueva, A., Nilsson, S. & Fuentes, D. (2013). "Improving the performance of fl uidized bed biomass/waste gasi fi ers for distributed electricity : A new three-stage gasi fi cation system," *Appl. Therm. Eng.*, vol. *50*, no. 2, pp. 1453–1462.

[43] Basu, P. (2010). *Biomass gasification and pyrolysis - Practical design and theory. Estados Unidos: Academic Press*, 365 p.

[44] van der Meijden, C. M., Veringa, H. J. & Rabou, L. P. L. M. (2010). "The production of synthetic natural gas (SNG): A comparison of three wood gasification systems for energy balance and overall efficiency," *Biomass and Bioenergy*, vol. *34*, no. 3, pp. 302–311.

[45] E4TECH. (2009). "Review of Technologies for Gasification of Biomass and Wastes. Final report," *NNFCC*, no. June.

[46] Worley, M. & Yale, J. (2012). *Biomass Gasification Technology Assessment. Consolidated Report National Laboratory of the U.S. Departmant of Energy*. United States.

[47] Bridgwater, A. (2003). "Renewable fuels and chemicals by thermal processing of biomass.," *Chem. Eng. J.*, vol. *91*, pp. 87–102.

[48] Larson, E. D., Williams, R. H. & Leal, M. R. L. V. (2001). "A review of biomass integrated-gasifier/gas turbine combined cycle technology and its application in sugarcane industries, with an analysis for Cuba," *Energy Sustain. Dev.*, vol. *5*, no. 1, pp. 54–76, Mar.

[49] Pérez, N. P., et al. (2015). "Energetic and exergetic analysis of a new compact trigeneration system run with liquefied petroleum gas," *Energy*, vol. *90*, pp. 1411–1419.

[50] Fiomari, M. C. (2004). *Análise Energética E Exergética De Uma Usina Sucroalcooleira Do Oeste Paulista Com Sistema De Cogeração De Energia Em Expansão*, Universidade Estadual Paulista. Faculdade de Engenharia de Ilha Solteira. [*Energy and Exergetic Analysis of a Western São Paulo Sucroalcooleira Power Plant with Expansion Energy Cogeneration System*, São Paulo State University (Unesp), Faculdade de Engenharia de Ilha Solteira.]

[51] Cooperation, G. "Cogeneration & Trigeneration – How to Produce Energy Efficiently." *A Practical Guide for Experts in Emerging and Developing Economies* Published by: GIZ, 144 p.

[52] C. for S. Energy, "Combined Heat and Power (CHP)." *Center for Sustainable Energy.*, Accessed June 12. https://energycenter.org/self-generation-incentive-program/business/technologies/chp.

[53] Neto, V. C. (2001). *Análise de Viabilidade da Cogeração de Energia Eletrica em Ciclo Combinado com Gaseificação de Biomassa de Cana-de-açucar e Gás Natural*, Universidade Federal do Rio de Janeiro, COPPE.[*Feasibility Analysis of Combined Cycle Electric Energy Cogeneration with Biomass Gasification of Sugarcane and Natural Gas*, Universidade Federal do Rio de Janeiro, COPPE].

[54] B. of E. Efficiency. Bureau of Energy Efficiency." *Charter Cogeneration.*"., pp. 155-172.

[55] Rocha, G. J. M., Gonçalves, A. R., Oliveira, B. R., Olivares, E. G. & Rossell, C. E. V. (2012). "Steam explosion pretreatment reproduction and alkaline delignification reactions performed on a pilot scale with sugarcane bagasse for bioethanol production," *Ind. Crops Prod.*, vol. *35*, no. 1, pp. 274–279, Jan.

[56] RENAC - Renewables Academy. (2015). "*CHP – Cogeneration Power*," p. 74.

[57] Wu, D. W. & Wang, R. Z. (2006). "Combined cooling, heating and power: A review," *Prog. Energy Combust. Sci.*, vol. *32*, no. 5–6, pp. 459–495, Sep.

[58] Crossman, K. (2007). "*Biomass combined heat and power catalog of technologies*," no. September, p. 123.

[59] Williams, E. D. & Larson, R. H. (1996). "Biomass gasifier gas turbine power generating technology," *Biomass and Bioenergy*, vol. *10*, no. 2–3, pp. 149–166.

[60] Ibrahim, T. K. & Rahman, M. M. (2012). "Effect of Compression Ratio on Performance of Combined Cycle Gas Turbine," *Int. J. Energy Eng.*, vol. *2*, no. 1, pp. 9–14.

[61] "U.S. Environmental Protection Agency, Combined Heat and Power Partnership – *Catalogue of CHP Technologies – 2008*."

About the Editor

Michel Brienzo
Bioenergy Research Institute
São Paulo State University
Bairro Santana, Rio Claro
São Paulo, Brazil
mbrienzo@ipben.unesp.br

INDEX

C

D

E

F

G

H

I

J

K

L

M

N

O

T

U

V

W

X

Y

α

β